Power Systems Handbook

Load Flow Optimization and Optimal Power Flow

Volume 2

Power Systems Handbook

Series Author
J.C. Das
Power System Studies, Inc., Snellville, Georgia, USA

Volume 1: Short-Circuits in AC and DC Systems: ANSI, IEEE, and IEC Standards

Volume 2: Load Flow Optimization and Optimal Power Flow

Volume 3: Harmonic Generation Effects Propagation and Control

Volume 4: Power System Protective Relaying

Load Flow Optimization and Optimal Power Flow

Volume 2

J.C. Das

CRC Press is an imprint of the
Taylor & Francis Group, an **informa** business

CRC Press
Taylor & Francis Group
6000 Broken Sound Parkway NW, Suite 300
Boca Raton, FL 33487-2742

First issued in paperback 2022

ISBN-13: 978-1-498-74544-4 (hbk)
ISBN-13: 978-1-03-233944-3 (pbk)
DOI: 10.1201/9781351228299

Publisher's Note

The publisher has gone to great lengths to ensure the quality of this reprint but points out that some imperfections in the original copies may be apparent.

Visit the Taylor & Francis Web site at
http://www.taylorandfrancis.com

and the CRC Press Web site at
http://www.crcpress.com

Contents

Series Preface

This handbook on power systems consists of a set of four volumes. These are carefully planned and designed to provide the state-of-the-art material on major aspects of electrical power systems, short-circuit currents, load flow, harmonics, and protective relaying.

An effort has been made to provide a comprehensive coverage, with practical applications, case studies, examples, problems, extensive references, and bibliography.

The material is organized with sound theoretical base and its practical applications. The objective of creating this series is to provide the reader with a comprehensive treatise, which could serve as a reference and day-to-day application guide for solving the real-world problem. It is written for plasticizing engineers and academia, level of education upper undergraduate and graduate degrees.

Though there are published texts on similar subjects, this series provides a unique approach to the practical problems that an application engineer or consultant may face in conducting the system studies and applying it to varied system problems.

Some parts of the work are fairly advanced on a postgraduate level and get into higher mathematics. Yet, the continuity of the thought process and basic conceptual base are maintained. A beginner and advanced reader will equally benefit from the material covered. An undergraduate level of education is assumed, with fundamental knowledge of electrical circuit theory, rotating machines, and matrices.

Currently, power systems, large or small, are analyzed on digital computers with appropriate software packages. However, it is necessary to understand the theory and basis of these calculations to debug and decipher the results.

A reader may be interested only in one aspect of power systems and may like to purchase only one of the volumes of the series. Many aspects of power systems are transparent between different types of studies and analyses—for example, knowledge of short-circuit currents and symmetrical component is required for protective relaying, and knowledge of fundamental frequency load flow is required for harmonic analysis. Though appropriate references are provided, the material is not repeated from one volume to another.

The series is a culmination of the vast experience of the author in solving real-world problems in the industrial and utility power systems for the last more than 40 years.

Another key point is that the solutions to the problems are provided in Appendix D. A reader should be able to independently solve these problems after perusing the contents of a chapter, and then look back to the solutions provided, as a secondary help. The problems are organized, so that these can be solved with manual manipulations, without the help of any digital computer power system software.

It is hoped that the series will be a welcome addition to the current technical literature.

The author thanks Ms. Nora Konopka of CRC Press for her help and cooperation throughout the publication effort.

—J.C. Das

Preface to Volume 2: Load Flow Optimization and Optimal Power Flow

This volume discusses the major aspects of load flow, optimization, optimal load flow, and culminating in modern heuristic optimization techniques and evolutionary programming.

In the deregulated environment, the economic provision of electrical power to consumers requires knowledge of so many related aspects—maintaining a certain power quality and load flow being the important aspects. Chapter 1 provides basic structures: security assessments, load estimation, forecasting, and nonlinear nature of load flow, requiring iterative techniques for optimum solutions. This is followed by Chapter 2 which goes into AGC and AFC concept, controls, and underfrequency load shedding.

Load flow over AC and HVDC transmission lines is somewhat unique in the sense that it much depends upon line length and in HVDC system the system configurations.

Nodal analysis techniques that have developed over the past many years for solution of the load flow problem are covered in Chapters 5 and 6. These discuss and compare almost all available load flow algorithms in practical use on commercial software packages.

Maintaining an acceptable voltage profile on sudden load impacts and contingency load flow problems, large induction and synchronous motor starting, their models and characteristics, and stability considerations are discussed in Chapter 7.

The related topic of reactive power control and voltage instability and impact on the acceptable voltages in the electrical power system is the subject of Chapter 8, followed by FACTS and SVC applications for transmission and distribution systems in Chapter 9.

The distribution systems can have phase unbalances that cannot be ignored. The symmetrical component analysis cannot be applied with prior phase unbalance. The techniques of phase coordinate methods for three-phase modeling and advanced models of three-phase transformers, with optimum locations of capacitor banks using dynamic modeling concepts, are provided in Chapter 10.

Chapters 11 and 12 cover the classical methods of optimization. Gradient methods, linear programming, dynamic programming, barrier methods, security- and environmental-constrained OPF, generation scheduling considering transmission losses, and unit commitment are discussed.

Finally, Chapter 13 provides an introduction to evolutionary programming, genetic algorithms, particle swarm optimization, and the like.

Thus, the subject of load flow, optimization, and optimal load flow is completely covered.

Many case studies and practical examples are included. The problems at the end of a chapter can be solved by hand calculations without resort to any computer software. Appendix A is devoted to calculations of line and cable constants and Appendix B provides solutions to the problems.

—J.C. Das

Author

J.C. Das is an independent consultant, Power System Studies, Inc., Snellville, Georgia, USA. Earlier, he headed the Power System Analysis Department at AMEC Foster Wheeler for the last 30 years. He has varied experience in the utility industry, industrial establishments, hydroelectric generation, and atomic energy. He is responsible for power system studies, including short circuit, load flow, harmonics, stability, arc flash hazard, grounding, switching transients, and protective relaying. He conducts courses for continuing education in power systems and has authored or coauthored about 70 technical publications nationally and internationally. He is the author of the following books:

- *Arc Flash Hazard Analysis and Mitigation*, IEEE Press, 2012.
- *Power System Harmonics and Passive Filter Designs*, IEEE Press, 2015.
- *Transients in Electrical Systems: Analysis Recognition and Mitigation*, McGraw-Hill, 2010.
- *Power System Analysis: Short-Circuit Load Flow and Harmonics*, Second Edition, CRC Press, 2011.
- *Understanding Symmetrical Components for Power System Modeling*, IEEE Press, 2017.

These books provide extensive converge, running into more than 3000 pages and are well received in the technical circles. His interests include power system transients, EMTP simulations, harmonics, passive filter designs, power quality, protection, and relaying. He has published more than 200 electrical power system study reports for his clients.

He has published more than 200 study reports of the power systems analysis addressing one problem or the other.

Mr. Das is a Life Fellow of the Institute of Electrical and Electronics Engineers, IEEE (USA), a Member of the IEEE Industry Applications and IEEE Power Engineering societies, a Fellow of the Institution of Engineering Technology (UK), a Life Fellow of the Institution of Engineers (India), a Member of the Federation of European Engineers (France), a Member of CIGRE (France), etc. He is a registered Professional Engineer in the States of Georgia and Oklahoma, a Chartered Engineer (C.Eng.) in the UK and a European Engineer (Eur. Ing.) in Europe. He received a meritorious award in engineering, IEEE Pulp and Paper Industry in 2005.

He earned a PhD in electrical engineering at Atlantic International University, Honolulu, an MSEE at Tulsa University, Tulsa, Oklahoma, and a BA in advanced mathematics and a BEE at Panjab University, India.

1

Load Flow—Fundamental Concepts

Historically, the development of technology for predictive and control analysis of load flow problems was rather slow—the analogue computing techniques: network analyzers taking much time to model and were slow for a specific load flow problem. Another limitation is that large systems could not be modeled. With the advent of digital computers, much advancement has taken place. This is discussed in the chapters to follow.

The load flow problem arises from the fact that consumers must be provided with reliable source of electrical power for their varying needs, which involves much planning, like generation, renewables, transmission, and distribution and in the deregulated environment, this has become a very competitive field. The economic provisions of electrical power to the consumers require knowledge of so many interrelated functions; for example, adequately sizing the electrical equipment in the long chain of generation, transmission, and distribution: see Volume 1 for the nature of modern electrical power systems. Even in the isolated systems, same fundamental considerations arise, the renewables, microgrids, solar and wind generation, government subsidies for green power adding to the complexity. The load flow and optimization techniques are the tools to provide guidance through this maze.

1.1 Security Assessments

The 1965 blackout led to the various efforts for improving reliability. Several emergency guidelines and criteria were introduced by the Federal Power Commission and North American Power System Interconnection subcommittee. In this respect, "load flow" is not a stand-alone problem. The stability, protection, and transient behavior of a system plus the system management are important links. The power system security means the ability to withstand sudden disturbances such as short circuits and unexpected loss of system components. In terms of planning and operation, a power system should

- Ride through the transients and return to a steady-state operation without much impact on the consumers
- In the new steady state, no system component should be overloaded and should operate within their design parameters

These ideal conditions may not always be met in practice.

We can define three states as follows:

1. Normal
2. Emergency
3. Restorative

Under normal operation, all consumer demands are met and all equipment operates slightly below its rated capability. Mathematically, under normal state

$$\begin{aligned} &x_1(a_1, a_2, \ldots, a_n; b_1, b_2, \ldots, b_n) = 0 \\ &x_2(a_1, a_2, \ldots, a_n; b_1, b_2, \ldots, b_n) = 0 \\ &\ldots \\ &x_n(a_1, a_2, \ldots, a_n; b_1, b_2, \ldots, b_n) = 0 \end{aligned} \tag{1.1}$$

where a_1, a_2,... are a set of dependent state variables and b_1, b_2,... are a set of independent variables, representing demand, inputs, and control. These correspond to load flow equations.

Relative to equipment the constraints are as follows:

$$\begin{aligned} &y_1(a_1, a_2, \ldots, a_n; b_1, b_2, \ldots, b_n) \leq 0 \\ &y_2(a_1, a_2, \ldots, a_n; b_1, b_2, \ldots, b_n) \leq 0 \\ &\ldots \\ &y_n(a_1, a_2, \ldots, a_n; b_1, b_2, \ldots, b_n) \leq 0 \end{aligned} \tag{1.2}$$

The equipment constraints can be described, for example, as upper and lower limits of a generator reactive power limits, system plus minus voltage limits, and the like.

In the emergency state, for example, brought out by a short circuit, some of the inequality constraints may be violated—the frequency may deviate, the lines and machines may be overloaded, etc. Mathematically, we write

$$\begin{aligned} &x_1(a_1, a_2, \ldots, a_n; b_1, b_2, \ldots, b_n) = 0 \\ &x_2(a_1, a_2, \ldots, a_n; b_1, b_2, \ldots, b_n) = 0 \\ &\ldots \\ &x_n(a_1, a_2, \ldots, a_n; b_1, b_2, \ldots, b_n) = 0 \end{aligned} \tag{1.3}$$

and

$$\begin{aligned} &y_1(a_1, a_2, \ldots, a_n; b_1, b_2, \ldots, b_n) \neq\leq 0 \\ &y_2(a_1, a_2, \ldots, a_n; b_1, b_2, \ldots, b_n) \neq\leq 0 \\ &\ldots \\ &y_n(a_1, a_2, \ldots, a_n; b_1, b_2, \ldots, b_n) \neq\leq 0 \end{aligned} \tag{1.4}$$

In the restorative state, only some customers may be satisfied without overloading any equipment—thus, all equality constraints are not satisfied. We write

$$
\begin{aligned}
&x_1(a_1,a_2,\ldots,a_n;b_1,b_2,\ldots,b_n)\neq 0\\
&x_2(a_1,a_2,\ldots,a_n;b_1,b_2,\ldots,b_n)\neq 0\\
&\ldots\\
&x_n(a_1,a_2,\ldots,a_n;b_1,b_2,\ldots,b_n)\neq 0
\end{aligned}
\tag{1.5}
$$

and

$$
\begin{aligned}
&y_1(a_1,a_2,\ldots,a_n;b_1,b_2,\ldots,b_n)=0\\
&y_2(a_1,a_2,\ldots,a_n;b_1,b_2,\ldots,b_n)=0\\
&\ldots\\
&y_n(a_1,a_2,\ldots,a_n;b_1,b_2,\ldots,b_n)=0
\end{aligned}
\tag{1.6}
$$

1.2 Control Actions

A normal state is *secure* if following any one of the postulated disturbances, the system remains in the normal state; otherwise, it is *insecure.* An operator may intervene and manipulate the system variables so that the system is secure—this is called preventive control. When the system is in the emergency mode, a *corrective* control action is possible which will send the system to normal state. If corrective control fails, emergency control can be applied. Finally, a restorative control should put the system in normal state.

1.3 Consumer Loads

Before we discuss load types, it is useful to define certain established parameters with respect to consumer loads.

1.3.1 Maximum Demand

The load demand of a consumer is not constant. It varies with the time of the day, the season (i.e., summer or winter months). For industrial processes, the load demand may not be constant and vary with the process or production rate. For residential consumers, the utilities do not, generally, levy a demand charge, but for industrial consumers, the maximum demand registered over a period of 30 min or 1 h is built into the tariff rates.

The instantaneous peak is always the highest. A peaky load imposes constraints on the electrical supply system. Imagine if the peaks of all consumers were to occur simultaneously at a certain time of the day, then the complete supply system should be capable of supplying this peak, while at other times its installed capacity lies idle. When considering peaky loads, we must remember that

- The electrical equipment has limited short-duration overload capability, i.e., power transformers can take a certain amount of overload for certain duration without derating the life expectancy and these limits have been established in the standards. However, the circuit breakers may have only limited overload capability. The overload capability of a system will be established by the weakest link in the system.
- A peaky load may give rise to excessive voltage saga and power quality may become a problem.
- Protective devices may operate on sudden inrush currents. Starting of a large motor in an industrial system is an example of such peaky loads. Another example can be cited of a process plant, which generates most of its power, and has utility's system as a standby. Normally, the plant generators supply most of the power requirements, but on loss of a machine, the utility tie must be capable of supporting the load without voltage dips for continuity of the processes.
- All these scenarios are interrelated and are carefully analyzed before a tie with the utility's system is established.

1.3.2 Load Factor

Load factor can be defined as the ratio of the average power to the maximum demand *over a certain period.* The definition will be meaningless unless the time period is associated with it, i.e., 30 min, monthly, or yearly:

$$\text{Load factor} = \frac{\text{Kwh consumed}}{\text{Maximum demand} \times \text{Hours}} \tag{1.7}$$

1.3.3 Diversity Factor

Diversity factor is the ratio of the sums of the maximum demands of various consumers or subdivisions of a system or the part of a system to the maximum demand of the whole or part of the system under consideration:

$$DF = \frac{\sum d}{D} \tag{1.8}$$

where d is the sum of the maximum demands of the subdivisions or consumers and D is the maximum demand of the whole system. This is graphically shown in Figure 1.1.

Example 1.1

A power station supplies the following consumers with monthly maximum demands of: Industrial = 100 MW, Residential = 30 MW, and Commercial = 20 MW. The maximum demand of the power station is 120 MW and it generates 40 GWh units in 1 month. Calculate the monthly load factor and diversity factor.

Load factor = 46%
Diversity factor = 1.25

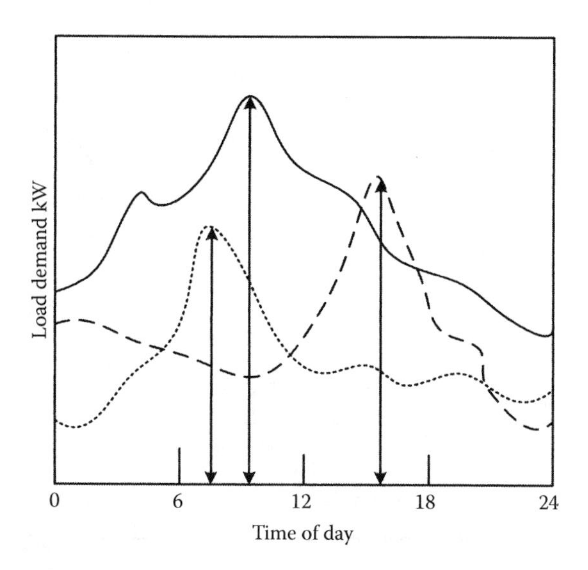

FIGURE 1.1
Illustration of diversity factor.

1.4 Load Types

The load types can be divided into the following categories:

Domestic: These consist of lighting, HVAC, household appliances, television, computers, etc. and have poor load factor. On a cold day, all consumers will switch on their heaters and on a hot day, their air conditioners, practically at the same time.

Commercial: These consist of mainly lighting, computer terminals, HVAC, neon signs, and displays as used in most commercial establishments. The power requirements for hospitals and health care facilities are special requirements.

Industrial: Industrial power systems vary considerably in size, complexity, and nature of loads. A 500 kVA outdoor pad-mounted transformer in a radial feed arrangement connected to a lineup of low voltage motor control center may be classed as an industrial distribution system. Conversely, a large industrial distribution system may consume tens of MW of power.

Transit systems, lighting of public buildings, entertainment parks, and roads are some other examples of specific load types.

Figure 1.2 shows the patterns of residential, commercial, and industrial loads.

1.4.1 Load Characteristics

Load characteristics are important. The voltages in power system swing and the various types of load will behave differently under voltage variations. We can divide the loads into

- Constant MVA type
- Constant impedance type
- Constant current type

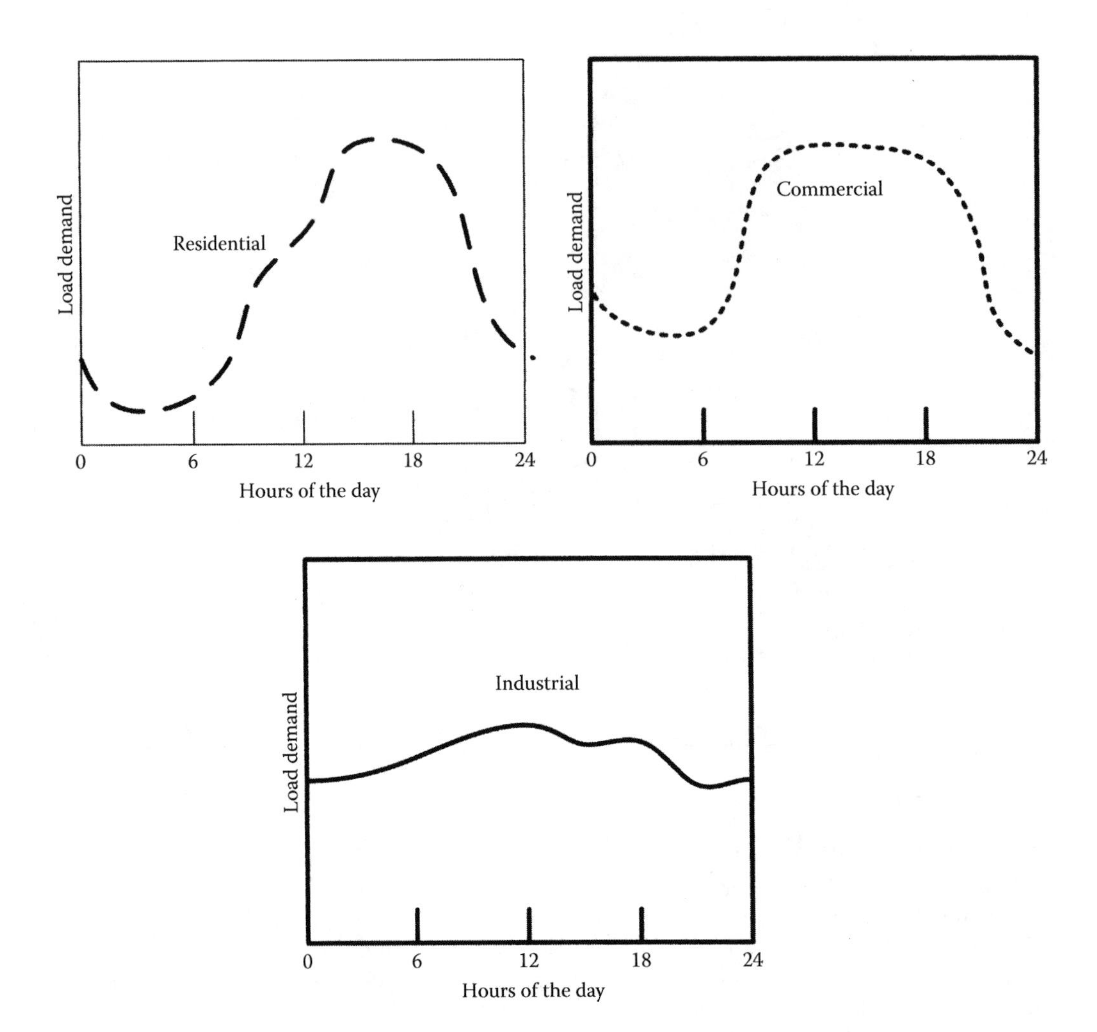

FIGURE 1.2
Daily load profiles of typical customers.

The effect of change of operating voltage on constant current, constant MVA, and constant impedance load types is discussed in Chapter 6. Heavy industrial motor loads are approximately constant MVA loads, while commercial and residential loads are mainly constant impedance loads. Classification into commercial, residential, and industrial is rarely adequate and one approach has been to divide the loads into individual load components. The other approach is based upon measurements. Thus, the two approaches are as follows:

- Component-based models
- Models based upon measurements

See Chapter 6 for further discussions and mathematical equations. Modeling the correct load type in a load flow problem is important to avoid errors and erroneous results. The nonlinar loads are discussed in Volume 3.

1.5 Effect on Equipment Sizing

Load forecasting has a major impact on the equipment sizing and future planning. If the forecast is too optimistic, it may lead to the creation of excess generating capacity, blocking of the capital, and uneconomical utilization of assets. On the other hand, if the forecast errs on the other side, the electrical power demand and growth will not be met. This has to be coupled with the following facts:

- Electrical generation, transmission, and distribution facilities cannot be added overnight and takes many years of planning and design engineering efforts.
- In industrial plant distribution systems, it is far easier and economical to add additional expansion capacity in the initial planning stage rather than to make subsequent modifications to the system, which are expensive and also may result in partial shutdown of the facility and loss of vital production. The experience shows that most industrial facilities grow in the requirements of power demand.
- Energy conservation strategies should be considered and implemented in the planning stage itself.
- Load management systems—an ineffective load management and load dispatch program can offset the higher capital layout and provide better use of plant, equipment, and resources.

Thus, load forecasting is very important in order that a electrical system and apparatus of the most economical size be constructed at the correct place and the right time to achieve the maximum utilization.

For industrial plants, the load forecasting is relatively easy. The data are mostly available from the similar operating plants. The vendors have to guarantee utilities within narrow parameters with respect to the plant capacity.

Something similar can be said about commercial and residential loads. Depending upon the size of the building and occupancy, the loads can be estimated.

Though these very components will form utility system loading, yet the growth is sometimes unpredictable. The political climate and the migration of population are not easy to forecast. The sudden spurs of industrial activity in a particular area may upset the past load trends and forecasts.

Load forecasting is complex, statistical, and econometric models are used [1, 2]. We will briefly discuss regression analysis in the following section.

1.6 Regression Analysis

Regression or trend analysis is the study of time series depicting the behavior pattern of process in the past and its mathematical modeling so that the future behavior can be extrapolated from it. *The fitting of continuous mathematical functions through actual data to achieve the least overall error is known as regression analysis.* The purpose of regression is to estimate one of the variables (dependent variable) from the other (independent variable). If y is estimated from x by some equation, it is referred to as regression. In other words, if the scatter diagram of two variables indicates some relation between these variables, the dots will be concentrated around a curve. This curve is called the curve of regression. When the curve is a straight line, it is called a line of regression.

Typical regression curves used in forecasting are as follows:

$$\text{Linear}: y = A + Bx \tag{1.9}$$

$$\text{Exponential}: y = A(1+bx)^2 \tag{1.10}$$

$$\text{Power}: y = Ax^B \tag{1.11}$$

$$\text{Polynomial}: A + Bx + Cx^2 \tag{1.12}$$

The limitations are that the past data may not be indicative of the future. They depend upon the standard of living, the GNP, and the geography of the terrain.

For short-term forecasting, a sequence of discontinuous lines or curves may be fitted.

Consider a total process represented by Figure 1.3. It can be broken down into the following subprocesses:

- *Basic trend, Figure 1.4a.*
- *Seasonal variations, Figure 1.4b*: The monthly or yearly variations of the load. The average over the considered period of time is zero. The long-term mean is zero.
- *Random variations, Figure 1.4c* These occur on account of day-to-day changes and are usually dependent upon the time of the week, weather, etc. The long-term mean is zero.

The interest is centered on yearly peaks and not on the whole load curve. A straight regression line can be drawn through the load peaks. The least square regression method is most suitable for long-term forecasts, Figure 1.5a and b.

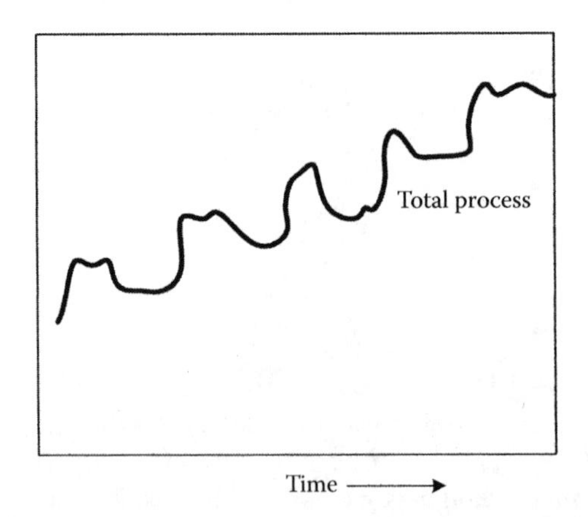

FIGURE 1.3
Total load growth.

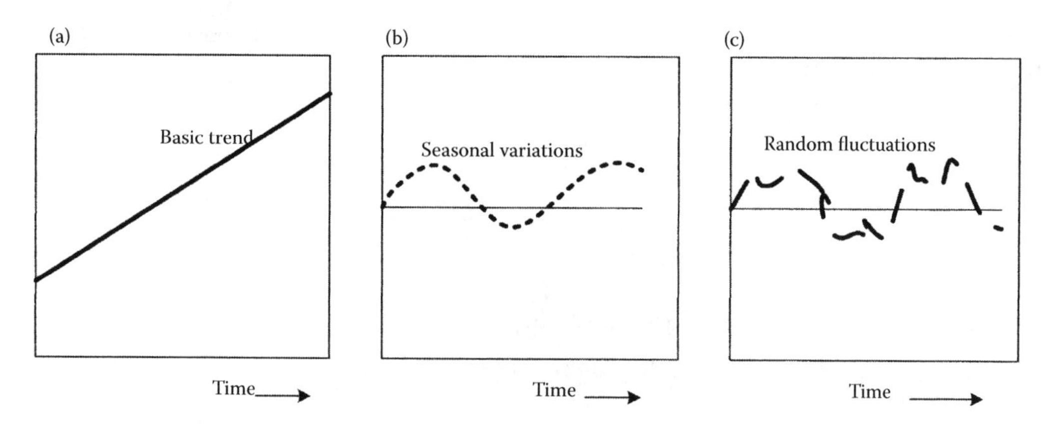

FIGURE 1.4
(a)–(c) Decomposition of total load growth in Figure 1.3.

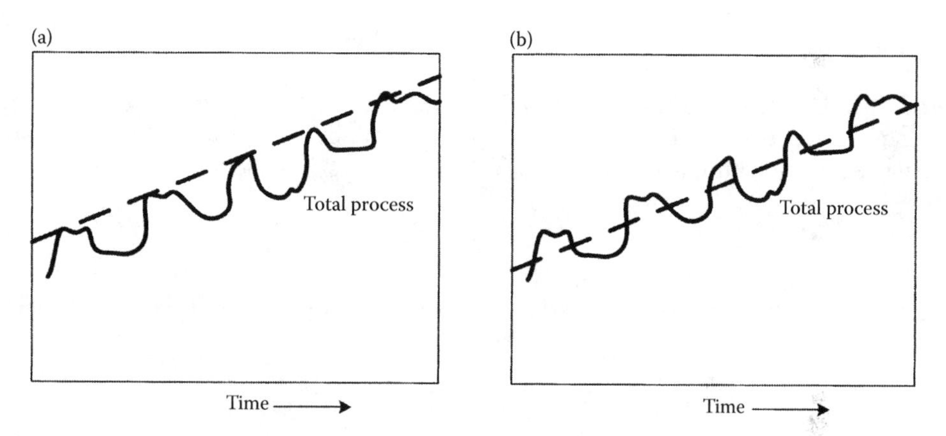

FIGURE 1.5
Regression curves through peak and through mean.

1.6.1 Linear Trend

This denotes a trend where increase in demand from year to year is constant, rarely true. Such a trend can be expressed as follows:

$$C_t = a + bt \tag{1.13}$$

C_t=consumption of electricity in some future year t
a=consumption for base year, t=0
b=annual increase in energy consumption
$t=T-1+n$, where T is the number of years for which statistical trend is studied and n is the number of years for which the forecast is required

1.6.2 Exponential Trend

An exponential trend is denoted by

$$C_t = C_0(1+m)^t \tag{1.14}$$

where m=mean rate of growth observed in T years. By taking logs to base 10, the graph becomes a straight line:

$$\log C_t = a + bt \tag{1.15}$$

1.7 Curve Fitting—Least Square Line

For some given data points, more than one curve may seem to fit. Intuitively, it will be hard to fit an appropriate curve in a scatter diagram and variation will exist.

Referring to Figure 1.6, a measure of goodness for the appropriate fit can be described as follows:

$$d_1^2 + d_2^2 + \cdots + d_n^2 = a = \min \tag{1.16}$$

A curve meeting these criteria is said to fit the data in the *least square sense* and is called a least square regression curve, or simply a least square curve—straight line or parabola. The least square line imitating the points $(x_1, y_1), \ldots, (x_n, y_n)$ has the following equation:

$$y = a + bx \tag{1.17}$$

The constants a and b are determined from solving simultaneous equations, which are called the normal equations for the least square line:

$$\begin{aligned} \sum y &= an + b\sum x \\ \sum xy &= a\sum x + b\sum x^2 \end{aligned} \tag{1.18}$$

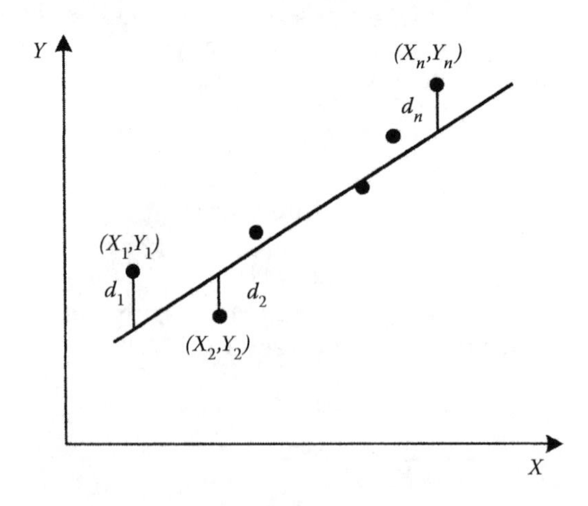

FIGURE 1.6
Fitting a least square line.

This gives

$$a = \frac{\left(\sum y\right)\left(\sum x^2\right) - \left(\sum x\right)\left(\sum xy\right)}{n\sum x^2 - \left(\sum x\right)^2}$$

$$b = \frac{n\sum xy - \left(\sum x\right)\left(\sum y\right)}{n\sum x^2 - \left(\sum x\right)^2} \tag{1.19}$$

The constant *b* can be written as follows:

$$b = \frac{\sum (x - x_{\text{mean}})(y - y_{\text{mean}})}{\sum (x - x_{\text{mean}}^2)} \tag{1.20}$$

This yields

$$y_{\text{mean}} = a + bx_{\text{mean}} \tag{1.21}$$

From the above equation, we can also write the least square line as follows:

$$y - y_{\text{mean}} = b(x - x_{\text{mean}}) = \frac{\sum (x - x_{\text{mean}})(y - y_{\text{mean}})}{\sum (x - x_{\text{mean}})^2}(x - x_{\text{mean}}) \tag{1.22}$$

Similarly, for *x* on *y*,

$$x - x_{\text{mean}} = \frac{\sum (x - x_{\text{mean}})(y - y_{\text{mean}})}{\sum (y - y_{\text{mean}})^2}(y - y_{\text{mean}}) \tag{1.23}$$

The least square line can be written in terms of variance and covariance. The sample variance and covariance are given by

$$S_x^2 = \frac{\sum (x - x_{\text{mean}})^2}{n}$$

$$S_y^2 = \frac{\sum (y - y_{\text{mean}})^2}{n} \tag{1.24}$$

$$S_{xy} = \frac{\sum (x - x_{\text{mean}})(y - y_{\text{mean}})}{n}$$

In terms of these, the least square lines of y on x and x on y

$$y - y_{\text{mean}} = \frac{S_{xy}}{S_x^2}(x - x_{\text{mean}})$$
$$x - x_{\text{mean}} = \frac{S_{xy}}{S_y^2}(y - y_{\text{mean}}) \quad (1.25)$$

A sample correlation coefficient can be defined as follows:

$$r = \frac{S_{xy}}{S_x S_y} \quad (1.26)$$

For further reading see References [3–5].

Example 1.2

Given the data points (x, y) as (1, 1), (3, 2), (4, 5), (6, 7), (7, 6), (9, 8), (12, 10), (15, 16), fit a least square line with x as independent variable, y as dependent variable.

Table 1.1 shows the various steps of calculations. Then, we have

$$8a + 57b = 55$$
$$57a + 561b = 543$$

where n=number of samples=8. Solving these equations, a=–0.0775 and b=0.975. Therefore, the least square line is

$$y = -0.0775 + 0.975x$$

If x is considered as the dependent variable and y as the independent variable, then

$$8c + 55d = 57$$
$$55c + 535d = 543$$

Solution of which gives, c=0.507 and d=0.963.

TABLE 1.1

Fitting the Least Square Line, Example 1.2

x	y	x^2	xy	y^2
1	1	1	1	1
3	2	9	6	4
4	5	16	20	25
6	7	36	42	49
7	6	49	42	36
9	8	81	72	64
12	10	144	120	100
15	16	225	240	256
$\sum x = 57$	$\sum y = 55$	$\sum x^2 = 561$	$\sum xy = 543$	$\sum y^2 = 535$

1.8 Load Estimation and Projection in Distribution Systems

The load estimation and forecasting is one key element in the distribution system planning. The metered loads at all points in the distribution system are rarely available. Thus, multi-scenario load studies become mandatory. For every MW of load, engineering analysis of roughly 100 nodes is required. The location of substations, routing of feeders, and sizing of equipment depend on it. Furthermore, there is an element of uncertainty—even the best estimates may leave some uncertainty about the future conditions.

Thus, forecasting involves projecting the number and types of customers, providing geographical or spatial information on future load growth required to identify future problem areas in advance and perform studies of substation sizing, locations, and feeder routings.

1.8.1 Small-Area Forecasting

The utility service area is divided into a number of small areas and the load forecasting is done in each area; this gives an idea where and how much the load will grow throughout the system. Two approaches are as follows:

- Equipment-area-oriented forecasting by feeder, substation, or other sets of areas defined by equipment—most utilities define small areas in the range of 10–160 acres (0.04–0.64 km^2)
- A uniform grid based on some mapping coordinate systems

See Figure 1.7a and b.

The planning is performed on an annual basis. The small-area forecasts concentrated on a projection of the annual peak load for each area, over the period of time for which planning is to be done. The concept is to provide a load forecast change over time. The average short-range period used by utilities is 5 years. Long-range planning involves 5–25 years. Figure 1.8 shows the load behavior with respect to area size (number of consumers).

1.8.2 Spatial Forecast Methods

Mainly, there are two categories:

Trending methods that extrapolate past trends in annual peaks—these are applied to equipment on small-area basis.

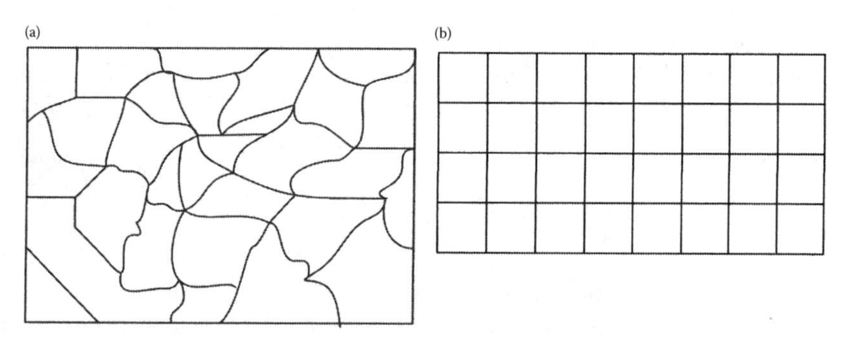

FIGURE 1.7
(a) Diffused small areas for load growth defined by equipment; (b) by an explicit grid.

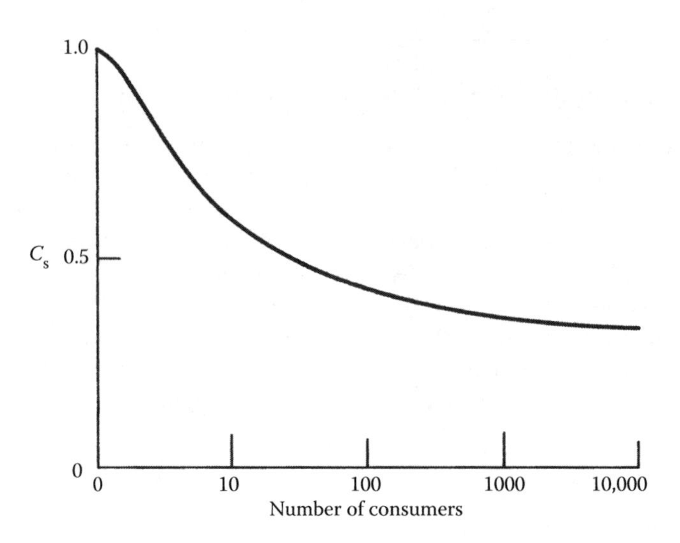

FIGURE 1.8
Coincidence of peak load of a number of consumers.

Land-use-based methods that forecast by analyzing zoning, municipal plans, and other land-use factors—these are applied on a grid basis.

The spatial forecast methods on digital computers were first applied in 1950 and land-use methods in the early 1960s. Development of computer-based forecast methods has accelerated during the last 20 years. EPRI project report RP-570 investigated a wide range of forecast methods, established a uniform terminology, and established many major concepts and priorities [6]. Almost all modern spatial forecast methods are allocation methods, on a small-area basis.

We have already observed that the load demands vary, and these will not occur simultaneously. Define a "coincidence factor":

$$C_s = \frac{\text{(Peak system load)}}{\text{(Sum of customers peak loads)}}$$

The more the number of customers, the smaller will be the value of C_s. For most power systems, it is 0.3–0.7.

One approach is to use coincidence factors and the equipment may be sized using a coincidence curve similar to Figure 1.8. But few spatial models explicitly address coincidence of the load as shown in Figure 1.8. An approach to include coincidence is given by

$$I_k(t) = \frac{C(e)}{C(a)} \times N_k \times L \tag{1.27}$$

where a is the average number of consumers in an area, e is the average number of consumers in a substation of feeder service area, N is the total consumers in the small area, and L is their average peak load. The load behavior as a function of the area size is shown in Figure 1.9. Figure 1.9a is the total system, Figure 1.9b is several km^2, and Figure 1.9c is small

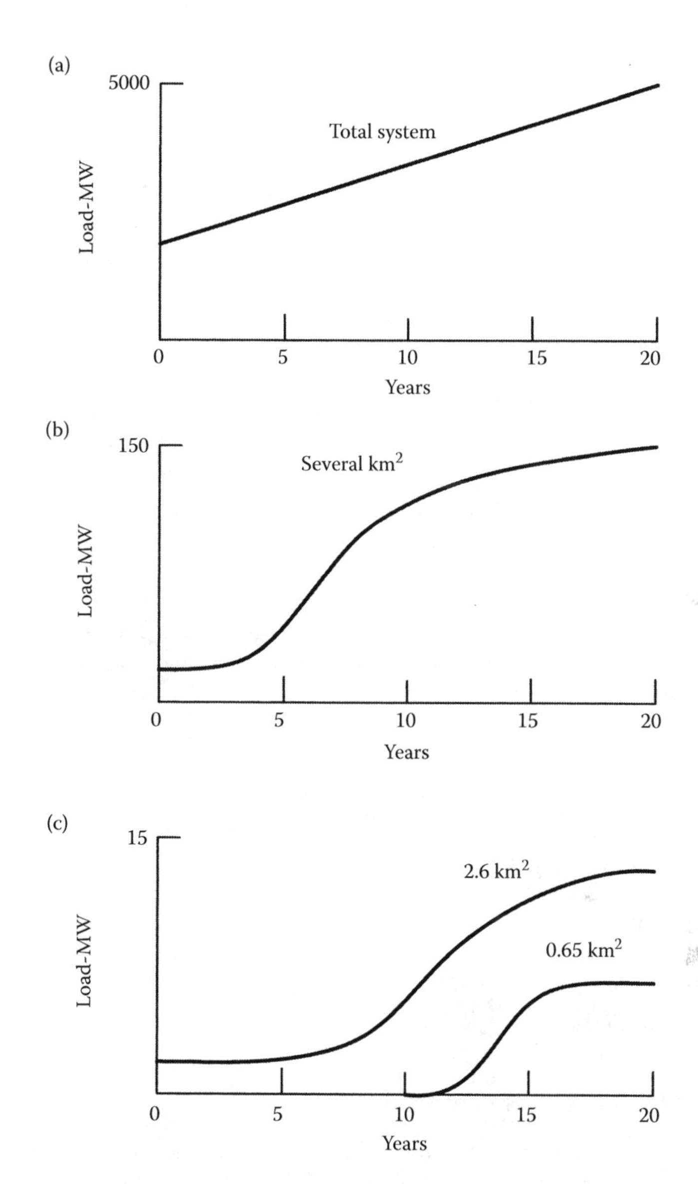

FIGURE 1.9
Average load growth profiles as a function of area: (a) total load growth; (b) area of several km^2; and (c) areas of 2.6 and 0.65 km^2.

areas. Furthermore, Figure 1.10 shows the behavior of load growth in a small area; the first dotted curve is so-called *S* curve. The majority of load growth occurs only in a few years and the curve flattens out. The *S* curve varies with respect to the size of the area.

In spite of all the advancements in computer modeling, the future load projections may have errors, which impact the distribution system planning. The political climate of an area may change, the industries may move in and out, and with that the number of consumers.

This led to the spatial frequency analysis. The errors can be broken into their spatial frequencies. The planning is not sensitive to that portion of the error that is composed of high spatial frequencies—like load forecast errors. The low-frequency spatial errors have a high impact. Forecasting algorithms can be studied in the spatial frequency

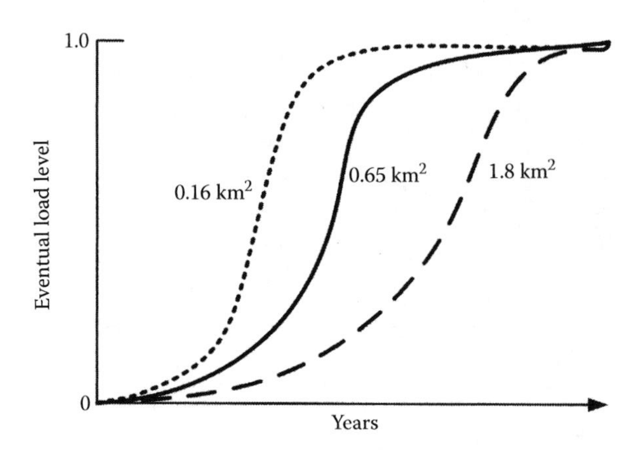

FIGURE 1.10
Typical load behavior in small areas, *S* curves, the sharpness depending on the area.

domain with respect to their impact on distribution system planning. The forecast of individual small-area loads is not as important as the assessment of overall spatial aspects of load distribution.

1.9 Load Forecasting Methods

We can categorize these as follows:

- Analytical methods
- Nonanalytical methods

1.9.1 Analytical Methods: Trending

There are various algorithms that interpolate past small-area load growth and apply regression methods as discussed in Section 1.7 These extrapolate based on past load values—these are simple, require minimum data, and are easy to apply. The best option is the multiple regression curve fitting of a cube polynomial to most recent, about 6 years of small-area peak history. However, this extrapolation can lead to inaccuracy. When a "horizon load estimate" is inputted, the accuracy can be much improved, see Figure 1.11 curves *A* and *B*.

Other sources of error in trending forecast are as follows:

Load transfer coupling (LTC) regression. The load is moved from one service area to another—these load shifts may be temporary or permanent. These create severe accuracy problems. The exact amount of load transfer may not be measured. The accuracy can be improved with a modification to the regression curve fit method, called LTC. The exact amount of load transfer need not be known. LTC is not described, see Reference [7].

The other problem is inability to trend vacant areas. Future growth in a vacant area can often be estimated by assuming that the continuous load growth in an area can continue

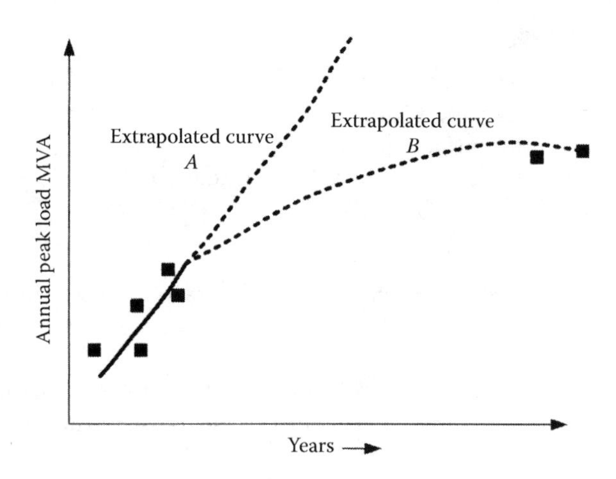

FIGURE 1.11
(a) Trending analysis and extrapolation with no ultimate or horizon year load, curve *A*; (b) with input of horizon year loads, curve *B*.

only if vacant areas within the region start growing. A technique called vacant area interference applies this concept in a repetitive and hierarchical manner [8].

Another variation of the trending methods is clustering template matching method—a set of about six typical "s" curves of various shapes called templates are used to forecast the load in a small area.

1.9.2 Spatial Trending

An improvement in trending can be made by including parameters other than load history. For example, the load density will be higher in the central core of the city. Small-area trending used a function that is the sum of one or more monotonically decreasing functions of distance from the central urban pole, see Figure 1.12.

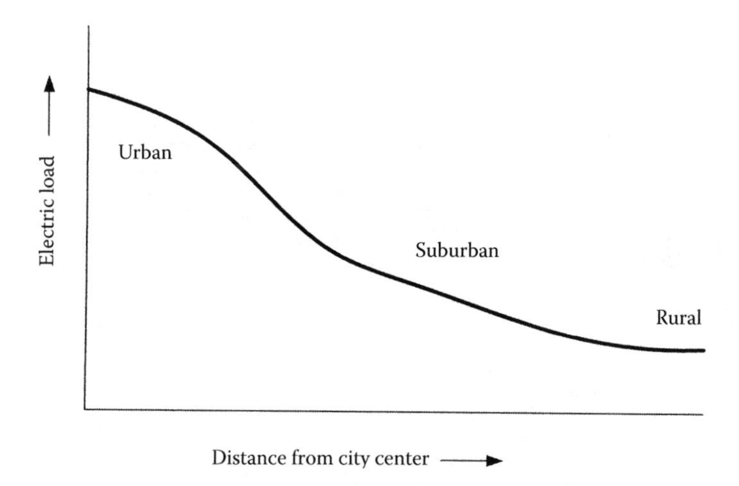

FIGURE 1.12
Spatial load distribution urban, suburban, and rural, as the distance from the city center increases.

1.9.3 Multivariate Trending

Multivariate trending used as many as 30 non-load-related measurements on each small area. These total variables are called "data vector." The computer program was developed under EPRI, known as "multivariate" [6]. This produced better results compared to other trending methods as per a series of test results conducted by EPRI.

1.9.4 Land-Use Simulations

Land-use spatial methods involve an intermediate forecast of growth of land-use type and density as a first step in electrical load forecast. The steps are listed as follows:

- The total growth and the regional nature of its geographical distribution are determined on a class-by-class basis. This means determining growth rated for each land-use class.
- Assign growth to small areas and determine how much of total growth in each class occurs in which small area.
- Determine the load of each small area based on its forecast land-use class composition and a class-based load model which converts class-based use to kW of load.

All land-use-based methods employ analyses in each of three categories, see Figure 1.13.

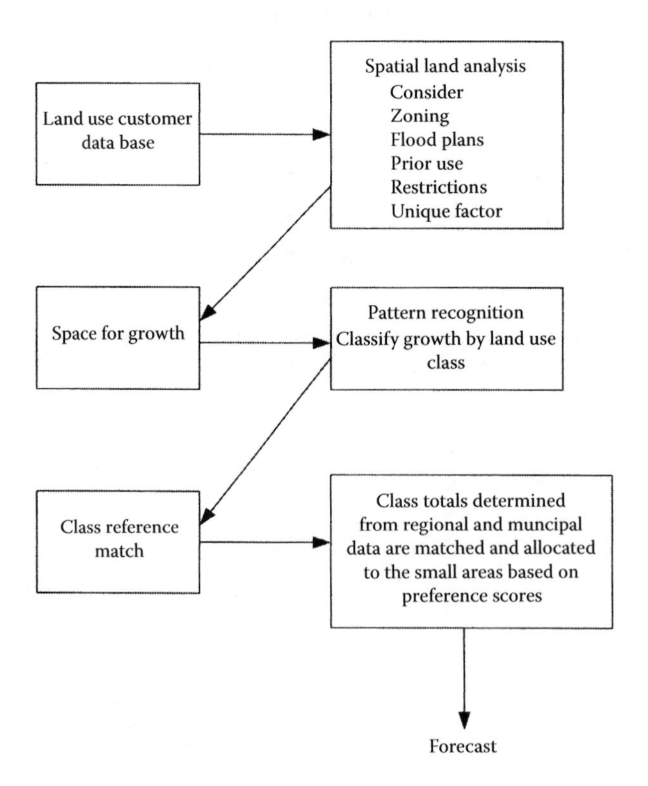

FIGURE 1.13
Land-use-based load estimation simulation steps.

1.9.5 Nonanalytical Methods

These rely on users intuition and do not use computer simulations. A "color-book" approach is described in Reference [9]. A map of utility area is divided into a series of grid lines. The amount of existing land use, for example, industrial, residential, and commercial is colored in on the map. Future land use based on user's intuition is similarly coded in to the map. An average load density for each land use is determined from experience, survey data, and utilities overall system load forecast.

Figure 1.14 shows the accuracy of major types of load forecasts over a period of time, also see Reference [10].

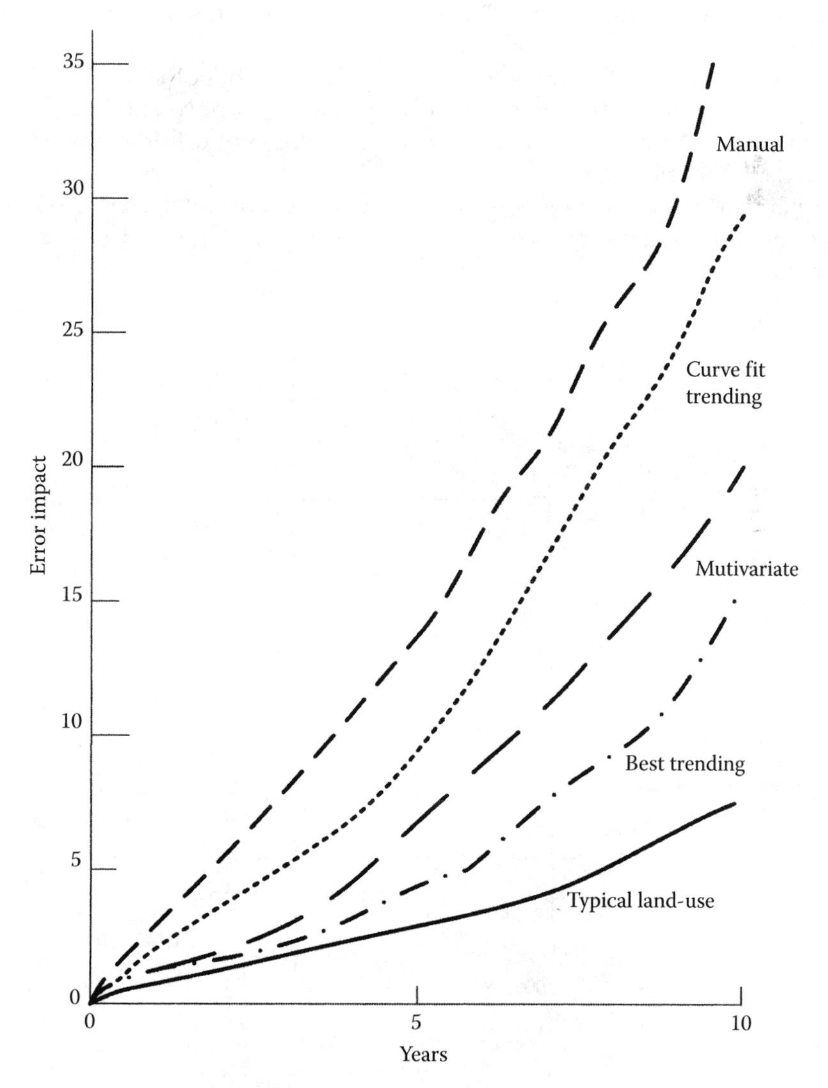

FIGURE 1.14
Estimated load forecasting errors with different methodologies.

1.10 Iterative Nature of Load Flow Problem

Load flow solutions require numerical techniques, discussed in the chapters to follow. To explain the nature of load flow, consider a simple circuit, as shown in Figure 1.15a, which depicts a 2.5 MVA 13.8–0.48 kV transformer, carrying a constant impedance type of 2000 kVA load at a power factor of 0.8 lagging. The transformer impedance and source short-circuit levels are as shown. It is required to calculate the transformer secondary voltage.

The loads on buses are invariably specified in terms of MW/Mvar or KVA and power factor. In this small problem, we need to find the current flowing through the system impedance, to calculate the voltage, assuming that the source voltage is known. However, the load current at the secondary of the transformer, in the above example, cannot be calculated, as the voltage is not known. Thus, there are two unknowns: current and voltage that depend on each other.

We conceive a "swing bus" or an "infinite bus," which is a hypothetical bus, for the load flow problem. It is an ideal Thévenin source as well as an ideal Norton's source. In other words, any amount of current taken from this source does not result in any voltage drop. This means zero Thévenin impedance.

Figure 1.15b shows the equivalent impedance circuit in ohms referred to 480 V side. Load flow programs work on a pu base, generally 100 or 10 MVA. All impedances are converted to this common base.

(a)

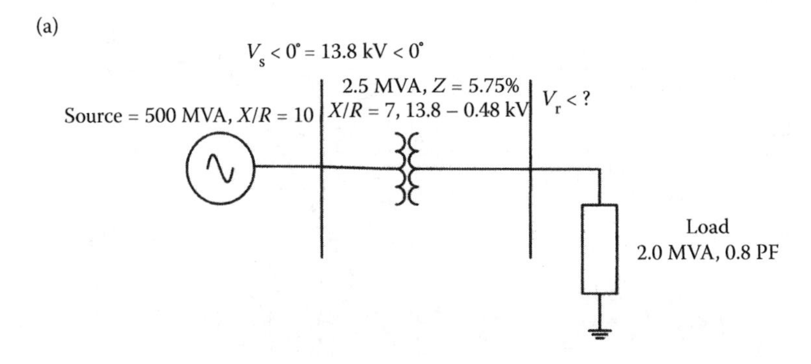

(b)

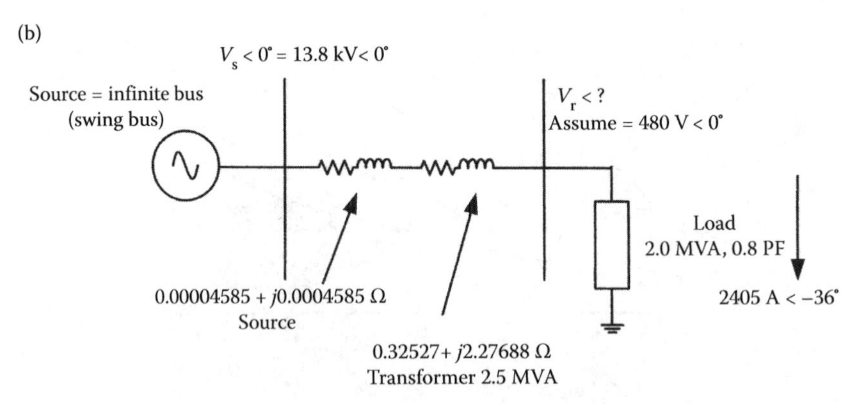

FIGURE 1.15
Hand calculations of secondary voltage on 2.5 MVA transformer carrying 2.0 MVA load at 0.8 PF: (a) system configuration; (b) step one of the calculations.

As a first step assume that

$$V_s < 0° = V_R < 0°$$

This means no voltage drop on load flow. Here, we have only one load bus, and practically, there may be hundreds or thousands of load buses interconnected through cables, transformer, or overhead lines. Initially, a voltage equal to 1.0 pu at an angle of 0° is assumed for all the buses.

As the secondary voltage is assumed as 480<0°, we can calculate the load current: 4520 < –36°.

As it is constant impedance load, the load impedance is 0.199<36°.

Then, the transformer secondary voltage is

$$480 - \left[\sqrt{3}(2405)(0.8 - j0.6)(0.000795 + j0.005704)\right] = 480 - 16.905 - j17.021$$

$$= |463.40| < -2.105°$$

This is the first estimate of the voltage. The final result can be close to it or very much different.

Based on this first estimate of voltage, recalculate the load current. As it is a constant impedance load, the current varies in proportion to the load. The new load current is

$$\frac{463.09 - j17.02}{1.99(0.8 + j0.6)} = 1810 - j1465$$

Then, the second estimate of the voltage is

$$480 - \left[\sqrt{3}(1810 - j1465)(0.000795 + j0.005704)\right] = 480 - 16.965 - j15.864$$

$$= |463.30| < -2.180°$$

This is fairly close to the first result. It is a simple one-bus system that we are studying to illustrate the procedure. The iterations are stopped when the differences between two successive calculations go on decreasing. The final result is printed when the difference goes as little as 0.00001 for the Gauss–Seidel method. (The tolerance limits are user adjustable.) In this case, we say that the load flow has converged.

Sometimes, the results in successive calculations may first narrow down and then increase; the results may swing and a convergence may not be achieved. The characteristics of various load flow algorithms are discussed in the chapters to follow.

Using a computer program with Gauss–Seidel method, the load flow converges in 14 iterations and the final value of the voltage is 46.392 V. The power input to the transformer is 1.5404 MW and 1.211 Mvar (power factor 77.90 lagging); the system losses are 0.013 MW and 0.092 Mvar.

The transformer is provided with off-load taps of +5%, +2.5%, rated tap, –2.5% and –5% on the 13.8 kV windings. By setting the taps to –5% (13,110 V), the turns ratio of the transformer is reduced and the secondary voltage rises by 5% at no-load. Table 1.2 shows the

TABLE 1.2

Effect of Off-Load Tap Settings on Load Flow

		Load on Transformer			Losses	
Tap	**Bus Voltage (%)**	**MW**	**Mvar**	**PF**	**MW**	**Mvar**
–5%	101.59	1.606	1.342	77.88 Lag	0.014	0.105
–2.5%	99.00	1.582	1.274	77.89 Lag	0.014	0.097
Rated	96.54	1.504	1.211	77.90 Lag	0.013	0.092
+2.5%	94.20	1.4323	1.152	77.91 Lag	0.012	0.088
+5%	91.97	1.365	1.098	77.91 Lag	0.012	0.083

effect of various tap settings on load flow. Note that the transformer load varies as we have modeled a constant impedance load. Various load types will behave differently. Transformers may be provided with under load tap changing and the appropriate taps to maintain a voltage close to the rated voltage can be automatically selected in the load flow. These aspects are further discussed in the chapters to follow.

Practically, the load flow can be complex. There may be a number of constraints and certain objectives are required to be met, for example,

- Active power cost minimization
- Active power loss minimization
- Minimize control operations
- Unit commitment

The control variables include the following:

- Real and reactive power generation
- Net interchange control, see Chapter 2
- Load scheduling, see Chapter 2
- Control voltage settings
- Load tap changer and phase-shifter controls

The equality and nonequality constraints are as follows:

- Generation and load balance
- Generator active and reactive power limits
- Branch and tie-line load flow limits
- Bus voltage limits
- Limits on control variables

The optimal power flow is a nonlinear numerical optimization problem, discussed in Chapters 11 and 12.

References

1. MC Douglas, LA Johnson. *Forecasting and Time Series Analysis*, McGraw-Hill, New York, 1976.
2. "Load Forecast bibliography," *IEEE Trans Power Appar Syst*, 99, 53–58, 1980.
3. FM Dekking, C Kraaikamp, HP Lopuhaä, LE Meester. *A Modern Introduction to Probability and Statistics (Springer Texts in Statistics)*, Springer, Berlin, New York, Heidelberg, 2007.
4. JL Devore. *Probability and Statistics for Engineering and Sciences*, Brooks/Cole Publishing Co., Duxbury Press, CA, 2007.
5. MR Spiegel, JJ Schiller, RA Srinivasan. *Theory and Problems of Probability and Statistics*, 2nd ed. Schaum's Outline Series, McGraw-Hill, New York, 2000.
6. EPRI Research into load forecasting and distribution planning, EPRI-EL-1198, 1979.
7. HL Willis, RW Powell, DL Wall. "Load transfer coupling regression curve fitting for distribution load forecasting," *IEEE Trans Power Appar Syst*, 5, 1070–1077, 1984.
8. HL Willis, AE Schauer, JED Northcote-green, TD Vismor. "Forecasting distribution system loads using curve shape clustering," IEEE PES Summer Meeting, Paper B2SM385–3, San Francisco, CA, July 18–23, 1982.
9. H Jung, K Smalling. "Determination of area load on LILCO system using kWh," Presented at the Penn. Electric Association Meeting, Long Island Lighting Co., May 7, 1962.
10. HL Willis, H.N. Tram. "Distribution system load forecasting," IEEE Tutorial Course 92-EHO 361–6-PWR 1992.

2

Automatic Generation and Frequency Control—AGC and AFC

2.1 Fundamental Concepts

We have seen that the load at any instant in a power system is not constant and is continuously varying. For the power system stability and control, it is necessary that the varying load demand is met by loading and unloading the generating sources, taking these out of service and bringing them in service as soon as practical. The contingency load flow adds further complexity to this fundamental problem of meeting the varying load demand. Under contingency conditions, say under a fault when a certain route of power flow is taken out of service, the load demand should be met by alternate routes of power flows. Automatic generation control (AGC) is defined as the automatic regulation of the mechanical power to synchronous generators within a predefined control area.

- Power system frequency is one criterion for such controls. For satisfactory operation, the power system frequency should remain nearly constant.
- The performance of generating units is dependent upon frequency. The frequency deviations can seriously impact the operations of auxiliaries like synchronous and induction motors.
- The steam turbines can be damaged due to operation at a reduced or higher frequency. The safe operating times at frequency other than the normal which are critical for operation are supplied by the manufacturers, and a guideline is provided in ANSI/IEEE standard 122, also see Volume 4.
- The frequency is dependent upon active power balance; a change in load cannot be suddenly met by rotating synchronous generators. On a sudden load on a synchronous generator, driven by a steam turbine, the governing system must act to open the steam valves, which can only occur with a certain time constant. Thus, the frequency of the generator will dip. Conversely, on load rejection, the frequency will rise. The synchronous generators may be tripped out with a short-time delay, if the frequency happens to be beyond a certain range (generally indicative of fault conditions).
- As so many generators are paralleled in a grid system supplying power, some means to allocate changes in demand must be provided.
- The operations of all loads vary with the frequency. See EPRI load models with respect to variations in voltage and frequency in Chapter 6.
- The undervoltage due to sudden reactive power demand can give rise to voltage instability; see Volume 4, and also Chapter 9. The active and reactive power flows

can be decoupled, as further discussed. While reactive power flow mainly impacts voltage stability, a sudden increase or decrease in the active power impacts frequency.

- The under frequency load shedding is resorted to, in extreme cases, limit the area of shutdown and counteract voltage instability.
- A speed governor on each generating unit provides primary speed control function, while a central control allocates generation priorities.
- Each area has to be controlled to maintain a scheduled power interchange. The control of generation and frequency is commonly referred to as load-frequency control (LFC).
- The extensive use of timing clocks and timing devices, based on frequency, requires accurate maintenance of synchronous time, which is proportional to integral of frequency.
- It may take 11–12h or more to start a cold turbine unit and bring it on line. Thus, there should be some spinning reserves in the system which can be quickly loaded on an increase in the demand. Some units are designated "unit on regulation" or there may be more units to meet the changing load demand which are regulated. The base units operate continuously.
- This indicates the problem of unit commitment. How to allocate units most efficiently and economically to meet the load demand? This is further discussed in Chapter 12 considering fuel costs, operating costs, and transmission line losses.

2.2 Control Centers

There are two levels of controls. One is for bulk power generation and transmission (system control center) and the other for distribution (distribution control center). There may be many distribution centers depending on the area served.

The system control functions are as follows:

Automatic generation control (AGC)

Load-frequency control (LFC)

Economic dispatch control (EDC)

These functions overlap each other to some extent.

2.2.1 Scheduling and Forecasting

Table 2.1 describes these functions that are discussed in the following sections.

2.2.1.1 Hourly Interchange

The projected load for hour of interchange must be determined in advance (generally 1–4h in advance). The production cost of interchange is the cost of total committed load including previous commitments and the new interchange minus the previous commitments. These are determined by assuming that units are economically utilized.

TABLE 2.1

Scheduling and Forecasting

Time Frame		Required Data
Hourly interchange	Hours	Current unit status
Transmission maintenance scheduling	Days	Projected unit status and typical bus load data
Unit commitment	Days or weeks	Long-term interchange forecast Projected unit availability
Generator maintenance scheduling	Weeks or months	Overall unit and load data

2.2.1.2 Unit Commitment

A unit commitment program simulates the generation system over a period of time from a couple of days to a couple of weeks. Economic dispatch, reserve and reliability evaluation, and operating costs are simulated considering unit priority list and other such criteria. The renewable, solar, and wind generation (Volume 1) add to this complexity of programming. The objective is to optimize the total operating cost including startups, within all the imposed constraints. The basic inputs are the daily projected peak loads and long-term interchange schedules already agreed.

2.2.1.3 Transmission and Generator Maintenance Scheduling

As the transmission and generating systems are becoming more complex, these functions are now taken by complex computer simulations. It is not so simple to take a unit or section of the system under maintenance. A study is required, which simulates the system at the time of desired outage. This involves load forecasting, economic dispatch, investigations of bus voltages, projected units on line and standby, and keeping the load demand uninterrupted without excessive risks.

2.3 Controls in Real Time

Controls in real time signify "now" and in the present. The various aspects interconnections are shown in Figure 2.1. Note the interdependence and interrelations between various components. A block wise description is provided.

2.3.1 AGC and LFC

Fundamental concepts have been outlined in Section 2.1. The LFC system operates to regulate the system generation under various load conditions for proper performance of interconnected systems and optimize the economy of dispatch. Earlier, a constant frequency mode of LFC was used; today tie-line bias mode is most common. It strives to keep net interchange and frequency at prescheduled values.

First total generation required to meet area requirements are ascertained. Then, this generation can be apportioned to various units, including renewable sources (Volume 1) to

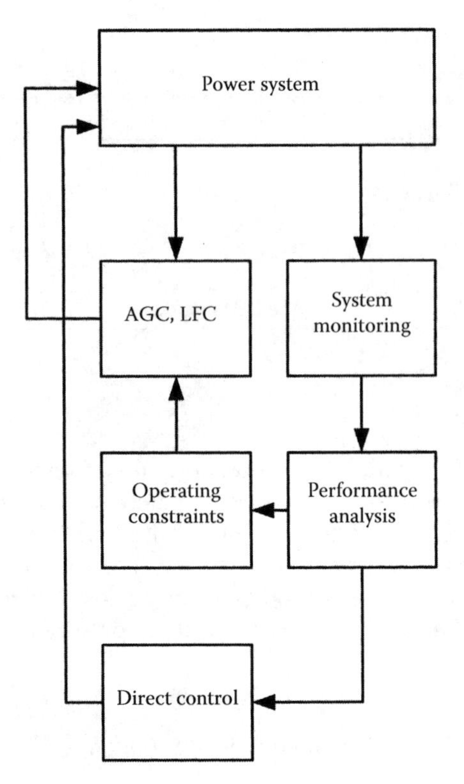

FIGURE 2.1
Forecasting and scheduling functions.

optimize the production costs, transmission losses, relative efficiencies, and environmental constraints like nitrous oxide emissions. The data for tie-line bias regulation are power flow and system frequency. With incremental power loading, the power output for each generator must also be monitored.

The performance should meet the following objectives:

- Minimize fuel costs.
- Avoid sustained operation of generation in undesirable modes, like overloads, out of range set points and avoid unnecessary operation on the generating units, for example, transient overloads.

2.3.2 System Monitoring

The complexity of power systems has grown and the data acquisition is an important link in the control strategies. The data are too voluminous to indicate on meters. Large CRT screens, CRT displays, and color one line diagrams are the modern tools with data updated every 10 s or so. Any switching or fault operation has the priority over the data communication systems.

The supervisory control means monitoring and control of circuit breaker status. The communication facilities for data acquisition and supervisory control are common.

2.3.3 Performance Analysis

A deterministic approach is required to evaluate the security of the system. At a time and periodically, a group of contingencies are selected and the system response monitored for any weaknesses. Undervoltages, overloads, and frequency deviations may be detected, and in some preprogrammed situations, possible corrective actions are displayed to the operating personnel. The manual controls can override the displayed solutions. A very extensive data collection system is required to support this activity (Volume 4). A real-time load flow situation is displayed, and topology of transmission systems and substations, and circuit breaker monitoring can be examined.

2.3.4 Operating Constraints

The constraints are applied say to prevent overloads and other anomalies during operation. Security controlled dispatch systems signify that additional constraints are added to economic dispatch. In real time, the magnitude and phase angle of bus voltages should be known and also MW and Mvar flows. This requires accurate contingency simulations. Allowance has to be made in the measurement and transmission errors of data, and techniques are available for exploiting the redundancy in the measurements. To obtain a better estimate, state estimation (Volume 1) is being accepted as a necessary tool for security monitoring. Human error and judgments are taking a back seat in the modern automation processes.

2.3.5 Direct Control

Much effort is being directed for directly controlling some power system operations, like voltage regulations, exciter controls, control of turbine steam valves, under-load tap changing from a centralized station rather than from a localized location. The objective is to optimize the transient stability performance of the power system. This requires the following:

- Fast data acquisition systems
- Fast actuators' and sensors
- High-speed computers

2.4 Past Data Logging

The future planning and current operation are dependent on the past data logging. The past data that should be available for easy access are as follows:

- Post-disturbance data—any post-disturbance data, say for a fault or overload situation or tripping, are captured and sampled at high speed for analysis and possible future remedial action.
- Dispatcher actions and the specific situations which give rise to these.
- Load survey is helpful for validating computer-based load flow study results. The monthly and yearly peaks and minimums are recorded.

- Energy recording consists of MWh in and out of the tie-line and system net interchange.
- Cost reconstruction provides the means for calculating production costs for a given combination of units operating under given constraints. This is used to determine the cost of interchange and energy billing.
- Intercompany billing data.
- Operating reports are statistical daily reports containing data such as hour of the peak load, total generation, and load interchange.
- Data index reports provide an indicator how well the AGC and LFC systems are operating; and if enough regulating capacity is available. These indexes may be used to improve the system regulation and controls.

2.5 Deregulated Market

Volume 1, Chapter 2, "Modern Electrical Systems" describes the new structure of utility companies GENCO, TRANSCO, DISTCO, and the deregulation of power industry, not repeated here. As discussed above, the interchanges of power can be from any source, from neighboring systems or the power can be *wheeled off* across intermediate systems, depending upon economics of such interchanges. There may be a central pool for dispatch which may economize the interchanges. Wheeling of power could result in transmission losses.

Other types of interchanges between utilities include capacity interchange, diversity interchange, energy banking, and inadvertent power exchange; and these could lead to economic benefits. To maximize these benefits, several utility companies have formed power pools that incorporate a common dispatch center.

2.5.1 Auction-Based Mechanism

In this competitive environment, an auction marker mechanism is one of the ways to price-based operations. It is a method of matching buyers and sellers through bids and offers. Each player generates bids (specified amount of electricity at a given price) and submits to the auctioneer, who matches the buy and sell bids to the approval of contract evaluator. This role of evaluation is played by an independent system operator. The mechanism allows for cash futures and planning markets.

A *forward contract* is an agreement in which the seller agrees to deliver a particular amount of electricity with a specified quality at a specified time to the buyer.

A *future contract* is a financial instrument that allows traders to lock-in a price in some future month—to manage their risks for future losses or gains.

A *future–option contract* is a form of insurance that gives the right (but not the obligation) to an option purchaser to buy or sell future contracts at a given price. Both the options and future contracts are financial instruments to minimize risk.

A reader may draw similarities with a commodity trading stock market.

With the competitive market structure, several technical issues arise, i.e., the capability of network to handle power flows reliably and securely. The Federal Energy Regulatory Commission (FERC) issued orders that specify the role of available transmission capacity (ATC). In using the transmission access for effective competition and established the Open

Access Same Time Information System (OASIS), operational since January 1977. Following it National Reliability Council (NERC) initiated establishment of ATC evaluations. It is a measure of ability of the interconnected electrical systems to transfer power reliability from one location to another considering transmission paths between two areas. Total transfer capability (TTC) determines it and is based on the following:

1. Thermal limits: All facility loadings are within normal ratings and all voltages are within normal limits.
2. Stability: The electrical system is capable of riding through dynamic power swings consequent to loss of any single electrical system element like a transmission line, transformer, or a generating unit.
3. Post-dynamic power swings: After the swings subside and a new operating state is restored, post-item 2, and after operation of any automatic systems, but before any operator initiated system adjustments are made, all transmission loadings are within emergency limits and all voltages within emergency limits, Chapter 9.
4. Post-contingency loadings: With reference to condition (1) when post-contingency loadings reach normal thermal limits, at a transfer level below that at which the first contingency transfer limits are reached, the transfer limits are the ones at which such normal ratings are reached.
5. Multiple contingencies: In some areas, multiple contingencies are required to determine transfer capability limits.

2.5.1.1 Transmission Reliability Margin

Transmission reliability margin is the transfer capability necessary to ensure that the interconnected transmission network is secure under a range of reasonable system conditions. This accounts for uncertainties in the system and operating conditions. Also the system needs to be flexible for secure operations.

2.5.1.2 Capacity Benefit Margin

This is the amount of transfer capability reserved to ensure access to generation from interconnected neighboring systems to meet generation reliability requirements.

Thus, we can write

$$\text{ATC} = \text{TTC} - \text{TRM} - \text{CBM} - \text{existing transmission commitments} \tag{2.1}$$

where,

TTC = min (thermal limits, voltage limits, and transient stability limits).

Transmission reliability margin (TRM): It is the amount of transfer capability necessary to ensure that the interconnected transmission network is secure under various sources of uncertainty under system operating conditions.

Capacity benefit margin (CBM): It is the amount of transfer capability reserved by load-serving entities to ensure access to generation from interconnected neighboring systems.

FERC requires the calculations and posting on OASIS, continuous ATC information for the next hour, month, and for following 12 months. The accuracy of these data depends on many factors, like use of accurate power flow methods and simulations to calculate ATC.

2.6 Load-Frequency Control

Normal governor control relates the load of a unit to the system frequency. Supplementary controls change the load versus frequency characteristics of the governor. Referring to Figure 2.2, a load change, ΔP, changes the electrical torque of the generator which causes a mismatch between the mechanical torque produced by the turbine and this results in speed variations. A rigorous method is solution of the swing equations of the generator using transient stability type of programs. As power is proportional to rotor speed in radians multiplied by the torque, the following relations exist.

2.6.1 Single Generator with Isochronous Governor

Isochronous means constant speed. Referring to Figure 2.2, P_0 is the active MW supplied and ΔP is the incremental load addition. The shaft speed or the frequency of the generator terminal voltage is sensed, and governor operates a valve that controls the flow of steam and therefore mechanical output of the turbine. Define the following:

T_m = mechanical torque (pu)
T_e = electrical torque (pu)
T_a = accelerating torque (pu)
P_m = mechanical power
P_e = electrical power
P_0 = initial active load
ω_r = rotor speed in radians
$\Delta\omega_r$ = rotor speed deviation

Then

$$P = P_0 + \Delta P$$
$$T = T_0 + \Delta T \tag{2.2}$$
$$\omega_r = \omega_0 + \Delta\omega_r$$

Then, we can write

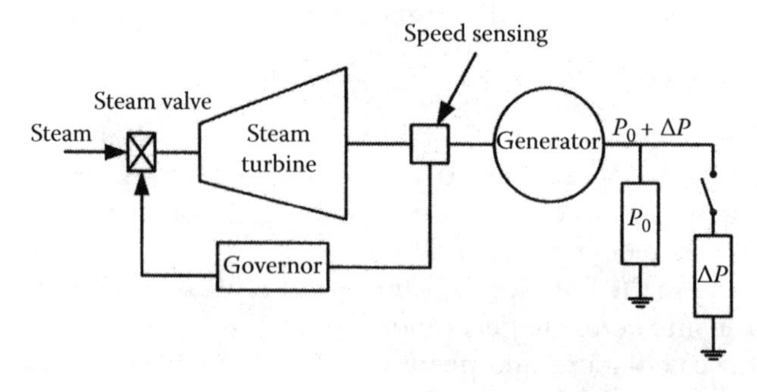

FIGURE 2.2
Step load addition to an isolated generator.

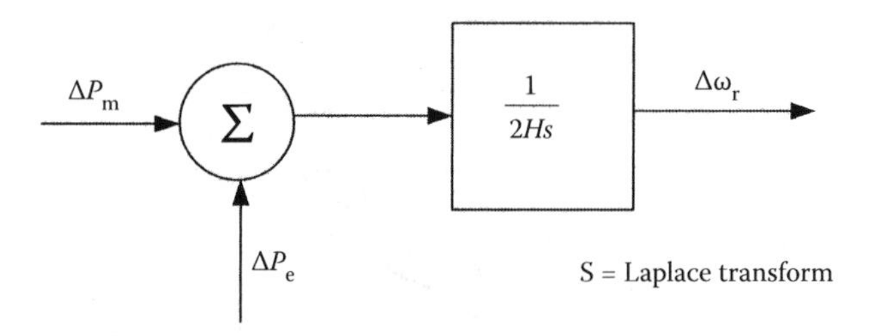

FIGURE 2.3
Transfer function relating speed and power.

$$P_0 + \Delta P = (\omega_0 + \Delta\omega_r)(T_0 + \Delta T) \quad (2.3)$$

from where

$$\Delta P \approx \omega_0 \Delta T + T_0 \Delta\omega_r \quad (2.4)$$

In steady state, electrical and mechanical torques are equal and $(\omega_0) = 1$ pu:

$$\Delta P_m - \Delta P_e = \Delta T_m - \Delta T_e \quad (2.5)$$

Equation 2.5 represents an obvious conclusion. The transfer function relating speed and power can be represented as in Figure 2.3. *H* is the inertia constant, see Chapter 7 for its definition and equation. The turbine mechanical power is a function of valve position: percentage open or closed.

It remains to define the load-frequency characteristics, which vary with the type of load, see Chapter 6. The loads can be divided into two parts; one part is not dependent on frequency and the other part is

$$\Delta P_e = \Delta P_L + D\Delta\omega_r \quad (2.6)$$

where,

ΔP_L = part of the load change that is not frequency sensitive
$D\Delta\omega_r$ = part of the load change that is frequency sensitive
D = damping factor

The response of an isochronous governor is shown in Figure 2.4. As speed drops, the turbine power begins to increase; ultimately, the turbine picks up the additional load and the speed returns to 60 Hz.

2.6.2 Two (or Multiple) Generators with Isochronous Governor

An isochronous governor works well with a single turbine. When more than one turbine and governor are present, they will fight with each other, Figure 2.5. This will be obvious if two governors with slightly different speed controls operate in parallel. Trying to control with two different set points will cause the power to oscillate between the units. This is often referred to as hunting.

The transfer function of a governor with steady-state feedback is illustrated in Figure 2.6. Define *R* as

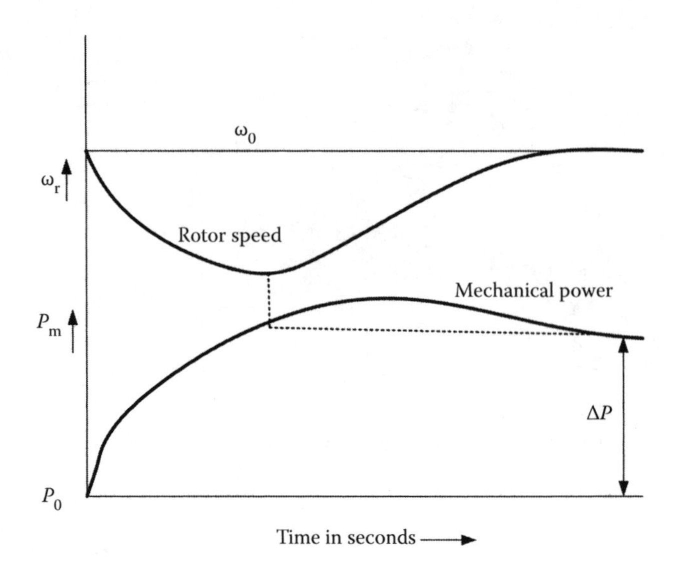

FIGURE 2.4
Response of an isochronous governor.

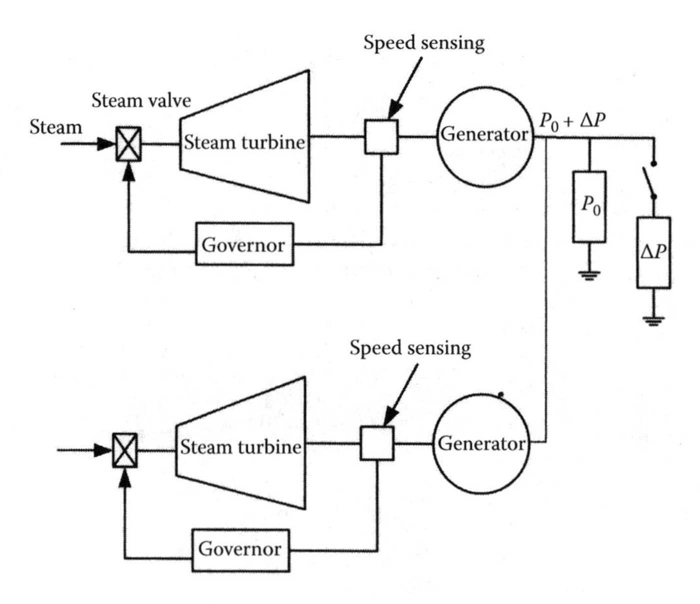

FIGURE 2.5
Multiple generators with step load addition.

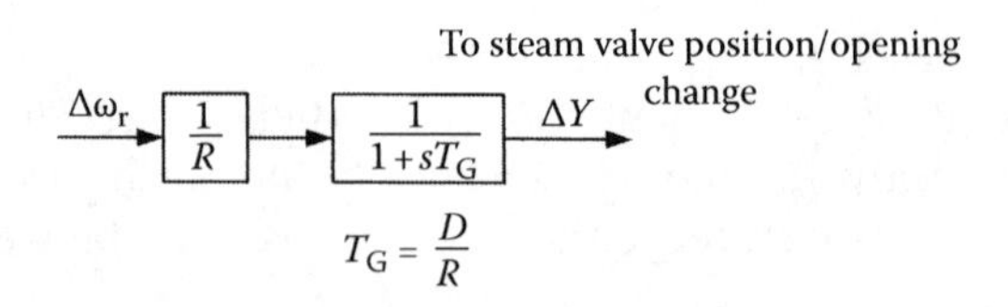

FIGURE 2.6
Block control circuit diagram of an isochronous governor with droop.

$$R = \left(\frac{\omega_{NL} - \omega_{FL}}{\omega_0} \right) = \frac{\Delta f}{\Delta P} = \frac{f - f_0}{\Delta P} \tag{2.7}$$

The ideal steady-state characteristics of the governor are shown in Figure 2.7 with speed droop. Figure 2.8 depicts the time response.

This figure shows that on an isolated system, the turbine will pick up the additional load ΔP but will not restore the speed to original value. A governor having these characteristics will parallel well with similar units since each governor has a definite steady-state valve position for each value of speed error. If the speed should be 60 Hz to deliver a power P_0, the speed will be lesser to deliver a power $P_0 + \Delta P$. Typically, a 5%–25% change in generation will occur for a 1% change in frequency, depending on governor characteristic, valve response, load point, etc. This may be expressed in another way; if $1/R$ is 20, a 5% change in speed will cause a 100% change in generation. Another definition of R is as follows:

> The change in steady state speed, when the power output of the turbine operating isolated is gradually reduced from rated power output to zero power output with unchanged settings of all adjustments of the speed governing system.

Consider now two generators operating in parallel, their speed has to be the same as these are in parallel. A speed regulation characteristic representing the second unit can be drawn parallel to the one shown in Figure 2.7, as depicted in Figure 2.9. The additional load is now shared between the two units according to the regulation characteristics. It is likely that the resulting split of power between the units is unacceptable from standpoint of overall economy, security, and operation. *Also there is no way of getting back to 60 Hz operation.* This points to the necessity of supplemental control. The load sharing in case of two units in parallel is determined by

$$\frac{\Delta P_1}{\Delta P_2} = \frac{R_2}{R_1} \tag{2.8}$$

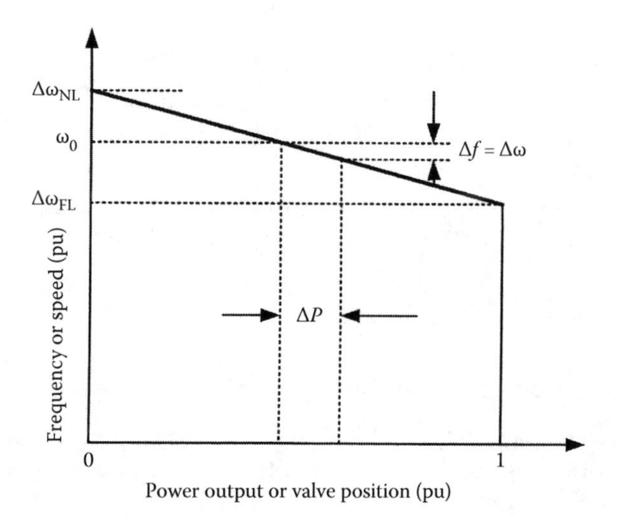

FIGURE 2.7
Ideal characteristics of an isochronous governor.

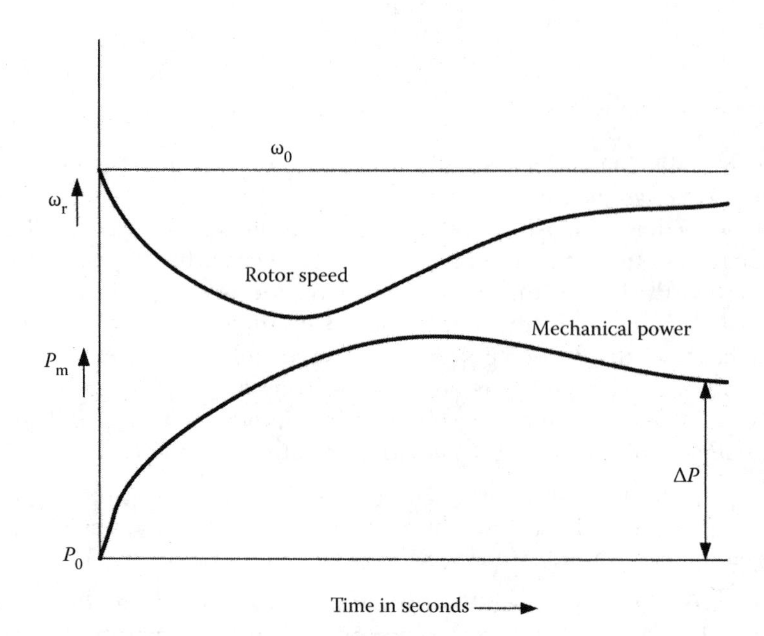

FIGURE 2.8
Response of a generating unit with governor having droop characteristics.

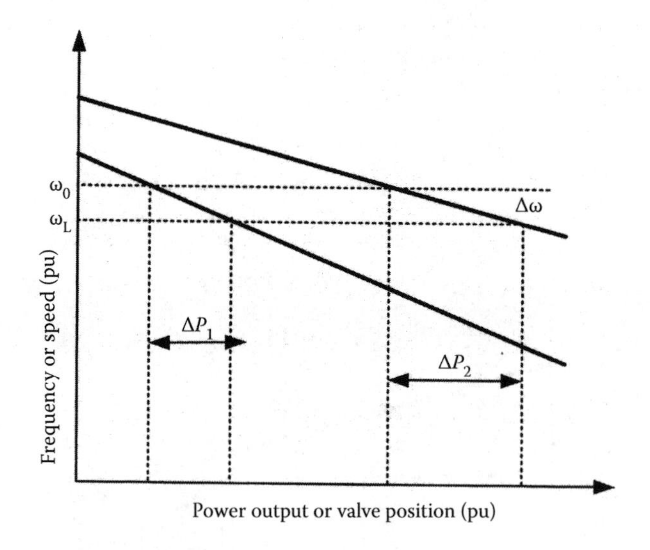

FIGURE 2.9
Load sharing between parallel units with governor droop characteristics.

2.6.3 Supplementary Constant Frequency Control

In order to restore frequency to normal and desired load split between units, some common control is required, Figure 2.10. It looks at the frequency and signals the individual governors to reset their operating points to the desired levels. The control shifts the operating line parallel to itself so that the new steady-state operating point is the intersection of the rated speed and new desired power.

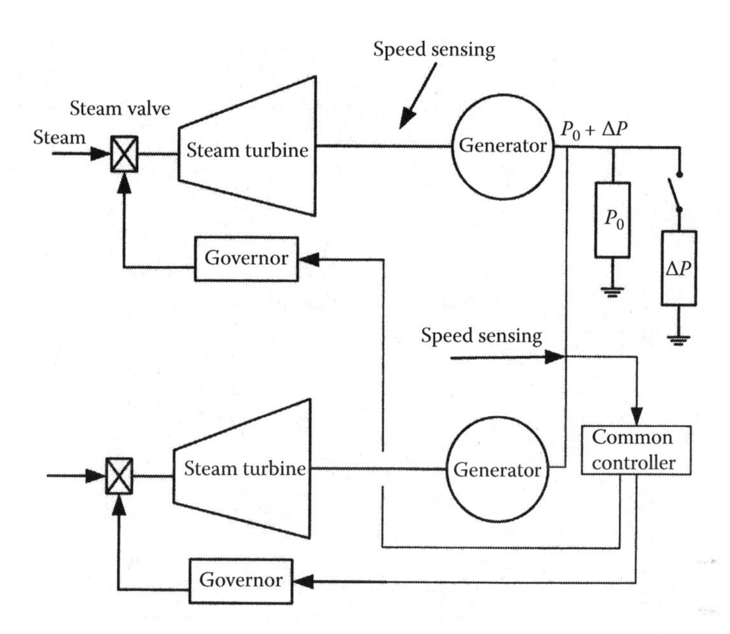

FIGURE 2.10
A typical power system with supplementary speed controller.

Referring to Figure 2.11, the impact of centralized supplementary control is seen. As load fluctuates, the governor responds by raising/lowering along the steady-state characteristics for one speed changer setting, labeled initial setting. When the frequency is high or low, the supplementary control shifts the complete governor characteristic to right or left parallel to the steady-state characteristics.

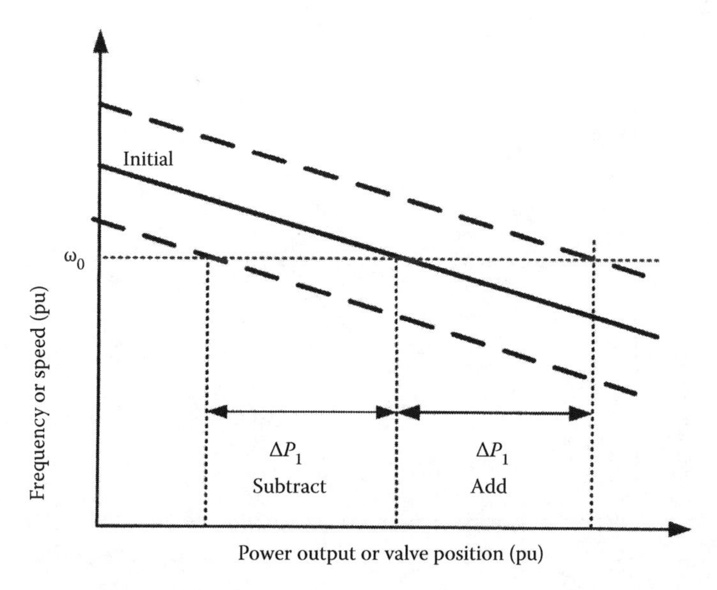

FIGURE 2.11
Steady-state characteristics of governor with speed changer.

We discussed the frequency dependence of the load and introduced damping factor D in Equation 2.6. The loads were divided into two categories, one non-frequency dependent and the other frequency dependent. Motor loads, generally, drop 2% or more for 1% drop in frequency. Heating and lighting loads are insensitive to frequency. The composite load may be approximated by 1% drop in load for 1% drop in frequency. Figure 2.12 illustrates this frequency-dependent load characteristic. The dotted line toward the right applies with additional load, ΔP.

When the effect of load is not considered, the frequency drop is proportional to R, the steady-state regulation. When the frequency dependence characteristics of the load are considered, the frequency will not drop that far:

$$\frac{\Delta P - \Delta P_L}{\Delta f_2} = \frac{1}{R}$$

$$\frac{\Delta P}{\Delta f_2} = \frac{1}{R} + \frac{\Delta P_L}{\Delta f_2} = \frac{1}{R} + D \tag{2.9}$$

A further depiction is provided in Figure 2.13. The operating point in the absence of a central controller is the intersection of two sloping lines at the bottom. The system regulation characteristics change as per Equation 2.9. The D is comparatively small, equal to 1 as stated before.

A number of generators, n, can be represented by their equivalent inertia constant. Also all the loads on the generators can be lumped together with a composite load damping factor D. The composite power frequency characteristics of a power system, therefore, depend on combined effect of droops of all generators speed governors. In such a system, the steady-state frequency deviation following a load change that impacts n generator is as follows:

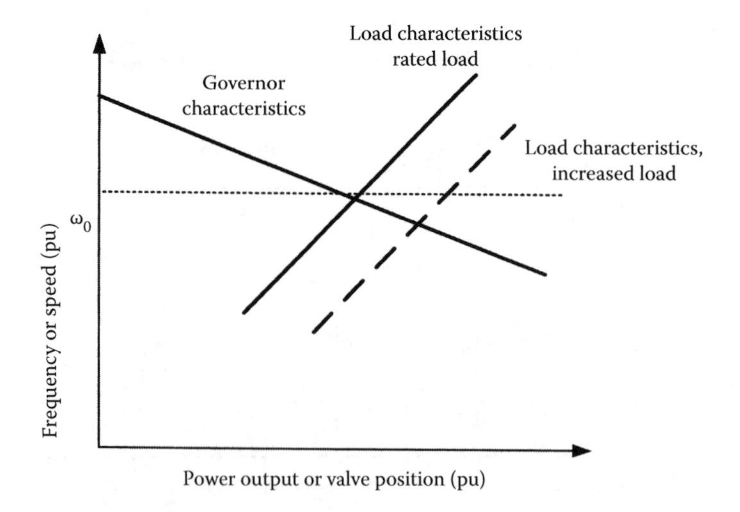

FIGURE 2.12
Steady-state characteristics of a typical power system.

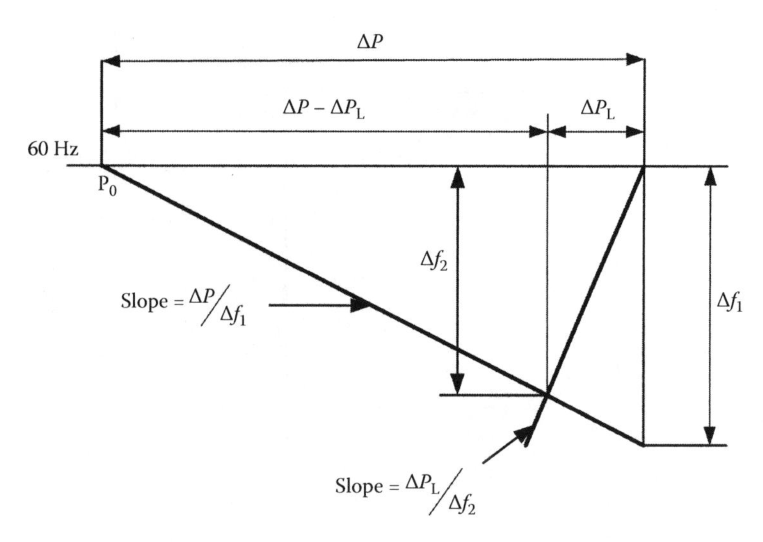

FIGURE 2.13
Expanded steady-state characteristics, see the text.

$$\Delta f_{\text{steadystate}} = \frac{-\Delta P_L}{(1/R_1 + 1/R_2 + \cdots + 1/R_n) + D}$$

$$= \frac{-\Delta P_L}{1/R_{eq} + D} \tag{2.10}$$

Composite frequency response characteristic is represented as follows:

$$\beta = \frac{-\Delta P_L}{\Delta f_{\text{steadystate}}} = \frac{1}{R_{eq}} + D \tag{2.11}$$

where β is normally expressed in MW/Hz. It is sometimes referred to as the stiffness of the system. The composite regulating characteristics are 1/β.

2.7 Interconnected System Control

Consider two areas, A and B, with their respective generations G_A and G_B and loads L_A and L_B. ΔL is the additional load applied at area A. The two areas are connected through a tie-line, Figure 2.14. The operation is described in a number of steps.

1. Before the load is added in area A, there is no flow in the tie-line and the generations in areas A and B serve their respective loads and the frequency is 60 Hz. The initial operating point is shown in Figure 2.15. This assumes that generations at either end have the same governor characteristics, given by the diagonal solid line. Thus, tie-line flow is zero. If R is not same, then see step 2

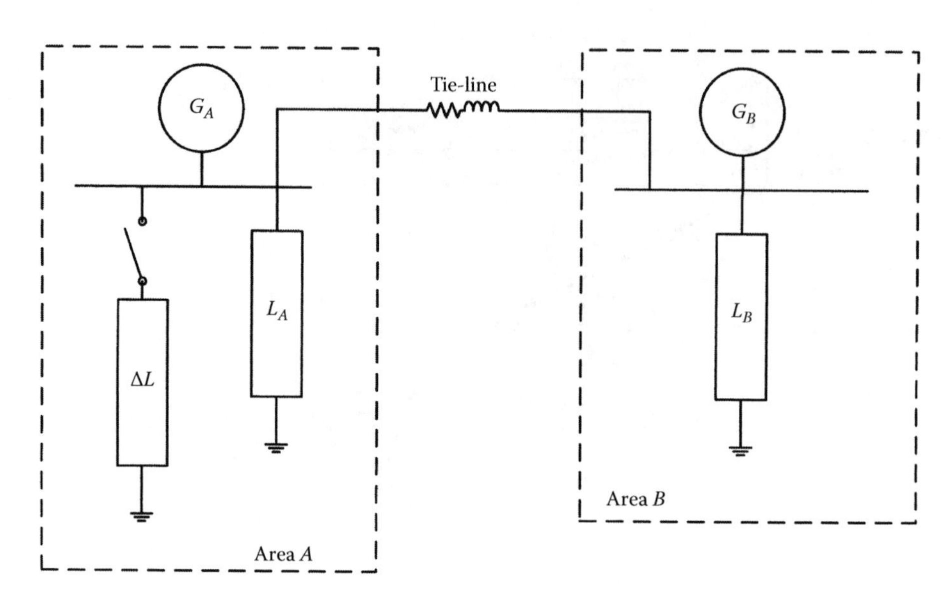

FIGURE 2.14
Schematic representation of two control areas interconnected together.

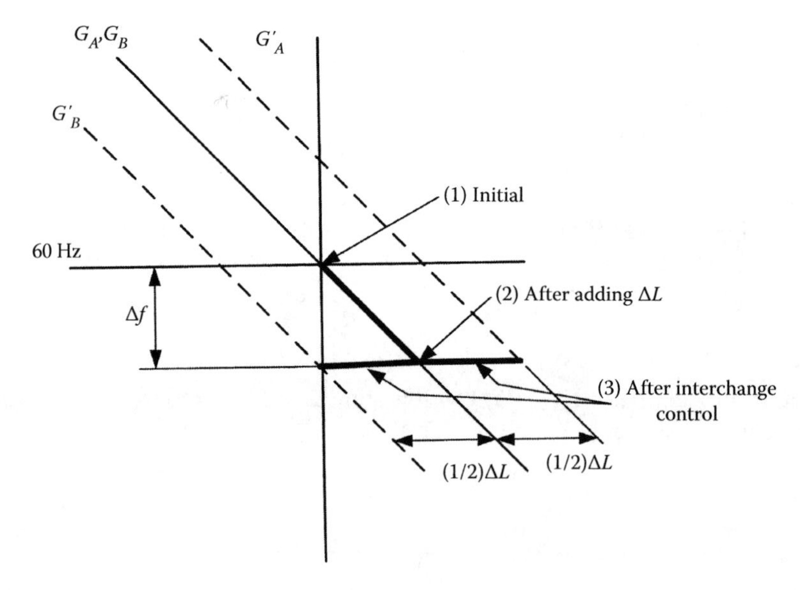

FIGURE 2.15
Net interchange control.

$$\Delta f = \frac{-\Delta P}{(1/R_1 + D_1) + (1/R_2 + D_2)} = \frac{-\Delta P}{\beta_1 + \beta_2} \tag{2.12}$$

2. When a load is added in area *A*, the frequency drops, yet the frequency in both areas *A* and *B* is the same, i.e., the drop in frequency in area *A* is equal to the drop in frequency in area *B*. As the governor characteristics are identical, it is fair to assume that each area has picked up 50% of the load applied in area *A*. Now the

tie-line flow is also equal to 0.5 ΔL, from area B to area A. The tie-line flow with different governor and load characteristics will be

$$\Delta P_{BA} = \frac{-\Delta P(1/R_2 + D_2)}{(1/R_1 + D_1) + (1/R_2 + D_2)} = \frac{-\Delta P\beta_2}{\beta_1 + \beta_2} \tag{2.13}$$

3. Ideally, no power interchange should occur and the area A load addition must be supplied by area A itself. Supplemental control is required. Prioritize that first the interchange should be corrected and then the frequency. Define area control error as the actual interchange minus the schedule interchange. Then, ACE for areas A and B are as follows:

$$\text{ACE}_{\text{area}A} = -\frac{1}{2}\Delta L$$

$$\text{ACE}_{\text{area}B} = \frac{1}{2}\Delta L \tag{2.14}$$

That is, that area A is deficient in generation by $0.5\Delta L$; while area B is having excess generation of the same amount. Figure 2.15 shows that the generation in area A raises at the same rate as the generation in area B lowers. The new governor characteristics, parallel to solid line, are shown in this figure. When ACE returns to zero in both the areas by supplemental control, this is shown as step 3 in Figure 2.15.

It has been assumed that the generation in area A raises at the same rate that generation in area B lowers to maintain constant frequency. That may not be the case. Generation in area A or B can be faster than in the other area. When generator B lowers faster than generator A raises, the system frequency is lowered by supplementary control lower than governor action, which is not desirable. From this viewpoint, the priorities may be reversed, correct frequency first and then the interchange, see Figure 2.16a and b.

4. Next the frequency control is depicted in Figure 2.17. Supplementary control action corrects the frequency, area A raises generation in area A, and, in area B, the same action occurs. Generation B returns to its original position, and generation in area A moves so that its governor characteristic is the last dotted line to the right in Figure 2.17. The 60 Hz operating point is ΔL greater than it was in the beginning, before additional load was applied.
5. Steps 1 through 4 may look okay, but these are not. In this process, the generation at B was unnecessarily lowered and then raised. If the generation at A had moved from

$$G_A \rightarrow G_A' \rightarrow G_A'' \tag{2.15}$$

Then, the frequency is restored and generation in area B would have moved by natural governor action to the origin as shown in Figure 2.18.

The disturbance (additional load) was added in area A; thus, it is area A's responsibility to correct it, *without impacting area B*. As the loads are not continuously being monitored,

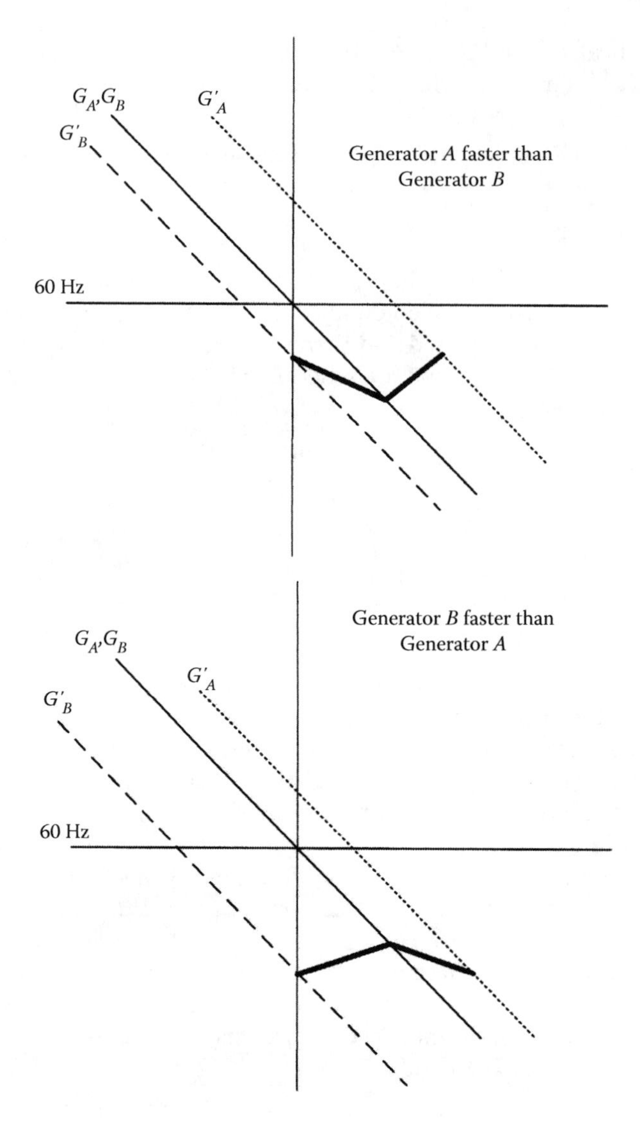

FIGURE 2.16
(a) Unequal response rates, area *A* faster than area *B*; (b) area *B* faster than area *A*.

how area *B* could be blocked and area *A* permitted to control? This is further discussed in the following section.

2.8 Tie-Line Bias Control

Consider that area *A* is much smaller generation compared to area *B*. The parameters of these two areas and the step load of 100 MW applied in area *A* are illustrated in Table 2.2. Thus, area *A* ACE should be –100 MW and area *B* ACE should be zero. The control should accomplish the following:

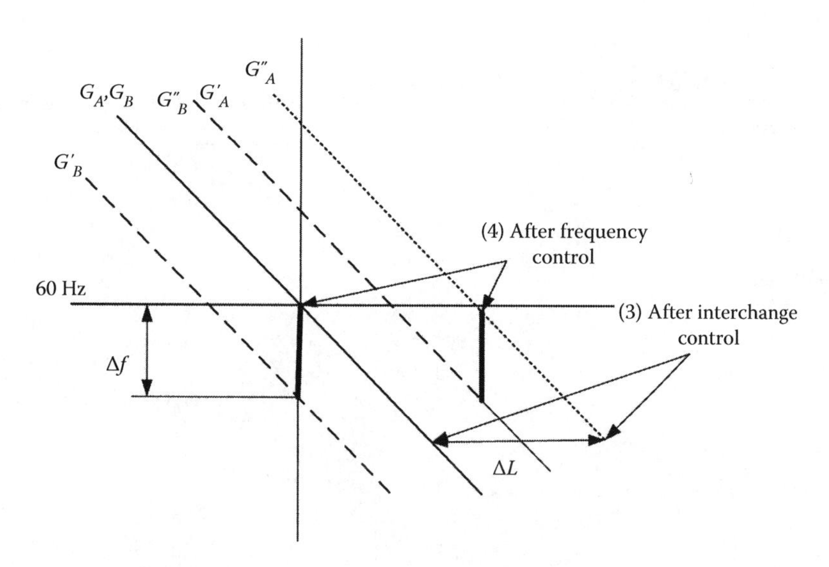

FIGURE 2.17
Frequency control.

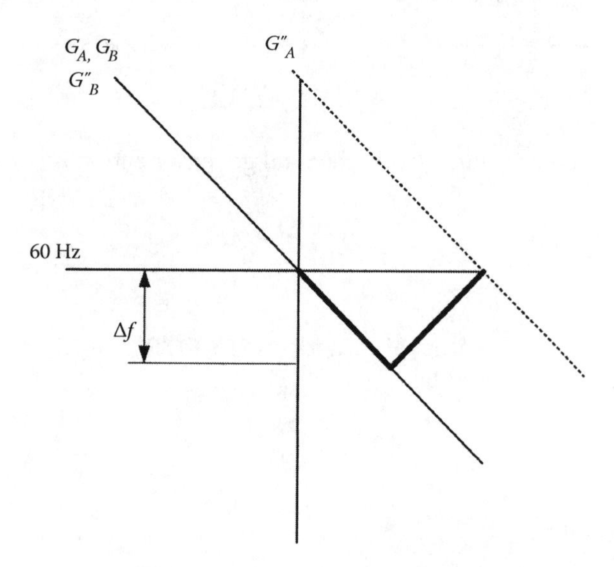

FIGURE 2.18
Supplemental control in one area only.

- Area A should pick up the entire load.
- Net interchange should make area B supply zero power, i.e., no tie-line flow.
- System frequency should return to normal.

As the load is not monitored, ACE should be calculated in a different manner. Neglecting damping the load change was met by governor action. The amount of governor action in area B should be determined from net interchange.

TABLE 2.2

System Conditions Following Increase in Load in Area *A*

System Parameters	Area *A*	Area *B*	Total
$1/R$	200	1000	1200
D	10	80	90
$1/R+D$	210	1080	1290
Starting Conditions			
Connected generation	1000 MW	8000 MW	9000 MW
Initial connected load	900 MW	7000 MW	7900 MW
Added load	100 MW		100 MW
Total connected load	1000 MW	7000 MW	8000 MW
Results after Governor Response			
Δf, steady state (100/1290)	(100/1290) = 0.07752		
$\Delta \text{GEN} = \Delta f(1/R)\text{MW}$	15.50	77.52	93.02
$\Delta \text{Load} = \Delta f(D)$	0.775	6.20	6.975
New generation MW	1000 + 15.50	8000 + 77.52	9000 + 93.02
New load (MW)	1000 − 0.775	7000 − 6.20	8000 − 6.975
New interchange (MW) = generation − load	−100 + 15.5 + 0.775 = −83.72	77.52 + 6.20 = 83.72	

If the interchange is represented by Δ_i, the total governor action in area *A* is

$$\Delta_i + \frac{\Delta f}{R_1} \tag{2.16}$$

Area *B* will observe a positive Δ_i. If this is added to $(\Delta f)(1/R)$, the result will be zero. Thus, the ACE for both systems should be

$$\Delta_i + \beta \Delta f \tag{2.17}$$

In Table 2.2, ACE in area *A* is

$$-90.9 + 210 \times (-0.07752) = -100\,\text{MW} \tag{2.18}$$

and in area *B* is

$$-90.9 + 1080 \times (-0.07752) = 0\,\text{MW} \tag{2.19}$$

In the above description, the scheduled interchange is zero and the frequency is 60 Hz. It is common for the utilities to buy and sell power to the neighboring systems when both parties find it economically desirable under emergency. To transfer *X* amount of MW, the seller increases the value of net exchange by *X* MW and the buyer decreases by the same amount. Both systems will have area control error and take supplementary control action.

When the area is interconnected to more than one area, which is usually the case, the interchanges between them do not occur from one area to another. The actual flows will split over parallel paths through other areas depending upon the impedances of the parallel paths.

2.9 Practical Implementation of AGC

2.9.1 Performance Criteria

The minimum performance control standards set up by NERC are as follows:

Under normal conditions, the following criteria apply:

- A1 criterion. The ACE must return to zero within 10 min of previously reaching zero. Violations count for each subsequent 10 min period during which ACE fails to return to zero.
- A2 criterion: The average ACE for each of the six 10 min periods must be within specified limits, which is referred to L_d. It is determined from

$$L_d = 5 + 0.025\Delta L \text{ MW} \tag{2.20}$$

where ΔL is the greatest hourly change in the net system load of a control area on the day of its maximum summer or winter peak load.

Under disturbance conditions, like sudden loss of generation or increase of load, the following criteria apply:

- B1 criterion: The ACE must return to zero within 10 min following the start of the disturbance.
- B2 criterion: The ACE must start to return to zero within 1 min following the start of the disturbance.

A disturbance is defined when a sampled area of ACE exceeds $3L_d$.

2.9.2 AGC and EDC

In economic dispatch control (EDC) each power source is loaded most economically, Chapter 12. For tie-line control, it is necessary to send signals to generating units to control generation. It is possible to use these signals to control generation to satisfy EDC. Thus, EDC can be accomplished as a part of AGC.

In EDC calculation, two factors are used: (1) base points that represent the most economical output of each generating unit and (2) participation factor that is the rate of change of unit output with respect to change in total generation. Then, the new generation required is power of base units plus participation factor multiplied by total generation.

2.9.3 Smooth ACE

The random variation in load will cause too many operations in the governor actions wearing these out. AGC schemes using filtering to filter out the random variations and smoothed ACE (SACE) is used for generation control. This reduces the speed of response.

2.9.4 Response Times

The rate at which the generating outputs can be changed depends on the type of generation.

- For thermal units, the initial load that can be picked up without causing much thermal stresses is about 10%, followed by a slow increase of 2% MCR (maximum continuous rating) per minute while for hydro units it is of the order of 100% MCR. This considers that the units are already in operation, and not operating at their rated outputs.
- The ability of a boiler to pick up a significant amount of load is limited. As the steam valve opens admitting more steam into the turbine, a pressure drop occurs and increased fuel input to the boiler is required to restore pressure, this may take several minutes and is inconsequential for capturing speed drop.
- The response time of governors is of the order of 3–5 s.

2.9.5 Time Deviation

In USA and Canadian interconnected systems the practice is to assign one area the maintenance of a time standard. Through communication channels, the information on the system time deviation is relayed to all control areas, and certain periods are designated as time correction periods. All areas are expected to simultaneously offset their frequency schedules by an amount related to accumulated system time deviation.

Execution of AGC every 2–4 min results in good performance.

2.9.6 Speed Governor Dead Band

Dead bands are caused by friction and backlash in the governor mechanical linkages. In some modern applications, dead bands have been eliminated. The effect of dead band depends on frequency deviation—a small deviation may remain within dead band and no correction will occur, and in an interconnected system of generation, the response of the units will be random. The random fluctuations in frequency in the large systems are of the order of 0.01 Hz. One effect of the dead band is to reduce the area response characteristic, β. It may cause cycling of the AGC system with periods ranging from 30 to 90 s.

2.10 Under Frequency Load Shedding

Under frequency load shedding is resorted to when the loads exceed the generation and the interchanges. Customers in some areas lose power, but this is better than tripping more generators on overload and jeopardizing their stability. Load shedding studies will show how much load should be shed and in what steps:

Frequency decay: 0.5 Hz/s: shed a certain percentage of load.

If the frequency decay is not arrested, shed more loads. The decay rates and loads to be shed are known from prior simulations.

2.11 Transient Response

The transient response will depend on a number of factors: step load change, type of turbine-thermal or hydro, type of governor, inertia constant *H*, and generator characteristics.

Consider a 500 MW generator of the technical parameters as shown in Table 2.3. A steam governor control circuit diagram is in Figure 2.19.

TABLE 2.3

500 MW, 2-pole, 50 Hz, 18 kV, 0.85 PF Steam Generator, High Resistance Grounded, Parameters

Description	Symbol	Value in pu at Generator MVA Base
Saturated subtransient reactance direct axis	X''_{dv}	0.19
Transient reactance direct axis, unsaturated	X'_d	0.28
Transient reactance quadrature axis, unsaturated	X'_q	0.55
Negative sequence reactance	X_2	0.18
Zero sequence reactance	X_0	0.07
Leakage reactance	X_L	0.15
Synchronous reactance, direct axis, unsaturated	X_d	1.55
X/R		100
Transient direct axis open-circuit time constant	T'_{do}	6.5 s
Transient quadrature axis open-circuit time constant	T'_{qo}	1.25 s

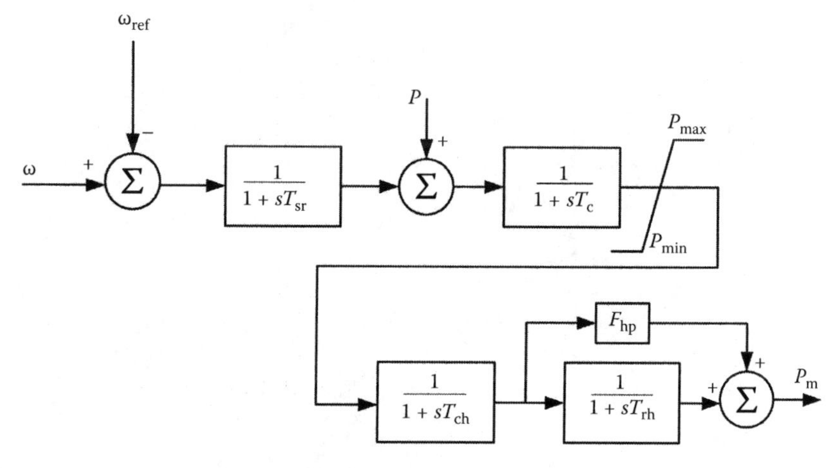

FIGURE 2.19
A steam governor control circuit diagram.

Droop = 5%
F_{hp} = Total shaft capacity
P_{max} = Maximum shaft power (rated MW)
P_{min} = Minimum shaft power (≥0)
T_c = control amplifier, servomotor, time constant
T_{ch} = Steam chest time constant
T_{rh} = Reheat time constant
T_{sr} = Speed relay time constant

The generator is carrying a load of 350 MW and a step load of 50 MW is applied at 50 ms.

Figure 2.20 shows the speed, the electrical power, and the mechanical power. Note that for an acceleration to occur, mechanical power should be greater than the electrical power. Though a step load of 50 MW is applied, the generator electrical power swings to approximately 405 MW on application of the step load. The drop in frequency is 0.75% recovering in approximately 8 s.

A simulation when a step load of 150 MW is applied; while the generator is supplying 350 MW is shown in Figure 2.21. There is a drop of 2.2%.

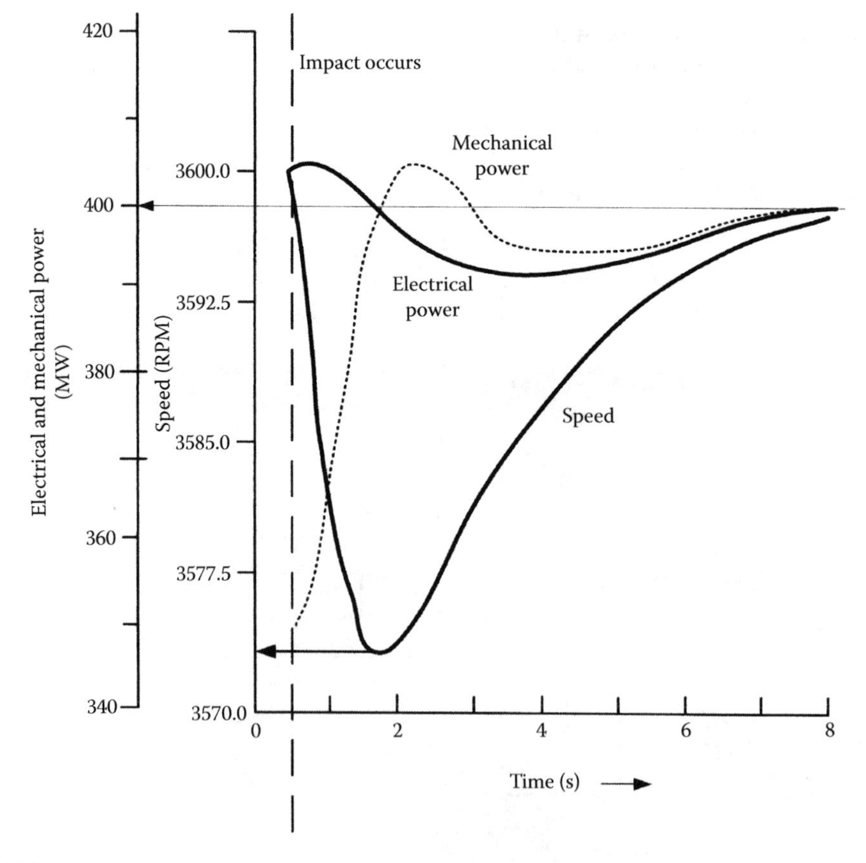

FIGURE 2.20
Transient response of a step load of 50 MW applied to a 500 MW generator, operating at base loads of 350 MW.

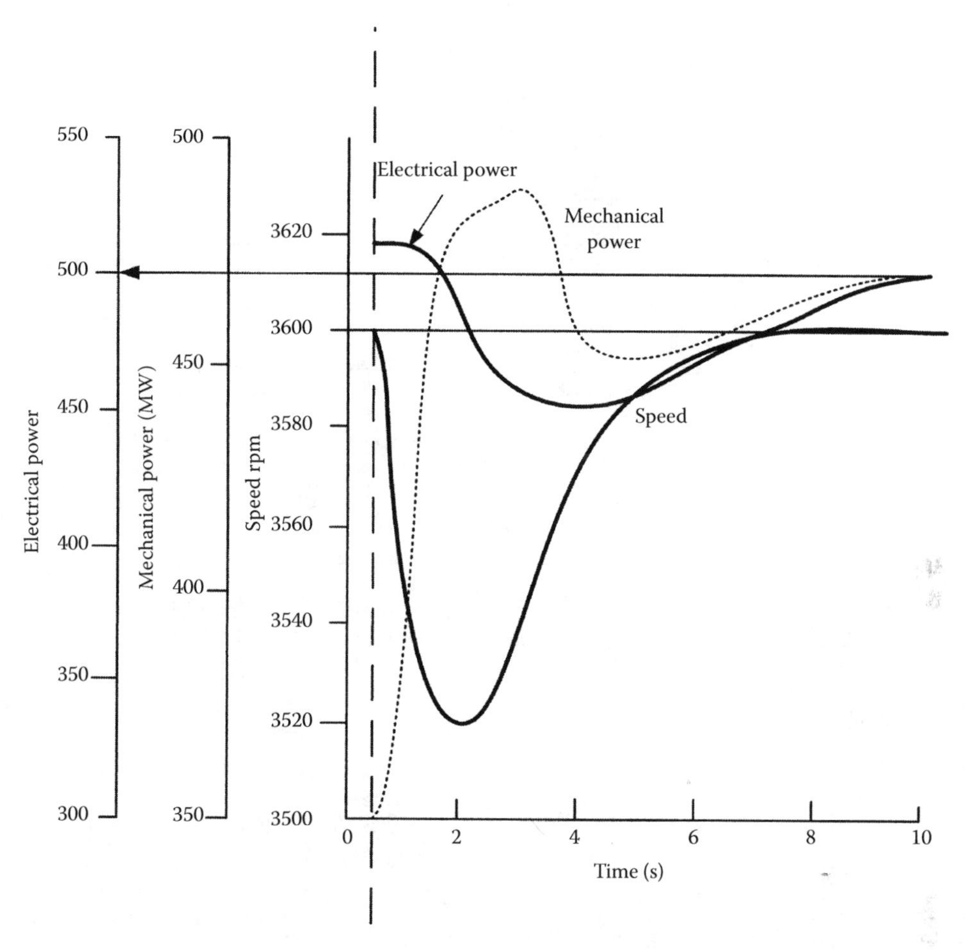

FIGURE 2.21
Transient response of a step load of 150 MW applied to a 500 MW generator, operating at base load of 350 MW.

Figure 2.22 shows the control circuit diagram of a tandem compounded double reheat steam turbine:

Droop = 5%
F_{hp} = total shaft capacity
F_{ip} = intermediate pressure turbine power fraction
F_{lp} = low pressure turbine power fraction
F_{vhp} = very high pressure turbine power fraction
P_{max} = maximum shaft power (rated MW)
P_{min} = minimum shaft power (≥0)
T_c = control amplifier, servomotor, time constant
T_{ch} = steam chest time constant
T_{co} = cross over time constant
T_{rh1} = first reheat time constant

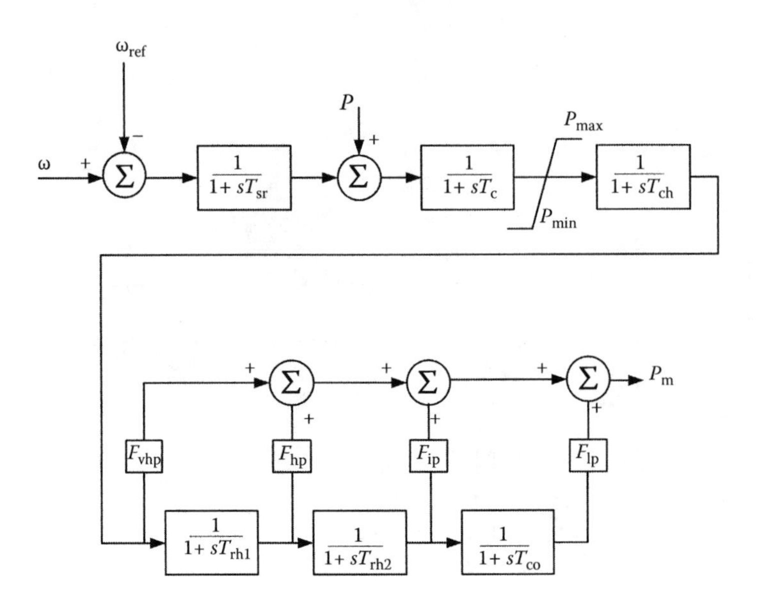

FIGURE 2.22
Control circuit diagram of a tandem compounded double reheat steam turbine.

T_{rh2} = second reheat time constant
T_{sr} = speed relay time constant

A turbine with reheat will, generally, have a larger frequency dip.

Problems

2.1 Considering no governor action, plot the frequency deviation as a function of time for $H = 5$ s, $D = 1\%$, and step load change of 0.02 pu.

2.2 A power system supplies 1500 MW and $D = 1$. Find the steady-state frequency deviation when 50 MW load is suddenly applied.

2.3 Two areas are interconnected, area 1; generation = 5000 MW and load = 4000 MW. Area 2 has a generation of 15,000 MW and a load of 13,000 MW. $R = 5$ and $D = 1$. Find the steady-state frequency, with a loss of 500 MW in area 1 with no supplementary control, and with supplementary control with frequency bias setting of 500 MW/0.1 Hz for both the areas.

Bibliography

1. IEEE Standard 122. Recommended Practice for Functional and Performance Characteristics of Control Systems for Steam Turbine-Generator Units, 1992.

2. IEEE Standard 125. Recommended Practice for Functional and Performance Characteristics of Equipment Specifications for Speed-Governing of Hydraulic Turbines Intended to Drive Electrical Generators, 1988.
3. IEEE AGC Task Force Report. "Understanding automatic generation control," *IEEE Trans*, PWRS-7(3), 1106–1122, 1992.
4. NERC Operating Guide 1: System Control, 1991.
5. OI Elgard, CE Fosha. "Optimum megawatt-frequency control of multiarea electrical energy systems," *IEEE Trans*, PAS-89, 556–563, 1970.
6. CW Ross. "Error adaptive control computer for interconnected power systems," *IEEE Trans*, PAS-85, 742–749, 1996.
7. C Concordia. "Effect of prime-mover speed control characteristics on electrical power system performance," *IEEE Trans*, PAS-88, 752–756, 1969.
8. CW Taylor, KY Lee, DP Dave. "AGC analysis with governor deadband effects," *IEEE Trans*, PAS-98, 2030–2036, 1979.

3

Load Flow over AC Transmission Lines

Load flow is a solution of the steady-state operating conditions of a power system. It presents a "frozen" picture of a scenario with a given set of conditions and constraints. This can be a limitation, as the power system's operations are dynamic. In an industrial distribution system, the load demand for a specific process can be predicted fairly accurately, and a few load flow calculations will adequately describe the system. For bulk power supply, the load demand from hour to hour is uncertain, and winter and summer load flow situations, though typical, are not adequate. A moving picture scenario could be created from static snapshots, but it is rarely adequate in large systems having thousands of controls and constraints. Thus, the spectrum of load flow (power flow) embraces a large area of calculations, from calculating the voltage profiles and power flows in small systems to problems of on-line energy management and optimization strategies in interconnected large power systems.

Load flow studies are performed using digital computer simulations. These address operation, planning, running, and development of control strategies. Applied to large systems for optimization, security, and stability, the algorithms become complex and involved. While the treatment of load flow, and finally optimal power flow, will unfold in the following chapters, it can be stated that there are many load flow techniques and there is a historical background to the development of these methods.

In this chapter, we will study the power flow over power transmission lines, which is somewhat distinct and a problem by itself. The characteristics and performance of transmission lines can vary over wide limits, mainly dependent on their length. Maintaining an acceptable voltage profile at various nodes with varying power flow is a major problem. We will consider two-port networks, i.e., a single transmission line, to appreciate the principles and complexities involved.

3.1 Power in AC Circuits

The concepts of instantaneous power, average power, apparent power, and reactive power are fundamental and are briefly discussed here. Consider a lumped impedance $Ze^{j\theta}$, excited by a sinusoidal voltage $E < 0°$ at constant frequency. The linear load draws a sinusoidal current. The time varying power can be written as

$$\begin{aligned} p(t) &= \mathrm{Re}\left(\sqrt{2}E\varepsilon^{j(\omega t-\theta)}\right) \\ &= 2EI\cos\omega t \cdot \cos(\omega t - \theta) \end{aligned} \tag{3.1}$$

$$p(t) = EI\cos\theta + EI\cos(2\omega t - \theta) \tag{3.2}$$

where E and I are the rms voltage and current, respectively. The first term is the average time-dependent power, when the voltage and current waveforms consist only of fundamental components. The second term is the magnitude of power swing. Equation 3.2 can be written as

$$EI\cos\theta(1+\cos 2\omega t)+EI\sin\theta\cdot\sin 2\omega t \tag{3.3}$$

The first term is the power actually exhausted in the circuit and the second term is power exchanged between the source and circuit, but not exhausted in the circuit. The active power is measured in watts and is defined as

$$P = EI\cos\theta(1+\cos 2\omega t)\approx EI\cos\theta \tag{3.4}$$

The reactive power is measured in var and is defined as

$$Q = EI\sin\theta\sin 2\omega t \approx EI\sin\theta \tag{3.5}$$

These relationships are shown in Figure 3.1; cos θ is called the *power factor* (PF) of the circuit, and θ is the PF angle.

The apparent power in VA (volt–ampères) is given by

$$S=\sqrt{P^2+Q^2}=\sqrt{P^2+(P\tan\theta)^2}=P\sqrt{(1+\tan^2\theta)}=P\sec\theta=P/\cos\theta \tag{3.6}$$

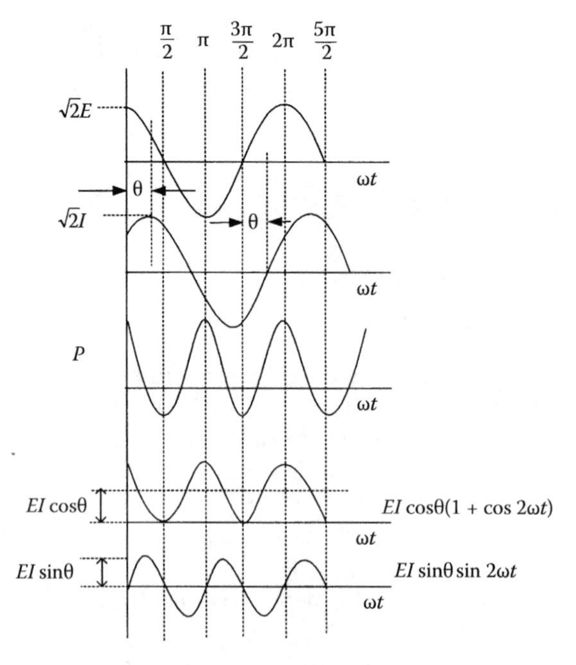

FIGURE 3.1
Active and reactive power in AC circuits.

The PF angle is generally defined as

$$\theta = \tan^{-1}\left(\frac{Q}{P}\right) \tag{3.7}$$

If $\cos\theta = 1$, $Q = 0$. Such a load is a unity PF load. Except for a small percentage of loads, i.e., resistance heating and incandescent lighting, the industrial, commercial, or residential loads operate at lagging PF. As the electrical equipment is rated on a kVA basis, a lower PF derates the equipment and limits its capacity to supply active power loads. The reactive power flow and control is one important aspect of power flow and Chapter 8 is devoted to it. The importance of PF and reactive power control can be broadly stated as follows:

- Improvement in the active power handling capability of transmission lines.
- Improvement in voltage stability limits.
- Increasing capability of existing systems: the improvement in PF for release of a certain per unit kVA capacity can be calculated from Equation 3.6:

$$\text{PF}_{\text{imp}} = \frac{\text{PF}_{\text{ext}}}{1 - \text{kVA}_{\text{ava}}} \tag{3.8}$$

  where PF_{imp} is improved PF, PF_{ext} is existing PF, and kVA_{ava} is kVA made available as per unit of existing kVA.
- Reduction in losses: the active power losses are reduced as these are proportional to the square of the current. With PF improvement, the current per unit for the same active power delivery is reduced. The loss reduction is given by the expression:

$$\text{Loss}_{\text{red}} = 1 - \left(\frac{\text{PF}_{\text{ext}}}{\text{PF}_{\text{imp}}}\right)^2 \tag{3.9}$$

  where Loss_{red} is reduction in losses in per unit with improvement in PF from PF_{ext} to PF_{imp}. An improvement of PF from 0.7 to 0.9 reduces the losses by 39.5%.
- Improvement of transmission line regulation: the PF improvement improves the line regulation by reducing the voltage drops on load flow.

All these concepts may not be immediately clear and are further developed.

The active power can also be written as

$$P = \frac{1}{T}\int_0^T p(t)\,\mathrm{d}t = EI\cos\theta = S\cos\theta \tag{3.10}$$

That is the active power is the average value of instantaneous power over one period T. The source current can be expressed as

$$\begin{aligned} I &= \frac{P}{E}\sin\omega t + \frac{Q}{E}\cos\omega t \\ &= \sqrt{\left(\frac{P}{E}\right)^2 + \left(\frac{Q}{E}\right)^2} \end{aligned} \tag{3.11}$$

In a three-phase circuit, the currents and voltages are represented as column vectors. The three-phase apparent power can be expressed as

$$S = E_a I_a + E_b I_b + E_c I_c$$
$$S = \sqrt{P^2_{3\text{-phase}} + Q^2_{3\text{-phase}}} \tag{3.12}$$
$$S = \sqrt{E_a^2 + E_b^2 + E_c^2} \cdot \sqrt{I_a^2 + I_b^2 + I_c^2}$$

where I_a, I_b, and I_c are the currents in the three phases and E_a, E_b, and E_c are the voltages. So long as the voltages are sinusoidal and loads are *linear* and balanced, all the three equations of S will give the same results.

3.1.1 Complex Power

If the voltage vector is expressed as $A+jB$ and the current vector as $C+jD$, then by convention the volt-ampères in AC circuits are vectorially expressed as

$$\begin{aligned} EI^* &= (A + jB)(C - jD) \\ &= AC + BD + j(BC - AD) \\ &= P + jQ \end{aligned} \tag{3.13}$$

where $P = AC + BD$ is the active power and $Q = BC - AD$ is the reactive power; I^* is the conjugate of I. This convention makes the imaginary part representing reactive power negative for the leading current and positive for the lagging current. *This is the convention used by power system engineers.* If a conjugate of voltage, instead of current, is used, the reactive power of the leading current becomes positive. The PF is given by

$$\cos\theta = \frac{AC + BD}{\sqrt{A^2 + B^2}\sqrt{C^2 + D^2}} \tag{3.14}$$

3.1.2 Conservation of Energy

The conservation of energy concept (Tellegen's theorem) is based on Kirchoff laws and states that the power generated by the network is equal to the power consumed by the network (inclusive of load demand and losses). If $i_1, i_2, i_3, \ldots, i_n$ are the currents and $v_1, v_2, v_3, \ldots, v_n$ are the voltages of n single-port elements connected in any manner then,

$$\sum_{k=1}^{k=n} V_k I_k = 0 \tag{3.15}$$

This is an obvious conclusion.

Also, in a linear system of passive elements, the complex power, active power, and reactive power should summate to zero:

$$\sum_{k=1}^{k=n} S_n = 0 \tag{3.16}$$

$$\sum_{k=1}^{k=n} P_n = 0 \tag{3.17}$$

$$\sum_{k=1}^{k=n} Q_n = 0 \tag{3.18}$$

3.2 Power Flow in a Nodal Branch

The modeling of transmission lines is unique in the sense that capacitance plays a significant role and cannot be ignored, except for short lines of length less than approximately 50 miles (80 km). Let us consider power flow over a short transmission line. As there are no shunt elements, the line can be modeled by its series resistance and reactance, load, and terminal conditions. Such a system may be called a nodal branch in load flow or a two-port network. The sum of the sending end and receiving end active and reactive powers in a nodal branch is not zero, due to losses in the series admittance Y_{sr} (Figure 3.2). Let us define Y_{sr}, the admittance of the series elements $= g_{sr} - jb_{sr}$ or $Z = zl = l(r_{sr} + jx_{sr}) = R_{sr} + X_{sr} = 1/Y_{sr}$, where l is the length of the line. The sending end power is

$$S_{sr} = V_s I_s^* \tag{3.19}$$

where I_s^* is conjugate of I_s. This gives

$$\begin{aligned} S_{sr} &= V_s[Y_{sr}(V_s - V_r)]^* \\ &= [V_s^2 - V_s V_r \varepsilon^{j(\theta_s - \theta_r)}](g_{sr} - jb_{sr}) \end{aligned} \tag{3.20}$$

where sending end voltage is $V_s < \theta_s$ and receiving end voltage is $V_r < \theta_r$. The complex power in Equation 3.20 can be divided into active and reactive power components. At the sending end,

$$P_{sr} = [V_s^2 - V_s \cos(\theta_s - \theta_r)]g_{sr} - [V_s V_r \sin(\theta_s - \theta_r)]b_{sr} \tag{3.21}$$

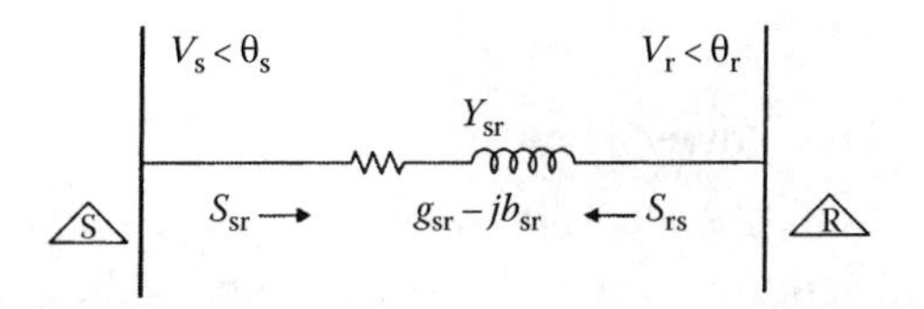

FIGURE 3.2
Power flow over a two-port line.

$$Q_{sr} = [-V_s V_r \sin(\theta_s - \theta_r)]\, g_{sr} - \left[V_s^2 - V_s V_r \cos(\theta_s - \theta_r)\right] b_{sr} \tag{3.22}$$

and at the receiving end,

$$P_{rs} = [V_r^2 - V_r V_s \cos(\theta_r - \theta_s)] g_{sr} - [V_r V_s \sin(\theta_r - \theta_s)] b_{sr} \tag{3.23}$$

$$Q_{rs} = [-V_r V_s \sin(\theta_r - \theta_s)] g_{sr} - [V_r^2 - V_r V_s \cos(\theta_r - \theta_s)] b_{sr} \tag{3.24}$$

If g_{sr} is neglected,

$$P_{rs} = \frac{|V_s||V_r| \sin\delta}{X_{sr}} \tag{3.25}$$

$$Q_{rs} = \frac{|V_s||V_r| \cos\delta - |V_r|^2}{X_{sr}} \tag{3.26}$$

where δ in the difference between the sending end and receiving end voltage vector angles=$(\theta_s - \theta_r)$. For small values of delta, the reactive power equation can be written as

$$Q_{rs} = \frac{|V_r|}{X_{sr}}\left(|V_s| - |V_r|\right) = \frac{|V_r|}{X_{sr}}|\Delta V| \tag{3.27}$$

where $|\Delta V|$ is the voltage drop. For a short line it is

$$|\Delta V| = I_r Z = (R_{sr} + jX_{sr})\frac{(P_{rs} - jQ_{rs})}{V_r} \approx \frac{R_{sr}P_{rs} + X_{sr}Q_{rs}}{|V_r|} \tag{3.28}$$

Therefore, the transfer of real power depends on the angle δ, called the *transmission* angle, and the relative magnitudes of the sending and receiving end voltages. As these voltages will be maintained close to the rated voltages, it is mainly a function of δ. The maximum power transfer occurs at δ=90° (steady-state stability limit).

The reactive power flow is in the direction of lower voltage and it is independent of δ. The following conclusions can be drawn:

1. For small resistance of the line, the real power flow is proportional to sin δ. It is maximum at δ = 90°. For stability considerations, the value is restricted to below 90°. The real power transfer rises with the rise in the transmission voltage.
2. The reactive power flow is proportional to the voltage drop in the line and is independent of δ. The receiving end voltage falls with increase in reactive power demand.

3.2.1 Simplifications of Line Power Flow

Generally, the series conductance is less than the series susceptance, the phase angle difference is small, and the sending end and receiving end voltages are close to the rated voltage as follows:

$$
\begin{aligned}
&g_{sr} \prec b_{sr} \\
&\sin(\theta_s - \theta_r) \approx \theta_s - \theta_r \\
&\cos(\theta_s - \theta_r) \approx 1 \\
&V_s \approx V_r \approx 1 \text{ per unit}
\end{aligned}
\tag{3.29}
$$

If these relations are used,

$$
\begin{aligned}
&P_{sr} \approx (\theta_s - \theta_r) b_{sr} \\
&Q_{sr} \approx (V_s - V_r) b_{sr} \\
&P_{rs} \approx -(\theta_r - \theta_s) b_{sr} \\
&Q_{rs} \approx -(V_r - V_s) b_{sr}
\end{aligned}
\tag{3.30}
$$

3.2.2 Voltage Regulation

The voltage regulation is defined as the rise in voltage at the receiving end, expressed as a percentage of full-load voltage when the full load at a specified PF is removed. The sending end voltage is kept constant. The voltage regulation is expressed as a percentage or as per unit of the receiving end full-load voltage:

$$
VR = \frac{V_{rnl} - V_{rfl}}{V_{rfl}} \times 100 \tag{3.31}
$$

where V_{rnl} is the no-load receiving end voltage and V_{rfl} is the full load voltage at a given PF.

3.3 *ABCD* Constants

A transmission line of any length can be represented by a four-terminal network, Figure 3.3a. In terms of *A*, *B*, *C*, and *D* constants, the relation between sending and receiving end voltages and currents can be expressed as

$$
\begin{vmatrix} V_s \\ I_s \end{vmatrix} = \begin{vmatrix} A & B \\ C & D \end{vmatrix} \begin{vmatrix} V_r \\ I_r \end{vmatrix} \tag{3.32}
$$

In the case where sending end voltages and currents are known, the receiving end voltage and current can be found by

$$
\begin{vmatrix} V_r \\ I_r \end{vmatrix} = \begin{vmatrix} D & -B \\ -C & A \end{vmatrix} \begin{vmatrix} V_s \\ I_s \end{vmatrix} \tag{3.33}
$$

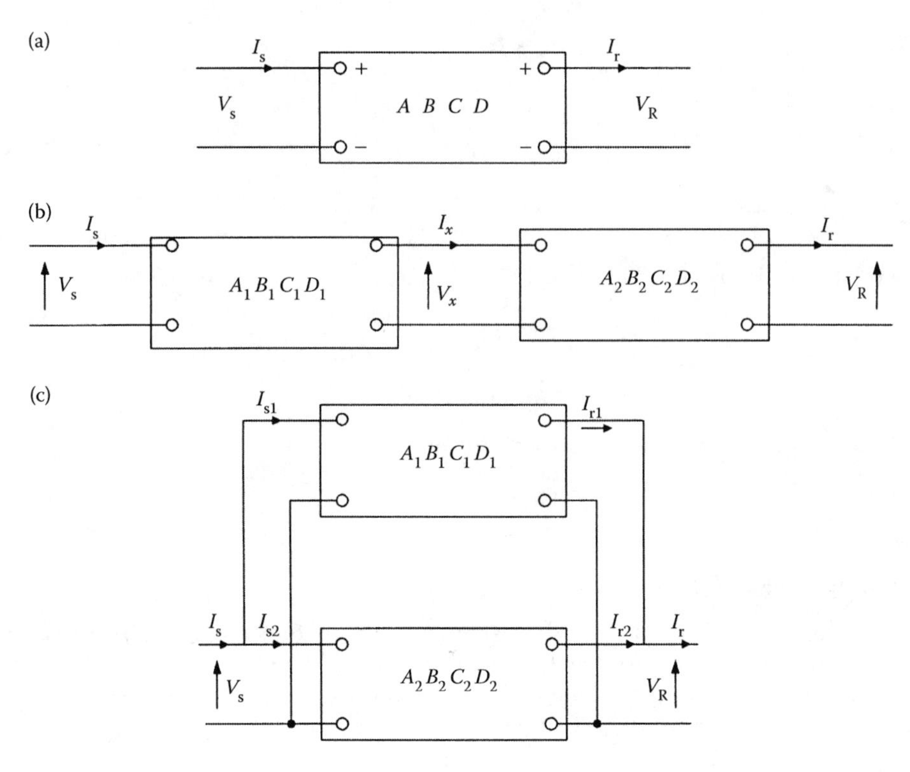

FIGURE 3.3
(a) Schematic representation of a two-terminal network using *ABCD* constants, (b) two networks in series, and (c) two networks in parallel.

Also,

$$AD - BC = 1 \tag{3.34}$$

The significance of these constants can be stated as follows:

$$\begin{aligned}
&A = V_s/V_r, \text{ when } I_r = 0, \text{ i.e., the receiving end is open-circuited. It is the ratio of two voltages and is dimensionless.}\\
&B = V_r/I_r, \text{ when } V_r = 0, \text{ i.e., the receiving end is short-circuited. It has the dimensions of an impedance and specified in ohms.}\\
&C = I_r/V_r, \text{ when the receiving end is open-circuited and } I_r \text{ is zero. It has the dimensions of an admittance.}\\
&D = I_s/I_r, \text{ when } V_r = 0, \text{ i.e., the receiving end is short-circuited. It is the ratio of two currents and dimensionless.}
\end{aligned} \tag{3.35}$$

Two *ABCD* networks in series, Figure 3.3b, can be reduced to a single equivalent network:

$$\begin{vmatrix} V_s \\ I_s \end{vmatrix} = \begin{vmatrix} A_1 & B_1 \\ C_1 & D_1 \end{vmatrix} \begin{vmatrix} A_2 & B_2 \\ C_2 & D_2 \end{vmatrix} \begin{vmatrix} V_r \\ I_r \end{vmatrix} = \begin{vmatrix} A_1A_2 + B_1C_2 & A_1B_2 + B_1D_2 \\ C_1A_2 + D_1C_2 & C_1B_2 + D_1D_2 \end{vmatrix} \begin{vmatrix} V_r \\ I_r \end{vmatrix} \tag{3.36}$$

For parallel *ABCD* networks (Figure 3.3c), the combined *ABCD* constants are as follows:

$$\begin{aligned} A &= (A_1B_2 + A_2B_1)/(B_1 + B_2) \\ B &= B_1B_2/(B_1 + B_2) \\ C &= (C_1 + C_2) + (A_1 - A_2)(D_2 - D_1)/(B_1 + B_2) \\ D &= (B_2D_1 + B_1D_2)/(B_1 + B_2) \end{aligned} \tag{3.37}$$

Example 3.1

Calculate the *ABCD* constants of a short transmission line, voltage regulation, and load PF for zero voltage regulation.

In a short transmission line, the sending end current is equal to the receiving end current. The sending end voltage can be vectorially calculated by adding the *IZ* drop to the receiving end voltage. Considering a receiving end voltage $V_r < 0°$ and current $I_r < \phi$,

$$V_s = V_r < 0° + ZI_r < \phi$$

Therefore, from Equation 3.32, $A = 1$, $B = Z$, $C = 0$, and $D = 1$:

$$\begin{vmatrix} V_s \\ V_r \end{vmatrix} = \begin{vmatrix} 1 & Z \\ 0 & 1 \end{vmatrix} \begin{vmatrix} V_r \\ I_r \end{vmatrix}$$

The equation can be closely approximated as

$$V_s = V_r + IR_{sr}\cos\phi + IX_{sr}\sin\phi \tag{3.38}$$

For a short line, the no-load receiving end voltage equals the sending end voltage:

$$V_s = AV_r + BI_r$$

At no load I_r=0 and A=1; therefore, the no-load receiving end voltage=sending end voltage=V_s/A. Therefore, the regulation is

$$\frac{(V_s/A) - V_r}{V_r} = \frac{IR_{sr}\cos\phi + IX_{sr}\sin\phi}{V_r} \tag{3.39}$$

The voltage regulation is negative for a leading PF.

Example 3.2

A short three-phase, 13.8-kV line supplies a load of 10 MW at 0.8 lagging PF. The line resistance is 0.25 Ω and its reactance is 2.5 Ω. Calculate the sending end voltage, regulation, and value of a capacitor to be connected at the receiving end to reduce the calculated line regulation by 50%.

The line current at 10 MW and 0.8 PF = 523 < −36.87°. The sending end voltage is

$$V_s = V_r < 0° + \sqrt{3} \times 523(0.8 - j0.6)(0.25 + j2.5)$$

or this is closely approximated by Equation 3.38, which gives a sending end voltage of 15.34 kV. From Equation 3.39, the line regulation is 11.16%. This is required to be reduced by 50%, i.e., to 5.58%. The sending end voltage is given by

$$\frac{|V_s| - 13.8}{13.8} = 0.0538 \text{ or } |V_s| = 14.54\,\text{kV}$$

The line voltage drop must be reduced to 742 V. This gives two equations: 428 = 4(0.25cos ϕ_n + 2.5sin ϕ_n) and I_n = 418/cos ϕ_n (10 MW of power at three phase, 13.8 kV, and unity PF 418 A), where I_n is the new value of the line current and ϕ_n the improved PF angle. From these two equations, ϕ_n = 17.2°, i.e., the PF should be improved to approximately 0.955. The new current I_n = 437.5 < −17.2°. Therefore, the current supplied by the intended capacitor bank is

$$I - I_n = (417.9 - j129.37) - (418.4 - j313.8) = -j184\,\text{A (leading)}$$

Within the accuracy of calculation, the active part of the current should cancel out, as the capacitor bank supplies only a reactive component. The capacitor reactance to be added per phase $= 13{,}800/\left(\sqrt{3.184}\right)$ = 43.30 Ω, which is equal to 61.25 μF.

3.4 Transmission Line Models

3.4.1 Medium Long Transmission Lines

For transmission lines in the range 50–150 miles (80–240 km), the shunt admittance cannot be neglected. There are two models in use, the nominal Π- and nominal T-circuit models. In the T-circuit model, the shunt admittance is connected at the midpoint of the line, while in the Π model, it is equally divided at the sending end and the receiving end. The Π equivalent circuit and phasor diagram are shown in Figure 3.4a and b. The nominal T-circuit model and phasor diagram are shown in Figure 3.4c and d. The *ABCD* constants are shown in Table 3.1.

Example 3.3

Calculate the *ABCD* constants of the Π model of the transmission line shown in Table 3.1.

The sending end current is equal to the receiving end current, and the current through the shunt elements $Y/2$ at the receiving and sending ends is

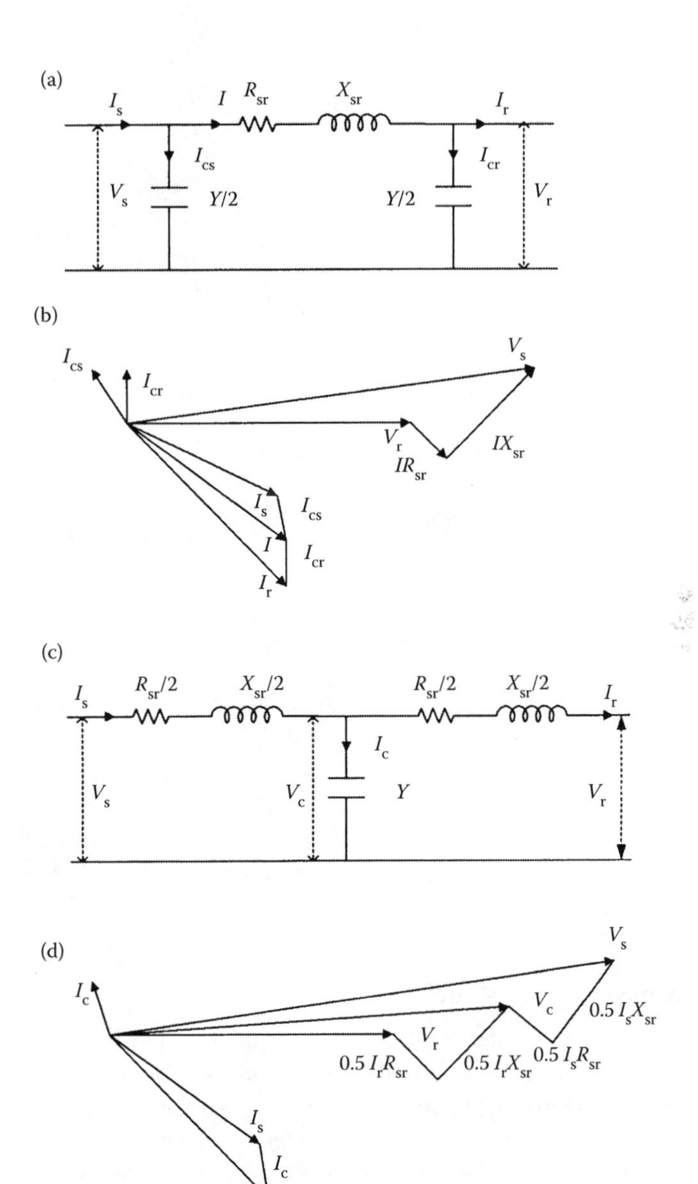

FIGURE 3.4
(a) Π representation of a transmission line, (b) phasor diagram of Π representation, (c) T representation of a transmission line, and (d) phasor diagram of T representation.

$$I_s = I_r + \frac{1}{2}V_rY + \frac{1}{2}V_sY$$

The sending end voltage is the vector sum of the receiving end voltage and the drop through the series impedance Z is

$$V_s = V_r + \left(I_r + \frac{1}{2}V_rY\right)Z = V_r\left(1 + \frac{1}{2}YZ\right) + I_rZ$$

TABLE 3.1

ABCD Constants of Transmission Lines

Line Length	Equivalent Circuit	A	B	C	D
Short	Series impedance only	1	Z	0	1
Medium	Nominal Π, Figure 10.4a	1 1+2 YZ	Z	Y[1+1/4(YZ)]	1 1+2 YZ
Medium	Nominal T, Figure 10.4b	1 1+2 YZ	Z[1+1/4(YZ)]	Y	1 1+2 YZ
Long	Distributed parameters	cosh γ1	Z_0 sinh γ1	(sinh γ1)/Z_0	cosh γ1

The sending end current can, therefore, be written as

$$I_s = I_r + \frac{1}{2}V_rY + \frac{1}{2}Y\left[V_r\left(1+\frac{1}{2}YZ\right)+I_rZ\right]$$

$$V_rY\left(1+\frac{1}{4}YZ\right)+I_r\left(1+\frac{1}{2}YZ\right)$$

or in matrix form

$$\begin{vmatrix} V_s \\ I_s \end{vmatrix} = \begin{vmatrix} \left(1+\frac{1}{2}YZ\right) & Z \\ Y\left(1+\frac{1}{4}YZ\right) & \left(1+\frac{1}{2}YZ\right) \end{vmatrix} \begin{vmatrix} V_r \\ I_r \end{vmatrix}$$

3.4.2 Long Transmission Line Model

Lumping the shunt admittance of the lines is an approximation and for line lengths over 150 miles (240 km), the distributed parameter representation of a line is used. Each elemental section of line has a series impedance and shunt admittance associated with it. The operation of a long line can be examined by considering an elemental section of impedance z per unit length and admittance y per unit length. The impedance for the elemental section of length $\mathrm{d}x$ is $z\,\mathrm{d}x$ and the admittance is $y\,\mathrm{d}x$. Referring to Figure 3.5, by Kirchoff's voltage law,

$$V = Iz\,\mathrm{d}x + V + \frac{\partial V}{\partial x}\mathrm{d}x$$
$$\frac{\partial V}{\partial x} = -IZ \tag{3.40}$$

Similarly, from the current law,

$$I = Vy\,\mathrm{d}x + I + \frac{\partial I}{\partial x}\mathrm{d}x$$
$$\frac{\partial I}{\partial x} = -Vy \tag{3.41}$$

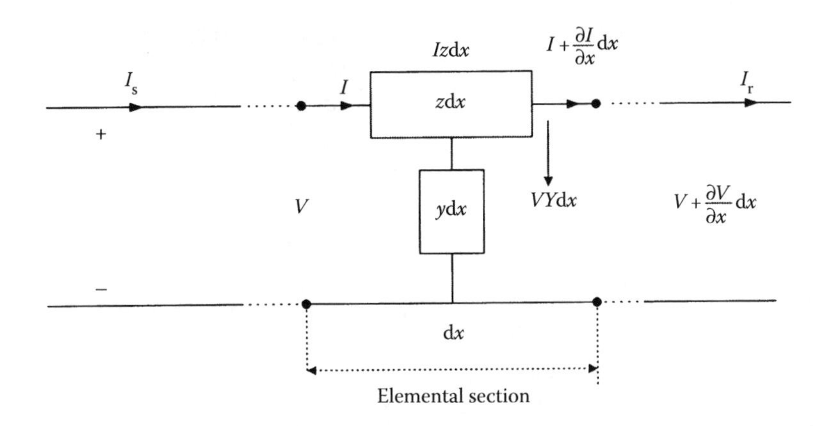

FIGURE 3.5
Model of an elemental section of a long transmission line.

Differentiating Equations 3.40 and 3.41

$$\frac{\partial^2 V}{\partial x^2} = yzV \tag{3.42}$$

$$\frac{\partial^2 I}{\partial x^2} = yzI \tag{3.43}$$

These differential equations have solutions of the form:

$$V = V_1 e^{\gamma x} + V_2 e^{-\gamma x} \tag{3.44}$$

where γ is the *propagation constant*. It is defined as

$$\gamma = \sqrt{zy} \tag{3.45}$$

Define the following shunt elements, to distinguish from line series elements: g_{sh} is the shunt conductance, b_{sh} is the shunt susceptance, x_{sh} is the shunt-capacitive reactance, and $Y = yl$, the shunt admittance, where l is the length of the line. The shunt conductance is small and ignoring it gives $y = j/x_{sh}$.

$$\begin{aligned} y &= jb_{sh} \\ yz &= -b_{sh}x_{sc} + jr_{sc}b_{sh} \\ |\gamma| &= \sqrt{b_{sh}}\,(r_{sc}^2 + x_{sc}^2)^{1/4} \end{aligned} \tag{3.46}$$

The complex propagation constant can be written as

$$\gamma = \alpha + j\beta \tag{3.47}$$

where α is defined as the *attenuation constant*. Common units are nepers per mile or per km.

$$\alpha = |\gamma| \cos\left[\frac{1}{2}\tan^{-1}\left(\frac{-r_{sc}}{x_{sc}}\right)\right] \tag{3.48}$$

$$\beta = |\gamma| \sin\left[\frac{1}{2}\tan^{-1}\left(-\frac{r_{sc}}{x_{sc}}\right)\right] \tag{3.49}$$

where β is the *phase constant*. Common units are radians per mile (or per km). The *characteristic impedance* is

$$Z_0 = \sqrt{\frac{z}{y}} \tag{3.50}$$

Again neglecting shunt conductance

$$\frac{z}{y} = \frac{r_{sc} + jx_{sc}}{jb_{sh}} = x_{sc}x_{sh} - jx_{sh}r_{sc}$$

$$Z_0 = \sqrt{x_{sh}}\,(r_{sc}^2 + x_{sc}^2)^{1/4} \tag{3.51}$$

$$< Z_0 = \frac{1}{2}\tan^{-1}\left(-\frac{r_{sc}}{x_{sc}}\right) \tag{3.52}$$

The voltage at any distance χ can be written as

$$V_x = \left|\frac{V_r + Z_0 I_r}{2}\right| e^{\alpha x + j\beta x} + \left|\frac{V_r - Z_0 I_r}{2}\right| e^{-\alpha x - j\beta x} \tag{3.53}$$

These equations represent traveling waves. The solution consists of two terms, each of which is a function of two variables, time and distance. At any instant of time, the first term, the incident wave, is distributed sinusoidally along the line, with amplitude increasing exponentially from the receiving end. After a time interval Δt, the distribution advances in phase by $\omega\,\Delta t/\beta$, and the wave is traveling toward the receiving end. The second term is the reflected wave, and after time interval Δt, the distribution retards in phase by $\omega\,\Delta t/\beta$, the wave traveling from the receiving end to the sending end.

A similar explanation holds for the current:

$$I_x = \left|\frac{V_r / Z_0 + I_r}{2}\right| e^{\alpha x + j\beta x} - \left|\frac{V_r / Z_0 - I_r}{2}\right| e^{-\alpha x - j\beta x} \tag{3.54}$$

These equations can be written as

$$V_x = V_r\left(\frac{e^{\gamma x} + e^{-\gamma x}}{2}\right) + I_r Z_0\left(\frac{e^{\gamma x} - e^{-\gamma x}}{2}\right)$$

$$I_x = \frac{V_r}{Z_0}\left(\frac{e^{\gamma x} - e^{-\gamma x}}{2}\right) + I_r\left(\frac{e^{\gamma x} + e^{-\gamma x}}{2}\right) \tag{3.55}$$

or in matrix form

$$\begin{vmatrix} V_s \\ I_s \end{vmatrix} = \begin{vmatrix} \cosh \gamma l & Z_0 \sinh \gamma l \\ \dfrac{1}{Z_0} \sinh \gamma l & \cosh \gamma l \end{vmatrix} \begin{vmatrix} V_r \\ I_r \end{vmatrix} \tag{3.56}$$

These *ABCD* constants are shown in Table 3.1.

3.4.3 Reflection Coefficient

The relative values of sending end and receiving end voltages, V_1 and V_2 depend on the conditions at the terminals of the line. The reflection coefficient at the load end is defined as the ratio of the amplitudes of the backward and forward traveling waves. For a line terminated in a load impedance Z_L,

$$V_2 = \left(\frac{Z_L - Z_0}{Z_L + Z_0} \right) V_1 \tag{3.57}$$

Therefore, the voltage reflection coefficient at the load end is

$$\rho_L = \frac{Z_L - Z_0}{Z_L + Z_0} \tag{3.58}$$

The current reflection coefficient is negative of the voltage reflection coefficient. For a short-circuit line, the current doubles and for an open-circuit line the voltage doubles. Figure 3.6 shows the traveling wave phenomenon. The reflected wave at an impedance discontinuity is a mirror image of the incident wave moving in the opposite direction. Every point in the reflected wave is the corresponding point on the incident wave multiplied by the reflection coefficient, but a mirror image. At any time, the total voltage is the sum of the incident and reflected waves. Figure 3.6b shows the reinforcement of the incident and reflected waves. The reflected wave moves toward the source and is again reflected. The source reflection coefficient, akin to the load reflection coefficient, can be defined as

$$\rho_s = \frac{Z_s - Z_0}{Z_s + Z_0} \tag{3.59}$$

A forward traveling wave originates at the source, as the backward traveling wave originates at the load. At any time, the voltage or current at any point on the line is the sum of all voltage or current waves existing at the line at that point and time.

Consider that the line is terminated with a reactor, and a step voltage is applied. Calculate the expressions for transmitted and reflected voltages as follows:

We can write $Z_2 = L_s$ in Laplace transform and let $Z_1 = Z$.

Then the reflection coefficient is

$$\rho_L = \frac{(s - Z/L)}{(s + Z/L)} \tag{3.60}$$

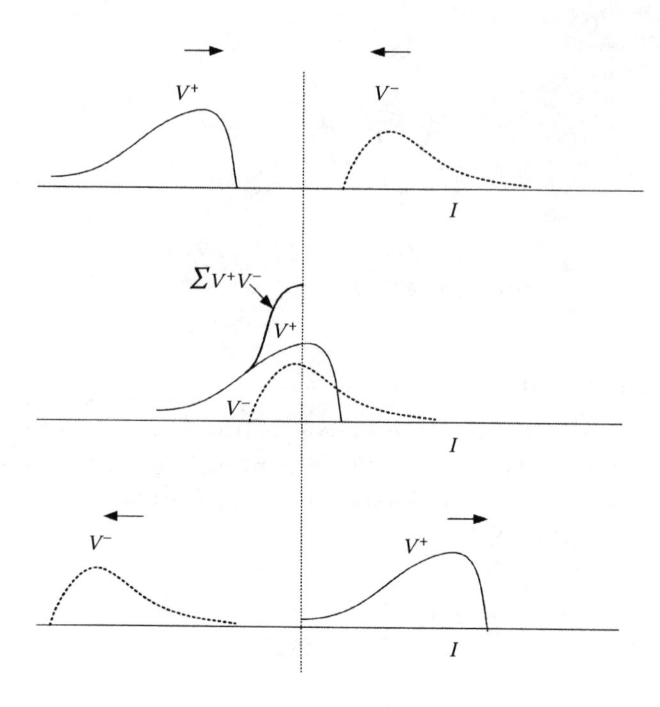

FIGURE 3.6
(a) Incident and reflected waves at an impedance change in a long transmission line, (b) reinforcement of incident and reflected waves, and (c) incident and reflected waves crossing each other.

The reflected wave is

$$V'(s) = \frac{(s - Z/L)}{(s + Z/L)} \frac{V}{s}$$
$$= \left[-\frac{1}{s} + \frac{2}{s + Z/L} \right] \tag{3.61}$$

Taking inverse transform,

$$V' = \left[-1 + 2e^{-(Z/L)t} \right] \tag{3.62}$$

Voltage of the transmitted wave is

$$V''(s) = (1 + \rho_L) V(s)$$
$$= \frac{2V}{(s + Z/L)} \tag{3.63}$$

Thus,

$$V'' = 2Ve^{-(Z/L)t} \tag{3.64}$$

The voltage across the inductor rises initially to double the incident voltage and decays exponentially.

3.4.4 Lattice Diagrams

The forward and backward traveling waves can be shown on a lattice diagram (Figure 3.7). The horizontal axis is the distance of the line from the source and the vertical axis is labeled in time increments, each increment being the time required for the wave to travel the line in one direction, i.e., source to the load. Consider a point P in the pulse shape of Figure 3.7b at time t'. The time to travel in one direction is l/u, where u is close to the velocity of light, as we will discuss further. The point P then reaches the load end at $t'+l/us$, and is reflected back. The corresponding point on the reflected wave is $P\rho_L$. At the sending end it is re-reflected as $\rho_L\rho_s$, Figure 3.7b.

Example 3.4

A lossless transmission line has a surge impedance of 300 Ω. It is terminated in a resistance of 600 Ω. A 120-V dc source is applied to the line at $t=0$ at the sending end. Considering that it takes t seconds to travel the voltage wave in one direction, draw the lattice diagram from $t=0$ to $7t$. Plot the voltage profile at the load terminals.

The sending end reflection coefficient from Equation 3.59 is –1 and the load reflection coefficient from Equation 3.58 is 0.33. The lattice diagram is shown in Figure 3.8a. At first reflection at the load, 120 V is reflected as $0.33\times120=39.6$ V, which is re-reflected from the sending end as $-1\times39.6=-39.6$ V. The voltage at the receiving end can be plotted from the lattice diagram in Figure 3.8a and is shown in Figure 3.8b.

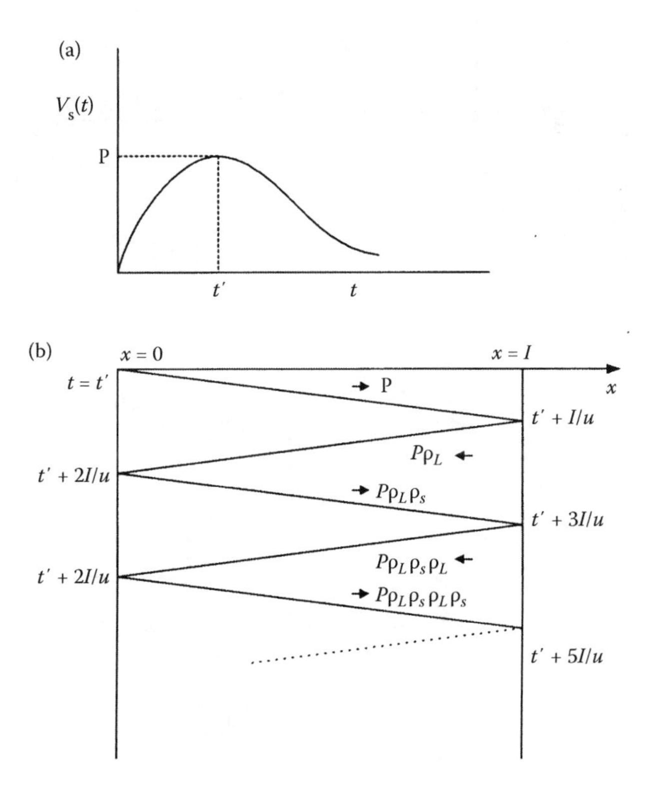

FIGURE 3.7
(a) A point P at time t' on a pulse signal applied to the sending end of a transmission line and (b) lattice diagram.

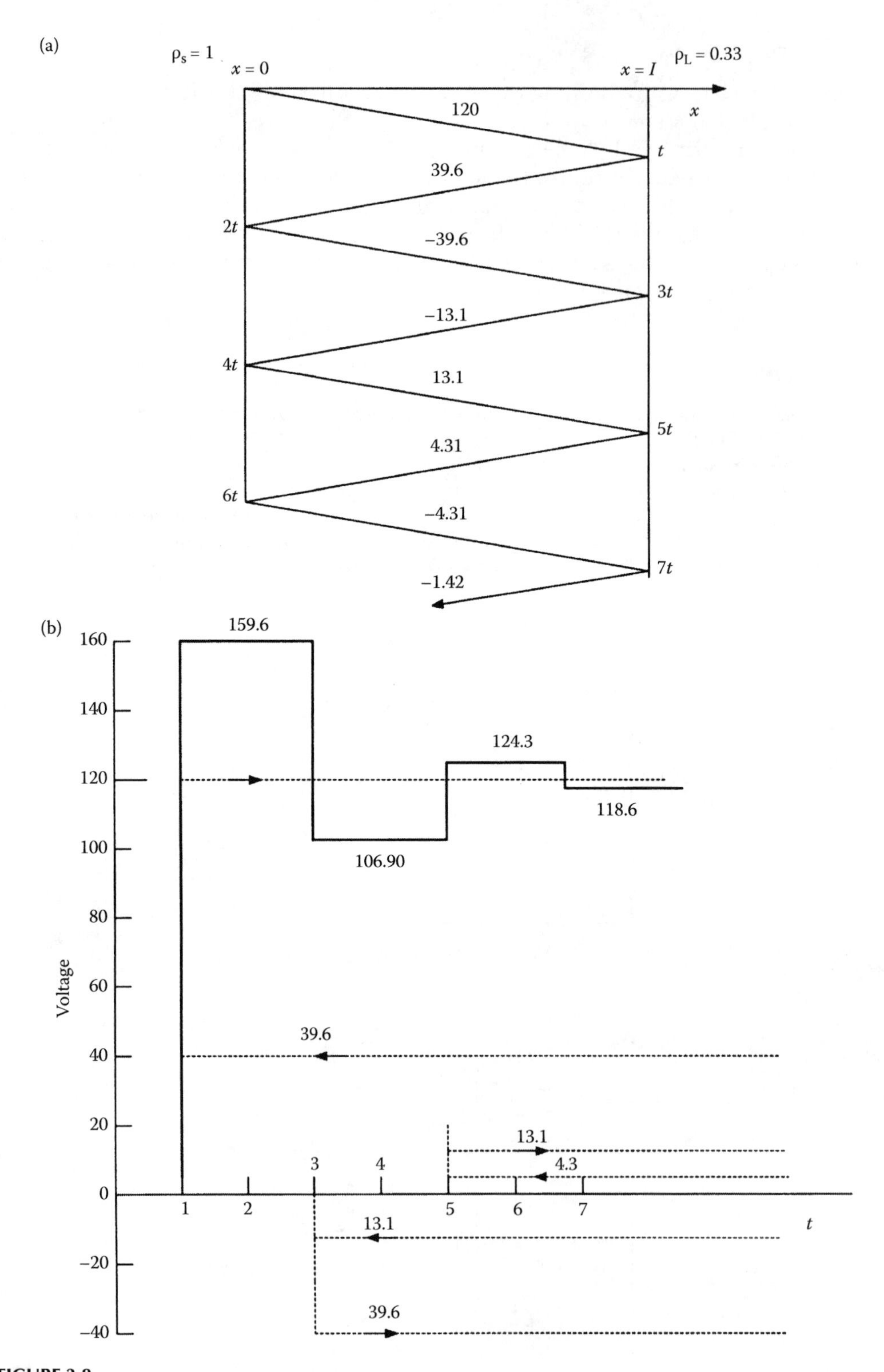

FIGURE 3.8
(a) Lattice diagram of Example 3.4 and (b) the receiving end voltage profile.

3.4.5 Infinite Line

When the line is terminated in its characteristic load impedance, i.e., $Z_L = Z_0$, the reflected wave is zero. Such a line is called an *infinite line* and the incident wave cannot distinguish between a termination and the continuation of the line.

The characteristic impedance is also called the surge impedance. It is approximately 400 Ω for overhead transmission lines and its phase angle may vary from 0° to 15°. For underground cables, the surge impedance is much lower, approximately 1/10 that of overhead lines.

3.4.6 Surge Impedance Loading

The surge impedance loading (SIL) of the line is defined as the power delivered to a purely resistive load equal in value to the surge impedance of the line:

$$\text{SIL} = \frac{V_r^2}{Z_0} \tag{3.65}$$

For 400 Ω surge impedance, SIL in kW is 2.5 multiplied by the square of the receiving end voltage in kV. The surge impedance is a real number and therefore the PF along the line is unity, i.e., no reactive power compensation (see Chapter 9) is required. The SIL loading is also called the natural loading of the transmission line.

3.4.7 Wavelength

A complete voltage or current cycle along the line, corresponding to a change of 2π radians in angular argument of β, is defined as the wavelength λ. If β is expressed in rad/mile or per km):

$$\lambda = \frac{2\pi}{\beta} \tag{3.66}$$

For a lossless line,

$$\beta = \omega\sqrt{LC} \tag{3.67}$$

Therefore,

$$\lambda = \frac{1}{f\sqrt{LC}} \tag{3.68}$$

and the velocity of propagation of the wave

$$v = f\lambda = \frac{1}{\sqrt{LC}} \approx \frac{1}{\sqrt{\mu_0 k_0}} \tag{3.69}$$

where $\mu_0 = 4\pi \times 10^{-7}$ is the permeability of the free space, and $k_0 = 8.854 \times 10^{-12}$ is the permittivity of the free space. Therefore, $1/[\sqrt{(\mu_0 k_0)}] = 3 \times 10^{10}$ cm/s or 186,000 miles/s velocity of light. We have considered a lossless line in developing the above expressions. The actual velocity of the propagation of the wave along the line is somewhat less than the speed of light.

3.5 Tuned Power Line

In the long transmission line model, if the shunt conductance and series resistance are neglected, then

$$\gamma = \sqrt{YZ} = j\omega\sqrt{LC} \tag{3.70}$$

$$\cosh \gamma l = \cosh j\omega l\sqrt{LC} = \cos \omega l\sqrt{LC} \tag{3.71}$$

$$\sinh \gamma l = \sinh j\omega l\sqrt{LC} = j \sin \omega l\sqrt{LC} \tag{3.72}$$

where l is the length of the line. This simplifies *ABCD* constants and the following relationship results:

$$\begin{vmatrix} V_s \\ I_s \end{vmatrix} = \begin{vmatrix} \cos \omega l\sqrt{LC} & jZ_0 \sin \omega l\sqrt{LC} \\ \dfrac{j}{Z_0} \sin \omega l\sqrt{LC} & \cos \omega l\sqrt{LC} \end{vmatrix} \begin{vmatrix} V_R \\ I_R \end{vmatrix} \tag{3.73}$$

If

$$\omega l\sqrt{LC} = n\pi \quad (n = 1, 2, 3\ldots) \tag{3.74}$$

then

$$|V_s| = |V_r| \quad |I_s| = |I_r| \tag{3.75}$$

This will be an ideal situation to operate a transmission line. The receiving end voltage and currents are equal to the sending end voltage and current. The line has a flat voltage profile.

$$\beta = \frac{2\pi f}{\upsilon} = \frac{2\pi}{\lambda} \tag{3.76}$$

As $1/\sqrt{LC} \approx$ velocity of light, the line length λ is 3100, 6200,... miles (4988, 9976,... km) or β = 0.116°/mile (0.0721°/km). The quantity β_1 is called the *electrical length* of the line. The length calculated above is too long to avail this ideal property and suggests that power lines can be tuned with series capacitors to cancel the effect of inductance and with shunt inductors to neutralize the effect of line capacitance. This compensation may be done by sectionalizing the line. For power lines, series and shunt capacitors for heavy load conditions and shunt reactors under light load conditions are used to improve power transfer and line regulation. We will revert to this in Chapter 9.

3.6 Ferranti Effect

As the transmission line length increases, the receiving end voltage rises above the sending end voltage, due to line capacitance. This is called the Ferranti effect. In a long line model, at no load (I_R=0), the sending end voltage is

$$V_s = \frac{V_r}{2} e^{\alpha l} e^{j\beta l} + \frac{V_r}{2} e^{-\alpha l} e^{-j\beta l} \tag{3.77}$$

At $l=0$, both incident and reflected waves are equal to $V_r/2$. As l increases, the incident wave increases exponentially, while the reflected wave decreases. Thus, the receiving end voltage rises. Another explanation of the voltage rise can be provided by considering that the line capacitance is lumped at the receiving end. Let this capacitance be Cl; then, on open circuit, the sending end current is

$$I_s = \frac{V_s}{\left(j\omega Ll - \frac{1}{j\omega Cl}\right)} \tag{3.78}$$

C is small in comparison with L. Thus, ωLl can be neglected. The receiving end voltage can then be written as

$$\begin{aligned} V_r &= V_s - I_s(j\omega Ll) \\ &= V_s + V_s\omega^2 CLl^2 \\ &= V_s(1+\omega^2 CLl^2) \end{aligned} \tag{3.79}$$

This gives a voltage rise at the receiving end of

$$|V_s|\omega^2 CLl^2 = |V_s|\omega^2 l^2 / v^2 \tag{3.80}$$

where v is the velocity of propagation. Considering that v is constant, the voltage rises with the increase in line length.

Also, from Equation 3.73, the voltage at any distance x terms of the sending end voltage, with the line open-circuited and resistance neglected, is

$$V_x = V_s \frac{\cos\beta(l-x)}{\cos\beta l} \tag{3.81}$$

and the current is

$$I_x = j\frac{V_s}{Z_0}\frac{\sin\beta(l-x)}{\cos\beta l} \tag{3.82}$$

Example 3.5

A 230-kV three-phase transmission line has 795 KCMIL (456.3 mm^2), ACSR conductors, one per phase. Neglecting resistance, $z=j0.8$ Ω/mile (= 0.4972 Ω/km) and $y=j5.4\times10^{-6}$ Siemens (same as mhos) per mile (= $j3.356\times10^{-6}$/km) Calculate the voltage rise at the receiving end for a 400 mile (644 km) long line.

Using the expressions developed above,

$$Z_0 = \sqrt{\frac{z}{y}} = \sqrt{\frac{j0.8}{j5.4\times10^{-6}}} = 385\,\Omega$$

$\beta = \sqrt{zy} = 2.078 \times 10^{-3}$ rad/mile = 0.119°/mile; β_l = 0.119×400 = 47.6°. (Same results with calculations on metric basis.) The receiving end voltage rise from Equation 3.81:

$$V_r = V_s \frac{\cos(l-l)}{\cos 47.6°} = \frac{V_s}{0.674} = 1.483V_s$$

The voltage rise is 48.3% and at 756 miles (1231 km), one-quarter wavelength, it will be infinite.

Even a voltage rise of 10% at the receiving end is not acceptable as it may give rise to insulation stresses and affect the terminal regulating equipment. Practically, the voltage rise will be more than that calculated above. As the load is thrown off, the sending end voltage will rise before the generator voltage regulators and excitation systems act to reduce the voltage, further increasing the voltages on the line. These point to the necessity of compensating the transmission lines.

The sending end charging current from Equation 3.82 is 1.18 per unit and falls to zero at the receiving end. This means that the charging current flowing in the line is 118% of the line natural load.

3.6.1 Approximate Long Line Parameters

Regardless of voltage, conductor size, or spacing of a line, the series reactance is approximately 0.8 Ω/mile (0.4972 Ω/km) and the shunt-capacitive reactance is 0.2 MΩ/mile (0.1243 MΩ/km). This gives a β of 1.998×10^{-3}/mile or 0.1145°/mile (0.07116°/km).

3.7 Symmetrical Line at No Load

If we consider a symmetrical line at no load, with the sending end and receiving end voltages maintained the same, these voltages have to be in phase as no power is transferred. Half the charging current is supplied from each end and the line is equivalent to two equal halves connected back-to-back. The voltage rises at the midpoint, where the charging current falls to zero and reverses direction. The synchronous machines at the sending end absorb leading reactive power, while the synchronous machines at the receiving end generate lagging reactive power Figure 3.9a, b, and c. The midpoint voltage is, therefore, equal to the voltage as if the line was of half the length.

On loading, the vector diagram shown in Figure 3.9d is applicable. By symmetry, the midpoint voltage vector exactly bisects the sending and receiving end voltage vectors, and the sending end and receiving end voltages are equal. The PF angle at both ends are equal but of opposite sign. Therefore, the receiving end voltage on a symmetric line of length $2l$ is the same as that of line of length l at unity PF load. From Equation 3.73, the equations for the sending end voltage and current for a symmetrical line can be written with βl replaced by $\beta l/2 = \theta/2$.

$$V_s = V_m \cos\left(\frac{\theta}{2}\right) + jZ_0 I_m \sin\left(\frac{\theta}{2}\right) \tag{3.83}$$

$$I_s = j\frac{V_m}{Z_0}\sin\left(\frac{\theta}{2}\right) + I_m \cos\left(\frac{\theta}{2}\right) \tag{3.84}$$

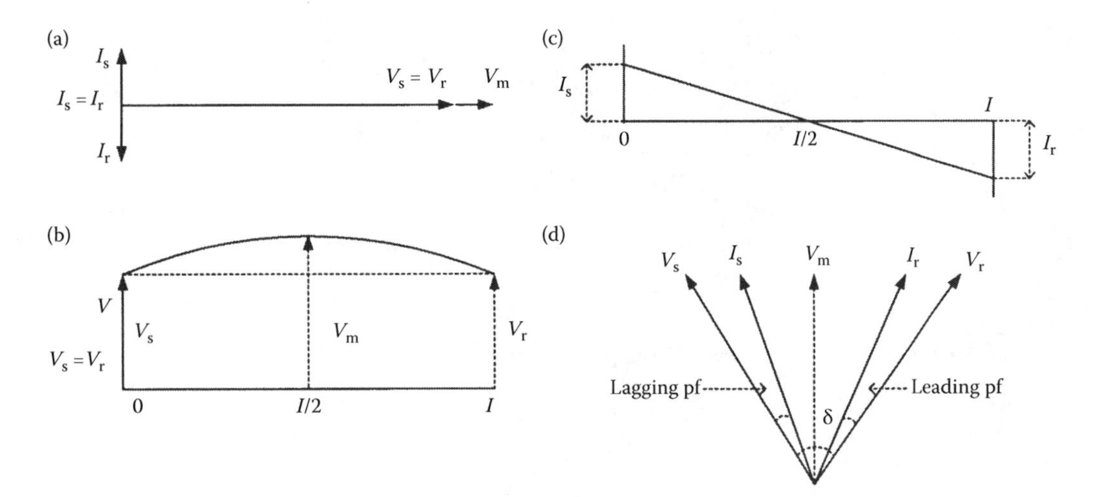

FIGURE 3.9
(a) Phasor diagram of a symmetrical line at no load, (b) the voltage profile of a symmetrical line at no load, (c) the charging current profile of a symmetrical line at no load, and (d) the phasor diagram of a symmetrical line at load.

At the midpoint,

$$P_m + jQ_m = V_m I_m^* = P \tag{3.85}$$

$$Q_s = \operatorname{Im} g V_s I_s^*$$

$$= j\frac{\sin\theta}{2}\left[Z_0 I_m^2 - \frac{V_m^2}{Z_0}\right] \tag{3.86}$$

where P is the transmitted power. No reactive power flows past the midpoint and it is supplied by the sending end.

3.8 Illustrative Examples

Example 3.6

Consider a line of medium length represented by a nominal Π circuit. The impedances and shunt susceptances are shown in Figure 3.10 in per unit on a 100-MVA base. Based on the given data calculate *ABCD* constants. Consider that one per unit current at 0.9 PF lagging is required to be supplied to the receiving end. Calculate the sending end voltage, currents, power, and losses. Compare the results with approximate power flow relations derived in Equation 3.30:

$$D = A = 1 + \frac{1}{2}YZ$$

$$= 1 + 0.5[j0.0538][0.0746 + j0.394]$$

$$= 0.989 + j0.002$$

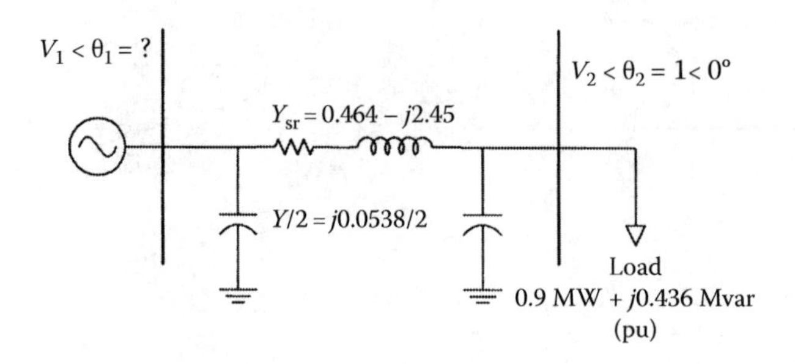

FIGURE 3.10
Transmission line and load parameters for Examples 3.6 and 3.7.

and

$$C = Y\left(1 + \frac{1}{4}YZ\right)$$
$$= (j0.0538)[1 + (-0.0053 + j0.9947)]$$
$$= -0.000054 + j0.0535$$
$$B = Z = 0.0746 + j0.394$$

The voltage at the receiving end bus is taken as equal to the rated voltage of one per unit at zero degree phase shift, i.e., $1 < 0°$. The receiving end power is, therefore,

$$V_2 I_2^*$$
$$(1 < 0°)(1_2 < 25.8°)$$
$$0.9 + j0.436$$

It is required to calculate the sending end voltage and current as follows:

$$V_1 = AV_2 + BI_2$$
$$= (0.989 + j0.002)(1 < 0°) + (0.0746 + j0.394)(1 < 25.8°)$$
$$= 1.227 + j0.234$$

The sending end voltage is

$$|V_1| = |1.269| < 14.79°$$

The sending end current is

$$I_1 = CV_2 + DI_2$$
$$(-0.000054 + j0.0535) + (0.989 + j0.002)(1 < -25.8°)$$
$$= 0.8903 - j0.3769$$
$$= 0.9668 < -22.944°$$

The sending end power is

$$\begin{aligned} V_1 I_1^* &= (1.269 < 14.79°)(0.9668 < 22.944°) \\ &= 0.971 + j0.75 \end{aligned}$$

Thus, the active power loss is 0.071 per unit and the reactive power loss is 0.314 per unit. The reactive power loss increases as the PF of the load becomes more lagging.

The power supplied based on known sending end and receiving end voltages can also be calculated using the following equations.

The sending end active power is

$$[V_1^2 - V_1V_2\cos(\theta_1 - \theta_2)]g_{12} - [V_1V_2\sin(\theta_1 - \theta_2)]b_{12}$$

$$[1.269^2 - 1.269\cos(14.79°)]0.464 - [1.269\sin(14.79°)](-2.450) = 0.971 \text{ as before}$$

In this calculation, a prior calculated sending end voltage of 1.269<14.79 for supply of per unit current at 0.9 lagging PF is used. *For a given load neither the sending end voltage nor the receiving end current is known.*

The sending end reactive power is

$$\begin{aligned} Q_{12} &= [-V_1V_2\sin(\theta_1 - \theta_2)]g_{12} - [V_1^2 - V_1V_2\cos(\theta_1 - \theta_2)]b12 - \frac{Y}{2}V_1^2 \\ &= [-1.269 \sin 14.79°]0.464 - [1.269^2 - 1.269\cos(14.79°)](-2.450) - (0.0269)(1.269^2) \\ &= 0.75 \text{ as before} \end{aligned}$$

The receiving end active power is

$$\begin{aligned} P_{21} &= [V_2^2 - V_2V_1\cos(\theta_2 - \theta_1)]g_{12} - [V_2V_1\sin(\theta_2 - \theta_1)]b_{12} \\ &= [1 - 1.269\cos(-14.79°)]0.464 - [1.269\sin(-14.79°)](-2.450) \\ &= 0.9 \end{aligned}$$

and the receiving end reactive power is

$$\begin{aligned} Q_{21} &= [-V_2V_1\sin(\theta_2 - \theta_1)]g_{12} - [V_2^2 - V_2V_1\cos(\theta_2 - \theta_1)]b_{21} - \frac{Y}{2}V_2^2 \\ &= [-1.269\sin(-14.79°)]0.464 - [1 - 1.269\cos(-14.79°)](-2.450) - (0.0269)(1) = 0.436 \end{aligned}$$

These calculations are merely a verification of the results. If the approximate relations of Equation 3.30 are applied,

$$\begin{aligned} P_{12} &\approx P_{21} = 0.627 \\ Q_{12} &\approx Q_{21} = 0.68 \end{aligned}$$

This is not a good estimate for power flow, especially for the reactive power.

Example 3.7

Repeat the calculation in Example 3.6, with a long line distributed parameter model, with the same data. Calculate the sending end current and voltage and compare the results.

$$Z = 0.0746 + j0.394$$
$$Y = 0.0538$$

The product ZY is

$$ZY = 0.021571 < 169.32°$$

α is the attenuation constant in Nepers:

$$\alpha = |\gamma| \cos\left[\frac{1}{2}\tan^{-1}\left(-\frac{R}{X}\right)\right]$$

and β is the phase shift constant in radians:

$$\beta = |\gamma| \sin\left[\frac{1}{2}\tan^{-1}\left(-\frac{R}{X}\right)\right]$$

Thus,

$$\alpha = 0.0136687 \text{ nepers}$$
$$\beta = 0.146235 \text{ radians}$$

The hyperbolic functions $\cosh\gamma$ and $\sinh\gamma$ are given by

$$\begin{aligned}\cosh\gamma &= \cosh\alpha\cos\beta + j\sinh\alpha\sin\beta \\ &= (1.00009)(0.987362) + j(0.013668)(0.14571) \\ &= 0.990519 < 0.1152° \\ \sinh\gamma &= \sinh\alpha\cos\beta + j\cosh\alpha\sin\beta \\ &= (0.013668)(0.987362) + j(1.00009)(0.14571) \\ &= 0.1463449 < 84.698°\end{aligned}$$

The sending end voltage is thus

$$\begin{aligned}V_1 &= \cosh\gamma V_2 + Z_0 \sinh\gamma I_2 \\ &= (0.990519 < 0.1152°)(1 < 0°) + (2.73 < -5.361°)(0.1463449 < 84.698°) \\ &\times (1 < -25.842°) \\ &\times 1.2697 < 14.73°\end{aligned}$$

This result is fairly close to the one earlier calculated by using the Π model.

The sending end current is

$$I_2 = \left(\frac{1}{Z_0}\right)(\sinh\gamma)V_2 + (\cosh\gamma)I_2$$

$$= (0.3663 < 5.361°)(0.146349 < 84.698°) + (0.990519 < 0.1152°)(1 < -25.84°)$$

$$= 0.968 < -22.865°$$

Again there is a good correspondence with the earlier calculated result using the equivalent Π model of the transmission line. The parameters of the transmission line shown in this example are for a 138-kV line of length approximately 120 miles (193 km). For longer lines the difference between the calculations using the two models will diverge. This similarity of calculation results can be further examined.

Equivalence between Π and Long Line Model

For equivalence between long line and Π-models, *ABCD* constants can be equated. Thus, equating the *B* and *D* constants,

$$Z_s = Z_0 \sinh\gamma l$$
$$1 + \frac{1}{2}YZ = \cosh\gamma l \tag{3.87}$$

Thus,

$$Z_s = \sqrt{\frac{z}{y}}\sinh\gamma l = zl\frac{\sinh\gamma l}{l\sqrt{yz}} = Z\left[\frac{\sinh\gamma l}{\gamma l}\right] \tag{3.88}$$

i.e., the series impedance of the Π network should be increased by a factor ($\sinh\gamma l/\gamma l$):

$$1 + \frac{1}{2}YZ_c\sinh\gamma l = \cosh\lambda l$$

This gives the shunt element Y_p as follows:

$$\frac{1}{2}Y = \frac{Y}{2}\left[\frac{\tanh\gamma l/2}{\gamma l/2}\right] \tag{3.89}$$

i.e., the shunt admittance should be increased by $[(\tanh\gamma l/2)/(\gamma l/2)]$. For a line of medium length both the series and shunt multiplying factors are ≈1.

3.9 Circle Diagrams

Consider a two-node two-bus system, similar to that of Figure 3.2. Let the sending end voltage be $V_s < \theta°$, and the receiving end voltage be $V_r < \theta°$. Then,

$$I_r = \frac{1}{B}V_s - \frac{A}{B}V_r \tag{3.90}$$

$$I_s = \frac{D}{B}V_s - \frac{1}{B}V_r \tag{3.91}$$

Constants *A*, *B*, C, and *D* can be written as

$$A=|A|<\alpha,\quad B=|B|<\beta,\quad D=|D|<\alpha\quad (A=D) \tag{3.92}$$

The receiving end and sending end currents can be written as

$$I_r=\left|\frac{1}{B}\right||V_s|<(\theta-\beta)-\left|\frac{A}{B}\right||V_r|<(\alpha-\beta) \tag{3.93}$$

$$I_s=\left|\frac{D}{B}\right||V_s|<(\alpha+\theta-\beta)-\left|\frac{1}{B}\right||V_r|<(\alpha-\beta) \tag{3.94}$$

The receiving end power is

$$\begin{aligned} S_r &= V_r I_r^* \\ &= \frac{|V_r||V_s|}{|B|}<(\beta-\theta)-\left|\frac{A}{B}\right||V_r|^2<(\beta-\alpha) \end{aligned} \tag{3.95}$$

The sending end power is

$$S_s=\left|\frac{D}{B}\right||V_s|^2<(\beta-\alpha)-\frac{|V_s||V_r|}{|B|}<(\beta+\theta) \tag{3.96}$$

The real and imaginary parts are written as follows:

$$P_r=\frac{|V_s||V_r|}{|B|}\cos(\beta-\delta)-\left|\frac{A}{B}\right||V_r|^2\cos(\beta-\alpha) \tag{3.97}$$

$$Q_r=\frac{|V_s||V_r|}{|B|}\sin(\beta-\delta)-\left|\frac{A}{B}\right|||V_r||^2\sin(\beta-\alpha) \tag{3.98}$$

Here, δ, the phase angle *difference*, is substituted for θ (the receiving end voltage angle was assumed to be 0°). Similarly, the sending end active and reactive powers are as follows:

$$P_s=\left|\frac{D}{B}\right||V_s|^2\cos(\beta-\alpha)-\frac{|V_s||V_r|}{|B|}\cos(\beta+\delta) \tag{3.99}$$

$$Q_s=\left|\frac{D}{B}\right||V_s|^2\sin(\beta-\alpha)-\frac{|V_s||V_r|}{|B|}\sin(\beta+\delta) \tag{3.100}$$

The received power is maximum at $\beta=\delta$:

$$P_r(\max)=\frac{|V_s||V_r|}{|B|}-\frac{|A||V_r|^2}{|B|}\cos(\beta-\alpha) \tag{3.101}$$

and the corresponding reactive power for maximum receiving end power is

$$Q_r=-\frac{|A||V_r|^2}{|B|}\sin(\beta-\alpha) \tag{3.102}$$

The leading reactive power must be supplied for maximum active power transfer. For a short line, the equations reduce to the following:

$$P_r = \frac{|V_s||V_r|}{|Z|}\cos(\varepsilon - \delta) - \frac{|V_r|^2}{|Z|}\cos\varepsilon \tag{3.103}$$

$$Q_r = \frac{|V_s||V_r|}{|Z|}\sin(\varepsilon - \delta) - \frac{|V_r|^2}{|Z|}\sin\theta \tag{3.104}$$

$$P_s = \frac{|V_s|^2}{|Z|}\cos\theta - \frac{|V_s||V_r|}{|Z|}\cos(\varepsilon + \delta) \tag{3.105}$$

$$Q_s = \frac{|V_s|^2}{|Z|}\sin\theta - \frac{|V_s||V_r|}{|Z|}\sin(\varepsilon + \delta) \tag{3.106}$$

where ε is the impedance angle.

The center of the receiving end circle is located at the tip of the phasor:

$$-\left|\frac{A}{B}\right||V_r|^2 < (\beta - \alpha) \tag{3.107}$$

In terms of the rectangular coordinates, the center of the circle is located at

$$-\left|\frac{A}{B}\right||V_r|^2\cos(\beta - \alpha) \quad \text{MW} \tag{3.108}$$

$$-\left|\frac{A}{B}\right||V_r|^2\sin(\beta - \alpha) \quad \text{Mvar} \tag{3.109}$$

The radius of the receiving end circle is

$$\frac{|V_s||V_r|}{|B|} \quad \text{MVA} \tag{3.110}$$

The receiving end circle diagram is shown in Figure 3.11. The operating point P is located on the circle by received real power P_r. The corresponding reactive power can be read immediately as Q_r. The torque angle can be read from the reference line as shown. For a constant receiving end voltage V_r, the center C is fixed and concentric circles result for varying V_s (Figure 3.12a). For constant V_s and varying V_r, the centers move along line OC and have radii in accordance with V_sV_r/B, Figure 3.12b.

The sending end circle diagram is constructed on a similar basis and is shown in Figure 3.13. The center of the sending end circle is located at the tip of the phasor:

$$\left|\frac{D}{B}\right||V_s|^2 < (\beta - \alpha) \tag{3.111}$$

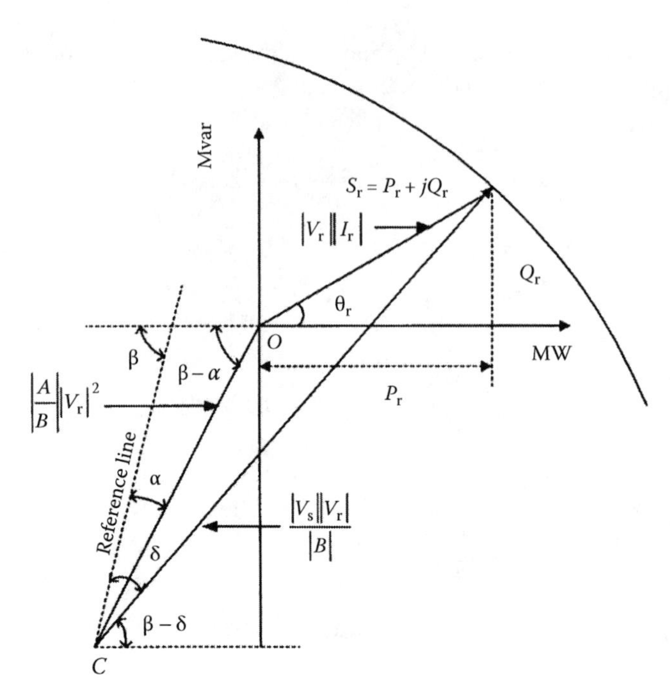

FIGURE 3.11
Receiving end power circle diagram of a transmission line.

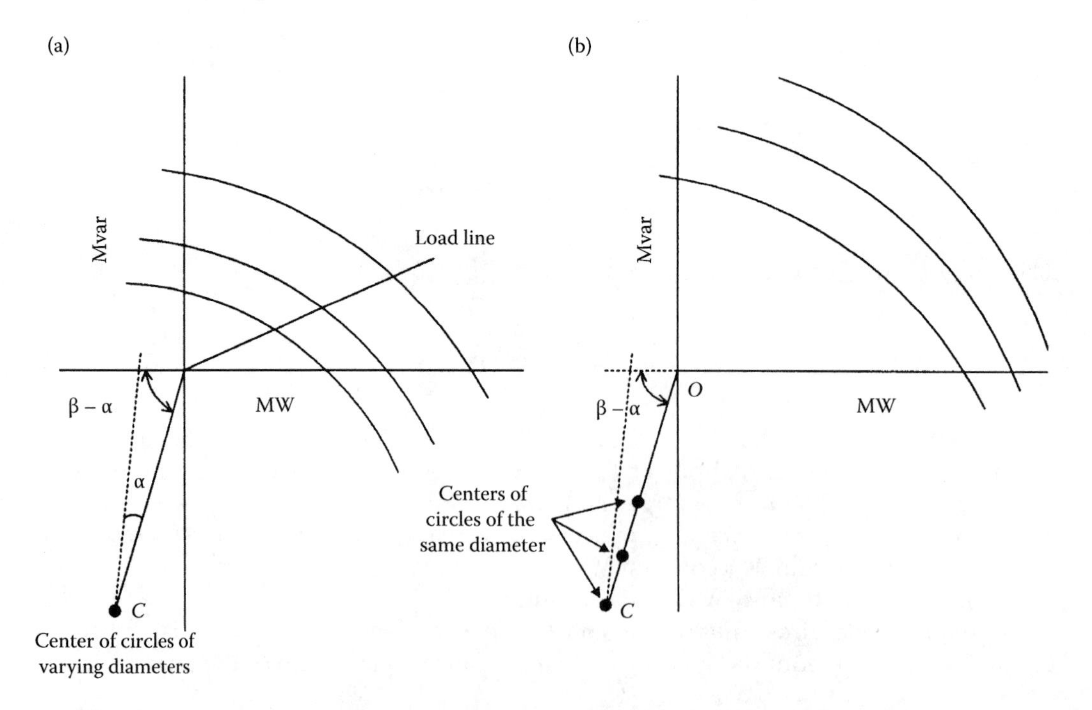

FIGURE 3.12
(a) Receiving end power circles for different sending end voltages and constant receiving end voltage and (b) receiving end power circles for constant sending end voltage and varying receiving end voltage.

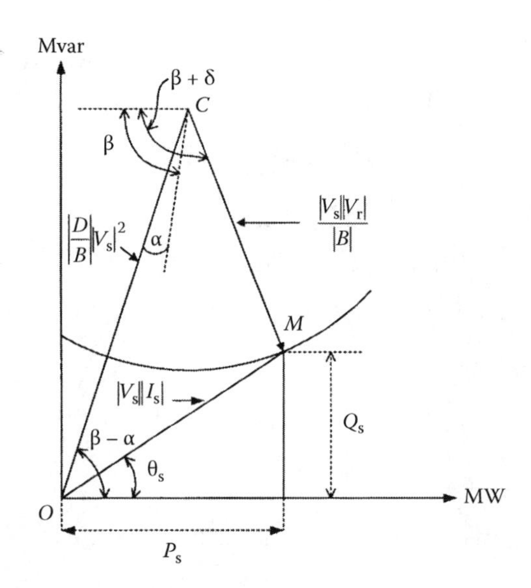

FIGURE 3.13
Sending end power circle diagram.

or in terms of rectangular coordinates at

$$\left|\frac{D}{B}\right||V_s|^2 \cos(\beta-\alpha) \quad \text{MW}$$
$$\left|\frac{D}{B}\right||V_s|^2 \sin(\beta-\alpha) \quad \text{Mvar} \tag{3.112}$$

The radius of the sending end circle is

$$\frac{|V_s||V_r|}{|B|} \tag{3.113}$$

Example 3.8

Consider Examples 3.6 and 3.7, solved for π and long line models. Draw a circle diagram and calculate (a) the sending end voltage, (b) the maximum power transferred in MW if the sending end voltage is maintained equal to the calculated voltage in (a), (c) the value of a shunt reactor required to maintain the receiving end voltage equal to the sending end voltage = 1 per unit at no load, (d) the leading reactive power required at the receiving end when delivering 1.4 per unit MW load at 0.8 PF, the sending end and receiving end voltages maintained as in (a).

The calculated A, and B constants from Example 3.6 are as follows:

$$A = 0.989 + j0.002 = |A| < \alpha = 0.989 < 0.116°$$

$$B = 0.076 + j0.394 = |B| < \beta = 0.40 < 79°$$

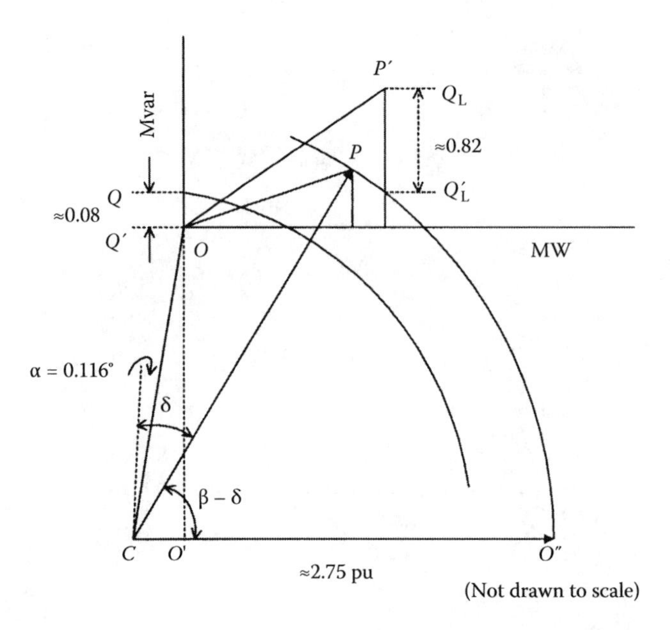

FIGURE 3.14
Circle diagram for solution of Example 3.8.

Thus, the coordinates of center C are

$$(-0.4732,-2.426)$$

(a) The receiving end power is $0.9+j0.436$. The circle diagram is shown in Figure 3.14. CP by measurement $=3.2=|V_s||V_r|/|B|$. This gives $V_s(=V_1$ in example 3.6$)=1.28$, as calculated in Example 3.6.
(b) The radius of the circle for $V_s=1.28$, $V_r=1$, is given by $|V_s||V_r|/|B|=3.2$, with center C. The maximum power transfer in MW is $=2.75$ per unit, equal to O′O″ in Figure 3.14.
(c) The diameter of the circle for per unit sending and receiving end voltages is 2.5, with C as the center. The shunt reactor required ≈ 0.08 per unit, given by QQ′ in Figure 3.14.
(d) Following the construction shown in Figure 3.14, a new load line is drawn and the leading reactive power required ≈ 0.82 per unit, given by Q_LQ_L'.

3.10 Modal Analysis

From Equation 3.69 for a perfect earth, we can write

$$\bar{L}\bar{C}=\frac{\bar{I}}{v^2}=\bar{M} \tag{3.114}$$

For kth conductor,

$$\frac{\partial V_k}{\partial x^2}=\frac{1}{v^2}\frac{\partial V_k}{\partial t^2} \tag{3.115}$$

This shows that the wave equation of each conductor is independent of the mutual influence of other conductors in the system—a result which in only partially true. However, for lightning surges and high frequencies, the depth of the current image almost coincides with the perfect earth. This may not be true for switching surges. The M matrix in Equation 3.114 is not a diagonal matrix. We will use decoupling to diagonalize the matrix, with respect to transmission lines; this is called *Modal Analysis* as it gives different modes of propagation.

Consider a matrix of modal voltages V_{m} and a transformation matrix T, transforming conductor voltage matrix V:

$$\bar{V} = \bar{T}\bar{V}_{\mathrm{m}} \tag{3.116}$$

Then the wave equation can be decoupled and can be written as

$$\begin{aligned}\frac{\partial^2 \bar{V}_{\mathrm{m}}}{\partial x^2} &= \bar{T}^{-1}\left[\bar{L}\bar{C}\right]\bar{T}\frac{\partial^2 \bar{V}_{\mathrm{m}}}{\partial t^2} \\ &= \bar{T}^{-1}\bar{M}^{-1}\bar{T}\frac{\partial^2 \bar{V}_{\mathrm{m}}}{\partial t^2} \\ &= \bar{\lambda}\frac{\partial^2 \bar{V}_{\mathrm{m}}}{\partial t^2}\end{aligned} \tag{3.117}$$

For decoupling matrix $\bar{\lambda} = \bar{T}^{-1}\bar{M}^{-1}\bar{T}$ must be diagonal. This is done by finding the eigenvalues of $\bar{M}$ from the solution of its characteristic equation. The significance of this analysis is that for n conductor system, the matrices are of order n, and n number of modal voltages are generated which are independent of each other. Each wave travels with a velocity:

$$v_n = \frac{1}{\lambda_k}, \quad k = 1, 2, \ldots, n \tag{3.118}$$

And the actual voltage on the conductors is given by Equation 3.117.

For a three-phase line,

$$\bar{T} = \begin{vmatrix} 1 & 1 & 1 \\ -1 & 0 & 1 \\ 0 & -1 & 1 \end{vmatrix} \tag{3.119}$$

Figure 3.15 shows modes for two conductor and three conductor lines. For a two conductor line, there are two modes. In the line mode, the voltage and current travel over one conductor returning through the other, none flowing through the ground. In the ground mode, modal quantities travel over both the conductors and return through the ground. For a three conductor line there are two line modes and one ground mode.

The line-to-line modes of propagation are close to the speed of light and encounter less attenuation and distortion compared to ground modes. The ground mode has more resistance and hence more attenuation and distortion. The resistance of the conductors and earth resistivity plays an important role.

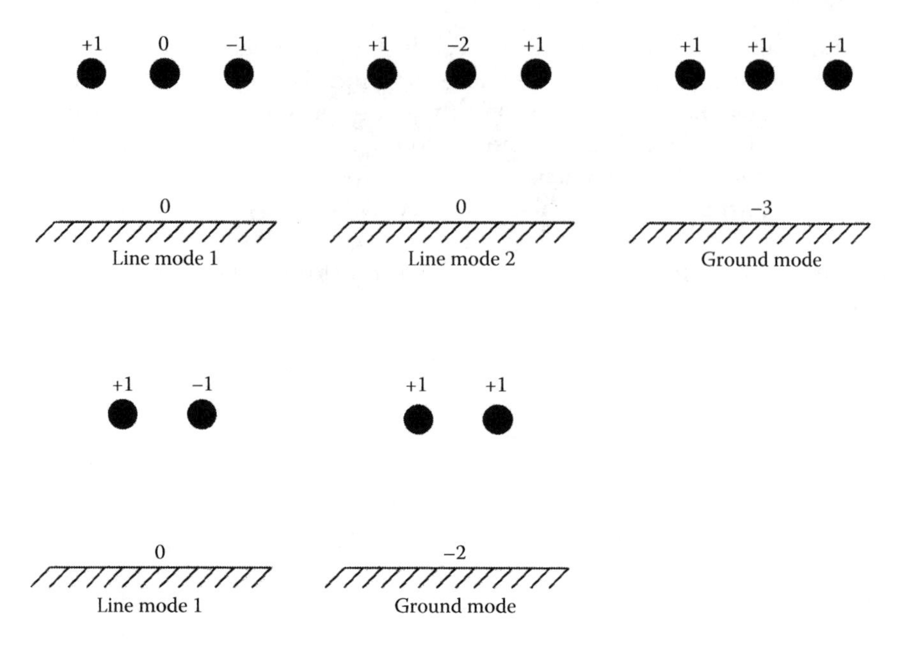

FIGURE 3.15
Mode propagation for two conductor and three conductor transmission lines.

Mode propagation gives rise to further distortion of multiconductor lines. For voltages below corona inception voltage, the velocity of propagation of the line components is close to the velocity of light and the distortion is negligible. For ground currents, the charges induced are near the surface, while the return current is well below the surface, at a depth depending upon the frequency and earth's resistivity. For a perfectly conducting earth, the current will flow at the earth's surface at velocity equal to the velocity of light. The concept of modal analysis is necessary for transient analysis and wave propagation.

3.11 Corona on Transmission Lines

The corona discharges form when the electrical field intensity exceeds the breakdown strength of air and local ionization occurs. There are a number of factors at which the corona forms, which include conductor diameter, conductor type (the smoother the conductor surface, the higher the critical disruptive voltage assuming the same diameter), line configuration and weather. Corona will form when an electrode is charged to sufficiently high voltage from a power frequency source. The mechanism of corona formation is different for the positive and negative half cycles. The negative cycle tends to produce short-duration pulses of current up to 1 mA, lasting for 0.03 μs. The corona noise generated by negative charged electrode tends to be of low level compared to the noise generated by a positive charged electrode. The corona inception voltage is given by

$$V_c = \frac{2\pi r \varepsilon_0 E_c}{C \times 10^6} \tag{3.120}$$

where r is the radius of conductor in m, ε_0 is the permittivity of free space, E_c is the corona inception electric field in kV/m on the conductor surface, and C is the capacitance of the overhead line in μF/m. The corona onset voltage is, generally, 30%–40% above the rated operating voltage.

Using Peek's equation, V_c can be expressed as

$$V_c = \frac{1.66\times10^{-3} nmk_d^{2/3} r\left(1+\frac{0.3}{\sqrt{r}}\right)}{k_h C} \tag{3.121}$$

where

n = number of sub conductors in a bundle

m = conductor surface factor which varies from 0.7 to 0.8

k_d = relative air density

r = conductor radius in cm

C = capacitance as above in μF/m

k_b = ratio of maximum to average surface gradient for bundle conductors. This is given by the expression:

$$k_b = 1+2(n-1)\sin\left(\frac{\pi}{n}\right)\left(\frac{r}{A}\right) \tag{3.122}$$

where A is the distance between adjacent sub conductors in cm.

The corona has three major effects:

1. Corona loss: A method of estimating the corona loss is established by Burgsdrof. If the line is designed to have appropriate radio interference voltage (RIV), the corona loss that forms the line will be relatively low in dry weather condition. Under rainfall or when the water droplets are still present on the conductors, this loss will increase substantially. As an example, a 500 kV line, with 4 bundle conductors, spaced 45 cm centers, and conductor radius 1.288 cm, gives a loss of 1.7 kW/km for fair weather conditions and a loss of 33.6 kW/km for wet weather conditions.
2. RIV: During design of a HV transmission line, RIV levels are considered taking into account the signal strengths of local transmissions and density of population.
3. Effect on wave propagations and flashovers: Corona can reduce overvoltages on open-circuited lines by attenuating the lightening and switching surges. The capacitance of a conductor increases due to corona effect and by increasing electrostatic coupling between shield wires and phase conductors, corona during lightning strikes to shield wires or towers reduces the voltage across the insulator strings and probability of flashovers.

3.12 System Variables in Load Flow

In the above analysis, currents have been calculated on the basis of system impedances or admittances and assumed voltages. The complex powers have been calculated on the basis of voltages. In load flow situations, neither the voltages nor the currents at load buses are known. The required power balance is known. A bus can connect system admittances, shunt admittances, generators, loads, or reactive power sources. The consumer demand at any instant is uncontrollable in terms of active and reactive power requirements. For a two-node two-port circuit, this is represented by a *four-dimensional disturbance* vector p:

$$p = \begin{vmatrix} p_1 \\ p_2 \\ p_3 \\ p_4 \end{vmatrix} = \begin{vmatrix} P_{D1} \\ Q_{D1} \\ P_{D2} \\ Q_{D2} \end{vmatrix} \tag{3.123}$$

where P_{D1}, P_{D2}, Q_{d1}, and Q_{d2} are the active and reactive power demands at the two buses.

The magnitudes of voltages and their angles at buses 1 and 2 can be called *state variables* and represented by

$$x = \begin{vmatrix} x_1 \\ x_2 \\ x_3 \\ x_4 \end{vmatrix} = \begin{vmatrix} \theta_1 \\ |V_1| \\ \theta_2 \\ |V_2| \end{vmatrix} \tag{3.124}$$

Lastly, the active and reactive power generation at buses 1 and 2 may be called *control variables*:

$$u = \begin{vmatrix} u_1 \\ u_2 \\ u_3 \\ u_4 \end{vmatrix} = \begin{vmatrix} P_{G1} \\ Q_{G1} \\ P_{G2} \\ Q_{G2} \end{vmatrix} \tag{3.125}$$

Thus, for a two-bus system, we have 12 variables. For large power systems, there will be thousands of variables and the load flow must solve for interplay between these variables. This will be examined as we proceed along with load flow calculations in the following chapters. In a steady-state operation,

$$f(x, p, u) = 0 \tag{3.126}$$

Problems

3.1 Mathematically derive *ABCD* constants for nominal T models shown in Table 3.1.

3.2 A 500 mile (800 km) long three-phase transmission line operating at 230 kV supplies a load of 200 MW at 0.85 PF. The line series impedance $Z = 0.35 + j0.476\ \Omega$/mile and shunt admittance $y = j5.2 \times 10^{-6}$ S/mile. Find the sending end current, voltage, power, and power loss, using short-line model, nominal Π model, and long line model. Find the line regulation in each case. What is the SIL loading of the line?

3.3 A 400-kV transmission line has the following A and B constants: $A = 0.9<2°$, $B = 120<75°$. Construct a circle diagram and ascertain (a) the sending end voltage for a receiving end load of 200 MW at 0.85 PF, (b) the maximum power that can be delivered for sending an end voltage of 400 kV, and (c) the Mvar required at the receiving end to support a load of 400 MW at 0.85 PF, the sending end voltage being held at 400 kV.

3.4 Plot the incident and reflected currents for the line of Example 3.4 at midpoint of the line.

3.5 A 230-kV transmission line has the following line constants: $A = 0.85<4°$, $B = 180<65°$. Calculate the power at unity PF that can be supplied with the same sending end and receiving end voltage of 230 kV. If the load is increased to 200 MW at 0.9 PF, with the same sending end and receiving end voltage of 230 kV, find the compensation required in Mvar at the receiving end.

3.6 Draw the current and voltage profile of a symmetrical 230 kV, 300 miles (483 km) long line at no load and at 100 MW, 0.8 PF load. Use the following line constants: $L = 1.98$ mH/mile, $C = 0.15$ μF/mile (1.609 km).

3.7 An underground cable has an inductance of 0.45 μH/m and a capacitance of 80 pF/m. What is the velocity of propagation along the cable?

3.8 Derive an expression for the maximum active power that can be transmitted over a short transmission line of impedance $Z = R + jX$. Find the maximum power for an impedance of $0.1 + j1.0$, line voltage 4160 V, and regulation not to exceed 5%.

3.9 A loaded long line is compensated by shunt power capacitors at the midpoint. Draw a general vector diagram of the sending end and receiving end voltages and currents.

3.10 Derive the *ABCD* constants of a transmission line having resistance 0.1 Ω/mile (1.609 km), reactance 0.86 Ω/mile, and capacitance 0.04 μF/mile. What is the electrical length of the line?

Bibliography

1. JG Anderson. *Transmission Reference Book, Edison Electric Company*, New York, 1968.
2. WA Blackwell, LL Grigsby. *Introductory Network Theory, PWS Engineering*, Boston, MA, 1985.
3. *Transmission Line Reference Book—345 kV and Above*. EPRI, Palo Alto, CA, 1975.
4. EHV Transmission. IEEE Trans 1966 (special issue), No. 6, PAS-85–1966, pp. 555–700.
5. Westinghouse Electric Corporation. *Electrical Transmission and Distribution Reference Book*, 4th ed. East Pittsburgh, PA, 1964.
6. EW Kimberk. *Direct Current Transmission*, vol. 1. John Wiley, New York, 1971.
7. B Stott. "Review of load-flow calculation methods," *Proc IEEE*, 62(7), 916–929, 1974.
8. C Gary, D Crotescu, G Dragon. Distortion and Attenuation of Traveling Waves Caused by Transient Corona, CIGRE Study Committee Report 33, 1989.
9. LV Beweley. *Traveling Waves on Transmission Systems*, 2nd ed. John Wiley, New York, 1951.
10. MS Naidu, V Kamaraju. *High Voltage Engineering*, 2nd ed. McGraw-Hill, New York, 1999.

11. T Gonen. *Electrical Power Transmission System Engineering, Analysis and Design*, 2nd ed. CRC Press, Boca Raton, 2009.
12. S Rao. *EHV-AC, HVDC Transmission and Distribution Engineering*, Khanna Publishers, New Delhi, 2004.
13. P Chowdhri. *Electromagnetic Transients in Power Systems*, Research Study Press, Somerset, 1996.
14. JC Das. *Transients in Electrical Systems*, McGraw-Hill, New York, 2010.

4

HVDC Transmission Load Flow

4.1 Fundamental Characteristics of HVDC Transmission

We introduced the concepts of high-voltage, direct current (HVDC) transmission in Volume 1, with some typical configurations and a terminal layout.

The first HVDC commercial transmission system was introduced in 1953. With developments in thyristor valves in the 1970s, the HVDC systems became a viable alternative to long-distance ac transmission, submarine cable transmission, and system interconnection of two ac systems. The earlier dc transmission systems were point-to-point transmission systems and the first multiterminal system was introduced in 1988.

The HVDC transmission system voltages have risen over the course of years, see Figure 4.1.

The 10 largest HVDC transmission systems are shown in Table 4.1.

The cumulative megawatts of HVDC systems around the world approach 100 GW. The major technology leap was in Brazil with the 3150 MW ±600 kV Itaipu project commissioned from 1984 to 1987. The overhead line is 800 km long and each 12-pulse converter is rated 790 MW, 300 kV. HVDC project list worldwide can be seen on the IEEE/PES website. Another website of interest is CIGRE Study Committee, HVDC, and Power Electronic Equipment (see bibliography).

4.1.1 Economics

The choice of ac transmission versus dc transmission depends upon the cost. The cost of conversion substations required for HVDC transmission overweigh the cost of savings in HV lines and lesser losses. The dc circuit requires two conductors. A bipolar line (discussed further) is equivalent to a double-circuit three-phase line. The corridor width of the dc line is approximately 50% that of ac line. As the distance over which the power to be transmitted increases, the economics swing in favor of HVDC transmission.

4.1.2 Cable Interconnections

AC cables take a considerable amount of charging current:

132 kV: 1250 kVA/circuit/km

230 kV: 3125 kVA/circuit/km

400 kV: 9375 kVA/circuit/km

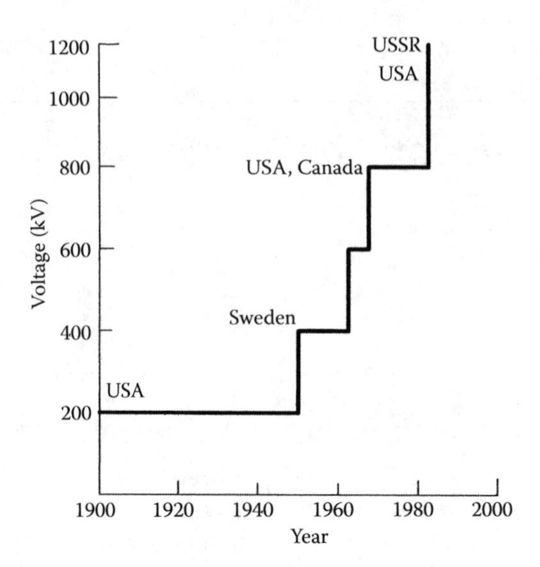

FIGURE 4.1
Increase in HVDC transmission voltages from 1900 to 2000.

TABLE 4.1
10 Largest DC Transmission Systems

Number	Link	Country	Year	MW	Voltage (kV)	Distance (km)
1.	Itaipu 2	Brazil	1984	3150	600	805
2.	Three Gorges-Changzhou	China	2003	3000	500	860
3.	Guizhou-Guangdong 1	China	2004	3000	500	980
4.	Ningxia-Shandong	China	2010	4000	±660	1348
5.	Yunnan-Guangdong	China	2010	5000	±800	1418
6.	Xiangjiaba-Shanghai	China	2010	6400	±800	2071
7.	Jinping-Sunana	China	2012	7200	±800	2100
8.	Xiluodu-Guangdong	China	2013	6400	±500	1286

These result in dielectric heating, and thermal limits can be reached even without loading the cables. Thus, power transfer capability of long ac cables is low. The cables can only be loaded to approximately 0.3 times their natural loading.

HVDC cables do not take charging current. Underwater transmission can interconnect two national grids separated by ocean waters.

4.1.3 HVDC Advantages

An HVDC interconnection properly planned and designed may achieve the following advantages:

1. An interlinking of two ac systems through interconnections which may have become overloaded during load development poses problems of trip-out and system separation and load/frequency management. There are limitations of load transfer with respect to deviations of frequency in the interconnected systems.

A dc link on the other hand is immune to such fluctuations and power transfer remains steady. Direction and magnitude of power can be changed quickly and accurately by controlling the inverter/rectifier. It may result in lesser overall installed capacity to meet peak load demand.

2. It is possible to improve the stability of the network as a whole by introducing control parameters from the ac network and disturbances in the ac network can be damped out by modulating power flow through dc link. Disturbances in one ac network can be damped out by modulating the power flow through HVDC interconnection.
3. An ac interconnection is a synchronous link and with increased generating capability, equipment of increasing short-circuit ratings are required. HVDC link is an *asynchronous* connection and the short-circuit levels of interconnected systems remain unchanged. Also frequency disturbances in one system do not impact the interconnected system. HVDC links are in operation interconnecting 40 or 50 Hz systems with 60 Hz systems.
4. A *synchronous* HVDC link is a combination of HVDC link in parallel with ac tie-lines. Power transfer through parallel dc line can be controlled to have a stabilizing effect on parallel ac tie-line. The stabilizing controls are initiated after first opening of ac line and continue during the dead time and subsequent return of the ac system to the equilibrium point.
5. The power through an HVDC link can be regulated more precisely and rapidly, change of the order of 30 MW/min or more. The limitations are imposed by the ac system reserve capacity and generation.
6. An HVDC link does not transmit reactive power (though converters require reactive power to operate). Thus, there is no loss of reactive power in the HVDC line itself. Continuous charging currents are absent. Thus, transmission losses are reduced.
7. The skin effect is absent in dc currents and the conductors can have a uniform current density through their cross section.
8. The phase-to-phase clearances, phase-to-ground clearances are smaller for dc transmission. HVDC towers are physically smaller for the same ac voltage level and ROW (right of way) is reduced.
9. With respect to line loading, the ac line remains loaded below its thermal limit due to limits of transient stability and conductors are not fully utilized. The voltage across the line varies due to absorption of reactive power, which is load power factor dependent. The voltage fluctuates with the load. An ac line cannot be loaded more than approximately 0.8 times the surge impedance loading. Such limitations do not exist in HVDC transmission. Line can be loaded up to the thermal limits of line or thyristor valves.
10. Intermediate substations are generally required in ac transmission for compensation located approximately 300 km apart—no such substations are required for HVDC transmission. It is point-to-point long-distance transmission, though multiterminal operation is possible.
11. An HVDC line can be operated with constant current (CC) or voltage regulation by suitable control of the thyristor valves and tap changer control.
12. For the same power transfer over the same distance, the corona losses and radio interference of dc systems are less than that of ac systems because the dc insulation level is lower than that of corresponding ac insulation.

4.1.4 Some Limitations of HVDC Transmission

- HVDC is generally a point-to-point transmission, though tee-offs are possible. These do not have parallel lines or mesh configurations—these do not have step-up or step-down transformers, and generally HVDC circuit breakers are not applied.
- For main distribution, subdistribution and subtransmission HVDC is not used; it is not economical.
- Cost of terminal substations is high and their configurations and controls are complex.
- Harmonic pollution due to converters requires ac filters and dc filters are required.

4.2 HVDC System Configurations

We will briefly discuss the following configurations:

- Bipolar
- Monopolar
- Homopolar
- Back-to-back coupling system
- Multiterminal HVDC system

The fundamental conceptual configurations of these systems are shown in Figure 4.2. Figure 4.3 shows a detailed bipolar terminal layout and Figure 4.4 shows back-to-back terminal connections.

Table 4.2 shows the characteristics of these systems.

4.2.1 Ground Electrodes and Ground Return

The current through ground line and electrodes depends on the HVDC configuration. In a bipolar circuit, it depends on the operation; in bipolar mode only a small differential current will flow. With monopolar operation at half the rated power, full dc current flows. In monopolar operation with metallic return, no ground current flows. In monopolar full-load dc current flows and in back-to-back connection no ground current flows.

The flow of ground current through ground gives rise to the following:

- Corrosion of buried metallic equipment, such as pipes, rails, cable sheaths, and fences
- Rapid consumption of electrode material
- Possible shocks to marine life and bathers
- Interference to communication circuits

The earth electrodes are installed away from the HVDC stations, 5–25 kM; the site is selected away from buildings and populations and buried metallic objects. The step potentials are controlled and cathodic protection provided. Before selecting a site, the

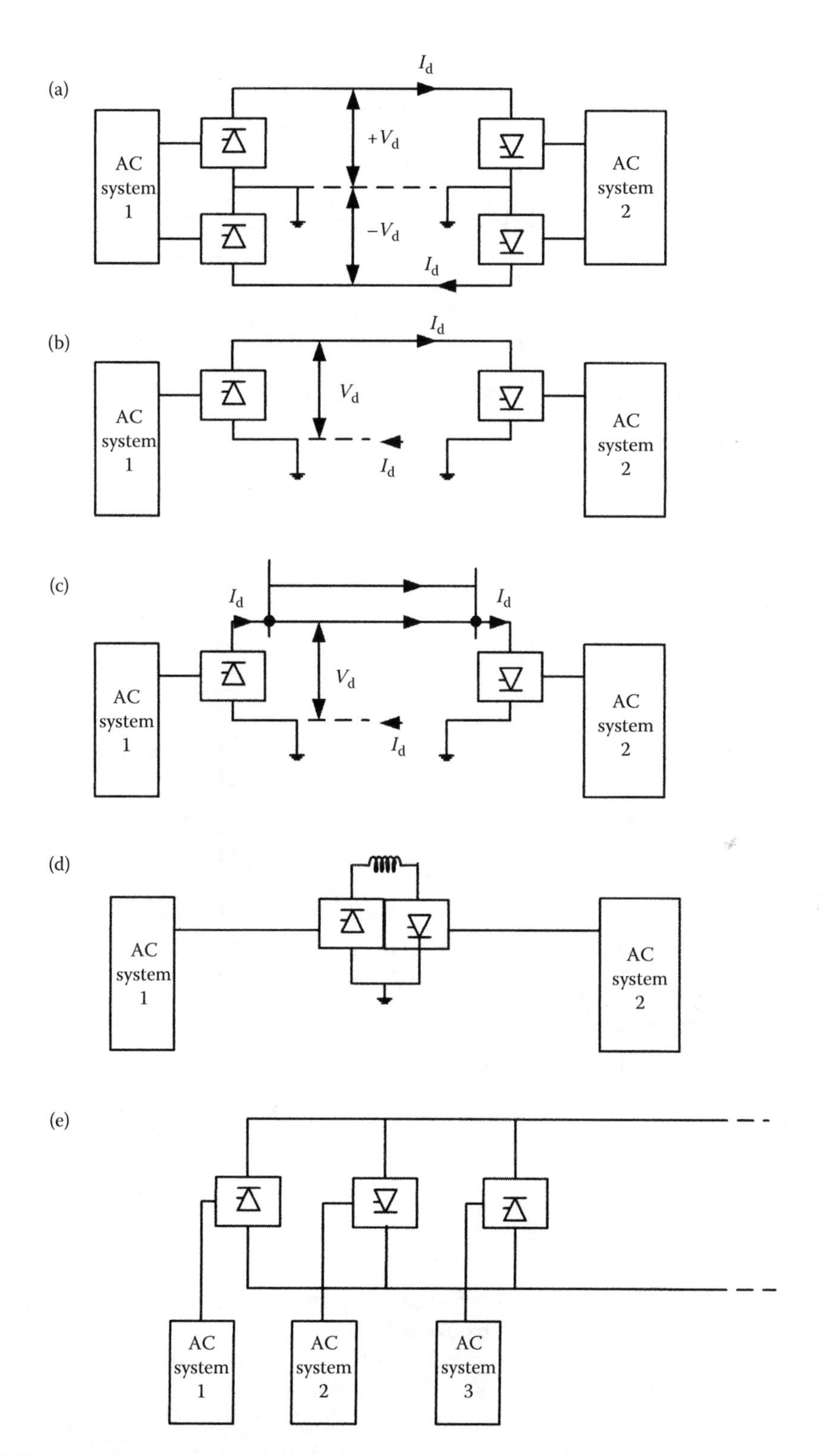

FIGURE 4.2
HVDC system configurations: (a) bipolar, (b) monopolar, (c) homopolar, (d) back-to-back, and (e) multipolar.

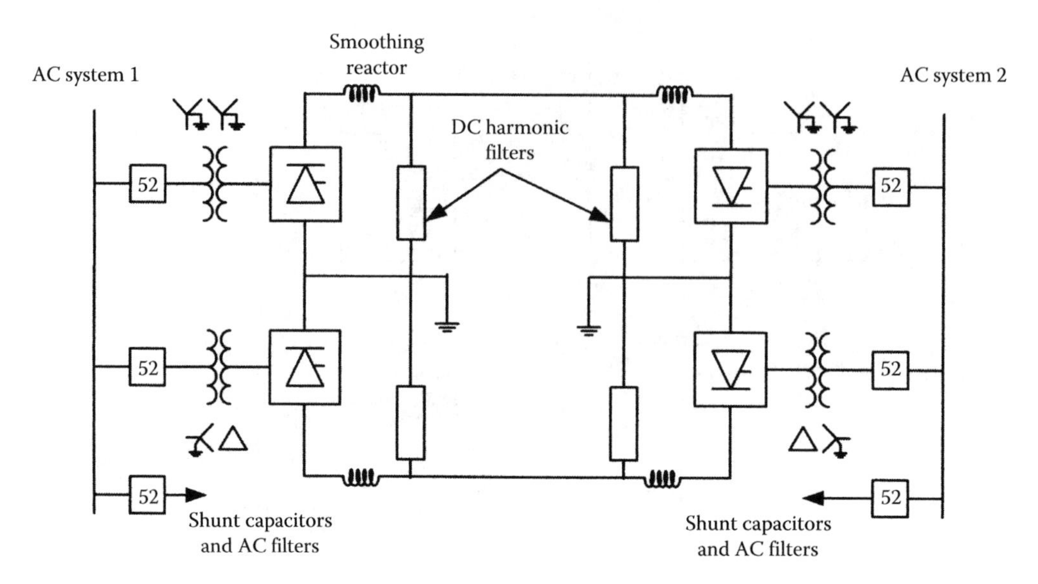

FIGURE 4.3
Terminal details of a bipolar connection.

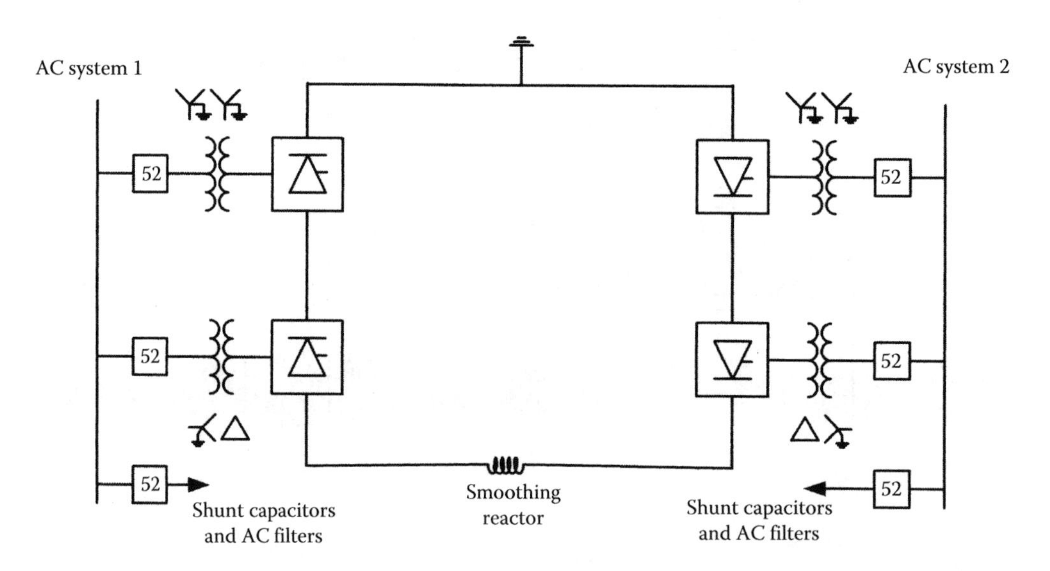

FIGURE 4.4
Terminal details of a back-to-back connection; there is no HVDC line.

mapping of buried metallic objects is carried out, and corrosion risk for the steel structures is assessed. Various formations of electrodes are used.

For submarine cable, the ground electrodes are in the form of sea or shore electrodes. The sea electrode is held in sea water near the shore, while shore electrode is buried near the sea on sea shore. A low ground resistance of the order of 0.01 Ω may be obtained; the soil near the sea has low resistivity. The electrodes are covered by nonconducting encapsulation of insulting materials like PVC.

TABLE 4.2

Characteristics of HVDC Configurations

Configuration	Characteristics
Bipolar	Two poles, positive and negative. The midpoint of two poles is grounded through an electrode line and ground electrodes, which are located up to 25 km away from the terminals Normal mode of operation is bipolar, with power flow through the line conductors and negligible return through ground. During a fault on one pole, the mode can be changed to monopolar, with power flow through one pole and return through earth
Monopolar	One pole and ground return, the pole is grounded through an electrode line and ground electrodes, located away from the terminals. The pole is normally negative with respect to ground The current monopolar systems are being transformed into bipolar. The power transfer is 50% of the bipolar configuration. Generally used for submarine cables
Homopolar	There are two poles of the same polarity and return ground. The poles are grounded through an electrode line and ground electrodes, located away from the terminals This system was used earlier for a combination of overhead lines and cables
Back-to-back connection	There are no transmission lines and two ac systems are linked through single HVDC back-to-back coupling station. Usually bipolar with no ground return. Reference ground provided for protection, control and measurements Provides asynchronous tie between two ac systems. Improvement in stability and power transfer can be in either direction depending on controls. A popular method of interconnection between adjacent ac networks and power exchange can be rapidly varied
Multiple terminals	Bipolar connection between three or more terminals. Some terminals feed power into while the others receive Exchange of power between terminals can be controlled rapidly and accurately. The system stability of ac networks can be improved

The three-phase converter transformers have, usually, grounded neutrals.

The electrode line is connected between the neutral bus of HVDC and the electrodes. For overhead HVDC, the electrode line is a separate overhead line, having two parallel conductors. This needs to have a reduced BIL level. Two conductors are used for reliability, each conductor is designed to carry the full value of return ground current. In case of submarine HVDC, the electrode lines are in the form of two separate insulated submarine cables.

4.3 HVDC Power Flow

It is necessary to have knowledge of the electronic converters, discussed in Volume 3. These details are not repeated here, except the following relevant extract:

1. In a 6-pulse full wave converter acting as a rectifier, the average dc voltage with a firing angle of α is

$$V_{dc} = \frac{3\sqrt{3}}{\pi} V_m \cos\alpha \tag{4.1}$$

where V_m is the peak ac line-to-neutral voltage.

2. The peak value of the fundamental frequency line current is

$$I_{\text{peak}} = \frac{2}{\pi}\sqrt{3}I_{\text{d}} = 1.11\, I_{\text{d}} \tag{4.2}$$

3. With overlap angle due to commutation

$$V_{\text{dc}} = V_{\text{do}} \cos\alpha - R_{\text{c}} I_{\text{d}} \tag{4.3}$$

   where R_c is called the equivalent commutating resistance

$$R_{\text{c}} = \frac{3}{\pi} X_{\text{c}} \tag{4.4}$$

   where X_c is the source reactance.

4. A 12-pulse operation can be obtained by applying transformers with 30° phase shift between their windings, see Figures 4.3 and 4.4. This is called phase multiplication and the lower order harmonics are eliminated; i.e., for a 12-pulse circuit, the lowest order characteristic harmonic will be 11th, while for a 6-pulse circuit it will be 5th, see Volume 3.
5. Back fire and commutation failures can occur.
6. Referring to Figure 4.5, the following angles are associated with rectifier and inverter operation.

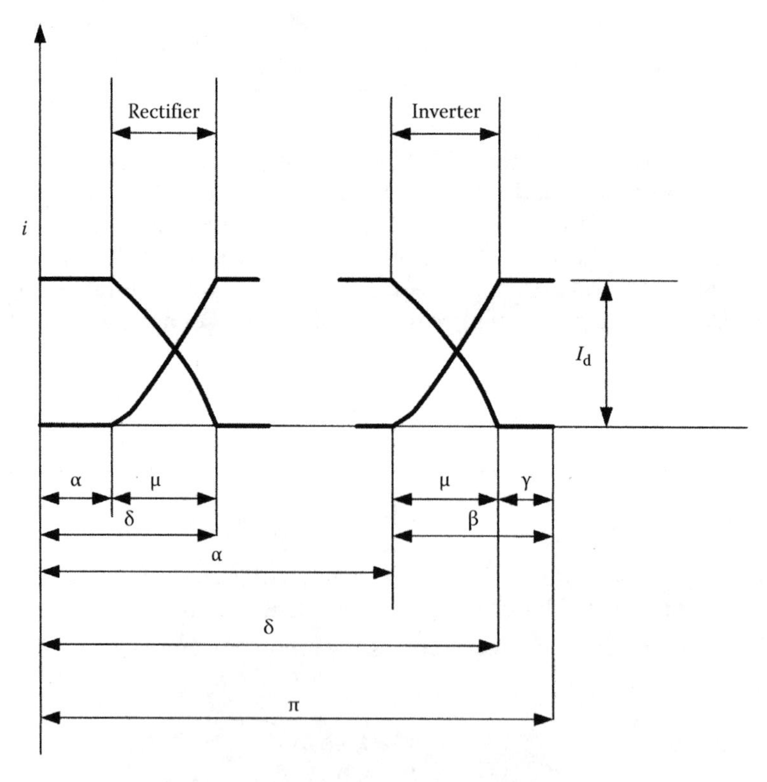

FIGURE 4.5
Definitions of angles in rectifier and inverter operation.

Rectifier Angles

α = ignition delay angle

μ = overlap angle

$\alpha+\mu$ = extinction delay angle

Inverter Angles

$B = \pi - \alpha$ = ignition advance angle

$\gamma = \pi - \delta$ = extinction advance angle

$\mu = \delta - \alpha = \beta - \gamma$ = overlap angle

7. For inverter operation

$$V_d = \frac{V_{do}}{2}(\cos\alpha + \cos\delta) \tag{4.5}$$

As $\cos\alpha = -\cos\beta$, and $\cos\delta = -\cos\gamma$

$$V_d = V_{do}\frac{\cos\gamma + \cos\beta}{2} = V_{do}\cos\beta + R_C I_d = V_{do}\cos\gamma - R_C I_d \tag{4.6}$$

See Volume 3 for further explanations of items (1) through (7).

The dc current is controlled by adjusting the rectifier and inverter voltages. Then, the dc current is

$$I_d = \frac{V_{dr} - V_{di}}{R_L} \tag{4.7}$$

where R_L is the circuit resistance. The dc power is

$$P_{dc} = V_d I_d \tag{4.8}$$

The voltages V_{di} and V_{dr} are adjusted by controlling the following:

- Tap changer on the rectifier transformers for slow variations of 8–10 s.
- Delay angle α (see Volume 3) of thyristors for fast variations of the order of a few ms.

V_{di} is held at a certain value by voltage control of the inverter and current I_d is controlled by the rectifier terminal.

The power at rectifier end is

$$P_R = \frac{V_{dr}^2}{R_L} - \frac{V_{dr} \times V_{di}}{R_L} \tag{4.9}$$

and at the inverter end is

$$P_i = -\frac{V_{di}^2}{R_L} + \frac{V_{dr} \times V_{di}}{R_L} \tag{4.10}$$

Power in the middle is

$$P_m = \frac{V_{dr}^2 \times V_{di}^2}{2R_L} \tag{4.11}$$

The losses are as follows:

$$P_L = \frac{(V_{dr} - V_{di})^2}{R_L} \tag{4.12}$$

Example 4.1

An HVDC bipolar HVDC transmission system, ±500 kV, 12-pulse, has rated power = 1500 MW, line resistance per pole = 5 Ω and dc voltage at rectifier end = ±500 kV. The following operating parameters are calculated.

Rated power per pole = 750 MW and rated dc current is = 1500 A. The dc voltage at inverter end = 492.5 kV/pole (500 − (5 × 5 × 1.5 × 10^3). Therefore, inverter end dc power = 738.75 MW, and bipolar power is 1447.5 MW (2 × 492.5 × 10^3 × 1.5 × 10^3). DC voltage of 6-pulse rectifier bridge = 250 kV and of inverter bridge = 246.25 kV. (Each pole of 12-pulse consists of two 6-pulse circuits.)

A dc link with rectifier and inverter angles is shown in Figure 4.6 schematically. Based on this figure, the rectifier to inverter current is

$$I_d = \frac{V_{dor} \cos\alpha - V_{doi} \cos\gamma}{R_{cr} + R_L - R_{ci}} \tag{4.13}$$

Note that the commutating resistance of the inverter is negative.

4.3.1 Rating of Converter Transformer

For a 6-pulse converter, it is shown in Volume 3 that the rms input ac current including harmonics is

$$I_{rms} = \sqrt{\frac{2}{3}} I_d = 0.8615\, I_d \tag{4.14}$$

The rms value of transformer line to neutral voltage is

$$V_{LN} = \frac{\pi}{3\sqrt{6}} V_{do} \tag{4.15}$$

Therefore, three-phase rating of the transformer is

$$3V_{LN} I_{rms} = \frac{\pi}{3} V_{do} I_d \tag{4.16}$$

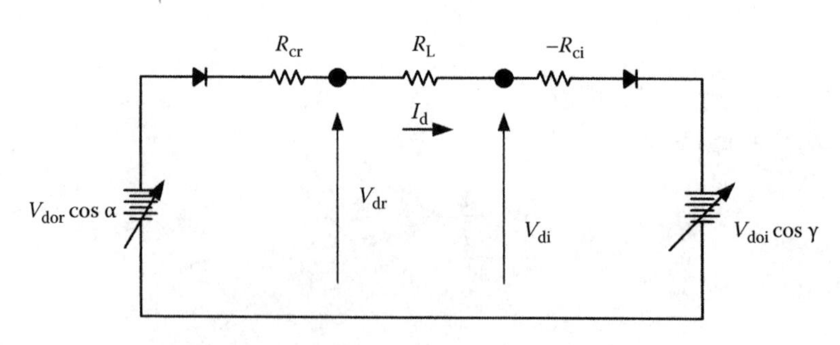

FIGURE 4.6
Equivalent circuit of an HVDC link.

4.4 HVDC Controls

The controls should achieve the following:

- Prevent large fluctuations in direct current due to variations in ac system voltages.
- Maintain dc voltage close to rated, which means reduced losses.
- Prevent commutation failures in inverters and arc-back in rectifiers.
- Maintain power factors to minimize reactive power supply.

The power factor is

$$\cos\varphi = 0.5[\cos\alpha + \cos(\alpha+\mu)]$$
$$= 0.5[\cos\gamma + \cos(\gamma+\mu)] \tag{4.17}$$

This means that α for the rectifier and μ for the inverter should be as low as possible.

Under normal operation, the rectifier maintains CC and the inverter maintains constant extinction angle (CEA) with adequate commutation margin to prevent commutation failure. The ideal VI characteristics are shown in Figure 4.7. The operating point is the intersection of the two characteristics.

From Figure 4.6,

$$V_d = V_{doi}\cos\gamma + (R_L - R_{ci})I_d \tag{4.18}$$

As the rectifier maintains CC, this gives inverter characteristics. If R_{ci} is slightly > R_L, the slope is slightly negative.

The rectifier maintains CC by changing the firing angle; but it cannot be less than the minimum value α_{min}. When the minimum angle is reached, no more voltage increase

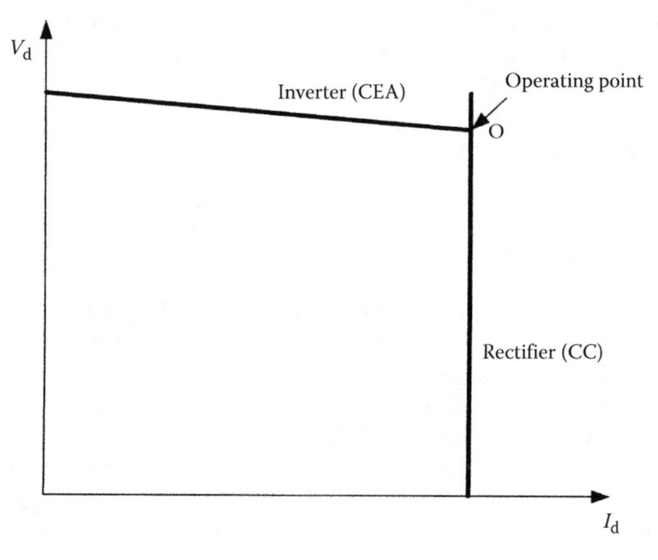

FIGURE 4.7
Ideal steady-state *V–I* characteristics.

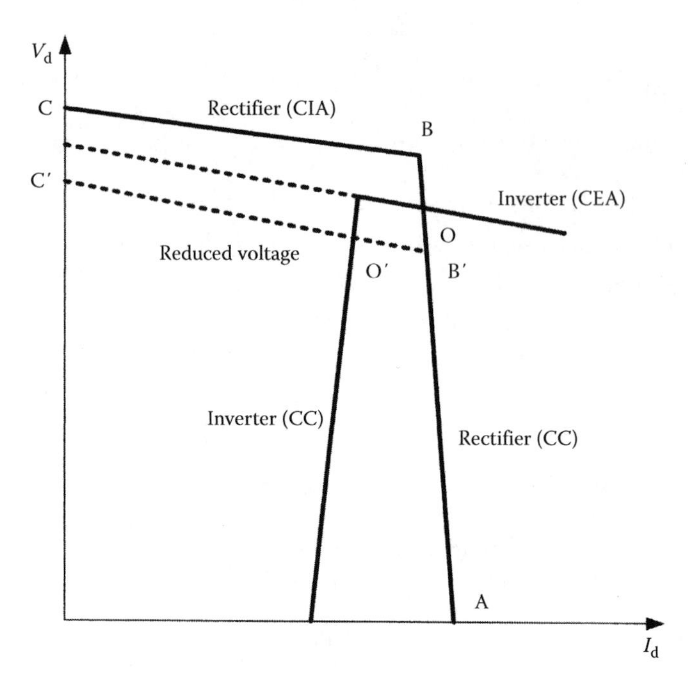

FIGURE 4.8
Practical converter control *V–I* characteristics.

is possible, and the rectifier operates at constant ignition angle (CIA). Thus, the rectifier has two segments as shown in Figure 4.8. The inverter characteristics can be raised or lowered by transformer tap control. When tap changer operates the CEA, regulator quickly restores desired γ. As a result, the dc current changes which is restored by current regulator of the rectifier.

Practically, the CC characteristic is not truly vertical. It has a negative slope due to finite gain of the current controller. The complete rectifier characteristic is ABC and at reduced voltage it is AB′C′, see Figure 4.8. O is the normal operating point. Note that the inverter characteristic does not intersect the rectifier characteristics at reduced voltage. Therefore, the inverter is also provided with a current controller. On reduced voltage, say due to a nearby fault, the operating point is O′. The inverter takes over current control and rectifier, the voltage.

4.4.1 Bidirectional Power Flow

For bidirectional power flow, the role of inverter and rectifier at each end must be interchangeable. Thus, each converter must be provided with combined characteristics.

This operation is shown in Figure 4.9.

The characteristics of each converter consist of three segments, CIA, CC, and CEA. Power is transferred from converter 1 to converter 2, with characteristics shown in solid lines, the operating point is O_1. The power reverse characteristics are shown in dotted lines, with operating point O_2. This is done by reversing the margin setting, I_m. This means that making the current order of setting of converter 2 exceed that of converter 1. The I_d is same as before, but the voltage polarity has changed.

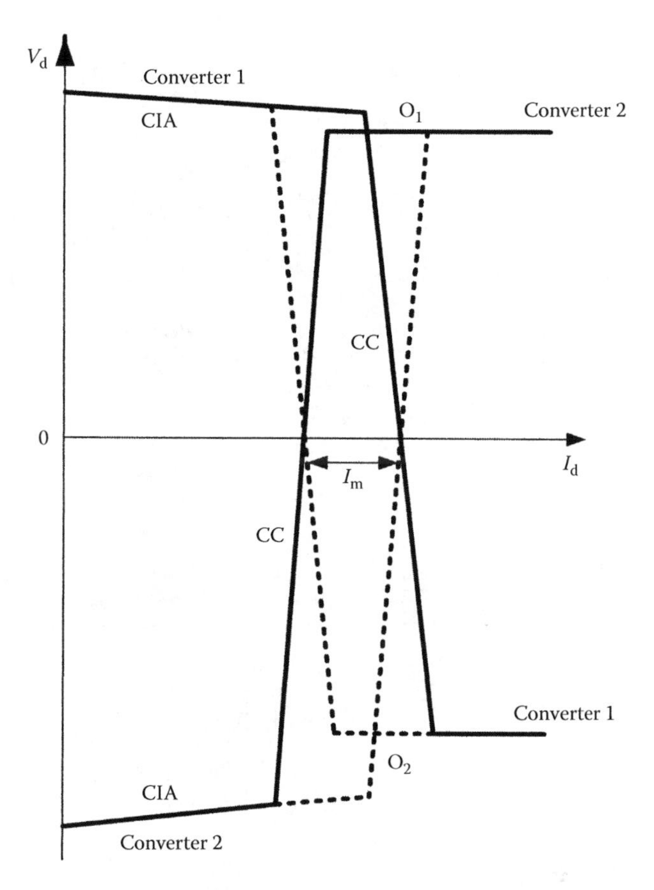

FIGURE 4.9
Rectifier and inverter characteristics at each end of the HVDC link.

4.4.2 Voltage-Dependent Current Order Limit

The voltage-dependent current order limit (VDCOL) addresses the problems with under-voltage conditions. It is not possible or desirable to maintain rated power flow under low-voltage conditions. The maximum allowable current is reduced when the voltage falls to a certain predetermined value. The VDCOL may be a function of ac or dc voltage. The rectifier–inverter characteristics with this control are shown in Figure 4.10. The inverter characteristic follows the rectifier characteristics and the current margin is maintained. The current order is reduced transiently through voltage-dependent current limit, see Figure 4.11.

4.4.3 Reactive Power Compensation

Reactive power is absorbed by the inverter and rectifier action, which must be supplied from ac filters or shunt compensation. Neglecting losses in the system the power factor is

$$\cos\varphi = \frac{V_d}{V_{do}} \tag{4.19}$$

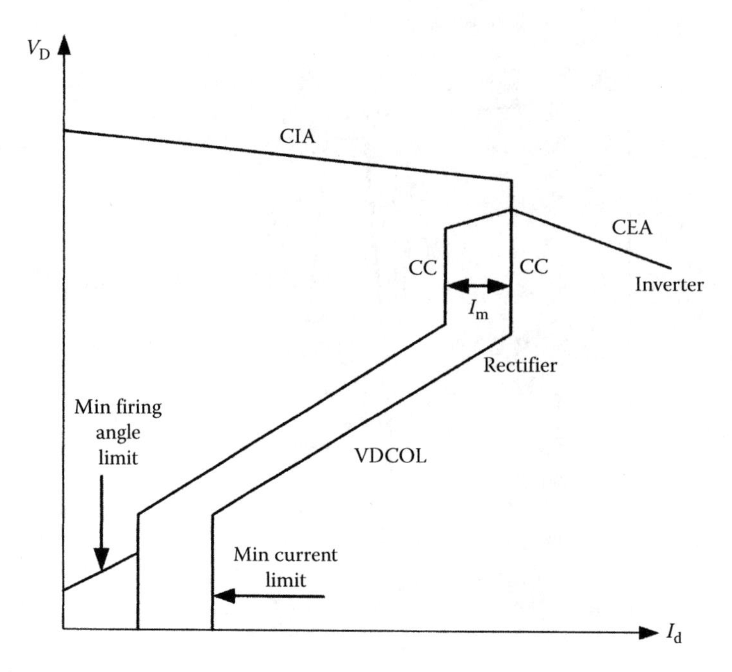

FIGURE 4.10
V–I characteristics in steady state with VDCOL control.

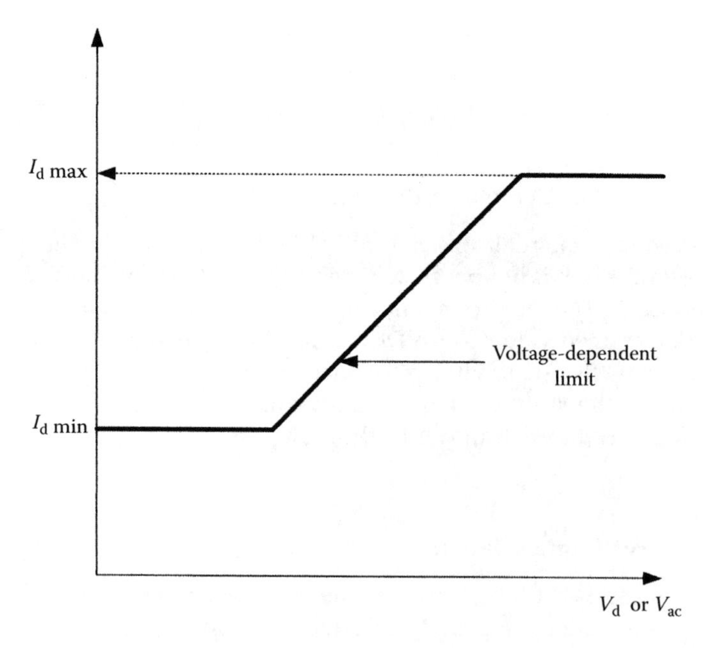

FIGURE 4.11
Current limit as a function of ac or dc voltage.

The power factor of ac current supplied to the converter is the ratio between actual dc voltage V_d and the ideal dc no load voltage, irrespective of whether the reduced dc voltage is due to reactance drop or delay angle.

The reactive power is given by

$$Q = V_d I_d \sqrt{\left[\frac{V_{do}}{V_d}\right]^2 - 1} = V_d I_d \tan\varphi \tag{4.20}$$

Example 4.2

A ±500 kV dc link has a rated power of 1500 MW, resistance of two conductors is 10 Ω, the converters are 12-pulse, converter transformers have 15% reactance on their ONAN base, firing angle = 22°, and overlap angle is 15°. The following parameters are calculated as follows:

$$\text{DC current} = 1500\ \text{A}$$

DC voltage at inverter is as follows:

$500 - (1500 \times 5 \times 10^{-3}) = 492.5\,\text{kV}$
Power at inverter end, per pole = 738.75 MW
DC voltage per pole rectifier = 250 kV
DC voltage per pole inverter = 246.52 kV

Converter transformer rating is as follows:

The dc voltage V_d each 6-pulse bridge, rectifier side = 250 kV. Then from

$$250 = V_{do}\frac{\cos\alpha + \cos\delta}{2} = V_{do}\frac{\cos 12^\circ + \cos 17^\circ}{2}$$

$$V_{do} = 256.9\ \text{kV}$$

Therefore, transformer rating

$$= \frac{\pi}{3} \times 256.9 \times 1.5 \times 10^3 = 403.53\ \text{MVA}$$

The same ratings are used at rectifier and inverter ends.

Secondary ac rms input current (including harmonics) = $0.8615 \times 1.5 \times 10^3 = 1.292$ kA.

The power factor is

$$\cos\varphi = \frac{250}{256.9} = 0.973$$

Therefore, the reactive power required at rectifier end

$$250 \times 2 \times 1.5 \times 10^3 \times 0.2372 = 177.9\ \text{Mvar}$$

This is per 6-pulse converter. Note that if the rectifier operates at a higher ignition angle, the reactive power required will increase. See Volume 3 for the reactive power requirement of converters.

At the inverter end, voltage is 246.52 kV:

$$V_{di} = V_{dio} \frac{\cos\gamma + \cos\beta}{2}$$

But β and γ have not been specified. Assume β = 35° and γ = 18°.
Then

$$246.52 = V_{dio} \frac{\cos 35° + \cos 18°}{2}, \; V_{dio} = 271.9 \text{ kV}$$

Therefore, the power factor at the inverter end is 0.9064.
The reactive power at the inverter end is

$$246{,}52 \times 2 \times 1.5 \times 10^3 \times 0.0466 = 177.9 \text{ Mvar}$$

4.5 Control Implementation

Different levels of control are shown in Figure 4.12. The controls are divided into four levels: the converter unit control, the pole control, master control, and overall control.

The bridge control determines the firing instants of the thyristors and defines α_{min} & γ_{min} limits.

The pole control coordinates the control of bridges in a pole. The conversion of current order to firing angle, tap changer, and protections are handled in this control.

The master control determines the current order and provided coordinated current signals to all poles.

System control is an interface with master control and determines power flow scheduling.

4.5.1 Equidistant Firing Control

The schematics of equidistant firing control (EFC) are shown in Figures 4.13 and 4.14. The pulses are derived from a pulse generator at normal frequency of f_c proportional to 6-pulse or 12-pulse operation. The pulses are separated in pulse distribution unit and supplied to the thyristors. The control pulses are equidistant and of constant frequency. The pulses are delivered by a ring counter.

The voltage-controlled oscillator gives a train of pulses of frequency f_c, proportional to V_c, which is derived from the feedback control system. Figure 4.14 shows a simplified diagram of this type of controller. The control function V_{cf} are pulses of a constant slope and initiated the control pulses at intersection of V_c (points F1, F2, F3, … , for V_{c1}) and F1′, F2′, F3′, … , for V_{c2}. The distance d is, therefore, determined by slope of control function V_{cf} and magnitude of V_c. When the point of intersection is shifted from F1 to F1′, it results in an increase in delay angle by $\Delta\alpha$.

The firing pulses of a 12-pulse converter are spaced 30° electrical, the terminal units—rectifier and inverter are provided with identical EFCs. During disturbances in power flow, the commutating voltage is sampled at a high rate and also the dc voltage and current. Based on these values, the actual firing instant of a thyristor is measured with a suitable commutation margin.

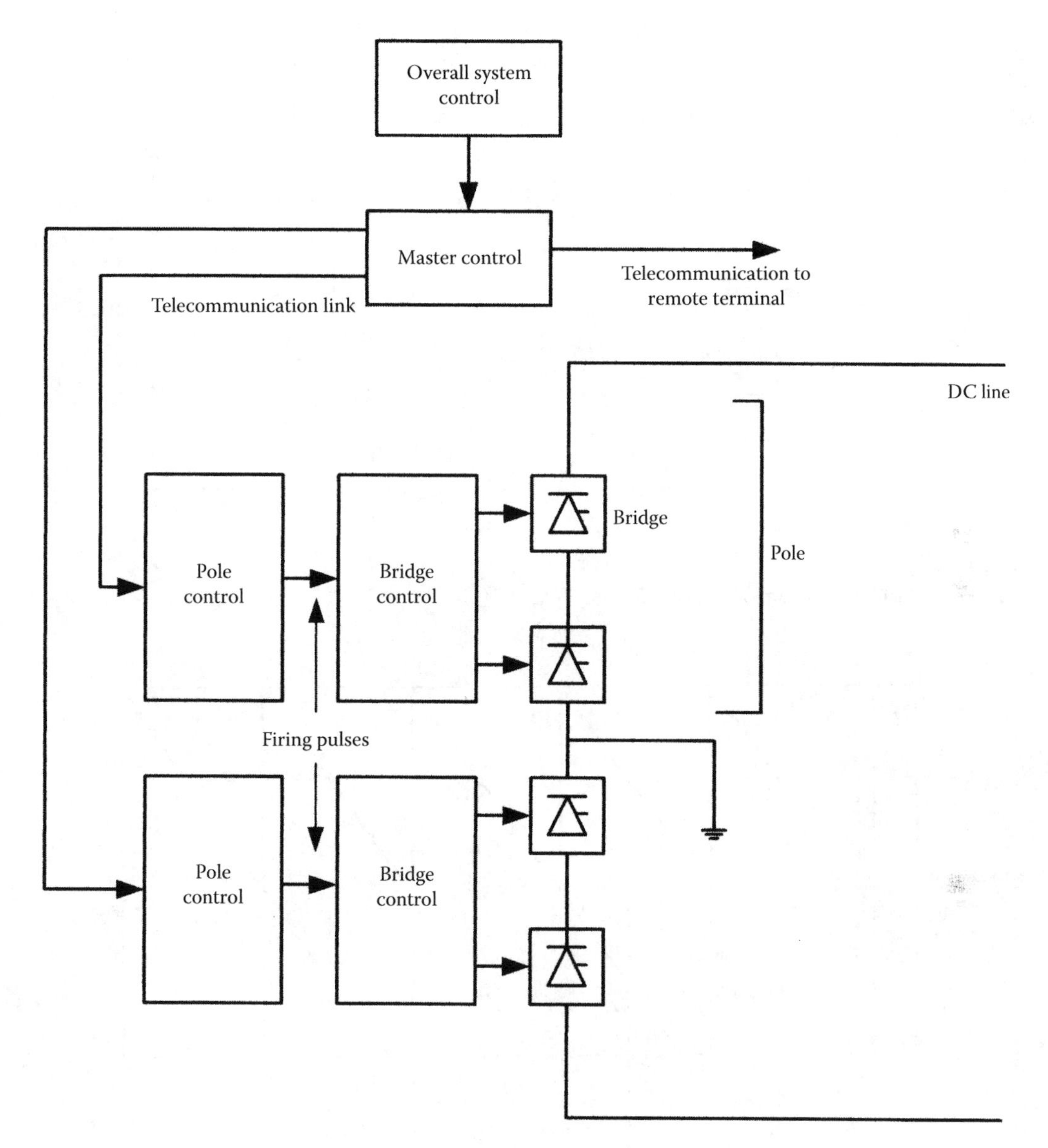

FIGURE 4.12
Different levels of control in an HVDC link.

4.6 Short-Circuit Ratio

The definition of SCR here should not be confused with SCR for synchronous generators described in Volume 1. Here, SCR signifies how strong the ac system with respect to the HVDC link is. It is defined as follows:

$$\mathrm{SCR} = \frac{S_{\mathrm{MVA,min}}}{P_{d,\max}} \tag{4.21}$$

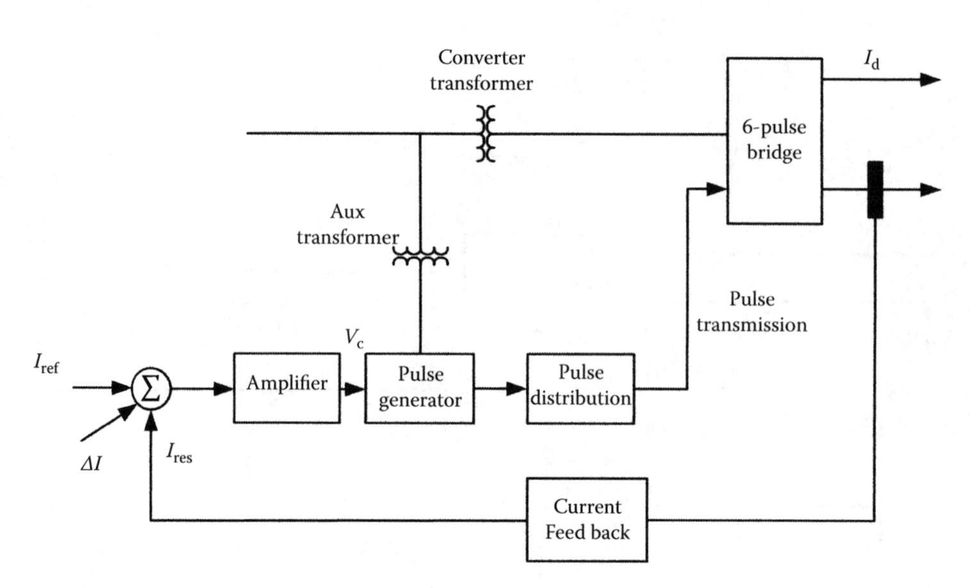

FIGURE 4.13
Control circuit block diagram of thyristor firing.

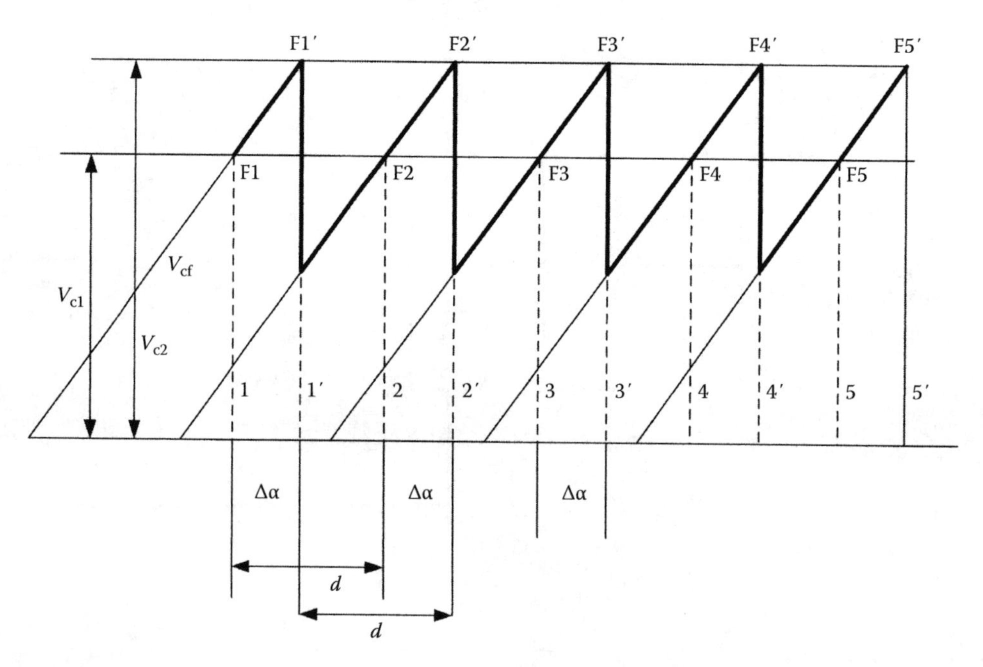

FIGURE 4.14
Pulse generation in equidistant firing control.

where $S_{MVA,min}$ is the minimum short-circuit MVA of the ac bus at the HVDC link and $P_{d,max}$ is the maximum rated bipolar power of the HVDC link. The minimum short-circuit level can be calculated based upon the minimum three-phase short-circuit current of the ac system at the point of interconnection, see Volume 1.

The equivalent SCR (ESCR) considers the reactive power of shunt capacitor filters and other shunt capacitors, if any:

$$\text{ESCR} = \frac{S_{\text{MVA,min}} - Q_{\text{c}}}{P_{d,\text{max}}} \tag{4.22}$$

An ac system with SCR < 2 is considered as weak, 2–4 as intermediate, and >4 as strong. The impacts of low SCR (weak electrical system) are as follows:

- High dynamic overvoltages
- Voltage instability
- Harmonic resonance and objectionable flicker

With low SCR, the $P_{\text{d}}/I_{\text{d}}$ curve has a stability limit above which increase *in* I_{d} *causes* a *decrease in* P_d. Control of voltage and recovery from disturbances becomes difficult. The dc system response may contribute to the collapse of the ac system. Synchronous condensers can be installed to increase SCR, as these contribute to the short-circuit currents.

The ac bus voltage depends upon SCR, active power, and reactive power supplied by the ac filters. For a given power flow, ac bus voltage can be raised by increasing the shunt compensation.

On sudden load rejection, there will be transient rise in ac voltage, which should be limited to 1.1 pu. This is controlled by rapidly reducing reactive power provided by shunt capacitors. SVCs have an obvious advantage, see Chapter 8:

- The harmonic resonance problems are due to parallel resonance between ac capacitors, filters, and ac system at lower harmonics, see Volume 3.
- The problems of flicker can arise as the switching of shunt capacitors and reactors may cause a large voltage swing in the vicinity of compensating equipment due to frequently switched compensating devices, see Volume 3.

4.7 VSC-Based HVDC, HVDC Light

Voltage converter-based HVDC, also called HVDC light, consists of a bipolar two-wire HVDC system with converters connected pole to pole. DC capacitors are used to provide a stiff dc voltage source. There is no earth return operation. The converters are coupled to the ac system through ac phase reactors and converter transformers. Harmonic filters are located between the phase reactors and converter transformers. This avoids the harmonic loading and dc stresses on the converter transformers. The harmonic filters are much smaller due to PWM modulation, see Volume 3.

The IGBT valves comprise series connected IGBT positions. An IGBT exhibits low forward voltage drop and has a voltage-controlled capacitive gate. The complete IGBT position consists of an IGBT, an anti-parallel diode, a voltage divider, and water-cooled heat sink. IGBTs may be connected in series to switch higher voltages, much alike thyristors in conventional HVDC.

Active power can be controlled by changing the phase angle of the converter ac voltage with respect to filter bus voltage, whereas the reactive power can be controlled by changing the magnitude of the phase voltage with respect to filter bus voltage. This allows separate active and reactive power control loops for HVDC system regulation. The active power loop can be set to control the dc voltage or the active power. In a dc link, one terminal may be set to control the active power while the other terminal to control the dc side voltage. Either of these two modes can be selected independently.

Problems

4.1 A 12-pulse bridge is fed from a three-phase transformer of voltage ration 220–230 kV. The ignition delay angle is 20°, and the commutation angle 15°. The dc current delivered by the rectifier is 2000 A. It is required to calculate: (1) dc output voltage, (2) effective commutating reactance, (3) rms fundamental current, (4) power factor, and (5) reactive power compensation.

4.2 A bipolar dc line operates at ±500 kV, rated power = 1000 MW, line resistance = 10 ohms, and 12-pulse bridge $Rc = 12\ \Omega$. It operated at ignition angle of 18°, inverter control with $\gamma = 18°$. Calculate power factor, inverter overlap angle, rms value of line-to-line voltage, and fundamental component of line current.

Bibliography

1. EW Kimbark. *Direct Current Transmission*, vol. 1. Wiley Interscience, New York, 1971.
2. KR Padiyar. *Power Transmission Systems*, John Wiley, New York, 1990.
3. J Arrillaga. *High Voltage Direct Current Transmission*, 2nd ed. IEEE Press, Piscataway, NJ, 1998.
4. VK Sood. *HVDC and FACTS Controllers*, Kluwer, Norwell, MA, 2004.
5. B Jacobson, Y Jiang-Hafner, P Rey, G Asplund. HVDC with voltage source converters, and extruded cables for up to ±300 kV and 1000 MW. In *Proceedings, CIGRE*, pp. B4–105, 2006.
6. A Ekstrom, G Liss. A refined HVDC control system. *IEEE Trans Power Syst*, PAS 89, 723–732, 1970.
7. IEEE/PES Transmission and Distribution Committee. http://www.ece.uidaho.edu/HVDCfacts.
8. CIGRE study committee B4, HVDC and Power Electronic Equipment. http://www.cigre-b4.org.
9. EPRI. HVDC Transmission Line Reference Book, EPRI Report TR-102764, 1993.
10. N Knudsen, F Iliceto. Contributions to the electrical design of HVDC overhead lines. *IEEE Trans*, PAS 93(1), 233–239, 1974.
11. EPRI. HVDC Electrode Design, EPRI Research Project 1467-1, Report EL-2020.
12. P Kundur, *Power System Stability and Control*, Chapter 10: HVDC transmission, EPRI, Palo Alto, CA, 1993.
13. CIGRE and IEEE Joint Task Force Report. Guide for planning dc links terminating at AC locations having low-short-circuit capabilities, Part 1: AC/DC interaction phenomena. CIGRE Publication 68, June 1992.
14. IEEE Committee Report. HVDC controls for system dynamic performance. *IEEE Trans*, PWRS-6(2), 743–752, 1991.

15. M Parker, EG Peattie. *Pipe Line Corrosion and Cathodic Protection*, 3rd ed. Gulf Professional Publishing, Houston, TX, 1995.
16. IEEE Standard 1031, IEEE Guide for the Functional Specifications of Transmission Line Static VAR Compensators-2000.
17. JJ Vithayathil, AL Courts, WG Peterson, NG Hingorani, S Nilsson, JW Porter. HVDC circuit breaker development and field tests. *IEEE Trans*, PAS-104, 2693–2705, 1985.
18. CIGRE Joint Working Group 13/14-08. Circuit breakers for meshed multi-terminal HVDC systems, Part 1: DC side substation switching under normal and fault conditions. *Electra*, 163, 98–122, 1995.
19. G Asplund, K Eriksson, H Jiang, J Lindberg, R Palsson, K Stevenson. *DC Transmission Based on Voltage Source Converters*, CIGRE, Paris, 1998.

5

Load Flow Methods: Part I

The *Y*-matrix iterative methods were the very first to be applied to load flow calculations on the early generation of digital computers. This required minimum storage, however, may not converge on some load flow problems. This deficiency in *Y*-matrix methods led to *Z*-matrix methods, which had a better convergence, but required more storage and slowed down on large systems. These methods of load flow are discussed in this chapter.

We discussed the formation of bus impedance and admittance matrices in Volume 1, not repeated here. For a general network with $n+1$ nodes, the admittance matrix is

$$\bar{Y} = \begin{vmatrix} Y_{11} & Y_{12} & \cdot & Y_{1n} \\ Y_{21} & Y_{22} & \cdot & Y_{2n} \\ \cdot & \cdot & \cdot & \cdot \\ Y_{n1} & Y_{n2} & \cdot & Y_{nn} \end{vmatrix} \tag{5.1}$$

where each admittance $Y_{ii}(i=1, 2, 3, 4, \ldots)$ is the self-admittance or driving point admittance of node i, given by the diagonal elements and it is equal to the algebraic sum of all admittances terminating at that node. $Y_{ik}(i, k=1, 2, 3, 4, \ldots)$ is the mutual admittance between nodes i and k or transfer admittance between nodes i and k and is equal to the negative of the sum of all admittances directly connected between those nodes. In Volume 1, we discussed how each element of this matrix can be calculated. The following modifications can be easily implemented in the bus admittance matrix:

1. Changing of a branch admittance from y_{sr} to $y_{sr}+\Delta y_{sr}$ between buses S and R leads to

$$\begin{aligned} Y_{ss} &\rightarrow Y_{ss} + \Delta y_{sr} \\ Y_{sr} &\rightarrow Y_{sr} + \Delta y_{sr} \\ Y_{rr} &\rightarrow Y_{rr} + \Delta y_{sr} \end{aligned} \tag{5.2}$$

2. Addition of a new branch of admittance y_{sr} between *existing* buses S and R gives

$$\begin{aligned} Y_{ss} &\rightarrow Y_{ss} + y_{sr} \\ Y_{sr} &\rightarrow Y_{sr} + y_{sr} \\ Y_{rr} &\rightarrow Y_{rr} + y_{sr} \end{aligned} \tag{5.3}$$

3. Addition of a new branch between an existing bus S and a *new bus* is shown in Figure 5.1. Let the new bus be designated N'. The order of the Y matrix is increased by one, from n to $n+1$. The injected current at bus N' is

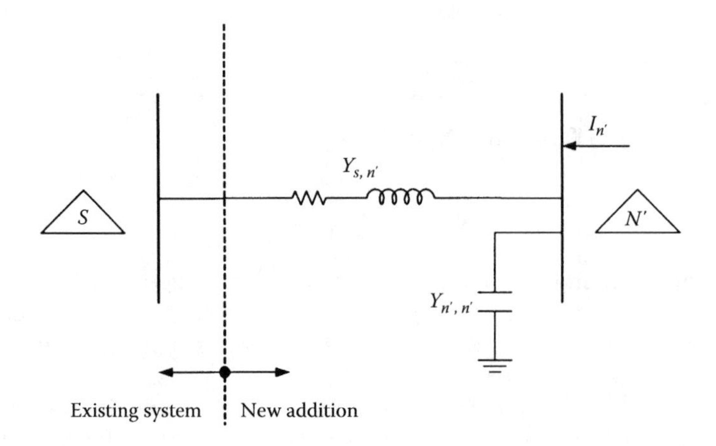

FIGURE 5.1
Adding a new branch between an existing bus S and a new bus N' in a Y matrix.

$$
\begin{aligned}
I_{n'} &= Y_{n's}V_s + Y_{n'n'}V_{n'} \\
Y_{n's} &= -y_{s,n'} \\
Y_{n'n'} &= y_{n'n'} + y_{sn'}
\end{aligned}
\tag{5.4}
$$

All equations remain unchanged except the bus S equation:

$$
I_s = Y_{s1}V_1 + \cdots + \left(Y_{ss} + y_{s,n'}\right)V_s + \cdots + Y_{sn'}V_{n'} \tag{5.5}
$$

Thus,

$$
\begin{aligned}
Y_{ss} &\rightarrow Y_{ss} + y_{sn'} \\
Y_{sn'} &\rightarrow -y_{sn'}
\end{aligned}
\tag{5.6}
$$

5.1 Modeling of a Two-Winding Transformer

A transformer can be modeled by its leakage impedance as in short-circuit calculations; however, in load flow calculations a transformer can act as a control element:

- Voltage control is achieved by adjustments of taps on the windings, which change the turns ratio. The taps can be adjusted under load, providing an automatic control of the load voltage (Volume 1). The under load taps generally provide ±10%–20% voltage adjustments around the rated transformer voltage, in 16 or 32 steps. Off-load taps provide ±5% V adjustment, two taps of 2.5% below the rated voltage and two taps of 2.5% above the rated voltage. These off-load taps must be set at optimum level before commissioning, as these cannot be adjusted under load.

- Transformers can provide phase-shift control to improve the stability limits.
- The reactive power flow is related to voltage change and voltage adjustments indirectly provide reactive power control.

We will examine each of these three transformer models at appropriate places. The model for a ratio-adjusting type transformer is discussed in this chapter. The impedance of a transformer will change with tap position. For an autotransformer there may be 50% change in the impedance over the tap adjustments. The reactive power loss in a transformer is significant, and the X/R ratio should be correctly modeled in load flow.

Consider a transformer of ratio 1:n. It can be modeled by an ideal transformer in series with its leakage impedance Z (the shunt magnetizing and eddy current loss circuit is neglected), as shown in Figure 5.2a. With the notations shown in this figure:

$$V_s = (V_r - ZI_r)/n \tag{5.7}$$

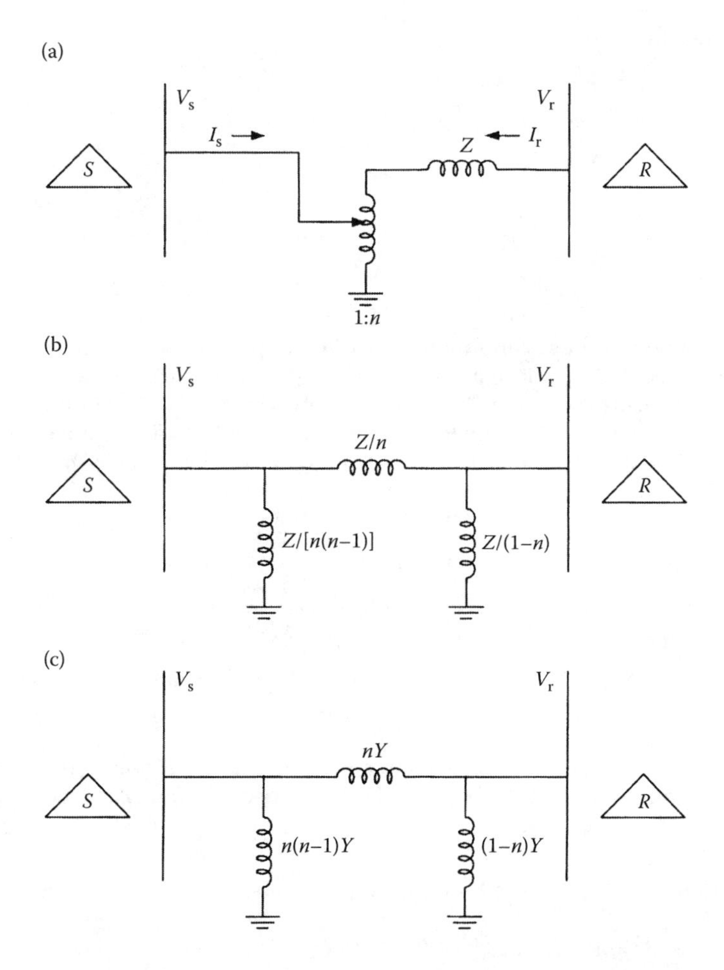

FIGURE 5.2
(a) Equivalent circuit of a ratio adjusting transformer, ignoring shunt elements, (b) equivalent Π impedance network, (c) equivalent Π admittance network.

As the power through the transformer is invariant:

$$V_s I_s^* + (V_r - ZI_r) I_r^* = 0 \tag{5.8}$$

Substituting $I_s/(-I_r)=n$, the following equations can be written:

$$I_s = \left[n(n-1)Y\right]V_s + nY(V_s - V_r) \tag{5.9}$$

$$I_r = nY(V_r - V_s) + \left[(1-n)Y\right]V_r \tag{5.10}$$

Equations 5.9 and 5.10 give the equivalent circuits shown in Figure 5.2b and c, respectively. For a phase-shifting transformer, a model is derived in Chapter 6.

The equivalent Π circuits of the transformer in Figure 5.2c leads to

$$Y_{ss} = n^2 Y_{sr} \qquad Y_{rs} = Y_{sr} = -nY_{sr} \qquad Y_{rr} = Y_{sr} \tag{5.11}$$

If the transformer ratio changes by Δn:

$$\begin{aligned} &Y_{ss} \to Y_{ss} + \left[(n+\Delta n)^2 - n^2\right]Y_{sr} \\ &Y_{sr} \to Y_{sr} - \Delta n Y_{sr} \\ &Y_{rr} \to Y_{rr} \end{aligned} \tag{5.12}$$

Example 5.1

A network with four buses and interconnecting impedances between the buses, on a per unit basis, is shown in Figure 5.3a. It is required to construct a Y matrix. The turns' ratio of the transformer should be accounted for according to the equivalent circuit developed above. All impedances are converted into admittances, as shown in Figure 5.3b, and the Y matrix is then written ignoring the effect of the transformer turns' ratio settings:

$$\bar{Y}_{bus} = \begin{vmatrix} 2.176 - j7.673 & -1.1764 + j4.706 & -1.0 + j3.0 & 0 \\ -1.1764 + j4.706 & 2.8434 - j9.681 & -0.667 + j2.00 & -1.0 + j3.0 \\ -1.0 + j3.0 & -0.667 + j2.00 & 1.667 - j8.30 & j3.333 \\ 0 & -1.0 + j3.0 & j3.333 & 1.0 - j6.308 \end{vmatrix}$$

The turns' ratio of the transformer can be accommodated in two ways: (1) by equations, or (2) by the equivalent circuit of Figure 5.2c. Let us use Equation 5.2 to modify the elements of the admittance matrix:

$$\begin{aligned} Y_{33} &\to Y_{33} + [(n+\Delta n)^2 - n^2]Y34 \\ &= 1.667 - j8.30 + [1.1^2 - 1][-j3.333] \\ &= 1.667 - j7.60 \end{aligned}$$

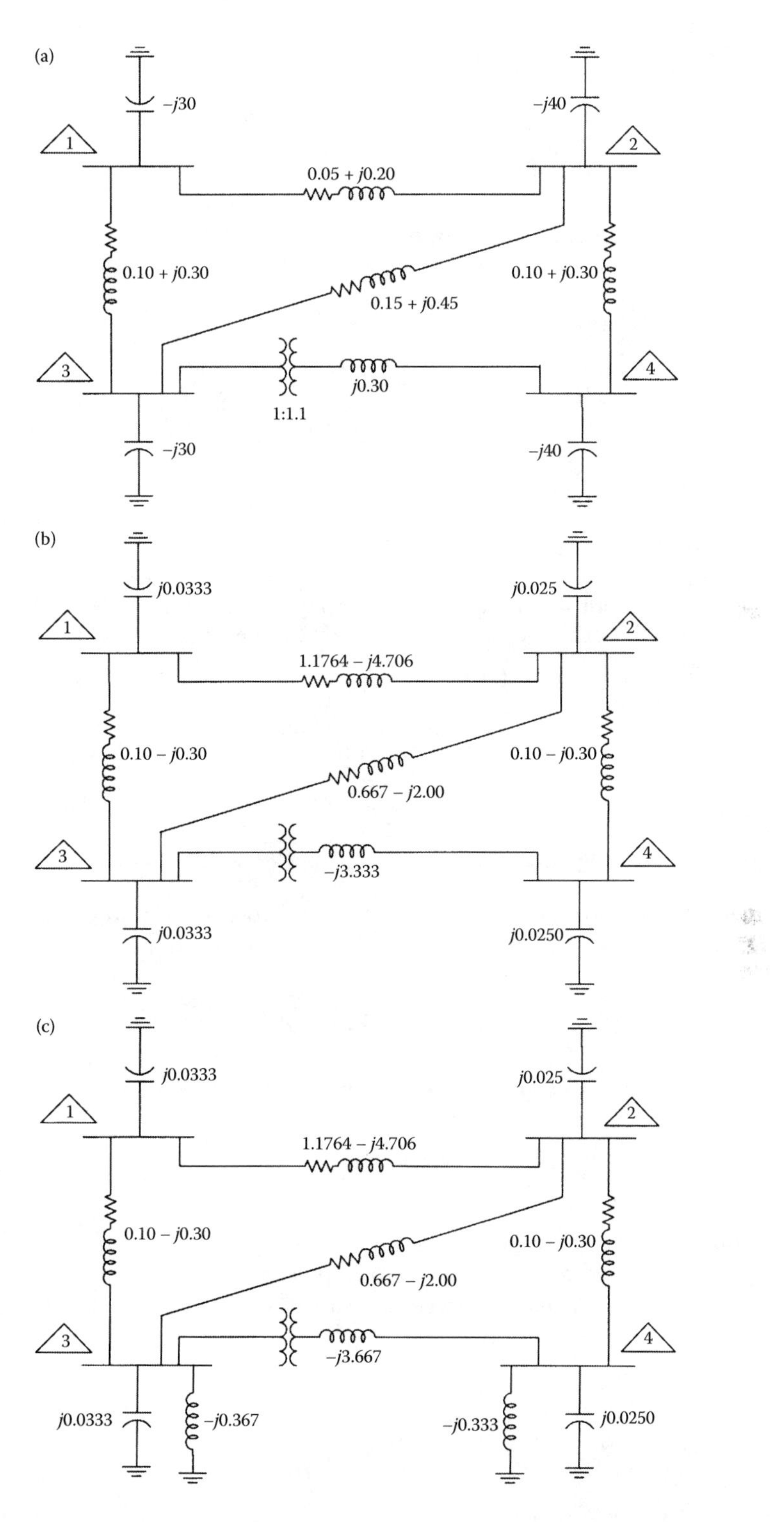

FIGURE 5.3
(a) Impedance data for a 4-bus network with a ratio-adjusting transformer, (b) the 4-bus system with impedance data converted to admittance, (c) equivalent circuit adjusted for transformer turns' ratio.

$$Y_{34} \rightarrow Y_{34} - \Delta n Y_{34}$$

$$j3.333 - (1.1 - 1)(-j3.333)$$

$$= j3.666$$

The modified Y bus matrix for the transformer turns' ratio is

$$\bar{Y}_{\text{bus, modified}} = \begin{vmatrix} 2.1764 - j7.673 & -1.1764 + j4.706 & -1.0 + j3.0 & 0 \\ -1.1764 + j4.706 & 2.8434 - j9.681 & -0.667 + j2.00 & -1.0 + j3.0 \\ -1.0 + j3.0 & -0.667 + j2.00 & 1.667 - j9.0 & j3.666 \\ 0 & -1.0 + j3.0 & j3.66 & 1.0 - j6.308 \end{vmatrix}$$

As an alternative, the equivalent circuit of the transformer is first modified for the required turns' ratio, according to Figure 5.2c. This modification is shown in Figure 5.3c, where the series admittance element between buses 3 and 4 is modified and additional shunt admittances are added at buses 3 and 4. After the required modifications, the admittance matrix is formed, which gives the same results as arrived at above.

5.2 Load Flow, Bus Types

In a load flow calculation, the voltages and currents are required under constraints of bus power. The currents are given by

$$\bar{I} = \bar{Y}\bar{V} \tag{5.13}$$

and the power flow equation can be written as

$$S_s = P_s + jQ_s = V_s I_s^* \quad s = 1, 2, \ldots, n \tag{5.14}$$

where n = number of buses

From Equation 5.13 and 5.14 there are $2n \times 2$ equations and there are $4n$ variables ($2n$ current variables and $2n$ V variables) presenting a difficulty in solution. The current can be eliminated in the above formation:

$$S_s = P_s + jQ_s = V_s \left(\sum_{r=1}^{n} Y_{sr} V_r \right)^* \quad r, s = 1, 2, \ldots, n \tag{5.15}$$

TABLE 5.1

Load Flow—Bus Types

Bus Type	Known Variable	Unknown Variable
PQ	Active and reactive power (*P*, *Q*)	Current and voltage (*I*, *V*)
PV	Active power and voltage (*P*, *V*)	Current and reactive power (*I*, *Q*)
Swing	Voltage	Current active and reactive power (*I*, *P*, and *Q*)

Now there are $2n$ equations and $2n$ variables. This does not lead to an immediate solution. The constant power rather than constant current makes the load flow problem nonlinear. The loads and generation may be specified but the sum of loads is not equal to the sum of generation as network losses are indeterminable until the bus voltages are calculated. Therefore, the exact amount of total generation is unknown. One solution is to specify the power of all generators except one. The bus at which this generator is connected is called the *swing bus* or slack bus. A utility source is generally represented as a swing bus or slack bus, as the consumers' system is much smaller compared to a utility's system. After the load flow is solved and the voltages are calculated, the injected power at the swing bus can be calculated.

Some buses may be designated as *PQ* buses while the others are designated as *PV* buses. At a *PV* bus the generator active power output is known and the voltage regulator controls the voltage to a specified value by varying the reactive power output from the generator. There is an upper and lower bound on the generator reactive power output depending on its rating, and for the specified bus voltage, these bounds should not be violated. If the calculated reactive power exceeds generator Q_{max}, then Q_{max} is set equal to Q. If the calculated reactive power is lower than the generator Q_{min}, then Q is set equal to Q_{min}.

At a *PQ* bus, neither the current, nor the voltage is known, except that the load demand is known. A "mixed" bus may have generation and also directly connected loads. The characteristics of these three types of buses are shown in Table 5.1.

The above description shows that the load flow problem is essentially non-linear and iterative methods are used to find a solution.

5.3 Gauss and Gauss–Seidel *Y*-Matrix Methods

The principle of Jacobi iteration is shown in Figure 5.4. The program starts by setting initial values of voltages, generally equal to the voltage at the swing bus. In a well-designed power system, voltages are close to rated values and in the absence of a better estimate all the voltages can be set equal to 1 per unit. From node power constraint, the currents are known and substituting back into the *Y*-matrix equations, a better estimate of voltages is obtained. These new values of voltages are used to find new values of currents. The iteration is continued until the required tolerance on power flows is obtained. This is diagrammatically illustrated in Figure 5.4. Starting from an initial estimate of x_0, the final value of x^* is obtained through a number of iterations. The basic flow chart of the iteration process is shown in Figure 5.5.

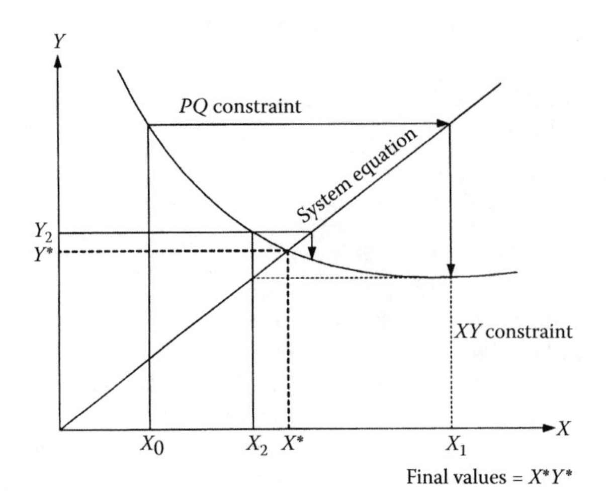

FIGURE 5.4
Illustration of numerical iterative process for final value of a function.

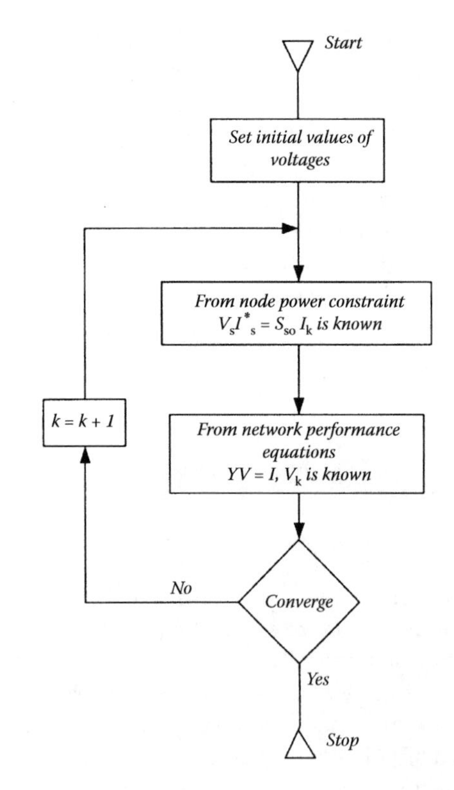

FIGURE 5.5
Flow chart of basic iterative process of Jacobi-type iterations.

5.3.1 Gauss Iterative Technique

Consider that n linear equations in n unknowns $(x_1, \ldots, x_n)$ are given. The a coefficients and b dependent variables are known:

$$\begin{aligned}
&a_{11}x_1 + a_{12}x_2 + \cdots + a_n x_n = b_1 \\
&a_{21}x_1 + a_{22}x_2 + \cdots + a_2 n x_n = b_2 \\
&\ldots \\
&a_{n1}x_1 + a_{n2}x_2 + \cdots + a_{nn}x_n = b_n
\end{aligned} \tag{5.16}$$

These equations can be written as

$$\begin{aligned}
&x_1 = \frac{1}{a_{11}}(b_1 - a_{12}x_2 - a_{13}x_3 - \cdots - a_{1n}x_n) \\
&x_2 = \frac{1}{a_{22}}(b_2 - a_{21}x_1 - a_{23}x_3 - \cdots - a_{2n}x_n) \\
&\ldots \\
&x_n = \frac{1}{a_{nn}}(b_n - a_{n1}x_1 - a_{n2}x_2 - \cdots - a_{n,n-1}x_{n-1})
\end{aligned} \tag{5.17}$$

An initial value for each of the independent variables $x_1, x_2, \ldots, x_n$ is assumed. Let these values be denoted by

$$x_1^0, x_2^0, x_3^0, \ldots, x_n^0 \tag{5.18}$$

The initial values are estimated as

$$\begin{aligned}
&x_1^0 = \frac{y_1}{a_{11}} \\
&x_2^0 = \frac{y_2}{a_{22}} \\
&\ldots \\
&x_n^0 = \frac{y_n}{a_{nn}}
\end{aligned} \tag{5.19}$$

These are substituted into Equation 5.17, giving

$$\begin{aligned}
&x_1^0 = \frac{1}{a_{11}}\left[b_1 - a_{12}x_2^0 - a_{13}x_3^0 - \cdots a_{1n}x_n^0\right] \\
&x_2^0 = \frac{1}{a_{22}}\left[b_2 - a_{21}x_1^0 - a_{23}x_3^0 - \cdots a_{2n}x_n^0\right] \\
&\ldots \\
&x_n^0 = \frac{1}{a_{nn}}\left[b_{nn} - a_{n1}x_1^0 - a_{n2}x_2^0 - \cdots a_{n,n-1}x_{n-1}^0\right]
\end{aligned} \tag{5.20}$$

These new values of

$$x_1^1, x_2^1, \ldots, x_n^1 \tag{5.21}$$

are substituted into the next iteration. In general, at the *k*th iteration:

$$\begin{aligned}
x_1^k &= \frac{1}{a_{11}}\left[b_1 - a_{12}x_2^{k-1} - a_{13}x_3^{k-1} - \cdots - a_{1n}x_n^{k-1}\right] \\
x_2^k &= \frac{1}{a_{22}}\left[b_2 - a_{21}x_1^{k-1} - a_{23}x_3^{k-1} - \cdots - a_{2n}x_n^{k-1}\right] \\
&\ldots \\
x_n^k &= \frac{1}{a_{nn}}\left[b_n - a_{n1}x_2^{k-1} - a_{n2}x_2^{k-1} - \cdots - a_{n,n-1}x_{n-1}^{k-1}\right]
\end{aligned} \tag{5.22}$$

Example 5.2

Consider a three-bus network with admittances as shown in Figure 5.6. The currents at each of the buses are fixed, which will not be the case in practice, as the currents are dependent on voltages and loads. Calculate the bus voltages for the first five iterations by the Gauss iterative technique.

The matrix equation is

$$\begin{vmatrix} 3 & -1 & 0 \\ -1 & 4 & -3 \\ 0 & -3 & 6 \end{vmatrix} \begin{vmatrix} v_1 \\ v_2 \\ v_3 \end{vmatrix} = \begin{vmatrix} 3 \\ 4 \\ 2 \end{vmatrix}$$

Thus,

$$\begin{aligned}
3v_1 - v_2 &= 3 \\
-v_1 + 4v_2 - 3v_3 &= 4 \\
-3v_2 + 6v_3 &= 2
\end{aligned}$$

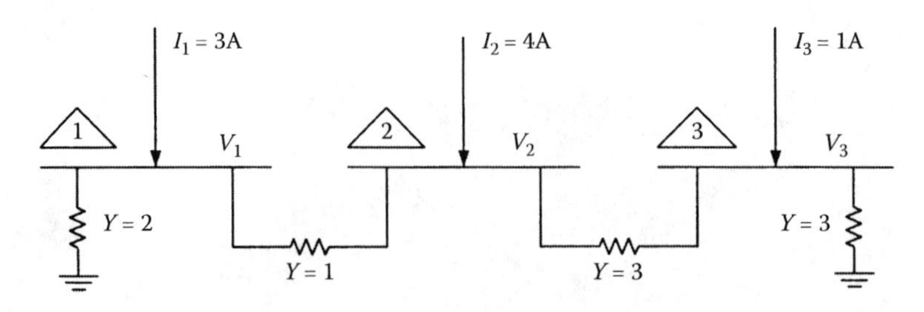

FIGURE 5.6
A three-bus system for Examples 5.2, 5.3, and 5.5.

TABLE 5.2

Gauss Iterative Solution of Example 5.2

Voltage	$k=0$	$k=1$	$k=2$	$k=3$	$k=4$	$k=5$
$\upsilon_{1(k)}$	1	1.333	1.667	1.653	1.806	1.799
$\upsilon_{2(k)}$	1	2.000	1.958	2.417	2.397	2.608
$\upsilon_{3(k)}$	1	0.8333	1.333	1.312	1.542	1.532

There are three equations and three unknowns. Let $\upsilon_1 = \upsilon_2 = \upsilon_3 = 1$ be the initial values. This relationship is generally valid as a starting voltage estimate in load flow, because the voltages throughout the network must be held close to the rated voltages. The following equations can then be written according to the Gauss iterative method:

$$\begin{aligned} \upsilon_1^{k+1} &= 1 + \frac{1}{3}\upsilon_2^k \\ \upsilon_2^{k+1} &= 1 + \frac{1}{4}\upsilon_1^k + \frac{3}{4}\upsilon_3^k \\ \upsilon_3^{k+1} &= \frac{1}{3} + \frac{1}{2}\upsilon_2^k \end{aligned} \tag{5.23}$$

The values of υ_1, υ_2, and υ_3 at $k=1$ are

$$k=1$$

$$\upsilon_1 = 1 + (0.3333)(1) = 1.333$$

$$\upsilon_2 = 1 + 0.25(1) + 0.75(1) = 2.000$$

$$\upsilon_3 = 0.3333 + (0.25)(1) = 0.833$$

$$k=2$$

$$\upsilon_1 = 1 + (0.3333)(2) = 1.667$$

$$\upsilon_2 = 1 + 0.25(1.333) + 0.75(0.8333) = 1.958$$

$$\upsilon_3 = 0.3333 + 0.5(2.000) = 1.333$$

The results for the first five iterations are presented in Table 5.2.

5.3.2 Gauss–Seidel Iteration

Instead of substituting the $k-1$ approximations into all the equations in the kth iteration, the kth iterations are immediately used as soon as these are found. This will hopefully reduce the number of iterations.

$$x_1^k = \frac{1}{a_{11}}\left[b_1 - a_{12}x_2^{k-1} - a_{13}x_3^{k-1} - \cdots - a_{1n}x_n^{k-1}\right]$$

$$
\begin{aligned}
x_1^k &= \frac{1}{a_{11}}\left[b_1 - a_{12}x_2^{k-1} - a_{13}x_3^{k-1} - \cdots - a_{1n}x_n^{k-1}\right] \\
x_2^k &= \frac{1}{a_{22}}\left[b_2 - a_{21}x_1^{k} - a_{23}x_3^{k-1} - \cdots - a_{2n}x_n^{k-1}\right] \\
x_3^k &= \frac{1}{a_{33}}\left[b_3 - a_{31}x_1^{k} - a_{32}x_2^{k} - \cdots - a_{3n}x_n^{k-1}\right] \\
&\cdots \\
x_n^k &= \frac{1}{a_{nn}}\left[b_n - a_{n1}x_1^{k} - a_{n2}x_2^{k} - \cdots - a_{n,n-1}x_{n-1}^{k}\right]
\end{aligned}
\tag{5.24}
$$

Example 5.3

Solve the three-bus system of Example 5.2 by Gauss–Seidel iteration.

We again start with the assumption of unity voltages at all the buses, but the new values of voltages are immediately substituted into the downstream equations. From Equation 5.23, the first iteration is as follows:

$$k = 1$$

$$\upsilon_1 = 1 + 0.3333(1) = 1.333$$

This value of υ_1 is immediately used:

$$\upsilon_2 = 1 = 0.25(1.333) + 0.75(1) = 2.083$$

This value of υ_2 is immediately used:

$$\upsilon_3 = 0.3333 + 0.5(2.083) = 1.375$$

The result of the first five iterations is shown in Table 5.3. A better estimate of the final voltages is obtained in fewer iterations.

5.3.3 Convergence

The convergence can be defined as

$$\varepsilon^k = \left|V_s^{k-1} - V_s^k\right| < \varepsilon \text{ (specified value)} \tag{5.25}$$

TABLE 5.3

Gauss–Seidel Iterative Solution of Network of Example 5.3

Voltage	$k = 0$	$k = 1$	$k = 2$	$k = 3$	$k = 4$	$k = 5$
υ_1	1	1.333	1.694	1.818	1.875	1.901
υ_2	1	2.083	2.455	2.625	2.703	2.739
υ_3	1	1.375	1.561	1.646	1.685	1.703

The calculation procedure is repeated until the specified tolerance is achieved. The value of ε is arbitrary. For Gauss–Seidel iteration, ε = 0.00001–0.0001 is common. This is the largest allowable voltage change on any bus between two successive iterations before the final solution is reached. Approximately 50–150 iterations are common, depending on the number of buses and the impedances in the system.

5.3.4 Gauss–Seidel *Y*-Matrix Method

In load flow calculations the system equations can be written in terms of current, voltage, or power at the *k*th node. We know that the matrix equation in terms of unknown voltages, using the bus admittance matrix for $n + 1$ nodes, is

$$\begin{vmatrix} I_1 \\ I_2 \\ \cdots \\ I_k \end{vmatrix} = \begin{vmatrix} Y_{11} & Y_{12} & \cdots & Y_{1n} \\ Y_{21} & Y_{22} & \cdots & Y_{2n} \\ \cdots & \cdots & \cdots & \cdots \\ Y_{n1} & Y_{n2} & \cdots & Y_{nn} \end{vmatrix} \begin{vmatrix} V_{01} \\ V_{02} \\ \cdots \\ V_{on} \end{vmatrix} \tag{5.26}$$

where 0 is the common node. The significance of zero or common node or slack bus is already discussed.

Although the currents entering the nodes from generators and loads are not known, these can be written in terms of *P*, *Q*, and *V*:

$$I_k = \frac{P_k - jQ_k}{V_k^*} \tag{5.27}$$

The convention of the current and power flow is important. Currents entering the nodes are considered positive, and thus the power *into* the node is also positive. A load *draws* power out of the node and thus the active power and inductive vars are entered as $-P-j(-Q) = -P + jQ$. The current is then $(-P + jQ)/V^*$. The nodal current at the *k*th node becomes

$$\frac{P_k - jQ_k}{V_k^*} = Y_{k1}V_1 + Y_{k2}V_2 + V_{k3}Y_3 + \cdots + Y_{kk}V_k + \cdots + Y_{kn}V_n \tag{5.28}$$

This equation can be written as

$$V_k = \frac{1}{Y_{kk}}\left[\frac{P_k - jQ_k}{V_k^*} - Y_{k1}V_1 - Y_{k3}V_3 - \cdots - Y_{kn}V_n\right] \tag{5.29}$$

In general, for the *k*th node:

$$V_k = \frac{1}{Y_{kk}}\left[\frac{P_k - jQ_k}{V_k^*} - \sum_{i=1}^{i=n} Y_{ki}V_i\right] \quad \text{for} \quad i \neq k \tag{5.30}$$

The kth bus voltage at $k+1$ iteration can be written as

$$V_k^{k+1} = \frac{1}{Y_{kk}}\left[\frac{P_k - jQ_k}{V_k^*} - \sum_{i=1}^{k-1} Y_{ki}V_i^{k+1} - \sum_{i=k+1}^{n} Y_{ki}V_i^k\right] \tag{5.31}$$

The voltage at the kth node has been written in terms of itself and the other voltages. The first equation involving the swing bus is omitted, as the voltage at the swing bus is already specified in magnitude and phase angle.

The Gauss–Seidel procedure can be summarized for PQ buses in the following steps:

1. Initial phasor values of load voltages are assumed, the swing bus voltage is known, and the controlled bus voltage at generator buses can be specified. Though an initial estimate of the phasor angles of the voltages will accelerate the final solution, it is not necessary and the iterations can be started with zero degree phase angles or the same phase angle as the swing bus. A *flat voltage start* assumes $1 + j0$ V at all buses, except the voltage at the swing bus, which is fixed.
2. Based on the initial voltages, the voltage at a bus in the first iteration is calculated using Equation 5.30, i.e., at bus 2:

$$V_{21} = \frac{1}{Y_{22}}\left[\frac{P_2 - jQ_2}{V_{2_0}^*} - Y_{21}V_{1_0} - \cdots - Y_{2n}V_{n0}\right] \tag{5.32}$$

3. The estimate of the voltage at bus 2 is refined by repeatedly finding new values of V_2 by substituting the value of V_2 into the right-hand side of the equation.
4. The voltages at bus 3 is calculated using the latest value of V_2 found in step 3 and similarly for other buses in the system.

This completes one iteration. The iteration process is repeated for the entire network till the specified convergence is obtained.

A generator bus is treated differently; the voltage to be controlled at the bus is specified and the generator voltage regulator varies the reactive power output of the generator within its reactive power capability limits to regulate the bus voltage:

$$Q_k = -I_m\left[V_k^*\left\{\sum_{i=1}^{i=n} Y_{ki}V_i\right\}\right] \tag{5.33}$$

where I_m stands for the imaginary part of the equation. The revised value of Q_k is found by substituting the most updated value of voltages:

$$Q_k^{k+1} = -I_m\left[V_k^{k*}\sum_{i=1}^{i=n} Y_{ki}V_i^{k+1} + V_k^{k*}\sum_{i=1}^{i=n} Y_{ki}V_i^k\right] \tag{5.34}$$

The angle δ_k is the angle of the voltage in Equation 5.30:

$$\delta_k^{k+1} = \angle \text{ of } V_k^{k+1}$$

$$= \angle \text{ of} \left[\frac{P_k - jQ_k^{k+1}}{Y_{kk}\left(V_k^k\right)^*} - \sum_{i=1}^{k-1} \frac{Y_{ki}}{Y_{kk}} V_i^{k+1} - \sum_{i=k+1}^{n} \frac{Y_{ki}}{Y_{kk}} V_i^k \right] \tag{5.35}$$

For a *PV* bus the upper and lower limits of var generation to hold the bus voltage constant are also given. The calculated reactive power is checked for the specified limits:

$$Q_{k(\min)} < Q_k^{k+1} < Q_{k(\max)} \tag{5.36}$$

If the calculated reactive power falls within the specified limits, the new value of voltage V_k^{k+1} is calculated using the specified voltage magnitude and δ_k^k. This new value of voltage V_k^{k+1} is made equal to the specified voltage to calculate the new phase angle δ_k^{k+1}.

If the calculated reactive power is outside the specified limits, then,

$$\text{If } Q_k^{k+1} > Q_{k(\max)} \quad Q_k^{k+1} = Q_{k(\max)} \tag{5.37}$$

$$\text{If } Q_k^{k+1} < Q_{k(\max)} \quad Q_k^{k+1} = Q_{k(\max)} \tag{5.38}$$

This means that the specified limits are not exceeded and beyond the reactive power bounds, the *PV* bus is treated like a *PQ* bus. A flow chart is shown in Figure 5.7.

Example 5.4

A three-bus radial system of distribution is shown in Figure 5.8 consisting of three branch impedances shown in per unit on a 100-MVA base. Bus 1 is a swing bus with voltage 1<0° in per unit. Buses 2 and 3 are load buses, loads as shown, and bus 4 is a generation bus, i.e., a *PV* bus with voltage of 1.05 per unit, phase angle unknown. The generator is a 25-MVA unit, rated power factor of 0.8 (15 Mvar output at rated MW output of 20). Though the generator will be connected through a step-up transformer, its impedance is ignored. Also, the capacitance to ground of the transmission lines is ignored. Calculate the load flow by the Gauss–Seidel iterative method.

The impedances are converted into admittances and a Y matrix is formed:

$$\bar{Y}_{\text{bus}} = \begin{vmatrix} 0.896 - j1.638 & -0.896 + j1.638 & 0 & 0 \\ -0.896 + j1.638 & 1.792 - j3.276 & -0.896 + j1.638 & 0 \\ 0 & -0.896 + j1.638 & 2.388 - j4.368 & -1.492 + j2.730 \\ 0 & 0 & -1.492 + j2.730 & 1.492 - j2.730 \end{vmatrix}$$

Let the initial estimate of voltages at buses 2 and 3 be 1<0°, 1<0° respectively. The voltage at bus 2, V_2 is

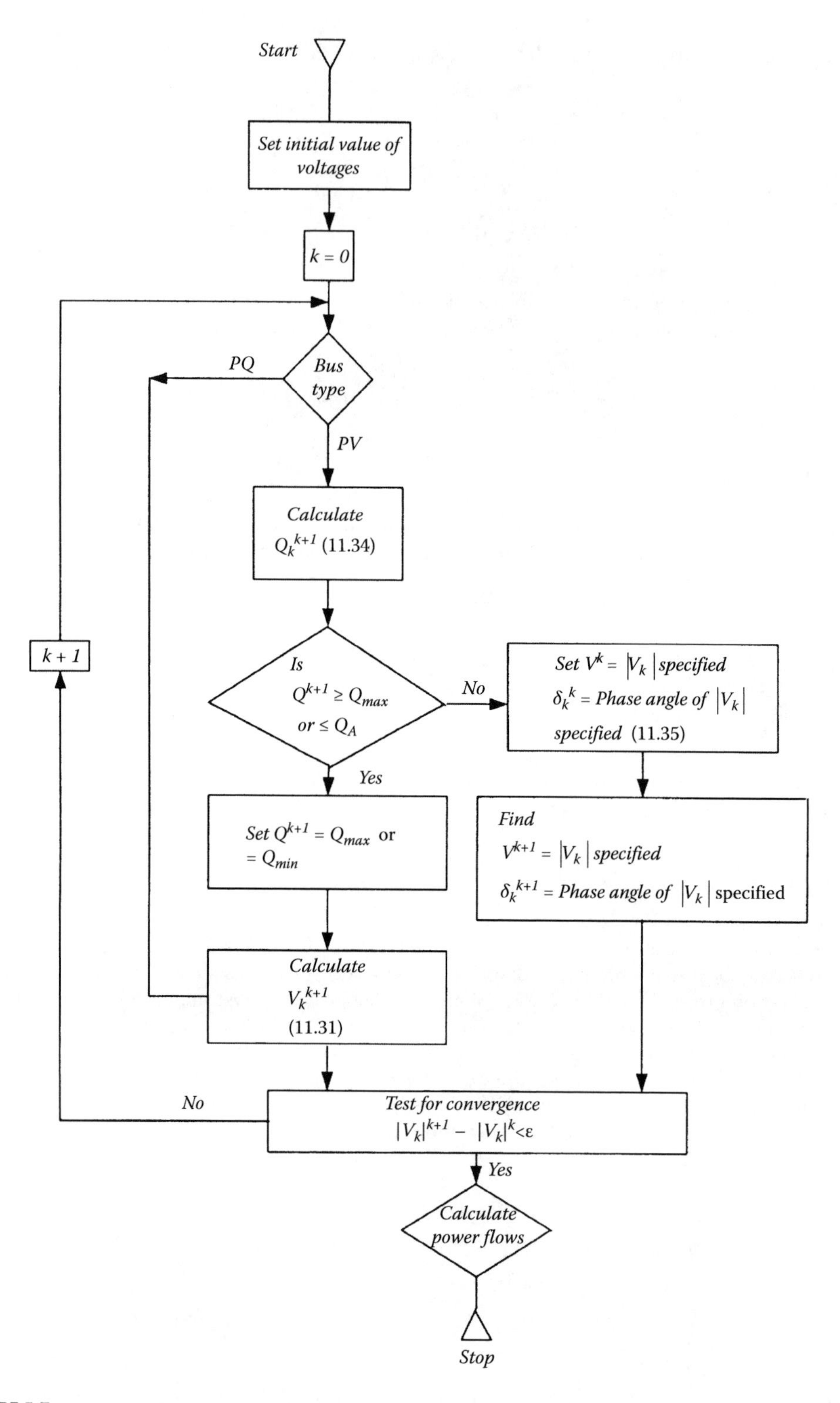

FIGURE 5.7
Flow chart for Gauss–Seidel method of load flow.

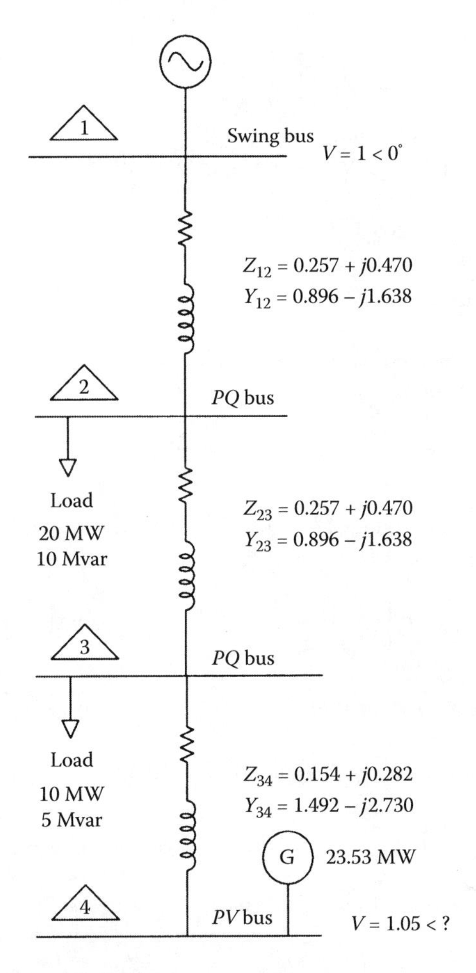

FIGURE 5.8
Four-bus network with generator for load flow (Examples 5.4, 5.6, and 5.9).

$$
\begin{aligned}
V_2^1 &= \frac{1}{Y_{22}}\left[\frac{P_2 - jQ_2}{(V_2^0)^*} - Y_{21}V_1^0 - Y_{23}V_3^0 - Y_{24}V_4^0\right] \\
&= \frac{1}{1.792 - j3.276}\left[\frac{-0.2 + j0.1}{1.0 + j0} - (-0.869 + j1.638)(1.0 < 0°) - (-0.896 + j1.638)(1.0 < 0°)\right] \\
&= 0.9515 < -2.059^0
\end{aligned}
\tag{5.39}
$$

Substitute this value back into Equation 5.39 and recalculate:

$$
\begin{aligned}
V_2^{1(2)} &= \frac{1}{1.792 - j3.276}\left[\frac{-0.2 + j0.1}{0.9515 < 2.059°} + 1.792 - j3.276\right] \\
&= 0.948 < -2.05°
\end{aligned}
$$

Another closer estimate of V_2 is possible by repeating the process. Similarly,

$$V_3^1 = \left[\frac{1}{2.388 - j4.368}\right]$$

$$\times\left[\frac{-0.1 + j0.05}{1.0 + j0} - (-0.896 + j1.638)(0.948 < -2.05°)\right.$$

$$\left. - (-1.492 + j2.730)(1.05 < 0°)\right]$$

$$0.991 < -1.445°$$

Note that the newly estimated value of bus 2 voltage is used in this expression. The bus 3 voltage after another iteration is

$$V_{3_{12}} = 0.993 < -1.453°$$

For the generator bus 4 the voltage at the bus is specified as 1.05<0°. The reactive power output from the generator is estimated as

$$Q_4^0 = -I_m\left[V_{4_0}^* \left\{Y_{41}V_1 + Y_{42}V_{2_{1(2)}} + Y_{43}V_{3_{1(2)}} + Y_{44}V_{4_0}\right\}\right] \tag{5.40}$$

Substituting the numerical values:

$$Q_4^0 = -I_m[1.05 < 0^0[(-1.492 + j2.730)(0.991 < -1.453^0) + (1.492 - j2.730)(1.05 < 0^0)]]$$

This is equal to 0.125 per unit. The generator reactive power output at rated load is 0.150 per unit.

The angle δ of the voltage at bus 4 is given by

$$\angle V_4^1 = \angle\frac{1}{Y_{44}}\left[\frac{P_4 - jQ_{4_0}}{V_4^{0*}} - Y_{41}V_1 - Y_{42}V_2^{1(2)} - Y_{43}V_3^{1(2)}\right] \tag{5.41}$$

The latest values of the voltage are used in Equation 5.41. Substituting the numerical values and the estimated value of the reactive power output of the generator:

$$\angle V_4^1 = \angle\frac{1}{1.492 - j2.730}\left[\frac{0.2 - j0.125}{1.05 < 0°} - (-1.492 + j2.730)(0.993 < -1.453°)\right]$$

This is equal to < 0.54°.

V_4^1 is given as 1.05<0.54°; the value of voltage found above can be used for further iteration and estimate of reactive power.

$$Q_4^1 = -I_m[1.05 < -0.54°]\left\{\begin{matrix}(-1.492 + j2.730)(0.993 < -1.453°) \\ +(1.492 - j2.730)(1.05 < 0.54°)\end{matrix}\right\}$$

This gives 0.11 per unit reactive power. One more iteration can be carried out for accuracy. The results of the first and subsequent iterations are shown in Table 5.4.

TABLE 5.4

Solution of Example 5.4, with Gauss–Seidel Iteration and Acceleration Factor of 1.6

	Bus 2			Bus 3		Bus 4			
Iteration	**Voltage**	**Load Current**	**Voltage**	**Load Current**	**Voltage**	**Generator Current**	**Generator Active Power**	**Generator Reactive Power**	**Converged for ε =**
1	0.948 < −2.05	0.236	0.993 < 1.453	0.1126	1.05 < 0.6	0.2179	0.2	0.12	
2	0.944 < −2.803	0.2370	0.991 < −1.268	0.1128	1.05 < 0.863	0.2194	0.2	0.1144	0.12
3	0.943 < −2.717	0.2373	0.991 < −1.125	0.1129	1.05 < 0.999	0.22	0.2	0.1155	0.0025
4	0.942 < −2.647	0.2373	0.990 < −1.012	0.1129	1.050 < 1.111	0.22	0.2	0.1157	0.002
10	0.942 < −2.426	0.2373	0.990 < −0.645	0.1129	1.050 < 1.482	0.2199	0.2	0.1155	0.0006

Now suppose that for the specified voltage of 1.05 per unit, the generator reactive power requirement exceeds its maximum specified limit of 0.15, i.e., a generator rated at 0.9 power factor will have a reactive capability at full load equal to 0.109 per unit. The 1.05 per unit voltage cannot then be maintained at bus 3. The maximum reactive power is substituted and the bus voltage calculated as in a *PV* bus.

5.4 Convergence in Jacobi-Type Methods

Example 5.4 shows that in the very first iteration the calculated results are close to the final results. Most power system networks are well conditioned and converge. Generally, the convergence is better ensured when the diagonal elements of the *Y* matrix are larger than the off-diagonal elements. This may not always be the case. In systems which have a wide variation of impedances, oscillations may occur without convergence.

5.4.1 Ill-Conditioned Network

Consider two linear equations:

$$1000x_1 + 20001x_2 = 40,003$$

$$x_1 + 2x_2 = 4$$

The solution is

$$x_1 = -2, x_2 = 3$$

Let the coefficient of x_1 in the first equation change by –0.1%, i.e., the coefficient is 999. The new equations are

$$999x_1 + 20001x_2 = 40,003$$

$$x_1 + 2x_2 = 4$$

The solution is

$$x_1 = -\frac{2}{3}, x_2 = \frac{7}{3}$$

Now let the coefficient of x_1 in the first equation change by +1%, i.e., it is 10001. The new equations are

$$10001x_1 + 20001x_2 = 40,003$$

$$x_1 + 2x_2 = 4$$

The solution of these equations is

$$x_1 = 2, x_2 = 1$$

This is an example of an *ill-conditioned* network. A small perturbation gives rise to large oscillations in the results.

5.4.2 Negative Impedances

Negative impedances may be encountered while modeling certain components, i.e., duplex reactors and sometimes the three-winding transformers. *When negative elements are present, the Gauss–Seidel method of load flow will not converge.*

5.4.3 Convergence Speed and Acceleration Factor

The convergence speed of the *Y*-matrix method corresponds to the indirect solution of linear simultaneous equations:

$$\bar{A}\bar{x} = \bar{b} \tag{5.42}$$

The matrix $\bar{A}$ can be written as

$$\bar{A} = D - \bar{L}_l - \bar{U}_u \tag{5.43}$$

The indirect process solution can be expressed as

$$\begin{aligned}
x_1^{k+1} &= \left(b_1 - a_{12}x_2^k - a_{13}x^k .... - a_{1n}x_n^k\right) / a_{11} \\
x_1^{k+1} &= x_1^k + \alpha\left(x_1^{k+1} - x_1^k\right) \\
x_2^{k+1} &= \left(b_2 - a_{21}x_1^{k+1} - a_{23}x_3^k .... - a_{2n}x_n^k\right) / a_{22} \\
x_2^{k+1} &= x_2^k + \alpha\left(x_2^{k+1} - x_2^k\right) \\
&\dots
\end{aligned} \tag{5.44}$$

In matrix form:

$$\begin{aligned}
\bar{x}^{k+1} &= \bar{D}^{-1}\left[\bar{b} + \bar{L}_l\bar{x}^{k+1} + \bar{U}_u\bar{x}^k\right] \\
\bar{x}^{k+1} &= \bar{x}^k + \alpha\left(\bar{x}^{k+1} - \bar{x}^k\right)
\end{aligned} \tag{5.45}$$

From these equations:

$$\bar{x}^{k+1} = \bar{M}\,\bar{x}^k + \bar{g}$$

where

$$\bar{M} = \left(\bar{D} - \alpha \bar{L}_l\right)^{-1}\left[(1-\alpha)\bar{L}_l + \bar{U}_u\right]$$
$$\bar{g} = \alpha\left(\bar{D} - \alpha \bar{L}_l\right)^{-1}\bar{b} \tag{5.46}$$

For Gauss method, $\alpha = 1$

Equation 5.46 can be transformed as

$$\begin{aligned}\bar{x}^{k+1} &= \bar{M}\bar{x}^k + \bar{g} \\ &= \bar{M}\left(\bar{M}\bar{x}^{k-1} + \bar{g}\right) + \bar{g} \\ &= \bar{M}^2\bar{x}^{k+1} + (\bar{M}+1)\bar{g} \\ &\ldots \\ &= \bar{M}^{k+1}\bar{x}^0 + \left(\bar{M}^k + \bar{M}^{k-1} + \ldots\bar{M} + 1\right)\bar{g}\end{aligned} \tag{5.47}$$

This can be written as

If $\lambda_1, \lambda_2, \lambda_3, \ldots$ are the eigenvalues of a square matrix $\bar{M}$, and $\lambda_i \neq \lambda_j\ (I \neq j)$, and x_1, x_2, x_3, … are the eigenvectors, then the eigenvectors are linearly independent. An arbitrary vector can be expressed as a linear combination of eigenvectors:

$$\bar{x} = c_1\bar{x}_1 + c_2\bar{x}_2 + \cdots + c_n\bar{x}_n \tag{5.48}$$

As $\bar{M}\,\bar{x} = \lambda\bar{x}$, the above equation can be written as

$$\begin{aligned}M\,\bar{x} &= C_1\lambda_1\bar{x}_1 + C_2\lambda_2\bar{x}_2 + \cdots + C_n\lambda_n\bar{x}_n \\ \bar{M}^2\bar{x} &= C_1\lambda_1^2\bar{x}_1 + C_2\lambda_2^2\bar{x}_2 + \cdots + C_n\lambda_n^2\bar{x}_n \\ \bar{M}^{k+1}\bar{x} &= C_1\lambda_1^{k+1}\bar{x}_1 + C_2\lambda_2^{k+1}\bar{x}_2 + \cdots + C_n\lambda_n^{k+1}\bar{x}_n\end{aligned} \tag{5.49}$$

If

$$|\lambda_i| < 1 (i = 1, 2, 3, \ldots, n)$$

Then the first term of Equation 5.49 converges to zero as $k \to \infty$. Otherwise it diverges. The speed of convergence is dominated by the maximum value of λ_i. One method of reducing the value of λ_i is to increase α, i.e., if the Δx between iterations is not small enough, it is multiplied by a numerical factor α to increase its value; α is called the acceleration factor. The value of α generally lies between:

$$0 < \alpha < 2$$

An α value of 1.4–1.6 is common. A value of $\alpha < 1$ is called a decelerating constant.

The larger the system, the more time it takes to solve the load flow problem, depending on the value of λ.

Example 5.5

Solve the system of Example 5.3 using the Gauss–Seidel method, with an acceleration factor of 1.6.

$$V_1^{(k+1)'} = 1 + \frac{1}{3}V_2^k$$

$$V_1^{(k+1)} = V_1^k + 1.6\left(V_1^{(k+1)'} - V_1^k\right)$$

$$V_2^{(k+1)'} = 1 + \frac{1}{4}V_1^{k+1} + \frac{3}{4}V_3^k$$

$$V_2^{(k+1)} = V_2^k + 1.6\left(V_2^{(k+1)'} - V_2^k\right)$$

$$V_3^{(k+1)'} = \frac{1}{3} + \frac{1}{2}V_2^{k+1}$$

$$V_3^{(k+1)} = V_3^k + 1.6\left(V_3^{(k+1)'} - V_3^k\right)$$

Using the above equations the values of voltages for $k=1$ are calculated as follows:

$$V_1^{1'} = 1 + \frac{1}{3}(1) = 1.333$$

$$V_1^1 = 1 + 1.6(1.333 - 1) = 1.6(0.333) = 1.538$$

$$V_2^{2'} = 1 + \frac{1}{4}(1.538) + \frac{3}{4}(1) = 2.1345$$

$$V_2^2 = 1 + 1.6(2.1345 - 1) = 2.8152$$

$$V_3^{3'} = \frac{1}{3} + \frac{1}{2}(2.8152) = 1.7406$$

$$V_3^3 = 1 + 1.6(1.7406 - 1) = 2.1849$$

The results of the first five iterations are shown in Table 5.5. The mismatch between bus voltages progressively reduces.

Example 5.6

Repeat the calculations of Example 5.4, with an acceleration factor of 1.6.

The voltage at bus 2 is calculated in the first two iterations as 0.948 < –2.05°, (Example 5.4). Using an acceleration factor of 1.6:

$$V_2^1 = V_2^0 + 1.6\left(V_2^{1'} - V_2^0\right)$$

$$= 1 < 0° + 1.6(0.948 < -2.05° - 1 < 0°)$$

$$= 0.916 - j0.054$$

TABLE 5.5

Solution of Example 5.5, with Gauss–Seidel Iteration and Acceleration Factor of 1.6

Voltage	$k=0$	$k=1$	$k=2$	$k=3$	$k=4$	$k=5$
$V_1^{k'}$	1	1.333	1.938	2.135	1.912	1.839
V_1^{k}	1	1.538	2.178	2.109	1.794	1.866
$V_2^{k'}$		2.135	3.183	2.986	2.598	2.686
V_2^{k}	1	2.815	3.404	2.735	2.516	2.788
$V_3^{k'}$		1.741	2.035	1.700	1.591	1.727
V_3^{k}		2.185	1.945	1.533	1.626	1.788

Use this voltage to calculate the initial voltage on bus 3:

$$V_3^{1'} = \frac{1}{2,388 - j4.368}\left[\begin{array}{l}\dfrac{-0.1+j0.05}{1.0+j0} - (-0.896+j1.638)(0.916-j0.054) \\ -(-1.493+j2.730)(1.05<0°)\end{array}\right]$$

$$= 0.979 - j0.032$$

Calculate the bus 3 voltage with the acceleration factor:

$$V_3^{1} = V_3^{0} + 1.6\left(V_3^{1'} - V_3^{0}\right)$$

$$= 1<0° + 1.6(0.979 - j0.032 - 1 + j0)$$

$$= 0.966 - j0.651$$

For bus 4 calculate the reactive power, as in Example 5.4, using the voltages found above. The results are shown in Table 5.6, which shows that there is initial oscillation in the calculated results, which are much higher than the results obtained in Table 5.4. However, at the 10th iteration, the convergence is improved compared to the results in Table 5.4. Selection of an unsuitable acceleration factor can result in a larger number of iterations and oscillations and a lack of convergence.

5.5 Gauss–Seidel Z-Matrix Method

From Volume 1, we know how a bus impedance matrix can be formed. We will revisit bus impedance matrix from load flow considerations. The voltage at the swing bus is defined and need not be included in the matrix equations. The remaining voltages are

$$\begin{vmatrix} V_2 \\ V_3 \\ \cdot \\ V_n \end{vmatrix} = \begin{vmatrix} Z_{22} & Z_{23} & \cdots & Z_{2n} \\ Z_{32} & Z_{33} & \cdots & Z_{3n} \\ \cdot & \cdot & \cdot & \cdot \\ Z_{n2} & Z_{n3} & \cdots & Z_{nn} \end{vmatrix}\begin{vmatrix} I_2 \\ I_3 \\ \cdot \\ I_n \end{vmatrix} \tag{5.50}$$

TABLE 5.6

Solution of Example 5.6, with an Acceleration Factor of 1.6

	Bus 2		Bus 3		Bus 4				
Iteration	Voltage	Load Current	Voltage	Load Current	Voltage	Generator Current	Generator Reactive Power	Generator Active Power	Convergence
1	0.917 < −3.403	0.2439	0.970 < −3.105	0.1153	1.023	0.23	0.124	0.2	−0.034
2	0.939 < −3.814	0.2380	0.974 < −3.134	0.1148	1.045	0.2252	0.124	0.2	0.0094
3	0.929 < −3.587	0.2406	0.988 < −2.049	0.1132	1.053	0.2236	0.124	0.2	0.01295
4	0.947 < −2.836	0.2360	0.998 < −1.083	0.1103	1.050	0.2100	0.0927	0.2	0.0140
10	0.943 < −.326	0.2372	0.991 < −0.485	0.1129	1.050<1.650	0.2193	0.1142	0.2	−0.00053

Therefore, we can write

$$V_k = Z_{k2}I_2 + Z_{k3}I_3 + \cdots + Z_{kn}I_n \tag{5.51}$$

The current at the kth branch can be written as

$$I_k = \frac{P_k - jQ_k}{V_k^*} - y_k V_k \tag{5.52}$$

where y_k is the admittance of bus k to a common reference bus. If the kth branch is a generator which supplies real and reactive power to the bus, P and Q are entered as positive values. If the kth branch draws power from the network, then P and Q are entered as negative values.

The general procedure is very similar to the Y admittance method:

$$\begin{aligned} V_k &= Z_{kl}I_1 + \cdots + Z_{kn}I_n \\ &= Z_{kl}\left[\frac{P_1 - jQ_1}{V_1^*} - y_1V_1\right] + \cdots + Z_{kn}\left[\frac{P_n - jQ_n}{V_n^*} - y_nV_n\right] \\ V_k &= \sum_{i=1}^{i-n} Z_{ki}\left[\frac{P_i - jQ_i}{V_i^*} - y_iV_i\right] \end{aligned} \tag{5.53}$$

The iteration process is

1. Assume initial voltages for the n buses:

$$V_1^0, V_2^0, \ldots, V_n^0 \tag{5.54}$$

2. Calculate V_1 from Equation 5.53 in terms of the initial assumed voltages and substitute back into the same equation for a new corrected value:

$$V_1^1 = \sum_{i=1}^{n} Z_{ki}\left(\frac{Pi - jQ_i}{V_i^{0*}} - yiV_i^0\right) \tag{5.55}$$

3. Repeat for bus 2, making use of corrected values of V_1 found in step 2. Iterate to find a corrected value of V_2 before proceeding to the next bus.
4. When all the bus voltages have been evaluated start all over again for the required convergence.

At a generator bus the reactive power is not known and an estimate of reactive power is necessary. This can be made from

$$Q_k = -I_m \frac{V_k^*}{Z_{kk}}\left[V_k - \sum_{i=1, i\neq k}^{i-n} Z_{ki}\frac{P_i - jQ_i}{V_i^*}\right] \tag{5.56}$$

A Z impedance matrix has strong convergence characteristics, at the expense of larger memory requirements and preliminary calculations. Modifications due to system changes are also comparatively difficult.

5.6 Conversion of *Y* to *Z* Matrix

A number of techniques for formation of a bus impedance matrix are outlined in Volume 1. We will examine one more method, by which a bus impedance matrix can be constructed from an admittance matrix by a step-by-step pivotal operation. The following equations are supported:

$$\text{new } Y_{kk} = \frac{1}{Y_{kk}} (Y_{kk} = \text{pivot}) \tag{5.57}$$

$$\text{new } Y_{kj} = \frac{Y_{kj}}{Y_{kk}} \quad j = 1, \ldots n(j \neq k) \tag{5.58}$$

$$\text{new } Y_{ik} = -\frac{Y_{ik}}{Y_{kk}} \quad i = 1, \ldots, n(j \neq k) \tag{5.59}$$

$$\text{new } Y_{ij} = Y_{ij} - \left[\frac{Y_{ik}Y_{kj}}{Y_{kk}}\right] \quad (i \neq k, j \neq k) \tag{5.60}$$

The choice of a pivot is arbitrary. The new nonzero elements can be avoided by a proper choice (Volume 1 Appendix B). The procedure can be illustrated by an example.

Example 5.7

Consider a hypothetical four-bus system, with the admittances as shown in Figure 5.9. Form a *Y* matrix and transform to a *Z* matrix by pivotal manipulation. Check the results with the step-by-step buildup method of the bus impedance matrix described in Volume 1.

The *Y*-bus matrix is easily formed by examination of the network:

$$\bar{Y} = \begin{vmatrix} 7 & -2 & 0 & -3 \\ -2 & 5 & -3 & 0 \\ 0 & -3 & 4 & 0 \\ -3 & 0 & 0 & 6 \end{vmatrix}$$

Pivot (3, 3):

$$Y_{31} = \frac{Y_{31}}{Y_{33}} = 0$$

$$Y_{32} = \frac{Y_{32}}{Y_{33}} = -3/4 = -0.75$$

$$Y_{33} = \frac{1}{Y_{33}} = 0.25$$

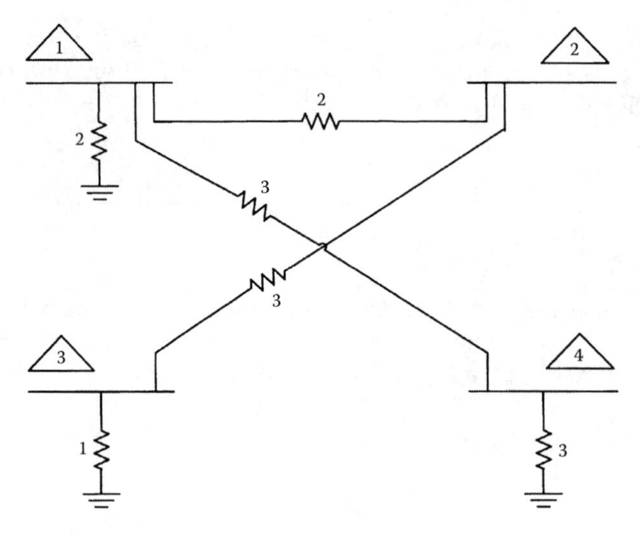

FIGURE 5.9
Four-bus network for Example 5.7.

$$Y_{34} = \frac{Y_{34}}{Y_{33}} = 0$$

$$Y_{13} = -\frac{Y_{13}}{Y_{33}} = 0$$

$$Y_{23} = -\frac{Y_{23}}{Y_{33}} = -(-3/4) = 0.75$$

$$Y_{43} = -\frac{Y_{43}}{Y_{33}} = 0$$

$$Y_{22} = Y_{22} - \frac{Y_{23}Y_{32}}{Y_{33}} = 5 - (-3)(-3)/4 = 2.75$$

etc.

The transformed matrix is

$$\begin{vmatrix} 7 & -2 & 0 & -3 \\ -2 & 2.75 & 0.75 & 0 \\ 0 & -0.75 & 0.25 & 0 \\ -3 & 0 & 0 & 6 \end{vmatrix}$$

Next, use the pivot (4, 4). The transformed matrix is

$$\begin{vmatrix} 5.5 & -2 & 0 & 0.5 \\ -2 & 2.75 & 0.75 & 0 \\ 0 & -0.75 & 0.25 & 0 \\ -0.5 & 0 & 0 & 0.167 \end{vmatrix}$$

Next, use the pivot (2, 2):

$$\begin{vmatrix} 4.045 & 0.727 & 0.545 & 0.5 \\ -0.727 & 0.364 & 0.273 & 0 \\ 0.545 & 0.273 & 0.455 & 0 \\ -0.5 & 0 & 0 & 0.167 \end{vmatrix}$$

and finally the pivot (1, 1) gives the Z matrix:

$$\bar{Z} = \begin{vmatrix} 0.2470 & 0.1797 & 0.1350 & 0.1236 \\ 0.1797 & 0.4950 & 0.3709 & 0.0898 \\ 0.1350 & 0.3709 & 0.528 & 0.0674 \\ 0.1236 & 0.0898 & 0.0674 & 0.2289 \end{vmatrix}$$

Step-by Step Buildup
Volume 1 showed a step-by-step buildup method for a Z matrix.
Node 1 to ground and adding branch 1–2 and 2–3 gives the Z matrix as

$$\begin{vmatrix} 0.5 & 0.5 & 0.5 \\ 0.5 & 1 & 1 \\ 0.5 & 1 & 1.33 \end{vmatrix}$$

Add the 1-Ω branch between node 3 and ground. The new Z matrix is

$$\begin{vmatrix} 0.5 & 0.5 & 0.5 \\ 0.5 & 1 & 1 \\ 0.5 & 1 & 1.33 \end{vmatrix} - \frac{1}{2.33} \begin{vmatrix} 0.5 \\ 1 \\ 1.33 \end{vmatrix} \begin{vmatrix} 0.5 & 1 & 1.33 \end{vmatrix} = \begin{vmatrix} 0.393 & 0.285 & 0.215 \\ 0.285 & 0.571 & 0.429 \\ 0.215 & 0.429 & 0.571 \end{vmatrix}$$

Add branch 1–4:

$$\bar{Z} = \begin{vmatrix} 0.393 & 0.285 & 0.215 & 0.393 \\ 0.285 & 0.571 & 0.429 & 0.285 \\ 0.215 & 0.429 & 0.571 & 0.215 \\ 0.393 & 0.285 & 0.215 & 0.726 \end{vmatrix}$$

Finally, add the 0.333-Ω branch between node 4 and ground. The final Z matrix is

$$\begin{vmatrix} 0.393 & 0.285 & 0.215 & 0.393 \\ 0.285 & 0.571 & 0.429 & 0.285 \\ 0.215 & 0.429 & 0.571 & 0.215 \\ 0.393 & 0.285 & 0.215 & 0.726 \end{vmatrix} - \frac{1}{1.059} \begin{vmatrix} 0.393 \\ 0.285 \\ 0.215 \\ 0.726 \end{vmatrix}$$

$$\begin{vmatrix} 0.393 & 0.285 & 0.215 & 0.726 \end{vmatrix} = \begin{vmatrix} 0.247 & 0.1797 & 0.135 & 0.1236 \\ 0.1797 & 0.495 & 0.3709 & 0.0898 \\ 0.1350 & 0.3709 & 0.528 & 0.0674 \\ 0.1236 & 0.0898 & 0.0674 & 0.2289 \end{vmatrix}$$

This is the same result as arrived at by pivotal operation.

Example 5.8

Solve the network of Example 5.2 by the Z-matrix method.

The Y matrix of Example 5.2 is

$$\bar{Y} = \begin{vmatrix} 3 & -1 & 0 \\ -1 & 4 & -3 \\ 0 & -3 & 6 \end{vmatrix}$$

Form the Z matrix:

$$\bar{Z} = \begin{vmatrix} 0.385 & 0.154 & 0.077 \\ 0.154 & 0.462 & 0.231 \\ 0.071 & 0.231 & 0.282 \end{vmatrix}$$

Then:

$$\begin{vmatrix} v_1 \\ v_2 \\ v_3 \end{vmatrix} = \bar{Z} \begin{vmatrix} 3 \\ 4 \\ 2 \end{vmatrix} = \begin{vmatrix} 1.923 \\ 2.769 \\ 1.718 \end{vmatrix}$$

No more iterations are required, as the currents are given. These values exactly satisfy the original equations. Compare these results obtained after five iterations in Tables 5.1, 5.2, and 5.5, which are still approximate and require more iterations for better accuracy.

Example 5.9

Solve Example 5.4 with a Z bus matrix of load flow.

Using the same initial voltage estimate on the buses as in Example 5.4:

$$\begin{aligned} I_2^0 &= \left[P_2 + jQ_2 / V_2^0 \right]^* - Y_{21}V_1 \\ &= \left[(-0.2 - j0.1) / 1.0 \right]^* - (-0.896 + j1.638) \times 1 < 0^\circ \\ &= 0.696 - j1.538 \end{aligned}$$

The equation for the swing bus need not be written.

$$\begin{aligned} I_3^0 &= \left[\frac{P_3 - jQ_3}{V_3^0} \right]^* - Y_{31}V_1 \\ &= \left[\frac{-0.1 - j0.05}{1 < 0^\circ} \right]^* - (0)(1 < 0^\circ) = -0.1 + j0.05 \end{aligned}$$

The reactive power at bus 4 is

$$Q_4^0 = \text{Im}\left[V_4^0\left(Y_{41}V_1^0 + Y_{42}V_2^0 + Y_{43}V_3^0 + Y_{44}V_4^0\right)^*\right]$$

$$I_m\left[1.05 < 0°\left[(-1.492 + j2.730)(1 < 0°) + (1.492 - j2.730)(1.05 < 0°)\right]^*\right] = j0.144$$

This is within the generator rated reactive power output.

$$I_4^0 = \left[(P_4 + jQ_4 / V_4^0)\right]^* - Y_{41}V_1$$
$$= 0.19 - j0.137$$

Thus, the voltages are given by

$$\begin{vmatrix} V_2 \\ V_3 \\ V_4 \end{vmatrix} = \bar{Z} \begin{vmatrix} I_2 \\ I_3 \\ I_4 \end{vmatrix}$$

where the Z matrix is formed from the Y matrix of Example 5.4. The equation for the swing bus is omitted.

$$\bar{Z} = \begin{vmatrix} 0.257 + j0.470 & 0.257 + j0.470 & 0.257 + j0.470 \\ 0.257 + j0.470 & 0.514 + j0.940 & 0.514 + j0.940 \\ 0.257 + j0.470 & 0.514 + j0.940 & 0.668 + j1.222 \end{vmatrix}$$

Therefore;

$$\begin{vmatrix} V_2 \\ V_3 \\ V_4 \end{vmatrix} = \bar{Z} \begin{vmatrix} 0.696 - j1.538 \\ -0.1 + j0.05 \\ 0.19 - j0.137 \end{vmatrix} = \begin{vmatrix} 0.966 - j0.048 \\ 1.03 - j0.028 \\ 1.098 + j0.004 \end{vmatrix} = \begin{vmatrix} 0.983 < -2.84° \\ 1.030 < -1.56° \\ 1.098 < 0.21° \end{vmatrix}$$

The calculations are repeated with the new values of voltages. Note that V_4 is higher than the specified limits in the first iteration. Since V_4 is specified as 1.05, in the next iteration we will use $V_4 = 1.05 < 0.21°$ and find the reactive power of the generator.

A comparison with the Y-matrix method is shown in Table 5.7.

5.7 Triangular Factorization Method of Load Flow

A commercial program of load flow used triangular factorization. A matrix can be factored into $\bar{L}$ and $\bar{U}$, Appendix A. Here we factor $\bar{Y}$ bus matrix in to $\bar{L}$ and $\bar{U}$ form:

$$\bar{L}\bar{U} = \bar{Y}_{\text{bus}} \tag{5.61}$$

TABLE 5.7

Comparison of Y- and Z-Matrix Methods for Load Flow

No.	Compared Parameter	Y Matrix	Z Matrix	Remarks
1	Digital computer memory requirements	Small	Large	Sparse matrix techniques easily applied to Y matrix
2	Preliminary calculations	Small	Large	Software programs can basically operate from the same data input
3	Convergence characteristics	Slow, may not converge at all	Strong	Both methods may slow down on large systems
4	System modifications	Easy	Slightly difficult	See text

Then

$$\bar{L}\bar{U}\bar{V} = \bar{I} \tag{5.62}$$

As an intermediate step define a new voltage $\bar{V}'$, so that

$$\begin{aligned} \bar{L}\bar{V}' &= \bar{I} \\ \bar{U}\bar{V} &= \bar{V}' \end{aligned} \tag{5.63}$$

The lower triangular system is first solved for the vector $\bar{V}'$ by forward substitution. Then $\bar{V}$ is found by backward substitution. When $\bar{Y}$ is symmetric only $\bar{L}$ is needed as it can be used to produce the back substitution as well as the forward substitution. When the sparsity is exploited by optimal ordering (Appendix B in vol. 1), and the computer program processes and stores only the non-zero terms, the triangular factorization can give efficient direct solution.

Example 5.10

To illustrate the principle, consider a bus admittance matrix:

$$\bar{Y}_B = \begin{vmatrix} -j & 2j & j \\ 2j & -j3 & j3 \\ j & j3 & -4j \end{vmatrix}$$

The current injection vector is

$$\begin{vmatrix} j1.75 \\ j2.0 \\ 0 \end{vmatrix}$$

Then the voltages at each of the three nodes can be calculated:

$$\begin{vmatrix} V_1 \\ V_2 \\ V_3 \end{vmatrix} = \bar{Y}_B^{-1}\bar{I} = \begin{vmatrix} 0.973 \\ 0.902 \\ 0.92 \end{vmatrix} \tag{5.64}$$

Now calculate matrices $\bar{L}$ and $\bar{U}$ based upon the techniques discussed in Appendix A. These matrices are

$$\bar{L} = \begin{vmatrix} -j & 0 & 0 \\ 2j & j & 0 \\ j & j5 & -j28 \end{vmatrix}$$

$$\bar{U} = \begin{vmatrix} 1 & -2 & -1 \\ 0 & 1 & 5 \\ 0 & 0 & 1 \end{vmatrix}$$

The product can be verified to give the original bus admittance matrix. Then we can write

$$\begin{vmatrix} -j & 0 & 0 \\ j2 & j & 0 \\ j & j5 & -j28 \end{vmatrix} \begin{vmatrix} V_1' \\ V_2' \\ V_3' \end{vmatrix} = \begin{vmatrix} j1.75 \\ j2.0 \\ 0 \end{vmatrix}$$

This gives

$$\begin{vmatrix} V_1' \\ V_2' \\ V_3' \end{vmatrix} = \begin{vmatrix} -1.75 \\ 5.5 \\ 0.92 \end{vmatrix}$$

Then

$$\begin{vmatrix} 1 & -2 & -1 \\ 0 & 1 & 5 \\ 0 & 0 & 1 \end{vmatrix} \begin{vmatrix} V_1 \\ V_2 \\ V_3 \end{vmatrix} = \begin{vmatrix} -1.75 \\ 5.5 \\ 0.92 \end{vmatrix}$$

or

$$\begin{vmatrix} V_1 \\ V_2 \\ V_3 \end{vmatrix} = \begin{vmatrix} 0.973 \\ 0.902 \\ 0.92 \end{vmatrix}$$

This is the same as arrived in Equation 5.64. The calculated voltages V_1, V_2, V_3 are much below rated voltages. The example is for illustrative purpose only; the voltages at the buses are maintained close to the rated voltages. In a practical distribution system it is desirable to operate with a voltage slightly higher than the rated voltages.

Note that the bus admittance matrix in the above example is symmetrical. Referring to Appendix A, for a symmetric matrix the calculation can be further simplified. The matrix $\bar{L}$ can be written as

$$\bar{L} = \bar{U}^t\bar{D} = \begin{vmatrix} 1 & 0 & 0 & 0 \\ \frac{Y_{21}}{Y_{11}} & 1 & 0 & 0 \\ \frac{Y_{31}}{Y_{11}} & \frac{Y_{32}}{Y_{22}} & 1 & 0 \\ \frac{Y_{41}}{Y_{11}} & \frac{Y_{42}}{Y_{22}} & \frac{Y_{43}}{Y_{33}} & 1 \end{vmatrix} \begin{vmatrix} Y_{11} & . & . & . \\ . & Y_{22} & . & . \\ . & . & Y_{33} & . \\ . & . & . & Y_{44} \end{vmatrix} \tag{5.65}$$

Then

$$\bar{Y}_B = \bar{U}^t\bar{D}\bar{U} \tag{5.66}$$

The Equation 5.66 can be solved by denoting:

$$\begin{aligned} \bar{U}^t\bar{V}'' &= \bar{I} \\ \bar{D}\bar{V} &= \bar{V}'' \\ \bar{U}\bar{V} &= \bar{V}' \end{aligned} \tag{5.67}$$

Example 5.11

Solve example using Equation 5.65.
From Example 5.10:

$$\bar{U}^t\bar{V}'' = \bar{I}$$

or

$$\begin{vmatrix} 1 & 0 & 0 \\ -2 & 1 & 0 \\ -1 & 5 & 1 \end{vmatrix} \begin{vmatrix} V_1'' \\ V_2'' \\ V_3'' \end{vmatrix} = \begin{vmatrix} j1.75 \\ j2.0 \\ 0 \end{vmatrix}$$

This gives

$$\begin{vmatrix} V_1'' \\ V_2'' \\ V_3'' \end{vmatrix} = \begin{vmatrix} j1.75 \\ j5.5 \\ -j25.75 \end{vmatrix}$$

Also

$$\bar{D}\bar{V}' = \bar{V}''$$

or

$$\begin{vmatrix} -j & 0 & 0 \\ 0 & j & 0 \\ 0 & 0 & -28j \end{vmatrix} \begin{vmatrix} V_1' \\ V_2' \\ V_3' \end{vmatrix} = \begin{vmatrix} V_1'' \\ V_2'' \\ V_3'' \end{vmatrix} = \begin{vmatrix} j1.75 \\ j5.5 \\ -j25.75 \end{vmatrix}$$

From which:

$$\begin{vmatrix} V_1' \\ V_2' \\ V_3' \end{vmatrix} = \begin{vmatrix} -1.75 \\ 5.5 \\ 0.92 \end{vmatrix}$$

This is the same result as of Example 5.10.
Finally

$$\bar{U}\bar{V} = \bar{V}'$$

As in Example 5.10. Thus, it is a three step calculation.

Problems

(The problems can be solved without modeling the systems on a digital computer.)

5.1 In Example 5.1 consider that the line between buses 2 and 3 is removed. Form the Y-impedance matrix. Modify this matrix by reconnecting the removed line.

5.2 Figure P5.1 shows a distribution system with data as shown. Convert all impedances to per unit base and form a Y matrix.

5.3 Calculate the currents and voltages in the system of Figure P5.1 by (1) Gauss iterative method, (2) Gauss–Seidel iterative method, and (3) Gauss–Seidel iterative method with an acceleration factor of 1.6. Calculate to first two iterations in each case.

5.4 Convert the Y matrix of Problem 2 into a Z matrix by (1) pivotal manipulation, and (2) step-by-step buildup.

5.5 Calculate currents and voltages by the Z-matrix method in Problem 5.3.

5.6 Figure P5.2 is a modification of the circuit in Example 5.5, with a tap adjusting transformer. Calculate load flow with the transformer at rated voltage tap and at 1:1.1 tap. How does the reactive power flow change with the transformer tap adjustment? Calculate to first iteration.

5.7 Solve for voltages in examples 5.8 and 5.9 using triangular factorization method and compare the results.

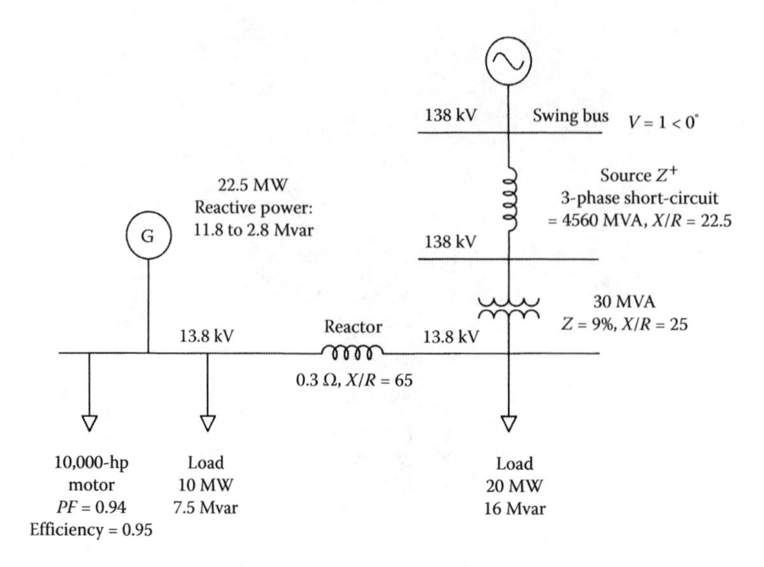

FIGURE P5.1
Network with impedance and generation data for load-flow Problems 2.5.

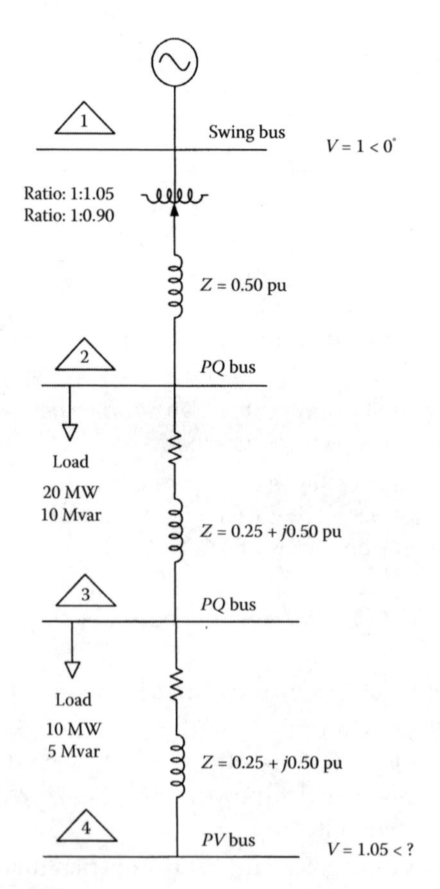

FIGURE P5.2
Four-bus system with ratio-adjusting transformer for Problem 6.

Bibliography

1. RB Shipley. *Introduction to Matrices and Power Systems,* John Wiley, New York, 1976.
2. J Arrillaga, JCP Arnold, BJ Harker. *Computer Modeling of Electrical Power Systems,* John Wiley, New York, 1983.
3. GW Stagg, AH El-Abiad. *Computer Methods in Power System Analysis,* McGraw-Hill, New York, 1968.
4. FJ Hubert, DR Hayes. A rapid digital computer solution for power system network load flow. *IEEE Trans Power App Syst,* 90, 934–940, 1971.
5. HE Brown, GK Carter, HH Happ, CE Person. Power flow solution by impedance iterative methods. *IEEE Trans Power App Syst,* 2, 1–10, 1963.
6. HE Brown. *Solution of Large Networks by Matrix Methods,* John Wiley, New York, 1972.
7. MA Laughton. Decomposition techniques in power systems network load flow analysis using the nodal impedance matrix. *Proc Inst Electri Eng,* 115(4), 539–542, 1968.
8. IEEE. Power System Analysis. Standard 399–1990.
9. AR Bergen, V. Vittal. *Power System Analysis,* 2nd ed., Prentice Hall, Upper Saddle River, NJ, 1999.
10. JJ Granger, WD Stevenson. *Power System Analysis,* McGraw-Hill, New York, 1994.
11. L Powell. *Power System Load Flow Analysis,* McGraw-Hill, New York, 2004.
12. Xi-Fan Wang, Y Song, M Irving. *Modern Power System Analysis.* Springer, New York, 2008.
13. WF Tinny, JW Walker. Direct solution of sparse network equations by optimally ordered triangular factorization. *Proc IEEE,* 55, 1801–1809, 1967.

6

Load Flow Methods: Part II

The Newton–Raphson (NR) method has powerful convergence characteristics, though computational and storage requirements are heavy. The sparsity techniques and ordered elimination (discussed in Volume 1, Appendix B) led to its earlier acceptability, and it continues to be a powerful load flow algorithm even in today's environment for large systems and optimization [1]. A lesser number of iterations are required for convergence, as compared to the Gauss–Seidel method, provided that the initial estimate is not far removed from the final results, and these do not increase with the size of the system [2]. The starting values can even be first estimated using a couple of iterations with the Gauss–Seidel method for load flow and the results input into the NR method as a starting estimate. The modified forms of the NR method provide even faster algorithms. Decoupled load flow and fast decoupled solution methods are offshoots of the NR method.

6.1 Functions with One Variable

Any function of x can be written as the summation of a power series, and Taylor's series of a function $f(x)$ is

$$y = f(x) = f(a) + f'(a)(x-a) + \frac{f''(a)}{2!}(x-a)^2 + \cdots + \frac{f^n(a)(x-a)^n}{n!} \tag{6.1}$$

where $f'(a)$ is the first derivative of $f(a)$. Neglecting the higher terms and considering only the first two terms, the series is

$$y = f(x) \approx f(a) + f'(a)(x-a) \tag{6.2}$$

The series converges rapidly for values of x near to a. If x_0 is the initial estimate, then the tangent line at $(x_0, f(x_0))$ is

$$y = f(x_0) + f'(x_0)(x_1 - x_0) \tag{6.3}$$

where x_1 is the new value of x, which is a closer estimate. This curve crosses the x-axis (Figure 6.1) at the new value x_1. Thus,

$$0 = f(x_0) + f'(x_0)(x_1 - x_0)$$

$$x_1 = x_0 - \frac{f(x_0)}{f'(x_0)} \tag{6.4}$$

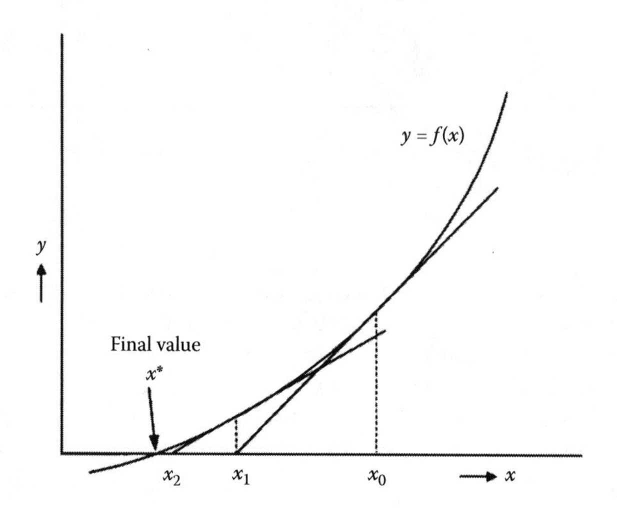

FIGURE 6.1
Progressive tangents to a function $y=f(x)$ at x_0, x_1, and x_2, showing the crossings with the X axis, each crossing approaching closer to the final value x^*.

In general

$$x_{k+1} = x_k - \frac{f(x_k)}{f'(x_k)} \tag{6.5}$$

Example 6.1

Consider a function

$$f(x) = x^3 - 2x^2 + 3x - 5 = 0$$

Find the value of x.
The derivative is

$$f'(x) = 3x^2 - 4x + 3$$

Let the initial value of $x=3$; then $f(x)=13$ and $f'(x)=18$, for $k=1$. From Equation 6.4, $x_1=3-(13/18)=2.278$. Table 6.1 is compiled to $k=4$ and gives $x=1.843$. As a verification, substituting this value into the original equation, the identity is approximately satisfied.

TABLE 6.1
Iterative Solution of a Function of One Variable (Example 6.1)

K	x_k	$f(x_k)$	$f'(x_k)$
0	3	13	18
1	2.278	3.277	9.456
2	1.931	0.536	6.462
3	1.848	0.025	5.853
4	1.844	≈0	

6.2 Simultaneous Equations

The Taylor series, Equation 6.1, is applied to n nonlinear equations in n unknowns, x_1, x_2, …, x_n:

$$\begin{aligned} y_1 &\cong f_1(x_1,x_2,\ldots,x_n)+\Delta x_1\frac{\partial f_1}{\partial x_1}+\Delta x_2\frac{\partial f_1}{\partial x_2}+\cdots+\Delta x_n\frac{\partial f_1}{\partial x_n} \\ &\ldots \\ y_n &\cong f_n(x_1,x_2,\ldots,x_n)+\Delta x_1\frac{\partial f_n}{\partial x_1}+\Delta x_2\frac{\partial f_n}{\partial x_2}+\cdots+\Delta x_n\frac{\partial f_n}{\partial x_n} \end{aligned} \quad (6.6)$$

As a first approximation, the unknowns represented by the initial values $x_1^0, x_2^0, x_3^0, \ldots$ can be substituted into the above equations, i.e.,

$$y_n = f_n(x_1^0,x_2^0,\ldots,x_n^0)+\Delta x_1^0\left.\frac{\partial f_n}{\partial x_1}\right|_0+\cdots+\Delta x_n^0\left.\frac{\partial f_n}{\partial x_n}\right|_0 \quad (6.7)$$

where $x_1^0, x_2^0, x_3^0, \ldots, x_n^0$ are the first estimates of n unknowns. On transposing

$$\begin{aligned} y_1 - f_1^0 &= \left.\frac{\partial f_1}{\partial x_1}\right|_0 \Delta x_1^0 + \left.\frac{\partial f_1}{\partial x_2}\right|_0 \Delta x_2^0 + \cdots + \left.\frac{\partial f_1}{\partial x_n}\right|_0 \Delta x_n^0 \\ &\ldots \\ y_1 - f_1^0 &= \left.\frac{\partial f_1}{\partial x_1}\right|_0 \Delta x_1^0 + \left.\frac{\partial f_n}{\partial x_2}\right|_0 \Delta x_2^0 + \cdots + \left.\frac{\partial f_n}{\partial x_{n(0)}}\right|_0 \Delta x_n^0 \end{aligned} \quad (6.8)$$

where

$$f_n(x_1^0, x_2^0, \ldots, x_n^0)$$

is abbreviated as f_n^0.

The original nonlinear equations have been reduced to linear equations in

$$\Delta x_1^0, \Delta x_2^0, \ldots, \Delta x_n^0 \quad (6.9)$$

The subsequent approximations are

$$\begin{aligned} x_1^1 &= x_1^0 + \Delta x_1^0 \\ x_2^1 &= x_2^0 + \Delta x_2^0 \\ &\ldots \\ x_n^1 &= x_n^0 + \Delta x_n^0 \end{aligned} \quad (6.10)$$

or in matrix form

$$\begin{vmatrix} y_1 - f_1^0 \\ y_2 - f_2^0 \\ \cdots \\ y_n - f_n^0 \end{vmatrix} = \begin{vmatrix} \frac{\partial f_1}{\partial x_1} & \cdots & \frac{\partial f_1}{\partial x_n} \\ \frac{\partial f_2}{\partial x_1} & \cdots & \frac{\partial f_2}{\partial x_n} \\ \cdots & \cdots & \cdots \\ \frac{\partial f_n}{\partial x_1} & \cdots & \frac{\partial f_n}{\partial x_n} \end{vmatrix} \begin{vmatrix} \Delta x_1^0 \\ \Delta x_2^0 \\ \cdots \\ \Delta x_n^0 \end{vmatrix} \tag{6.11}$$

The matrix of partial derivatives is called a Jacobian matrix. This result is written as

$$x^{k+1} = x^k - \left[J(x^k)\right]^{-1} f(x^k) \tag{6.12}$$

This means that determination of unknowns requires inversion of the Jacobian.

Example 6.2

Solve the two simultaneous equations

$$f_1(x_1, x_2) = x_1^2 + 2x_2 - 3 = 0$$

$$f(x_1, x_2) = x_1 x_2 - 3x_2^2 + 2 = 0$$

Let the initial values of x_1 and x_2 be 3 and 2, respectively:

$$\bar{J}(x) \stackrel{\bullet}{=} \begin{vmatrix} \frac{\partial f_1}{\partial x_1} & \frac{\partial f_1}{\partial x_2} \\ \frac{\partial f_2}{\partial x_1} & \frac{\partial f_2}{\partial x_2} \end{vmatrix} = \begin{vmatrix} 2x_1 & 2 \\ x_2 & x_1 - 6x_2 \end{vmatrix}$$

Step 0

$$x^1 = x^0 - \begin{vmatrix} 6 & 2 \\ 2 & -9 \end{vmatrix}^{-1} f(x^0)$$

$$= \begin{vmatrix} 3 \\ 2 \end{vmatrix} - \begin{vmatrix} \frac{9}{58} & \frac{2}{58} \\ \frac{2}{58} & -\frac{6}{58} \end{vmatrix} \begin{vmatrix} 10 \\ 4 \end{vmatrix} = \begin{vmatrix} 1.586 \\ 1.241 \end{vmatrix}$$

Step 1

$$x^2 = x^1 - \begin{vmatrix} 3.172 & 2 \\ 1.241 & -5.86 \end{vmatrix}^{-1} f(x^1)$$

$$= \begin{vmatrix} 1.586 \\ 1.241 \end{vmatrix} - \begin{vmatrix} 0.278 & 0.095 \\ 0.095 & -0.151 \end{vmatrix} \begin{vmatrix} 1.997 \\ -0.652 \end{vmatrix} = \begin{vmatrix} 1.092 \\ 1.024 \end{vmatrix}$$

Step 2

$$x^3 = x^2 - \begin{vmatrix} 2.184 & 2 \\ 1.025 & -5.058 \end{vmatrix}^{-1} f(x^2)$$

$$= \begin{vmatrix} 1.092 \\ 1.025 \end{vmatrix} - \begin{vmatrix} 0.386 & 0.153 \\ 0.078 & -0.167 \end{vmatrix} \begin{vmatrix} 0.242 \\ -0.033 \end{vmatrix} = \begin{vmatrix} 1.004 \\ 1.001 \end{vmatrix}$$

Thus, the results are rapidly converging to the final values of x_1 and x_2 which are 1 and 1, respectively.

6.3 Rectangular Form of the NR Method of Load Flow

The power flow equation at a *PQ* node is

$$S_s^0 = P_s^0 + jQ_s^0 = V_s \overline{\sum_{r=1}^{r=n} (Y_{sr} V_r)}^{*} \tag{6.13}$$

Voltages can be written as

$$\begin{aligned} V_s &= e_s + jh_s \\ V_r &= e_r - jh_r \end{aligned} \tag{6.14}$$

Thus, the power is

$$\begin{aligned} &(e_s + jh_s)\left[\sum_{r=1}^{r=n} (G_{sr} - jB_{sr})(e_r - jh_r)\right] \\ &= (e_s + jh_s)\left[\sum_{r=1}^{r=n} (G_{sr}e_r - B_{sr}h_r) - j\sum_{r=1}^{r=n} (G_{sr}h_r - B_{sr}e_r)\right] \end{aligned} \tag{6.15}$$

Equating the real and imaginary parts, the active and reactive powers at a *PQ* node are

$$P_s = e_s\left[\sum_{r=1}^{r=n}(G_{sr}e_r - B_{sr}h_r)\right] + h_s\left[\sum_{r=1}^{r=n}(G_{sr}h_r + B_{sr}e_r)\right] \tag{6.16}$$

$$Q_s = e_s\left[\sum_{r=1}^{r=n}(-G_{sr}h_r - B_{sr}e_r)\right] + h_s\left[\sum_{r=1}^{r=n}(G_{sr}e_r - B_{sr}h_r)\right] \tag{6.17}$$

where P_s and Q_s are the functions of e_s, e_n h_s, and h_r. Starting from the initial values, new values are found which differ from the initial values by ΔP_s and Δq_s:

$$\Delta P_s^0 = P_s - P_s^0 \text{(first iteration)} \tag{6.18}$$

$$\Delta Q_s^0 = Q_s - Q_s^0 \text{(first iteration)} \tag{6.19}$$

For a *PV* node (generator bus), voltage and power are specified. The reactive power equation is replaced by a voltage equation:

$$|V_g|^2 = e_g^2 + h_g^2 \tag{6.20}$$

$$\Delta|V_g^0|^2 = |V_g|^2 - |V_g^0|^2 \tag{6.21}$$

Consider the four-bus distribution system of Figure 5.8. It is required to write the equations for voltage corrections. Buses 2 and 3 are the load (*PQ* buses), bus 1 is the slack bus, and bus 4 is the generator bus. In Equation 6.11, the column matrix elements on the left consisting of $(y_1 - f_1^0)\ldots(y_n - f_n^0)$ are identified as ΔP, ΔQ, etc. The unknowns of the matrix equation are Δe and Δh. The matrix equation for voltage corrections is written as follows:

$$\begin{vmatrix} \Delta P_2 \\ \Delta Q_2 \\ \Delta P_3 \\ \Delta Q_3 \\ \Delta P_4 \\ \Delta V_4^2 \end{vmatrix} = \begin{vmatrix} \partial P_2/\partial e_2 & \partial P_2/\partial h_2 & \partial P_2/\partial e_3 & \partial P_2/\partial h_3 & \partial P_2/\partial e_4 & \partial P_2/\partial h_4 \\ \partial Q_2/\partial e_2 & \partial Q_2/\partial h_2 & \partial Q_2/\partial e_3 & \partial Q_2/\partial h_3 & \partial Q_2/\partial e_4 & \partial Q_2/\partial h_4 \\ \partial P_3/\partial e_2 & \partial P_3/\partial h_2 & \partial P_3/\partial e_3 & \partial P_3/\partial h_3 & \partial P_3/\partial e_4 & \partial P_3/\partial h_4 \\ \partial Q_3/\partial e_2 & \partial Q_3/\partial h_2 & \partial Q_3/\partial e_3 & \partial Q_3/\partial h_3 & \partial Q_3/\partial e_4 & \partial Q_3/\partial h_4 \\ \partial P_4/\partial e_2 & \partial P_4/\partial h_2 & \partial P_4/\partial e_3 & \partial P_4/\partial h_3 & \partial P_4/\partial e_4 & \partial P_4/\partial h_4 \\ \partial V_4^2/\partial e_2 & \partial V_4^2/\partial h_2 & \partial V_4^2/\partial e_3 & \partial V_4^2/\partial h_3 & \partial V_4^2/\partial e_4 & \partial V_4^2/\partial h_4 \end{vmatrix} \begin{vmatrix} \Delta e_2 \\ \Delta h_2 \\ \Delta e_3 \\ \Delta h_3 \\ \Delta e_4 \\ \Delta h_4 \end{vmatrix} \tag{6.22}$$

In the rectangular form, there are two equations per load (*PQ*) and generator (*PV*) buses. The voltage at the swing bus is known and thus there is no equation for the swing bus.

Equation 6.22 can be written in the abbreviated form as follows:

$$\bar{g} = \bar{J}\bar{x} \tag{6.23}$$

$$
\begin{aligned}
\Delta e_s &= e_s(\text{new}) - e_s(\text{old}) \\
\Delta P_s &= P_{so} - P_s\left(e_2, h_2, e_3, h_3, e_4, h_4\right) \\
\Delta Q_s &= Q_{so} - Q_s\left(e_2, h_2, e_3, h_3, e_4, h_4\right) \\
\Delta V_s^2 &= V_{so}^2 - (e_s^2 + h_s^2) \\
\Delta h_s &= h_s(\text{new}) - h_s(\text{old})
\end{aligned}
\tag{6.24}
$$

where Δe_s and Δh_s are the voltage corrections.

Equation 6.12 can be written as follows:

$$
x^{k+1} - x^k = -\left[\bar{J}(x^k)\right]^{-1} h(x^k) \tag{6.25}
$$

or

$$
\begin{vmatrix} \Delta e_2 \\ \Delta h_2 \\ \cdot \\ \cdot \end{vmatrix} = \bar{J}^{-1} \begin{vmatrix} -P_2(e_2, h_2, \ldots) + P_{20} \\ -Q_2(e_2, h_2, \ldots) + Q_{20} \\ \cdot \\ \cdot \end{vmatrix} \tag{6.26}
$$

The partial coefficients are calculated numerically by substituting assumed initial values into partial derivative equations:

Off-diagonal elements, $s \neq r$

$$
\partial P_s / \partial e_r = -\partial Q_s / \partial h_r = G_{sr} e_s + B_{sr} h_s \tag{6.27}
$$

$$
\partial P_s / \partial h_r = \partial Q_s / \partial e_r = -B_{sr} e_s + G_{sr} h_s \tag{6.28}
$$

$$
\partial V_s^2 / \partial e_r = \partial V_s^2 / \partial h_r = 0 \tag{6.29}
$$

Diagonal elements

$$
\partial P_s / \partial e_s = \sum_{r=1}^{r=n} (G_{sr} e_r - B_{sr} h_r) + G_{ss} e_s + B_{ss} h_s \tag{6.30}
$$

$$
\partial P_s / \partial h_s = \sum_{r=1}^{r=n} (G_{sr} h_r - B_{sr} e_r) - B_{ss} e_s + G_{ss} h_s \tag{6.31}
$$

$$
\partial Q_s / \partial e_s = \sum_{r=1}^{r=n} -(G_{sr} h_r - B_{sr} e_r) - B_{ss} e_s + G_{ss} h_s \tag{6.32}
$$

$$
\partial Q_s / \partial h_s = \sum_{r=1}^{r=n} (G_{sr} e_r - B_{sr} h_r) - G_{ss} e_s - B_{ss} h_s \tag{6.33}
$$

$$\partial V_s^2/\partial e_s = 2e_s \tag{6.34}$$

$$\partial V_s^2/\partial h_s = 2h_s \tag{6.35}$$

6.4 Polar Form of Jacobian Matrix

The voltage equation can be written in polar form as

$$V_s = V_s(\cos\theta_s + j\sin\theta_s) \tag{6.36}$$

Thus, the power is

$$V_s\left(\sum_{r=1}^{r=n} Y_{sr}V_{sr}\right)^*$$
$$= V_s\left(\cos\theta_s - j\sin\theta_s\right)\sum_{r=1}^{r=n}(G_{sr} - jB_{sr})V_r\left(\cos\theta_r - j\sin\theta_r\right) \tag{6.37}$$

Equating real and imaginary terms

$$P_s = V_s\sum_{r=1}^{r=n} V_r\left[(G_{sr}\cos(\theta_s - \theta_r) + B_{sr}\sin(\theta_s - \theta_r)\right] \tag{6.38}$$

$$Q_s = V_s\sum_{r=1}^{r=n} V_r\left[(G_{sr}\sin(\theta_s - \theta_r) - B_{sr}\cos(\theta_s - \theta_r)\right] \tag{6.39}$$

The Jacobian in polar form for the same four-bus system is

$$\begin{vmatrix} \Delta P_2 \\ \Delta Q_2 \\ \Delta P_3 \\ \Delta Q_3 \\ \Delta P_4 \end{vmatrix} = \begin{vmatrix} \partial P_2/\partial\theta_2 & \partial P_2/\partial V_2 & \partial P_2/\partial\theta_3 & \partial P_2/\partial V_3 & \partial P_2/\partial\theta_4 \\ \partial Q_2/\partial\theta_2 & \partial Q_2/\partial V_2 & \partial Q_2/\partial\theta_3 & \partial Q_2/\partial V_3 & \partial Q_2/\partial\theta_4 \\ \partial P_3/\partial\theta_2 & \partial P_3/\partial V_2 & \partial P_3/\partial\theta_3 & \partial P_3/\partial V_3 & \partial P_3/\partial\theta_4 \\ \partial Q_3/\partial\theta_2 & \partial Q_3/\partial V_2 & \partial Q_3/\partial\theta_3 & \partial Q_3/\partial V_3 & \partial Q_3/\partial\theta_4 \\ \partial P_4/\partial\theta_2 & \partial P_4/\partial V_2 & \partial P_4/\partial\theta_3 & \partial P_4/\partial V_3 & \partial P_4/\partial\theta_4 \end{vmatrix} \begin{vmatrix} \Delta\theta_2 \\ \Delta V_2 \\ \Delta\theta_3 \\ \Delta V_3 \\ \Delta\theta_4 \end{vmatrix} \tag{6.40}$$

The slack bus has no equation, because the active and reactive powers at this bus are unspecified and the voltage is specified. At *PV* bus 4, the reactive power is unspecified and there is no corresponding equation for this bus in terms of the variable ΔV_4:

$$\Delta P_s = P_s^0 - P_s(\theta_2 V_2 \theta_3 V_3 \theta_4) \quad (6.41)$$

$$\Delta Q_s = Q_s^0 - Q_s(\theta_2 V_2 \theta_3 V_3 \theta_4) \quad (6.42)$$

The partial derivatives can be calculated as follows:
Off-diagonal elements: $s \neq r$

$$\partial P_s/\partial\theta_r = G_{sr} V_s V_r \sin(\theta_s - \theta_r) - B_{sr} V_s V_r \cos(\theta_s - \theta_r) \quad (6.43)$$

$$\partial P_s/\partial V_r = G_{sr} V_s \cos(\theta_s - \theta_r) + B_{sr} \sin(\theta_s - \theta_r)$$
$$= -(1/V_r)(\partial Q_s)/(\partial\theta_r) \quad (6.44)$$

$$\partial Q_s/\partial\theta_r = -G_{sr} V_s V_r \cos(\theta_s - \theta_r) - B_{sr} V_s V_r \sin(\theta_s - \theta_r) \quad (6.45)$$

$$\partial Q_s/\partial V_r = G_{sr} V_s \sin(\theta_s - \theta_r) - B_{sr} V_s \cos(\theta_s - \theta_r)$$
$$= (1/V_r)(\partial P_s/\partial\theta_r). \quad (6.46)$$

Diagonal elements

$$\partial P_s / \partial\theta_s = V_s \sum_{r=1}^{r=n} V_r \left[-G_{sr} \sin(\theta_s - \theta_r) + B_{sr} \cos(\theta_s - \theta_r)\right] - V_s^2 B_{ss}$$
$$= -Q_s - V_s^2 B_{ss} \quad (6.47)$$

$$\partial P_s / \partial V_s = \sum_{r=1}^{r=n} V_r \left[G_{sr} \cos(\theta_s - \theta_r) + B_{sr} \sin(\theta_s - \theta_r)\right] - V_s G_{ss}$$
$$= (P_s / V_s) + V_s G_{ss} \quad (6.48)$$

$$\partial Q_s / \partial\theta_s = V_s \sum_{r=1}^{r=n} V_r \left[G_{sr} \cos(\theta_s - \theta_r) + B_{sr} \sin(\theta_s - \theta_r)\right] - V_s^2 G_{ss}$$
$$= P_s - V_s^2 G_{ss} \quad (6.49)$$

$$\partial Q_s / \partial V_s = \sum_{r=1}^{r=n} V_r \left[G_{sr} \sin(\theta_s - \theta_r) - B_{sr} \cos(\theta_s - \theta_r)\right] - V_s B_{ss}$$
$$= (Q_s / V_s) - V_s B_{ss} \quad (6.50)$$

6.4.1 Calculation Procedure of the NR Method

The procedure is summarized in the following steps, and flowcharts are shown in Figures 6.2 and 6.3.

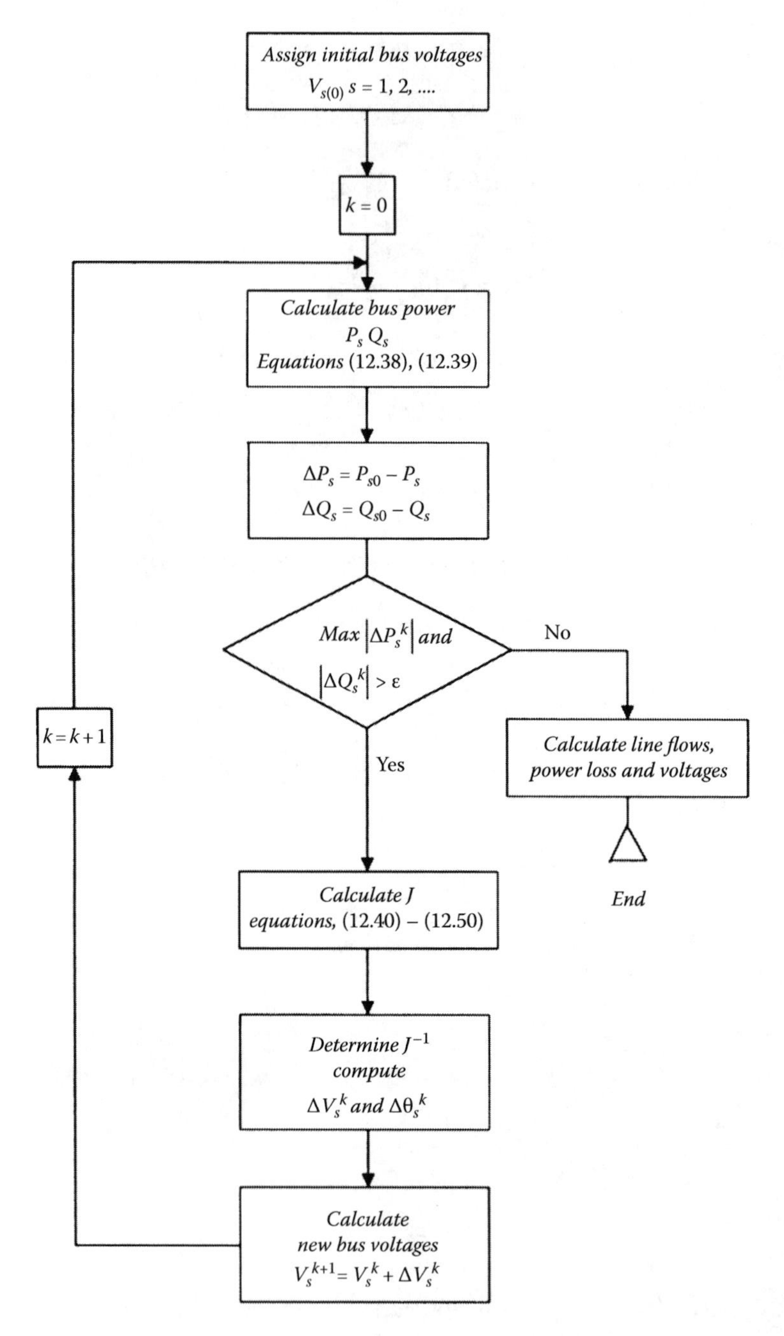

FIGURE 6.2
Flowchart for the NR method of load flow for *PQ* buses.

- Bus admittance matrix is formed.
- Initial values of voltages and phase angles are assumed for the load (*PQ*) buses. Phase angles are assumed for the *PV* buses. Normally, the bus voltages are set equal to the slack bus voltage, and phase angles are assumed equal to 0°, i.e., a *flat* start.
- Active and reactive powers, *P* and *Q*, are calculated for each load bus.

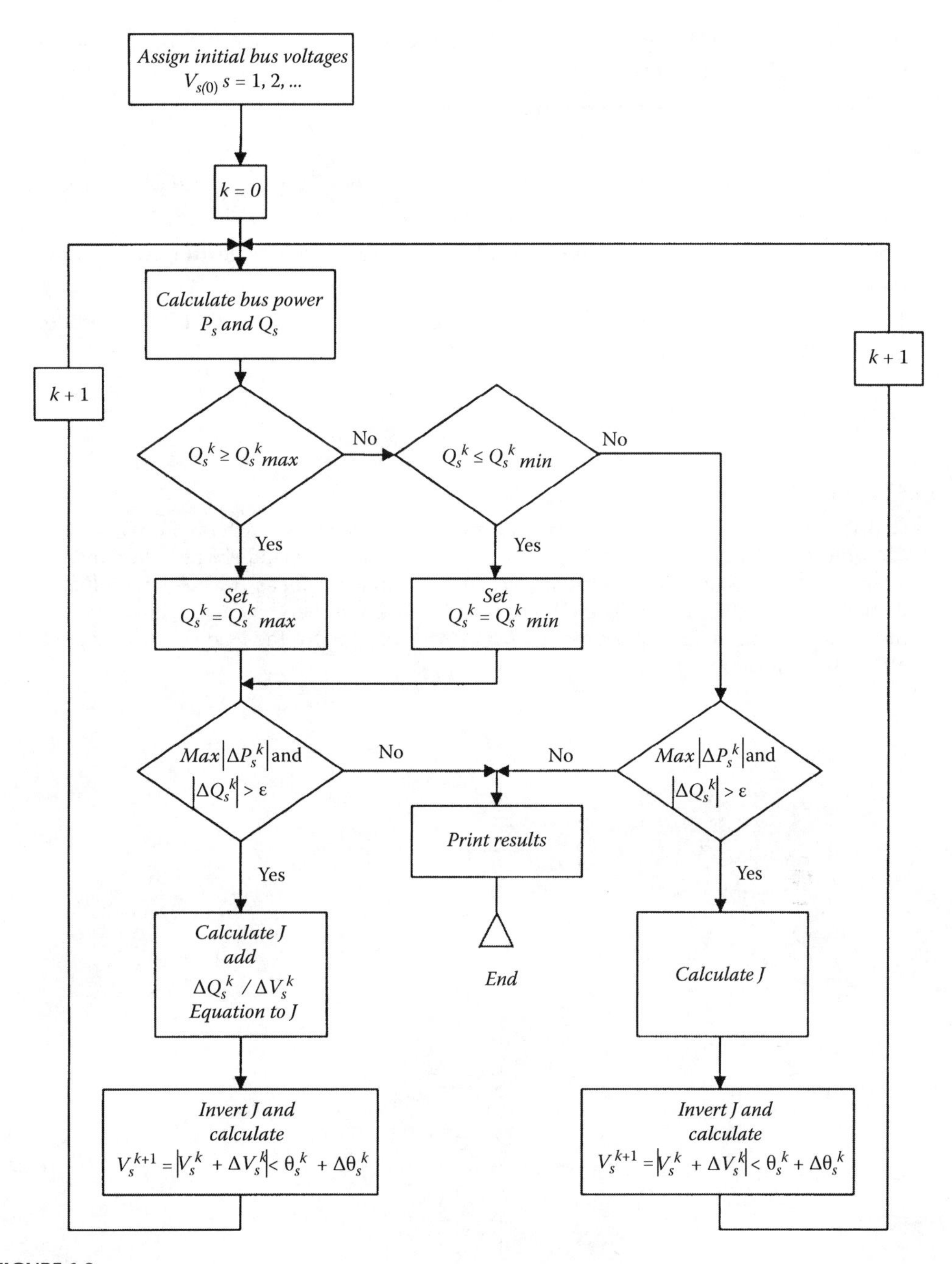

FIGURE 6.3
Flowchart for the NR method of load flow for *PV* buses.

- ΔP and ΔQ can, therefore, be calculated on the basis of the given power at the buses.
- For *PV* buses, the exact reactive power is not specified, but its limits are known. If the calculated value of the reactive power is within limits, only ΔP is calculated. If the calculated value of reactive power is beyond the specified limits, then an appropriate limit is imposed and ΔQ is also calculated by subtracting the

calculated value of the reactive power from the maximum specified limit. The bus under consideration is now treated as a *PQ* (load) bus.

- The elements of the Jacobian matrix are calculated.
- This gives $\Delta\theta$ and $\Delta|V|$.
- Using the new values of $\Delta\theta$ and $\Delta|V|$, the new values of voltages and phase angles are calculated.
- The next iteration is started with these new values of voltage magnitudes and phase angles.
- The procedure is continued until the required tolerance is achieved. This is generally 0.1 kW and 0.1 kvar.

See Reference [2].

Example 6.3

Consider a transmission system of two 138 kV lines, three buses, each line modeled by an equivalent Π network, as shown in Figure 6.4a, with series and shunt admittances as shown. Bus 1 is the swing bus (voltage 1.02 per unit), bus 2 is a *PQ* bus with a load demand of $0.25+j0.25$ per unit, and bus 3 is a voltage-controlled bus with a bus voltage of 1.02 and a load of 0.5 *j*0 per unit all on 100 MVA base. Solve the load flow using the NR method, polar axis basis.

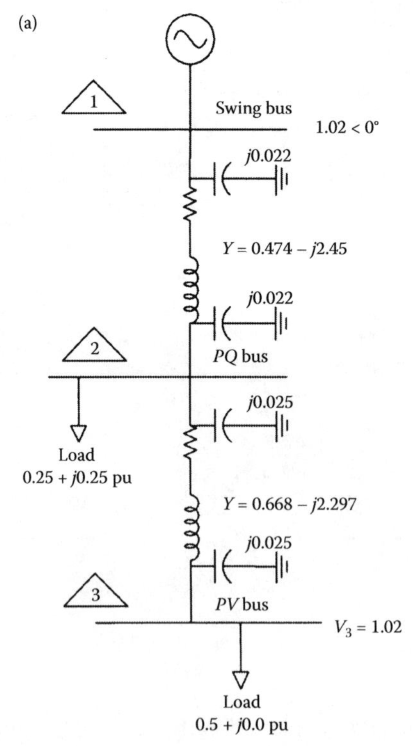

FIGURE 6.4
(a) System of Example 6.3 for load flow solution, (b) final converged load flow solution with reactive power injection at *PV* bus 3, and (c) converged load flow with bus 3 treated as a *PQ* bus.

(*Continued*)

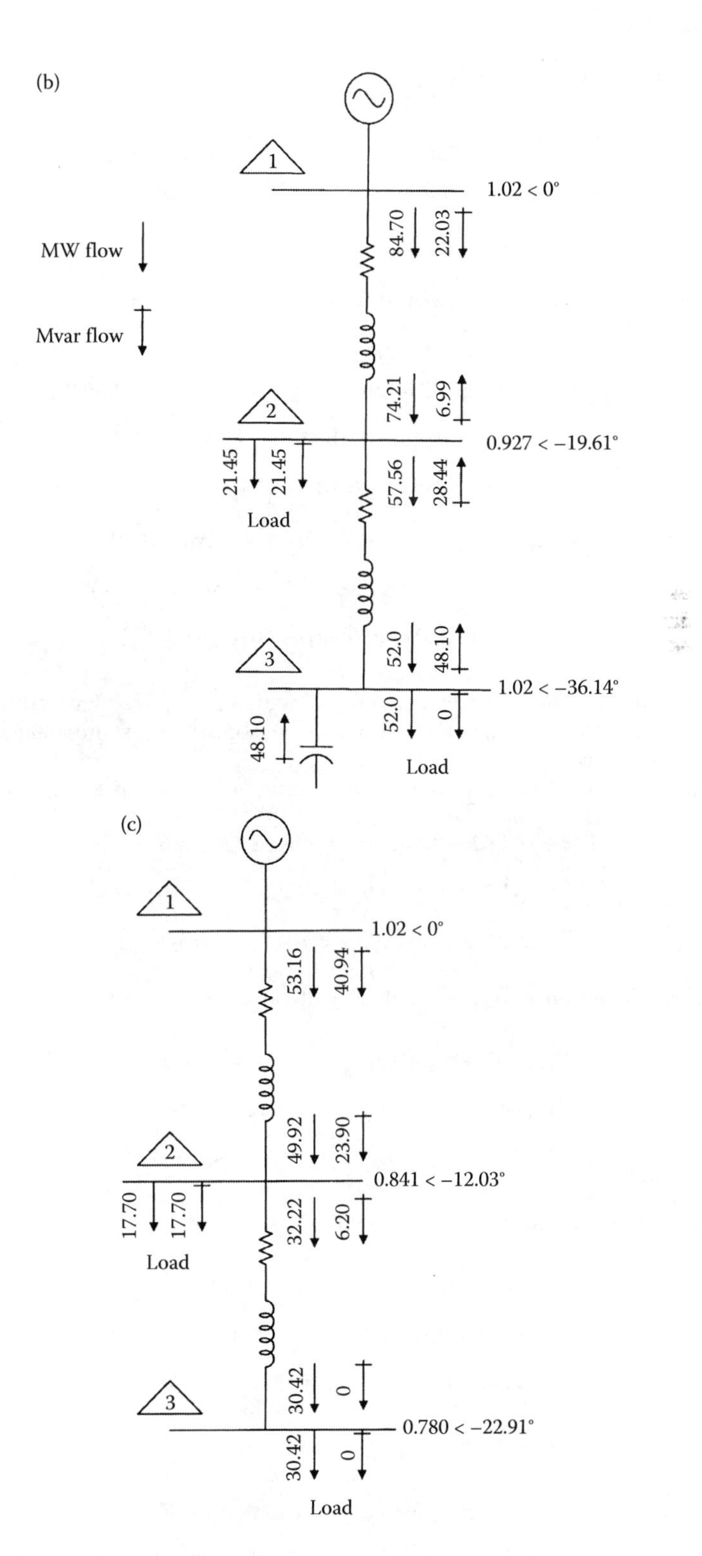

FIGURE 6.4 (CONTINUED)
(a) System of Example 6.3 for load flow solution, (b) final converged load flow solution with reactive power injection at *PV* bus 3, and (c) converged load flow with bus 3 treated as a *PQ* bus.

First, form a Y matrix as follows:

$$\bar{Y} = \begin{vmatrix} 0.474 - j2.428 & -0.474 + j2.428 & 0 \\ -0.474 + j2.428 & 1.142 - j4.70 & -0.668 + j2.272 \\ 0 & -0.668 + j2.272 & 0.668 - j2.272 \end{vmatrix}$$

The active and reactive powers at bus 1 (swing bus) can be written from Equations 6.38 and 6.39 as follows:

$$P_1 = 1.02 \times 1.02[0.474\cos(0.0 - 0.0) + (-2.428)\sin(0.0 - 0.0)]$$
$$+ 1.02V_2[(-0.474)\cos(0.0 - \theta_2) + 2.428\sin(0.0 - \theta_2)$$
$$+ 1.02 \times 1.02[0.0\cos(0.0 - \theta_3) + 0.0\sin(0.0 - \theta_3)]$$

$$Q_1 = 1.02 \times 1.02[0.474\sin(0.0 - 0.0) - (-2.428)\cos(0.0 - 0.0)]$$
$$+ 1.02V_2[(-0.474)\sin(0.0 - \theta_2) - 2.428\cos(0.0 - \theta_2)$$
$$+ 1.02 \times 1.02[0.0\sin(0.0 - \theta_3) - 0.0\cos(0.0 - \theta_3)]$$

These equations for the swing bus are not immediately required for load flow, but can be used to calculate the power flow from this bus, once the system voltages are calculated to the required tolerance.

Similarly, the active and reactive powers at other buses are written as

$$P_2 = V_2 \times 1.02[(-0.474)\cos(\theta_2 - 0) + 2.428\sin(\theta_2 - 0)] + V_2$$
$$\times V_2[1.142\cos(\theta_2 - \theta_2) + (-4.70)\sin(\theta_2 - \theta_2)] + V_2$$
$$\times 1.02[(-0.668)\cos(\theta_2 - \theta_3) + 2.272\sin(\theta_2 - \theta_3)]$$

Substituting the initial values ($V_2=1$, $\theta_2=0°$), $P_2=-0.0228$:

$$Q_2 = V_2 \times 1.02[(-0.474)\sin(\theta_2 - 0.0) - 2.428\cos(\theta_2 - 0.0)] + V_2$$
$$\times V_2[1.142\sin(\theta_2 - \theta_2) - (-4.70)\cos(\theta_2 - \theta_2)] + V_2$$
$$\times 1.02[(-0.668)\sin(\theta_2 - \theta_3) + 2.272\cos(\theta_2 - \theta_3)]$$

Substituting the numerical values, $Q_2=-0.142$:

$$P_3 = 1.02 \times 1.02[0.0\cos(\theta_3 - 0.0) + 0.0\sin(\theta_3 - 0.0)] + 1.02$$
$$\times V_2[(-0.668)\cos(\theta_3 - \theta_2) + 2.272\sin(\theta_3 - \theta_2)] + 1.02$$
$$\times 1.02[0.668\cos(\theta_3 - \theta_3) + (-2.272)\sin(\theta_3 - \theta_3)]$$

Substituting the values, $P_3=0.0136$:

$$Q_3 = 1.02 \times 1.02[0.0\sin(\theta_3 - 0.0) - 0.0\cos(\theta_3 - 0.0)] + 1.02$$
$$\times V_2[(-0.668)\sin(\theta_3 - \theta_2) - 2.272\cos(\theta_3 - \theta_2) + 1.02$$
$$\times 1.02[0.668\sin(\theta_3 - \theta_3) - (-2.272)\cos(\theta_3 - \theta_3)]$$

Substituting initial values, Q_3=−0.213.

The Jacobian matrix is

$$\begin{vmatrix} \Delta P_2 \\ \Delta Q_2 \\ \Delta P_3 \end{vmatrix} = \begin{vmatrix} \partial P_2/\partial\theta_2 & \partial P_2/\partial V_2 & \partial P_2/\partial\theta_3 \\ \partial Q_2/\partial\theta_2 & \partial Q_2/\partial V_2 & \partial Q_2/\partial\theta_3 \\ \partial P_3/\partial\theta_2 & \partial P_3/\partial V_2 & \partial P_3/\partial\theta_3 \end{vmatrix} \begin{vmatrix} \Delta\theta_2 \\ \Delta V_2 \\ \Delta\theta_3 \end{vmatrix}$$

The partial differentials are found by differentiating the equations for P_2, Q_2, P_3, etc.

$$\partial P_2/\partial\theta_2 = 1.02V_2[(0.474)\sin\theta_2 + 2.45\cos\theta_2] + 1.02[V_2(0.688)\sin(\theta_2-\theta_3) + V_2(2.272)\cos(\theta_2-\theta_3)]$$

$$= 4.842$$

$$\partial P_2/\partial\theta_3 = V_2(1.02)[(-0.668)\sin(\theta_2-\theta_3) - 2.272\cos(\theta_2-\theta_3)]$$

$$= -2,343$$

$$\partial P_2/\partial V_2 = 1.02[(-0.474)\cos\theta_2 + 2.45\sin\theta_2]$$

$$+2V_2(1.142) + 1.02[(-0.608)\cos(\theta_2-\theta_3) + 2.272\sin(\theta_2-\theta_3)]$$

$$= 1.119$$

$$\partial Q_2/\partial\theta_2 = 1.02[-V_2(0.474)\cos\theta_2 + 2.45\sin\theta_2] + 1.02[V_2(-0.668)\cos(\theta_2-\theta_3)$$

$$+2.297\sin(\theta_2-\theta_3)]$$

$$= -1.1648$$

$$\partial Q_2/\partial V_2 = 1.02[(-0.474)\sin\theta_2 - 2.45\cos\theta_2] + 2V_2(4.70)$$

$$+1.02[(-0.668)\sin(\theta_2-\theta_3) - 2.272\cos(\theta_2-\theta_3)]$$

$$= 4.56$$

$$\partial Q_2/\partial\theta_3 = 1.02V_2[0.668\cos(\theta_2-\theta_3) - 2.272\sin(\theta_2-\theta_3)]$$

$$= 0.681$$

$$\partial P_3/\partial\theta_2 = 1.02V_2[(0.668)\sin(\theta_3-\theta_2) - 2.272\cos(\theta_3-\theta_2)$$

$$= -2.343$$

$$\partial P_3/\partial V_2 = 1.02[(-0.668)\cos(\theta_3-\theta_2) + 2.272\sin(\theta_3-\theta_2)]$$

$$= -0.681$$

$$\partial P_3/\partial\theta_3 = 1.02[0.668\sin(\theta_3-\theta_2) + 2.272\cos(\theta_3-\theta_2)]$$

$$= 2.343$$

Therefore, the Jacobian is

$$\bar{J} = \begin{vmatrix} 4.842 & 1.119 & -2.343 \\ -1.165 & 4.56 & 0.681 \\ -2.343 & -0.681 & 2.343 \end{vmatrix}$$

The system equations are

$$\begin{vmatrix} \Delta\theta_2^1 \\ \Delta V_2^1 \\ \Delta\theta_3^1 \end{vmatrix} = \left|\bar{J}\right|^{-1} \begin{vmatrix} -0.25-(-0.0228) \\ -0.25-(-0.142) \\ -0.5-0.0136 \end{vmatrix}$$

Inverting the Jacobian gives

$$\begin{vmatrix} \Delta\theta_2^1 \\ \Delta V_2^1 \\ \Delta\theta_3^1 \end{vmatrix} = \begin{vmatrix} 0.371 & -0.153 & 0.4152 \\ 0.041 & 0.212 & -0.021 \\ 0.359 & -0.215 & 0.848 \end{vmatrix} \begin{vmatrix} -0.25-(-0.0228) \\ -0.25-(-0.142) \\ -0.5-0.0136 \end{vmatrix} = \begin{vmatrix} -0.293 \\ -0.021 \\ -0.518 \end{vmatrix}$$

The new values of voltages and phase angles are

$$\begin{vmatrix} \theta_2^1 \\ V_2^1 \\ \theta_3^1 \end{vmatrix} = \begin{vmatrix} 0 \\ 1 \\ 0 \end{vmatrix} - \begin{vmatrix} -0.293 \\ -0.021 \\ -0.518 \end{vmatrix} = \begin{vmatrix} -0.293 \\ -0.979 \\ -0.518 \end{vmatrix}$$

This completes one iteration. Using the new bus voltages and phase angles, the power flow is recalculated. Thus, at every iteration, the Jacobian matrix changes and has to be inverted.

In the first iteration, we see that the bus 2 voltage is 2.1% lower than the rated voltage; the angles are in radians. The first iteration is no indication of the final results. The hand calculations, even for a simple three-bus system, become unwieldy. The converged load flow is shown in Figure 6.4b. A reactive power injection of 43 Mvar is required at bus 3 to maintain a voltage of 1.02 per unit and supply the required active power of 0.5 per unit from the source. There is a reactive power loss of 48.65 Mvar in the transmission lines themselves, and the active power loss is 12.22 MW. The bus phase angles are high with respect to the swing bus, and the bus 2 operating voltage is 0.927 per unit, i.e., a voltage drop of 7.3% under load. Thus, even with voltages at the swing bus and bus 3 maintained above rated voltage, the power demand at bus 2 cannot be met and the voltage at this bus dips. The load demand on the system is too high, is lossy, and requires augmentation or reduction of load. A reactive power injection at bus 2 will give an entirely different result.

If bus 3 is treated as a load bus, the Jacobian is modified by adding a fourth equation of the reactive power at bus 3. In this case, the bus 3 voltage dips down to 0.78 per unit, i.e., a voltage drop of 22%; the converged load flow is shown in Figure 6.4c. At this lower voltage of 0.78 per unit, bus 3 can support an active load of only 0.3 per unit. This is not an example of a practical system, but it illustrates the importance of reactive power injection, load modeling, and its effect on the bus voltages.

The load demand reduces proportionally with reduction in bus voltages. This is because we have considered a constant impedance type of load, i.e., the load current varies directly with the voltage as the load impedance is held constant. The load types are discussed further.

6.5 Simplifications of the NR Method

The NR method has quadratic convergence characteristics; therefore, the convergence is fast and solution to high accuracy is obtained in the first few iterations. The number of iterations does not increase appreciably with the size of the system. This is in contrast to the Gauss–Seidel method of load flow that has slower convergence even with appropriately applied acceleration factors. The larger the system, the larger the number of iterations; 50–150 iterations are common.

The NR method, however, requires more memory storage and necessitates solving a large number of equations in each iteration step. The Jacobian changes at each iteration and must be evaluated afresh. The time required for one iteration in the NR method may be 5–10 times that of the Gauss–Seidel method. Some simplifications that can be applied are as follows.

From Equation 6.22, the first equation is

$$\Delta P_2 = (\partial P_2/\partial e_2)\Delta e_2 + (\partial P_2/\partial h_2)\Delta h_2 + (\partial P_2/\partial e_3)\Delta e_3 + (\partial P_2/\partial h_3)\Delta h_3$$
$$+(\partial P_2/\partial e_4)\Delta e_4 + (\partial P_2/\partial h_4)\Delta h_4 \quad (6.51)$$

The second term of this equation $(\partial P_2/\partial h_2)\Delta h_2$ denotes the change in bus 2 active power for a change of h_2 to $h_2+\Delta h_2$. Similarly, the term $(\partial P_2/\partial e_2)\Delta e_2$ indicates the change in bus 2 active power for a change of e_2 to $e_2+\Delta e_2$.

The change in power at bus 2 is a function of the voltage at bus 2, which is a function of the voltages at other buses. Considering the effect of e_2 only

$$\Delta P_2 \doteq (\partial P_2/\partial e_2)\Delta e_2 \quad (6.52)$$

Thus, the Jacobian reduces to only diagonal elements:

$$\begin{vmatrix} \Delta P_2 \\ \Delta Q_2 \\ \Delta P_3 \\ \Delta Q_3 \\ \Delta P_4 \\ \Delta V_4 \end{vmatrix} = \begin{Vmatrix} \partial P_2/\partial e_2 & \cdot & \cdot & \cdot & \cdot & \cdot \\ \cdot & \partial Q_2/\partial h_2 & \cdot & \cdot & \cdot & \cdot \\ \cdot & \cdot & \partial P_3/\partial e_3 & \cdot & \cdot & \cdot \\ \cdot & \cdot & \cdot & \partial Q_3/\partial h_3 & \cdot & \cdot \\ \cdot & \cdot & \cdot & \cdot & \partial P_4/\partial e_4 & \cdot \\ \cdot & \cdot & \cdot & \cdot & \cdot & \partial v_4^2/\partial h_4 \end{Vmatrix} \begin{vmatrix} \Delta e_2 \\ \Delta h_2 \\ \Delta e_3 \\ \Delta h_3 \\ \Delta e_4 \\ \Delta h_4 \end{vmatrix} \quad (6.53)$$

Method 2 reduces the Jacobian to a lower triangulation matrix:

$$\Delta P_2 \doteq (\partial P_2/\partial e_2)\Delta e_2$$
$$\Delta Q_2 \doteq (\partial Q_2/\partial e_2)\Delta e_2 + (\partial Q_2/\partial h_2)\Delta h_2 \quad (6.54)$$

Thus, the Jacobian matrix is

$$\begin{vmatrix} \Delta P_2 \\ \Delta Q_2 \\ \Delta P_3 \\ \Delta Q_3 \\ \Delta P_4 \\ \Delta V_4^2 \end{vmatrix} = \begin{vmatrix} \partial P_2/\partial e_2 & \cdot & \cdot & \cdot & \cdot & \cdot \\ \partial Q_2/\partial e_2 & \partial Q_2/\partial h_2 & \cdot & \cdot & \cdot & \cdot \\ \partial P_3/\partial e_2 & \partial P_3/\partial h_2 & \partial P_3/\partial 3e_3 & \cdot & \cdot & \cdot \\ \partial Q_3/\partial e_2 & \partial Q_3/\partial h_2 & \partial Q_3/\partial h_3 & \partial Q_3/\partial h_3 & \cdot & \cdot \\ \partial P_4/\partial e_2 & \partial P_4/\partial h_2 & \partial P_4/\partial e_3 & \partial P_4/\partial h_3 & \partial P_4/\partial e_4 & \cdot \\ \partial V_4^2/\partial e_2 & \partial V_4^2/\partial h_2 & \partial V_4^2/\partial e_3 & \partial V_4^2/\partial h_3 & \partial V_4^2/\partial e_4 & \partial V_4^2/\partial h_4 \end{vmatrix} \times \begin{vmatrix} \Delta e_2 \\ \Delta h_2 \\ \Delta e_3 \\ \Delta h_3 \\ \Delta e_4 \\ \Delta h_4 \end{vmatrix} \quad (6.55)$$

Method 3 relates P_2 and Q_2 to e_2 and h_2, P_3 and Q_3 to e_3 and h_3, etc. This is the Ward–Hale method. The Jacobian is

$$\begin{vmatrix} \Delta P_2 \\ \Delta Q_2 \\ \Delta P_3 \\ \Delta Q_3 \\ \Delta P_4 \\ \Delta Q_4 \end{vmatrix} = \begin{vmatrix} \partial P_2/\partial e_2 & \partial P_2/\partial h_2 & \cdot & \cdot & \cdot & \cdot \\ \partial Q_2/\partial e_2 & \partial Q_2/\partial h_2 & \cdot & \cdot & \cdot & \cdot \\ \cdot & \cdot & \partial P_3/\partial e_3 & \partial P_3/\partial h_3 & \cdot & \cdot \\ \cdot & \cdot & \partial Q_3/\partial e_3 & \partial Q_3/\partial h_3 & \cdot & \cdot \\ \cdot & \cdot & \cdot & \cdot & \partial P_4/\partial e_4 & \partial P_4/\partial h_4 \\ \cdot & \cdot & \cdot & \cdot & \partial Q_4/\partial e_4 & \partial Q_4/\partial h_4 \end{vmatrix} \times \begin{vmatrix} \Delta e_2 \\ \Delta h_2 \\ \Delta e_3 \\ \Delta h_3 \\ \Delta e_4 \\ \Delta h_4 \end{vmatrix} \tag{6.56}$$

Method 4 is a combination of Methods 2 and 3:

$$\begin{vmatrix} \Delta P_2 \\ \Delta Q_2 \\ \Delta P_3 \\ \Delta Q_3 \\ \Delta P_4 \\ \Delta V_4^2 \end{vmatrix} = \begin{vmatrix} \partial P_2/\partial e_2 & \partial P_2/\partial h_2 & \cdot & \cdot & \cdot & \cdot \\ \partial Q_2/\partial e_2 & \partial Q_2/\partial h_2 & \cdot & \cdot & \cdot & \cdot \\ \partial P_3/\partial e_2 & \partial P_3/\partial h_2 & \partial P_3/\partial e_3 & \partial P_3/\partial h_3 & \cdot & \cdot \\ \partial Q_3/\partial e_2 & \partial Q_3/\partial h_2 & \partial Q_3/\partial eh_3 & \partial Q_3/\partial h_3 & \cdot & \cdot \\ \partial P_4/\partial e_2 & \partial P_4/\partial h_2 & \partial P_4/\partial e_3 & \partial P_4/\partial h_3 & \partial P_4/\partial e_4 & \partial P_4/\partial h_4 \\ \partial V_4^2/\partial e_2 & \partial V_4^2/\partial h_2 & \partial V_4^2/\partial e_3 & \partial V_4^2/\partial h_3 & \partial V_4^2/\partial e_4 & \partial V_4^2/\partial h_4 \end{vmatrix} \times \begin{vmatrix} \Delta e_2 \\ \Delta h_2 \\ \Delta e_3 \\ \Delta h_3 \\ \Delta e_4 \\ \Delta h_4 \end{vmatrix} \tag{6.57}$$

Method 5 may give the least iterations for a value of $\beta < 1$, a factor somewhat akin to the acceleration factor in the Gauss–Seidel method (>1). The Jacobian is of the form lower, diagonal, and upper (LDU) (Appendix A):

$$\begin{vmatrix} \Delta P_2 \\ \Delta Q_2 \\ \Delta P_3 \\ \Delta Q_3 \\ \Delta P_4 \\ \Delta V_4 \end{vmatrix} = (L+D) \begin{vmatrix} \Delta e_2^k \\ \Delta h_2^k \\ \Delta e_3^k \\ \Delta h_3^k \\ \Delta e_4^k \\ \Delta h_4^k \end{vmatrix} + \beta U \begin{vmatrix} \Delta e_2^{k-1} \\ \Delta h_2^{k-1} \\ \Delta e_3^{k-1} \\ \Delta h_3^{k-1} \\ \Delta e_4^{k-1} \\ \Delta h_4^{k-1} \end{vmatrix} \tag{6.58}$$

6.6 Decoupled NR Method

It has already been demonstrated that there is strong interdependence between active power and bus voltage angle and between reactive power and voltage magnitude. The active power change ΔP is less sensitive to changes in voltage magnitude, and changes in reactive power ΔQ are less sensitive to changes in angles. In other words, the coupling between P and bus voltage magnitude is weak and between reactive power and phase angle is weak.

The Jacobian in Equation 6.22 can be rearranged as follows:

$$\begin{vmatrix} \Delta P_2 \\ \Delta P_3 \\ \Delta P_4 \\ \Delta Q_2 \\ \Delta Q_3 \end{vmatrix} = \begin{vmatrix} \partial P_2/\partial\theta_2 & \partial P_2/\partial\theta_3 & \partial P_2/\partial\theta_4 & \partial P_2/\partial V_2 & \partial P_2/\partial V_3 \\ \partial P_3/\partial\theta_2 & \partial P_3/\partial\theta_3 & \partial P_3/\partial\theta_4 & \partial P_3/\partial V_2 & \partial P_3/\partial V_3 \\ \partial P_4/\partial\theta_2 & \partial P_4/\partial\theta_3 & \partial P_4/\partial\theta_4 & \partial P_4/\partial V_2 & \partial P_4/\partial V_3 \\ \partial Q_2/\partial\theta_2 & \partial Q_2/\partial\theta_3 & \partial Q_2/\partial\theta_4 & \partial Q_2/\partial V_2 & \partial Q_2/\partial V_3 \\ \partial Q_3/\partial\theta_2 & \partial Q_3/\partial\theta_3 & \partial Q_3/\partial\theta_4 & \partial Q_3/\partial V_2 & \partial Q_3/\partial V_3 \end{vmatrix} \begin{vmatrix} \Delta\theta_2 \\ \Delta\theta_3 \\ \Delta\theta_4 \\ \Delta V_2 \\ \Delta V_3 \end{vmatrix} \tag{6.59}$$

Considering that

$$G_{sr} <<< B_{sr} \tag{6.60}$$

$$\sin(\theta_s - \theta_r) <<< 1 \tag{6.61}$$

$$\cos(\theta_s - \theta_r) \simeq 1 \tag{6.62}$$

The following inequalities are valid:

$$|\partial P_s/\partial\theta_r| >>> |\partial P_s/\partial V_r| \tag{6.63}$$

$$|\partial Q_s/\partial\theta_r| <<< |\partial Q_s/\partial V_r| \tag{6.64}$$

Thus, the Jacobian is

$$\begin{vmatrix} \Delta P_2 \\ \Delta P_3 \\ \Delta P_4 \\ \Delta Q_2 \\ \Delta Q_3 \end{vmatrix} = \begin{vmatrix} \partial P_2/\partial\theta_2 & \partial P_2/\partial\theta_3 & \partial P_2/\partial\theta_4 & \cdot & \cdot \\ \partial P_3/\partial\theta_2 & \partial P_3/\partial\theta_3 & \partial P_3/\partial\theta_4 & \cdot & \cdot \\ \partial P_4/\partial\theta_2 & \partial P_4/\partial\theta_3 & \partial P_4/\partial\theta_4 & & \\ \cdot & \cdot & \cdot & \partial Q_2/\partial V_2 & \partial Q_2/\partial V_3 \\ \cdot & \cdot & \cdot & \partial Q_3/\partial V_2 & \partial Q_3/\partial V_3 \end{vmatrix} \begin{vmatrix} \Delta\theta_2 \\ \Delta\theta_3 \\ \Delta\theta_4 \\ \Delta V_2 \\ \Delta V_3 \end{vmatrix} \tag{6.65}$$

This is called *P–Q* decoupling.

6.7 Fast Decoupled Load Flow

Two synthetic networks, *P*–θ and *P*–*V*, are constructed. This implies that the load flow problem can be solved separately by these two networks, taking advantage of *P*–*Q* decoupling [3].

In a *P*–θ network, each branch of the given network is represented by conductance, the inverse of series reactance. All shunt admittances and transformer off-nominal voltage taps which affect the reactive power flow are omitted, and the swing bus is grounded. The bus conductance matrix of this network is termed $\overline{B}^{\theta}$.

The second model is called a Q–V network. It is again a resistive network. It has the same structure as the original power system model, but voltage-specified buses (swing bus and PV buses) are grounded. The branch conductance is given by

$$Y_{sr} = -B_{sr} = \frac{x_{sr}}{x_{sr}^2 + r_{sr}^2} \tag{6.66}$$

These are equal and opposite to the series or shunt susceptance of the original network. The effect of phase-shifter angles is neglected. The bus conductance matrix of this network is called $\overline{B}^{\upsilon}$.

The equations for power flow can be written as follows:

$$P_s/V_s = \sum_{r=1}^{r-n} V_r \left[G_{sr}\cos(\theta_s - \theta_r) + B_{sr}\sin(\theta_s - \theta_r)\right] \tag{6.67}$$

$$Q_s/V_s = \sum_{r=1}^{r-n} V_r \left[G_{sr}\sin(\theta_s - \theta_r) - B_{sr}\cos(\theta_s - \theta_r)\right] \tag{6.68}$$

and partial derivatives can be taken as before. Thus, a single matrix for load flow can be split into two matrices as follows:

$$\begin{vmatrix} \Delta P_2/V_2 \\ \Delta P_3/V_3 \\ \cdot \\ \Delta P_n/V_n \end{vmatrix} = \begin{vmatrix} B_{22}^{\theta} & B_{23}^{\theta} & \cdot & B_{2n}^{\theta} \\ B_{32}^{\theta} & B_{33}^{\theta} & \cdot & B_{3n}^{\theta} \\ \cdot & \cdot & \cdot & \cdot \\ B_{n2}^{\theta} & B_{n3}^{\theta} & \cdot & B_{nn}^{\theta} \end{vmatrix} \begin{vmatrix} \Delta\theta_2 \\ \Delta\theta_3 \\ \cdot \\ \Delta\theta_n \end{vmatrix} \tag{6.69}$$

The correction of phase angle of voltage is calculated from this matrix:

$$\begin{vmatrix} \Delta Q_2/V_2 \\ \Delta Q_3/V_3 \\ \cdot \\ \Delta Q_n/V_n \end{vmatrix} = \begin{vmatrix} B_{22}^{\upsilon} & B_{23}^{\upsilon} & \cdot & B_{2n}^{\upsilon} \\ B_{32}^{\upsilon} & B_{33}^{\upsilon} & \cdot & B_{3n}^{\upsilon} \\ \cdot & \cdot & \cdot & \cdot \\ B_{n2}^{\upsilon} & B_{n3}^{\upsilon} & \cdot & B_{nn}^{\upsilon} \end{vmatrix} \begin{vmatrix} \Delta V_2 \\ \Delta V_3 \\ \cdot \\ \Delta V_n \end{vmatrix} \tag{6.70}$$

The voltage correction is calculated from this matrix.

These matrices are real, sparse, and contain only admittances; these are constants and do not change during successive iterations. This model works well for $R/X \ll 1$. If this is not true, this approach can be ineffective. If phase shifters are not present, then both the matrices are symmetrical. Equations 6.69 and 6.70 are solved alternately with the most recent voltage values. This means that one iteration implies one solution to obtain $|\Delta\theta|$ to update θ and then another solution for $|\Delta V|$ to update V.

Example 6.4

Consider the network of Figure 6.5a. Let bus 1 be a swing bus, buses 2 and 3 *PQ* buses, and bus 4 a *PV* bus. The loads at buses 2 and 3 are specified as is the voltage magnitude at bus 4. Construct *P*–θ and *Q*–*V* matrices.

$P-\theta$ Network

First construct the *P*–θ network shown in Figure 6.5b. Here, the elements are calculated as follows:

$$B_{sr}^{\theta}=-1/X_{sr}=-B_{sr}\left(1+(G_{sr}/B_{sr})^2\right)$$

$$B_{ss}^{\theta}=\sum_r(1/X_{sr})=\sum_r B_{sr}\left(1+(G_{sr}/B_{sr})^2\right)$$

All shunt susceptances are neglected and the swing bus is connected to ground.
The associated matrix is

$$\overline{B}^{\theta}=\begin{vmatrix} 6.553 & -2.22 & -0.333 \\ -2.22 & 5.886 & -3.333 \\ -0.333 & -3.333 & 3.666 \end{vmatrix}\begin{vmatrix} \Delta\theta_2 \\ \Delta\theta_3 \\ \Delta\theta_4 \end{vmatrix}=\begin{vmatrix} \Delta P_2/V_2 \\ \Delta P_3/V_3 \\ \Delta P_4/V_4 \end{vmatrix}$$

$Q-V$ Network

This has the structure of the original model, but voltage specified buses, that is swing bus and *PV* bus, are directly connected to ground:

$$B_{sr}^{v}=-B_{sr}=X_{sr}/(R_{sr}^2+X_{sr}^2)$$

The *Q*–*V* network is shown in Figure 6.5c. The associated matrix is

$$\overline{B}^{v}=\begin{vmatrix} 9.345 & -2.17 \\ -2.17 & 6.50 \end{vmatrix}\begin{vmatrix} \Delta V_2 \\ \Delta V_3 \end{vmatrix}=\begin{vmatrix} \Delta Q_2/V_2 \\ \Delta Q_3/V_3 \end{vmatrix}$$

The power flow equations can be written as in Example 6.3.

6.8 Model of a Phase-Shifting Transformer

A model of a transformer with voltage magnitude control through tap changing under load or off-load tap operation was derived in Section 5.1. The voltage drop in a transmission line is simulated in a line drop compensator, which senses the remote secondary voltage and adjusts the voltage taps. The voltage taps, however, do not change the phase angle of the voltages appreciably. A minor change due to change of the transformer impedance on account of tap adjustment and the resultant power flow through it can be ignored. The real power control can be affected through phase shifting of the voltage.

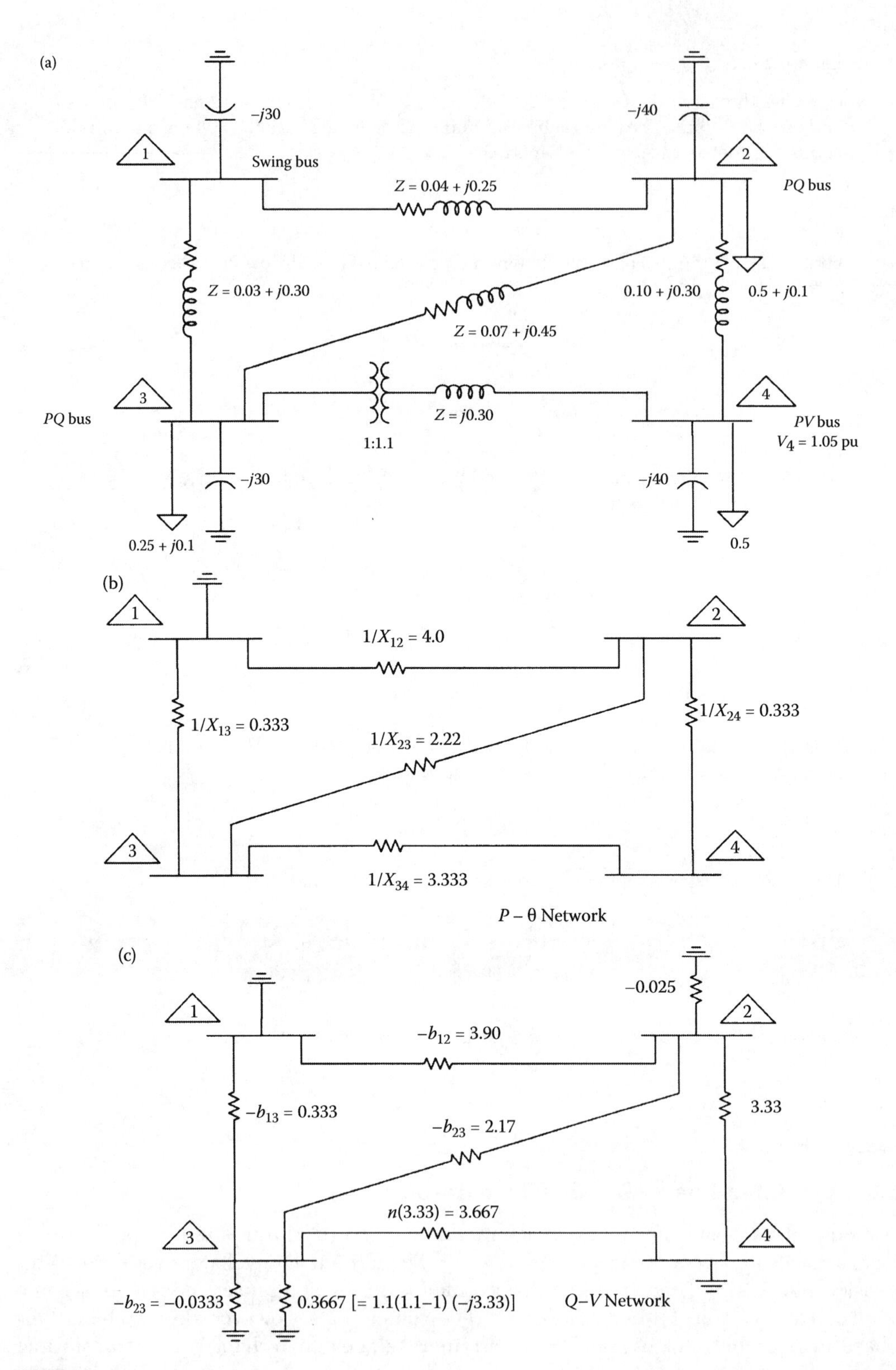

FIGURE 6.5
(a) Four-bus system with voltage tap adjustment transformer, (b) decoupled P–θ network, and (c) Q–V network.

A phase-shifting transformer changes the phase angle without appreciable change in the voltage magnitude; this is achieved by injecting a voltage at right angles to the corresponding line-to-neutral voltage, see Figure 6.6a.

The Y bus matrix for load flow is modified. Consider the equivalent circuit representation of Figure 6.6b. Let the regulating transformer be represented by an ideal transformer with a series impedance or admittance. Since it is an ideal transformer, the complex power input equals the complex power output, and for a voltage adjustment tap changing transformer, we have already shown that

$$I_s = n^2 yV_s - nyV_r \tag{6.71}$$

where n is the ratio of the voltage adjustment taps (or currents). Also

$$I_r = y(V_r - nV_s) \tag{6.72}$$

These equations cannot be represented by a bilateral network. The Y-matrix representation is

$$\begin{vmatrix} I_s \\ I_r \end{vmatrix} = \begin{vmatrix} n^2 y & -ny \\ -ny & y \end{vmatrix} \begin{vmatrix} V_s \\ V_r \end{vmatrix} \tag{6.73}$$

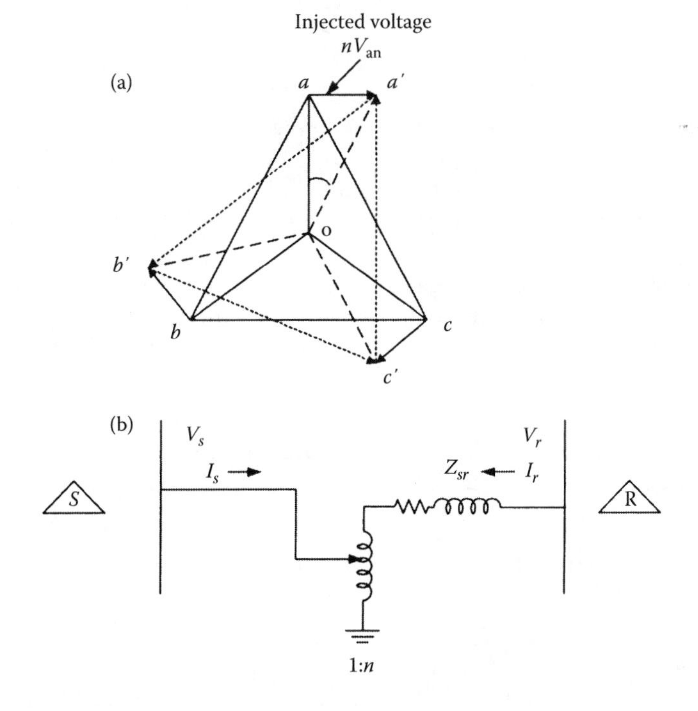

FIGURE 6.6
(a) Voltage injection vector diagram of a phase-shifting transformer and (b) schematic diagram of a phase-shifting transformer.

If the transformer has a phase-shifting device, n should be replaced with $n=Ne^{j}\varphi$ and the relationship is

$$\begin{vmatrix} I_s \\ I_r \end{vmatrix} = \begin{vmatrix} N^2y - N^*y \\ -Ny \quad y \end{vmatrix} \begin{vmatrix} V_s \\ V_r \end{vmatrix} \tag{6.74}$$

where N^* is a conjugate of N. In fact, we could write

$$N = n\varepsilon^{j0} \text{ for transformer without phase shifting} \tag{6.75}$$

$$N = n\varepsilon^{j\phi} \text{ for transformer with phase shifting} \tag{6.76}$$

The phase shift will cause a redistribution of power in load flow, and the bus angles solved will reflect such redistribution. A new equation must be added in the load flow to calculate the new phase-shifter angle based on the latest bus magnitudes and angles [4]. The specified active power flow is

$$P_{(\text{spec})} = V_s V_r \sin(\theta_s - \theta_r - \theta_\alpha) b_{sr} \tag{6.77}$$

where θ_α is the phase-shifter angle. As the angles are small,

$$\sin(\theta_s - \theta_r - \theta_\alpha) \approx \theta_s - \theta_r - \theta_\alpha \tag{6.78}$$

Thus,

$$\theta_\alpha = \theta_s - \theta_r - \frac{P_{(\text{spec})}}{V_s V_r b_{sr}} \tag{6.79}$$

θ_α is compared to its maximum and minimum limits. Beyond the set limits, the phase shifter is tagged nonregulating. Incorporation of a phase-shifting transformer makes the Y matrix nonsymmetric.

Consider the four-bus circuit of Example 11.1. Let the transformer in lines 3 and 4 be replaced with a phase-shifting transformer, phase shift $\varphi=-4°$. The Y bus can then be modified as follows.

Consider the elements related to lines 3 and 4, *ignoring the presence of the transformer.* The submatrix is

	3	4
3	1.667 – j8.30	j3.333
4	j3.333	1.0 – j6.308

With phase shifting, the modified submatrix becomes

	3	4
3	1.667 – j8.30	e^{j4} j3.333
4	e^{-j4} (j3.333)	1.0 – j6.308

The matrix is not symmetrical. See Reference [4].

6.9 DC Load Flow Models

The concept of decoupling and *P–θ and P–V models* is the key to the dc circuit models. All voltages are close to the rated voltages, the X/R ratio is >3.0, the angular difference between the tie-line flow is small, and the transformer taps are near 1.0 pu. With these assumptions, a high speed of calculation results, with some sacrifice of accuracy. The dc load flow was attractive to reduce the CPU time, and in modern times with high-speed computing, there is a demise of dc load flow. Its efficacy is to conduct examination of possible outage conditions and sensitivity analysis, for a very large number of system configurations.

6.9.1 *P–θ* Network

The *P–θ* circuit model of a series element is shown in Figure 6.7a. The active power flow between a bilateral node can be written as follows:

$$\Delta p_{ji} \approx \left(\Delta\theta_j - \Delta\theta_i\right) b_{ij} \tag{6.80}$$

When a phase adjustment element (like a phase-shifting transformer) is provided in series as in the model shown in Figure 6.7b, the relation is

$$\Delta p_{ji} \approx \left(\Delta\theta_j - \Delta\theta_i + \Delta\alpha\right) b_{ij} \tag{6.81}$$

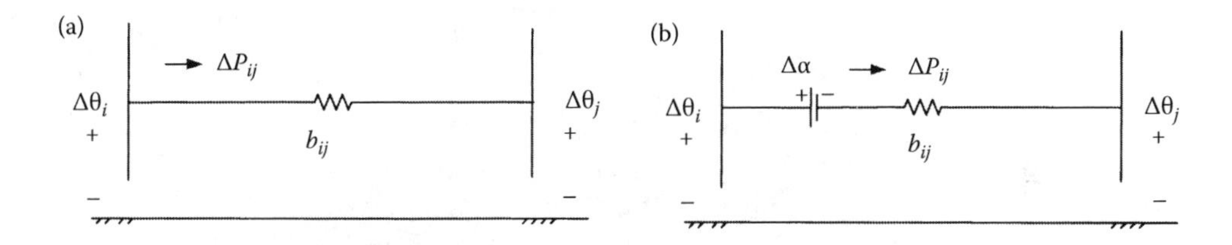

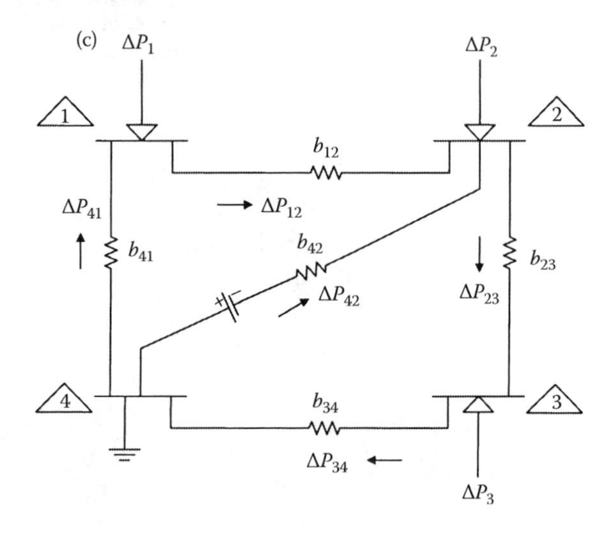

FIGURE 6.7
(a) DC circuit *P–θ* model of a tie-line, (b) dc circuit *P–θ* model of a voltage tap adjusting transformer, and (c) circuit diagram for *P–θ* network equations.

The partial derivatives $\partial P_{ij} / \partial V_i, \partial P_{ij} / \partial V_j$ are ignored.

Consider the four-bus circuit of Figure 6.7c. In terms of branch flows, the following equations can be written from branch relationships:

$$\begin{aligned}
\Delta\theta_2 &= \Delta\theta_1 - \Delta P_1 / b_{12} \\
\Delta\theta_3 &= \Delta\theta_2 - \Delta P_2 / b_{23} \\
\Delta\theta_4 &= \Delta\theta_3 - \Delta P_3 / b_{34} \\
\Delta\alpha + \Delta\theta_2 &= \Delta\theta_4 - \Delta P_4 / b_{24} \\
\Delta\theta_4 &= \Delta\theta_1 - \Delta P_5 / b_{14}
\end{aligned} \tag{6.82}$$

From the node relationships,

$$\begin{aligned}
\Delta P_1 + \Delta P_5 &= \Delta P_{1i} \\
-\Delta P_1 + \Delta P_2 - \Delta P_4 &= \Delta P_{2i} \\
-\Delta P_2 + \Delta P_3 &= \Delta P_{3i}
\end{aligned} \tag{6.83}$$

or in the matrix form,

$$\begin{vmatrix}
-1/b_{12} & 0 & 0 & 0 & 0 & \vdots & 1 & -1 & 0 \\
0 & -1/b_{23} & 0 & 0 & 0 & \vdots & 0 & 1 & -1 \\
0 & 0 & -1/b_{34} & 0 & 0 & \vdots & 0 & 0 & 1 \\
0 & 0 & 0 & -1/b_{24} & 0 & \vdots & 0 & -1 & 0 \\
0 & 0 & 0 & 0 & -1/b_{14} & \vdots & 1 & 0 & 0 \\
\cdots & \cdots & \cdots & \cdots & \cdots & \cdots & \cdots & \cdots & \cdots \\
1 & 0 & 0 & 0 & 1 & \vdots & 0 & 0 & 0 \\
-1 & 1 & 0 & -1 & 0 & \vdots & 0 & 0 & 0 \\
0 & -1 & 1 & 0 & 0 & \vdots & 0 & 0 & 0
\end{vmatrix}
\begin{vmatrix}
\Delta P_1 \\ \Delta P_2 \\ \Delta P_3 \\ \Delta P_4 \\ \Delta P_5 \\ \cdots \\ \Delta\theta_1 \\ \Delta\theta_2 \\ \Delta\theta_3
\end{vmatrix}
=
\begin{vmatrix}
0 \\ 0 \\ 0 \\ \Delta\alpha \\ 0 \\ \cdots \\ \Delta P_{1i} \\ \Delta P_{2i} \\ \Delta P_{3i}
\end{vmatrix} \tag{6.84}$$

This can be written as follows:

$$\begin{vmatrix} -\bar{z} & \bar{C} \\ \bar{C}^t & \bar{Y} \end{vmatrix} \begin{vmatrix} \Delta P \\ \Delta\theta \end{vmatrix} = \begin{vmatrix} \Delta\alpha \\ \Delta P_i \end{vmatrix} \tag{6.85}$$

or

$$\begin{aligned}
\begin{vmatrix} \Delta P \\ \Delta\theta \end{vmatrix} &= \begin{vmatrix} \bar{y} & \bar{T} \\ \bar{T}^t & \bar{Z} \end{vmatrix} \begin{vmatrix} \Delta\alpha \\ \Delta P_i \end{vmatrix} \\
&= \left|\bar{S}_\theta\right| \begin{vmatrix} \Delta\alpha \\ \Delta P_i \end{vmatrix}
\end{aligned} \tag{6.86}$$

where $\bar{S}_\theta$ is the sensitivity matrix for P–θ network, $\bar{C}$ is the connection matrix relating incidence nodes to branches, $\bar{Y}$ is the submatrix of shunt elements if present (normally=0), and $\bar{z}$ is the submatrix of branch impedances.

In Equation 6.86, $\bar{y}$ is the submatrix of terms relating a change in branch flow ΔP to a change in phase-shifter angle $\Delta\alpha$, $\bar{T}$ is the submatrix of terms relating a change in branch flow ΔP to a change in node power injection ΔP_i, $\bar{T}^t$ is the submatrix relating a change in phase angle $\Delta\theta$ to a change in phase-shifter angle $\Delta\alpha$, and $\bar{Z}$ is the submatrix of terms relating a change in node phase angle $\Delta\theta$ to node injection ΔP_i.

The sensitivity factor relating MW change to generation at bus i to an MW change in flow of line j is

$$\Delta P_j/\Delta P_{Gi} = (Z_{mi} - Z_{ni})b_j \tag{6.87}$$

where m and n are the respective terminal buses of line j.

The sensitivity factor relating a phase angle change in a phase shifter in line l to an MW change in flow of line j is

$$\Delta P_j/\Delta\alpha_l = \left[(Z_{ms} - Z_{mr} - Z_{ns} + Z_{nr})b_l\right]b_j \tag{6.88}$$

where buses m and n are the terminal buses of line j, and s and r are the terminal buses of phase-shifter circuit l.

6.9.2 Q–V Network

The dc formation of a transmission line, Figure 6.8a, is given by

$$\begin{aligned}\Delta q_{ij} &\approx (\Delta V_i - \Delta V_j)b_{ij} - 2b_{\text{sh}}\Delta V_i \\ &= \Delta q_{ij}b_{ij} - 2b_{\text{sh}}V_i\end{aligned} \tag{6.89}$$

In Equation 6.84, b_{sh} is one-half of the total shunt susceptance, i.e., the total line shunt susceptance=$2b_{\text{sh}}$. We obtain twice the normal susceptance in each leg of the Π model because of partial differentiation of the power flow equation with respect to voltage; ΔQ_{ijl} pertains to the line, see Figure 6.8a. Looking from the receiving end,

$$\begin{aligned}\Delta q_{ji} &\approx (\Delta V_j - \Delta V_i)b_{ji} - 2b_{\text{sh}}\Delta V_j \\ &= \Delta q b_{ji} - 2b_{\text{sh}}V_j\end{aligned} \tag{6.90}$$

The dc model of a voltage tap adjusting transformer, Figure 6.8b, is

$$\begin{aligned}\Delta q_{ij} &= (\Delta V_i - \Delta V_j - \Delta\alpha)b_{ij} \\ \Delta q_{ji} &= (\Delta V_j - \Delta V_i - \Delta\alpha)b_{ji}\end{aligned} \tag{6.91}$$

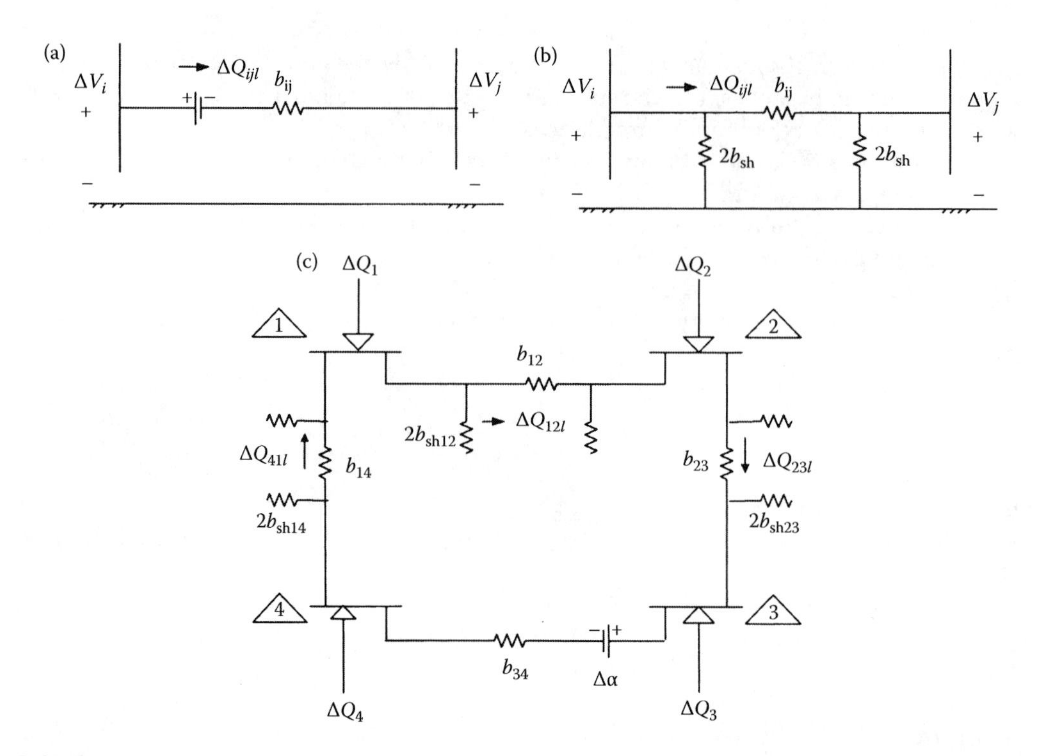

FIGURE 6.8
(a) DC circuit Q–V model of a tie-line, (b) dc circuit Q–V model of a phase-shifting transformer, and (c) circuit diagram for Q–V network equations.

In the circuit of Figure 6.8c, the generator terminal voltages ΔV_2 and ΔV_4 are considered as control parameters and ΔQ_2 and ΔQ_4 are the dependent parameters. Following equations can be written, based on branch relations:

$$
\begin{aligned}
&\Delta V_1 - \Delta V_2 = \Delta q_1 / b_1 \\
&\Delta V_2 - \Delta V_3 = \Delta q_2 / b_2 \\
&\Delta V_3 - \Delta V_4 - \Delta\alpha = \Delta q_3 / b_3 \\
&\Delta V_4 - \Delta V_1 = \Delta q_4 / b_4
\end{aligned}
\tag{6.92}
$$

From the node relations,

$$
\begin{aligned}
&\Delta q_1 + \Delta q_4 - 2(b_{\text{sh}1} + b_{\text{sh}4})\Delta V_1 = \Delta Q_1 \\
&-\Delta q_1 + \Delta q_2 - \Delta Q_2 = 2(b_{\text{sh}1} + b_{\text{sh}2})\Delta V_2 \\
&-\Delta q_2 + \Delta q_3 - 2b_{\text{sh}2}\Delta V_3 = \Delta Q_3 \\
&-\Delta q_4 - \Delta q_3 - \Delta Q_4 = 2b_{\text{sh}4}\, \Delta V_4
\end{aligned}
\tag{6.93}
$$

or in the matrix form,

$$\begin{bmatrix} -1/b_1 & 0 & 0 & 0 & \vdots & 1 & 0 & 0 & 0 \\ 0 & -1/b_2 & 0 & 0 & \vdots & 0 & 0 & -1 & 0 \\ 0 & 0 & -1/b_3 & 0 & \vdots & 0 & 0 & 1 & 0 \\ 0 & 0 & 0 & -1/b_4 & \vdots & 0 & 0 & 0 & 0 \\ \cdots & \cdots & \cdots & \cdots & \cdots & \cdots & \cdots & \cdots & \cdots \\ 1 & 0 & 0 & 1 & \vdots & -2b_{sh1}-2b_{sh4} & 0 & 0 & 0 \\ -1 & 1 & 0 & 0 & \vdots & 0 & -1 & 0 & 0 \\ 0 & -1 & 1 & 0 & \vdots & 0 & 0 & -2b_{sh2} & 0 \\ 0 & 0 & -1 & -1 & \vdots & 0 & 0 & 0 & -1 \end{bmatrix} \begin{bmatrix} \Delta q_1 \\ \Delta q_2 \\ \Delta q_3 \\ \Delta q_4 \\ \cdots \\ \Delta V_1 \\ \Delta Q_2 \\ \Delta V_3 \\ \Delta Q_4 \end{bmatrix}$$

$$= \begin{bmatrix} 1 & 0 & 0 & 0 & \vdots & 0 & 1 & 0 & 0 \\ 0 & 1 & 0 & 0 & \vdots & 0 & -1 & 0 & 0 \\ 0 & 0 & 1 & 0 & \vdots & 0 & 0 & 0 & 1 \\ 0 & 0 & 0 & 1 & \vdots & 0 & 0 & 0 & 1 \\ \cdots & \cdots & \cdots & \cdots & \cdots & \cdots & \cdots & \cdots & \cdots \\ 0 & 0 & 0 & 0 & \vdots & 1 & 0 & 0 & 0 \\ 0 & 0 & 0 & 0 & \vdots & 0 & 2b_{sh1}+2b_{sh2} & 0 & 0 \\ 0 & 0 & 0 & 0 & \vdots & 0 & 0 & 1 & 0 \\ 0 & 0 & 0 & 0 & \vdots & 0 & 0 & 0 & 2b_{sh4} \end{bmatrix} \begin{bmatrix} 0 \\ 0 \\ \Delta\alpha \\ 0 \\ \cdots \\ \Delta Q_1 \\ \Delta V_2 \\ \Delta Q_3 \\ \Delta V_4 \end{bmatrix} \tag{6.94}$$

or in the abbreviated form,

$$\bar{J}\Delta x = H\Delta u \tag{6.95}$$

The sensitivity factor relating an Mvar change in injection at bus i to an Mvar change in flow of line j is given by

$$\Delta q_j/\Delta q_i = (Z_{mi} - Z_{ni})b_j \tag{6.96}$$

where buses m and n are the respective buses of line j.

The sensitivity factor relating an Mvar change in injection at bus i to a voltage change at bus k is given by

$$\Delta V_k / \Delta Q_j = Z_{ki} \tag{6.97}$$

The sensitivity factor relating a change in voltage at the generator bus i to an Mvar change in flow of line j is

$$\Delta q_j / \Delta V_i = \left[(Z_{ms} - Z_{ns})b_{is} + (Z_{mr} - Z_{nr})b_{ir}\right]b_{mn} \tag{6.98}$$

where the generator is connected to bus i with terminal buses s and r.

The sensitivity factor relating a change in voltage at a generator bus i to a voltage change at a bus k is

$$\Delta V_k / \Delta V_i = Z_{ks} b_{is} + Z_{kr} b_{ir}$$

6.10 Second-Order Load Flow

In Equation 6.2 Taylor's series, we neglected second-order terms. Also from Equation 6.22 and including second-order terms, we can write

$$\begin{vmatrix} \Delta P \\ \Delta Q \end{vmatrix} = \bar{J} \begin{vmatrix} \Delta h \\ \Delta e \end{vmatrix} + \begin{vmatrix} SP \\ SQ \end{vmatrix} \tag{6.99}$$

Equation 6.99 has been written in rectangular coordinates. It can be shown that second-order terms can be simplified as [5]

$$\begin{aligned} SP_i &= P_i^{\text{cal}}(\Delta e, \Delta h) \\ SQ_i &= Q_i^{\text{cal}}(\Delta e, \Delta h) \end{aligned} \tag{6.100}$$

No third-order or higher order terms are present. If Equation 6.99 is solved, an exact solution or near to exact solution can be expected in the very first iteration. The second-order terms can be estimated from the previous iteration, and ΔP, ΔQ are corrected by the second-order term as

$$\begin{vmatrix} \Delta P - SP \\ \Delta Q - SQ \end{vmatrix} = \bar{J} \begin{vmatrix} \Delta h \\ \Delta e \end{vmatrix} \tag{6.101}$$

or

$$\begin{vmatrix} \Delta h \\ \Delta e \end{vmatrix} = \bar{J}^{-1} \begin{vmatrix} \Delta P - SP \\ \Delta Q - SQ \end{vmatrix} \tag{6.102}$$

Initially, the vectors SP and SQ are set to zero. And $P^{\text{cal}}, Q^{\text{cal}}$ are estimated. For a PV bus, the voltage magnitude is fixed; therefore, the increments must satisfy

$$e_i \Delta e_i + h_i \Delta h_i = V_i \Delta V_i \tag{6.103}$$

The voltages are updated

$$\begin{aligned} e^{k+1} &= e^k + \Delta e \\ h^{k+1} &= h^k + \Delta h \end{aligned} \tag{6.104}$$

The second-order terms SP and SQ are calculated using $\Delta P, \Delta Q$, and $P^{\text{cal}}, Q^{\text{cal}}$ are reestimated. The number of iterations and CPU time decrease with large systems over NR method.

Twenty-five years back there was, practically, no power system software available for PC use. Some of the commercial power system software available today claim unlimited bus numbers and the number of iterations is not so much of a consideration. The programs allow more than one load flow algorithm to be chosen. The computing requirements and storage for PC may be heavy; the project files may run into hundreds of Megabytes (MBs).

6.11 Load Models

Load modeling has a profound impact on load flow studies. Figure 6.9 shows the effect of change of operating voltage on constant current, constant MVA, and constant impedance load types. Heavy industrial motor loads are approximately constant MVA loads, while commercial and residential loads are mainly constant impedance loads. Classification into commercial, residential, and industrial is rarely adequate and one approach has been to divide the loads into individual load components. The other approach is based on measurements. Thus, the two approaches are

- Component-based models
- Models based on measurements

A component-based model is a *bottom-up* approach in the sense that different load components comprising the loads are identified. Each load component is tested to determine the relations between real and reactive power requirements versus voltage and frequency. A load model in exponential or polynomial form can then be developed from the test data.

The measurement approach is a *top-down* approach in the sense that the model is based on the actual measurement. The effect of the variation of voltage on active and reactive power consumption is recorded and, based on these, the load model is developed.

A composite load, i.e., a consumer load consisting of heating, air-conditioning, lighting, computers, and television, is approximated by combining load models in certain proportions based on load surveys. This is referred to as a *load window*.

Construction of the load window requires certain data, i.e., load saturation, composition, and diversity data. Any number of load windows can be defined.

The load models are normalized to rated voltage, rated power, and rated frequency and are expressed in per unit. The exponential load models are

$$\frac{P}{P_n} = \left|\frac{V}{V_n}\right|^{\alpha_v} \left|\frac{f}{f_n}\right|^{\alpha_f} \tag{6.105}$$

$$\frac{Q}{P_n} = \frac{Q_0}{P_0}\left|\frac{V}{V_n}\right|^{\beta_v} \left|\frac{f}{f_n}\right|^{\beta_f} \tag{6.106}$$

where V is the initial value of the voltage, P_0 is the initial value of the power, V_n is the adjusted voltage, and P_n is the power corresponding to this adjusted voltage. The exponential factors depend on the load type. The frequency dependence of loads is ignored for load flow studies and is applicable to transient and dynamic stability studies. Another form of load model is the polynomial model.

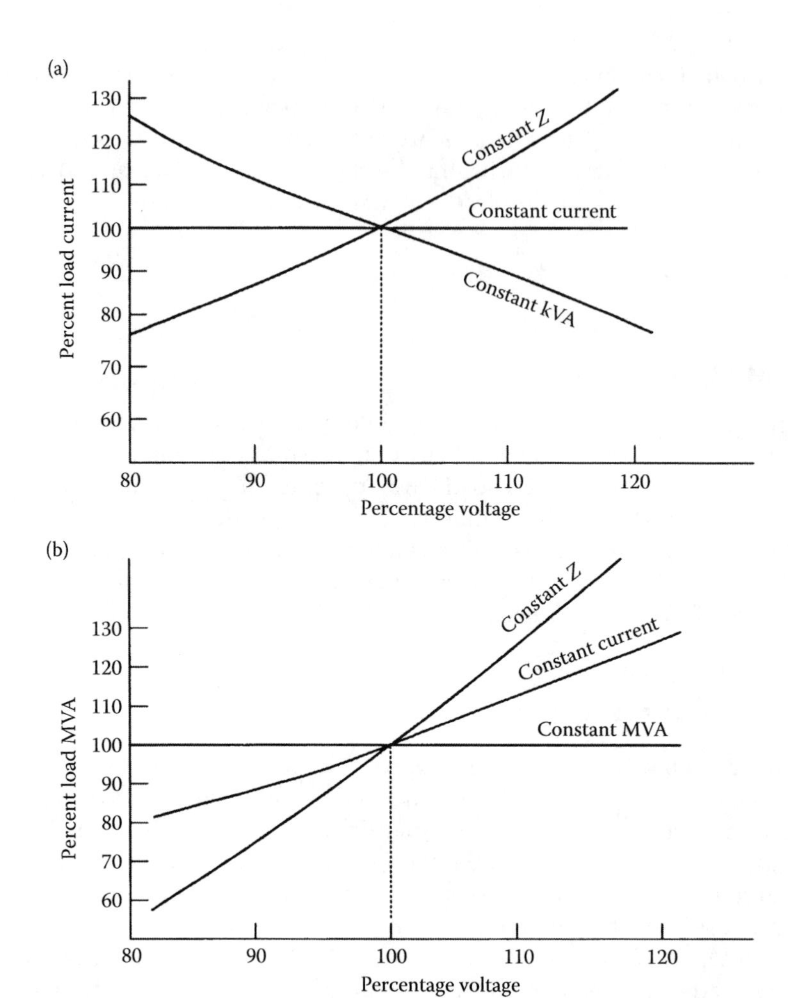

FIGURE 6.9
(a) Behavior of constant current, constant MVA, and constant impedance loads with respect to current loading as a function of voltage variations and (b) behavior of constant current, constant MVA, and constant impedance loads with respect to MVA loading as a function of voltage variations.

Consider the equation

$$P = V^n \tag{6.107}$$

For all values of n, $P=0$ when $V=0$, and $P=1$ when $V=1$, as it should be. Differentiating

$$\frac{dP}{dV} = nV^{n-1} \tag{6.108}$$

For $V=1$, $dP/dV=n$. The value of n can be found by experimentation if dP/dV, i.e., change in active power with change in voltage, is obtained. Also, by differentiating $P=VI$.

$$\frac{dP}{dV} = I + V\frac{dI}{dV} \tag{6.109}$$

For V and n equal to unity, the exponential n from Equations 6.108 and 6.110 is

$$n = 1 + \frac{\Delta I}{\Delta V} \tag{6.110}$$

The exponential n for a composite load can be found by experimentation if the change in current for a change in voltage can be established. For a constant power load n=0, for a constant current load n=1, and for a constant MVA load n=2. The composite loads are a mixture of these three load types and can be simulated with values of n between 0 and 2. The following are some quadratic expressions for various load types [6]. An EPRI report [7] provides more detailed models.

Air conditioning

$$P = 2.18 + 0.298V - 1.45V^{-1} \tag{6.111}$$

$$Q = 6.31 - 15.6V + 10.3V^2 \tag{6.112}$$

Fluorescent lighting

$$P = 2.97 - 4.00V + 2.0V^2 \tag{6.113}$$

$$Q = 12.9 - 26.8V + 14.9V^2 \tag{6.114}$$

Induction motor loads

$$P = 0.720 + 0.109V + 0.172V^{-1} \tag{6.115}$$

$$Q = 2.08 + 1.63V - 7.60V^2 + 4.08V^3 \tag{6.116}$$

Table 6.2 shows approximate load models for various load types. A synchronous motor is modeled as a synchronous generator with negative active power output and positive reactive power output, akin to generators, assuming that the motor has a leading rated

TABLE 6.2

Representation of Load Models in Load Flow

Load Type	*P* (Constant kVA)	*Q* (Constant kVA)	*P* (Constant Z)	*Q* (Constant Z)	Generator Power	Generator Reactive (min/max)
Induction motor running	$+P$	Q (lagging)				
Induction or synchronous motors starting			$+P$	Q (lagging)		
Generator					$+P$	$Q_{(max)}/Q_{(max)}$*
Synchronous motor					$-P$	Q (leading)
Power capacitor				Q (leading)		
Rectifier	$+P$			Q (lagging)		
Lighting			$+P$	Q (lagging)		

Q_{min}* for generator can be leading.

power factor. The reactive power supplied into the system depends on the type of excitation and control system. As a load flow provides a static picture, the time constants applicable with control devices are ignored, i.e., if a load demands a certain reactive power output, which is within the capability of a generator, it is available instantaneously. The drive systems can be assumed to have a constant active power demand with the reactive power demand varying with the voltage.

6.12 Practical Load Flow Studies

The requirements for load flow calculations vary over a wide spectrum, from small industrial systems to large automated systems for planning, security, reactive power compensation, control, and online management. The essential requirements are as follows:

- High speed, especially important for large systems.
- Convergence characteristics, which are of major consideration for large systems, and the capability to handle ill-conditioned systems.
- Ease of modifications and simplicity, i.e., adding, deleting, and changing system components, generator outputs, loads, and bus types.
- Storage requirement, which becomes of consideration for large systems.

The size of the program in terms of number of buses and lines is important. Practically, all programs will have data reading and editing libraries, capabilities of manipulating system variables, adding or deleting system components, generation, capacitors, or slack buses. Programs have integrated databases, i.e., the impedance data for short circuit or load flow calculations need not be entered twice, and graphic user interfaces. Which type of algorithm will give the speediest results and converge easily is difficult to predict precisely. Table 5.7 shows a comparison of earlier *Z*- and *Y*-matrix methods. Most programs will incorporate more than one solution method. While the Gauss–Seidel method with acceleration is still an option for smaller systems, for large systems, some form of the NR decoupled method and fast load flow algorithm are commonly used, especially for optimal power flow studies. Speed can be accelerated by optimal ordering (Appendix B in vol. 1). In fast decoupled load flow, the convergence is geometric, and less than five iterations are required for practical accuracies. If differentials are calculated efficiently, the speed of the fast decoupled method can be even five times that of the NR method. Fast decoupled load flow is employed in optimization studies and in contingency evaluation for system security.

The preparation of data, load types, extent of system to be modeled, and specific problems to be studied are identified as a first step. The data entry can be divided into four main categories: bus data, branch data, transformers and phase shifters, and generation and load data. Shunt admittances, i.e., switched capacitors and reactors in required steps, are represented as fixed admittances. Apart from voltages on the buses, the study will give branch power flows; identify transformer taps, phase-shifter angles, loading of generators and capacitors, power flow from swing buses, load demand, power factors, system losses, and overloaded system components.

Example 6.5

Consider a 15-bus network shown in Figure 6.10. It is solved by the fast decoupled load flow method, using a computer program. The impedance data are shown in Table 6.3. The bus types and loading input data are shown in Table 6.4. The bus voltages on *PV* buses and generator maximum and minimum var limits are specified in this table.

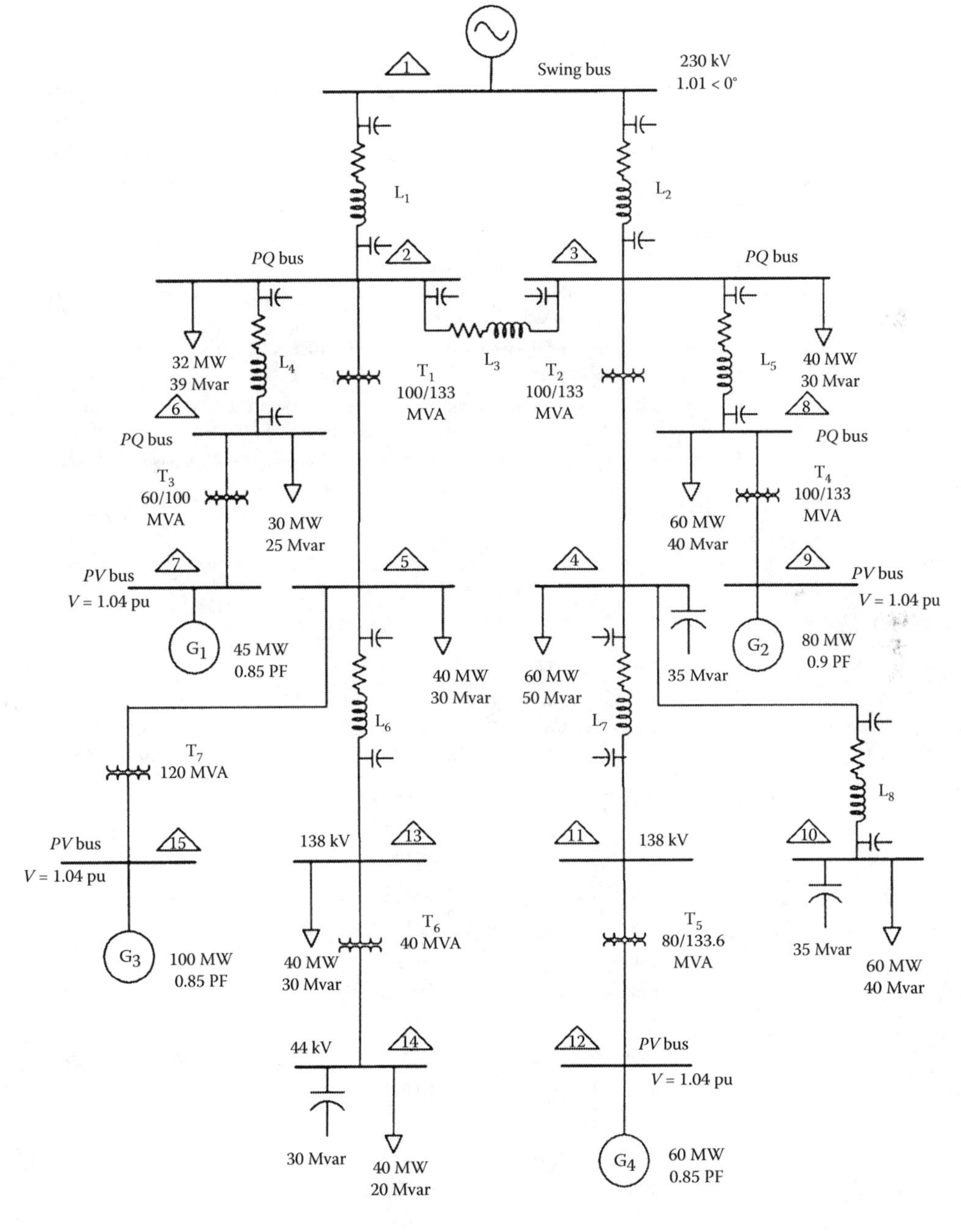

FIGURE 6.10
A 15-bus power system for Examples 6.5 and 6.6.

TABLE 6.3

Examples 6.5 and 6.6: Impedance Data

From Bus	To Bus	Connection Type	Impedance Data in per Unit on a 100 MVA Base, Unless Specified Otherwise
1	2	L1: 230 kV transmission line	$Z=0.0275+j0.1512$, Π model, $y=j0.2793$
1	3	L2: 230 kV transmission line	$Z=0.0275+j0.1512$, Π model, $y=j0.2793$
2	3	L3: 230 kV transmission line	$Z=0.0413+j0.2268$, Π model, $y=j0.4189$
2	6	L4: 230 kV transmission line	$Z=0.0157+j0.008$, Π model, $y=j0.1396$
3	8	L5: 230 kV transmission line	$Z=0.0157+j0.008$, Π model, $y=j0.1396$
4	11	L6: 138 kV transmission line	$Z=0.0224+j0.1181$, Π model, $y=j0.0301$
4	10	L7: 138 kV transmission line	$Z=0.0224+j0.1181$, Π model, $y=j0.0301$
5	13	L8: 138 kV transmission line	$Z=0.0373+j0.1968$, Π model, $y=j0.0503$
2	5	T1: 100/133 MVA two-winding transformer, 230–138 kV	$Z=10\%$ on a 100 MVA base, $X/R=34.1$, tap 1:1.02
3	4	T2: 100/133 MVA two-winding transformer, 230–138 kV	$Z=10\%$ on a 100 MVA base, $X/R=34.1$, tap 1:1.03
6	7	T3: 60/100.2 MVA transformer, 230–13.8 kV (generator step-up transformer)	$Z=10\%$ on a 60 MVA base, $X/R=34.1$, tap 1:1.03
8	9	T4: 100/133 MVA transformer, 230–13.8 kV (generator step up transformer)	$Z=10\%$ on a 100 MVA base, $X/R=34.1$, tap 1:1.04
11	12	T5: 80/133.6 MVA two-winding transformer, 138–13.8 kV (generator step-up transformer)	$Z=9\%$ on an 80 MVA base, $X/R=34.1$, tap 1:1.05
13	14	T6: 40 MVA two-winding transformer 138–44 kV	$Z=8\%$ on a 40 MVA base, $X/R=27.3$, tap 1:1.01
5	15	T7: 120 MVA two-winding transformer (generator step-up transformer), 138–13.8 kV	$Z=11\%$ on a 120 MVA base, $X/R=42$, tap 1:1.02

TABLE 6.4

Examples 6.5 and 6.6: Bus, Load, and Generation Data

Bus Identification	Type of Bus	Nominal Voltage (kV)	Specified Voltage (per unit)	Bus Load (MW/Mvar)	Shunt Capacitor (Mvar)	Generator Rating (MW) and Max/Min var Output (Mvar)
1	Swing	230	1.01<0	–		–
2	*PQ*	230	–	52/39		
3	*PQ*	230	–	40/30		
4	*PQ*	138	–	60/50	35	
5	*PQ*	138	–	40/30		
6	*PQ*	230	–	30/25		
7	*PV*	13.8	1.04 < ?	–	–	45 MW, 0.85 pF (28/3)
8	*PQ*	230	–	60/40		
9	*PV*	13.8	1.04 < ?	–		80 MW, 0.9 pF (38/3)
10	*PQ*	138	–	60/40	35	
11	*PQ*	138	–	–		
12	*PV*	13.8	1.04 < ?	–	–	60 MW, 0.85 pF (37/3)
13	*PQ*	138	–	40/30	–	–
14	*PQ*	44		40/30	30	
15	*PV*	13.8	1.04 < ?			100 MW, 0.85 pF (62/6)

All loads are modeled as *constant kVA loads.* There are 100 Mvar of shunt power capacitors in the system; this is always a constant impedance load.

The results of the load flow calculation are summarized in Tables 6.5 through 6.7. Table 6.5 shows bus voltages, Table 6.6 shows power flows, losses, bus voltages, and voltage drops, and Table 6.7 shows the overall load flow summary. The following observations are of interest:

- Voltages on a number of buses in the distribution are below acceptable level. Voltage on the 138 kV bus 10 is 12% below rated voltage and at bus 4 is 6.1% below rated voltage. There is a 5.85% voltage drop in transmission line L8. The voltages at buses 2 and 3 are 4.2 and 6.6%, respectively, below rated voltages. Bus 9 voltage cannot be maintained at 1.04 per unit with the generator supplying its maximum reactive power. When corrective measures are to be designed, it is not necessary to address all the buses having low operating voltages. In this example, correcting voltages at 230 kV buses 2 and 3 will bring up the voltages in the rest of the system. Bus 10 will require additional compensation.

TABLE 6.5

Example 6.5: Bus Voltages, Magnitudes, and Phase Angles (Constant kVA Loads)

Bus#	Voltage	Bus#	Voltage	Bus#	Voltage
1	1.01 < 0.0°	6	0.9650 < –5.0°	11	0.9696 < –6.4°
2	0.9585 < –5.7°	7	1.04 < –0.7°	12	1.04 < –2.4°
3	0.9447 < –6.6°	8	0.9445 < –5.5°	13	0.9006 < –13.0°
4	0.9393 < –10.7°	9	1.0213 < –0.6°	14	0.9335 < –18.7°
5	0.9608 < –7.1°	10	0.8808 < –18.6°	15	1.03.54 < –1.8°

TABLE 6.6

Example 6.5: Active and Reactive Power Flows, Losses, Bus Voltages, and Percentage Voltage Drops (Constant kVA Loads)

	Connected Buses		From-to Flow		To-from Flow		Losses		Bus Voltage (%)		
Circuit	From Bus#	To Bus #	MW	Mvar	MW	Mvar	kW	kvar	From	To	Voltage Drop (%)
L1	1	3	68.174	10.912	–66.748	–30.160	1426.0	–19248.6	101.00	95.85	5.15
L2	1	2	78.177	19.253	–76.219	–35.239	1957.7	–15986.0	101.00	94.47	6.53
L3	2	3	6.819	–14.602	–6.788	–23.169	30.6	–37770.9	95.85	94.47	1.38
L4	2	6	–14.794	–11.240	14.835	–1.466	41.3	–12705.9	95.85	96.50	0.65
T_1	2	5	22.732	17.002	–22.697	–16.126	25.7	876.3	95.85	96.08	0.23
L5	3	8	–19.786	–2.021	19.857	–10.072	71.9	–12092.4	94.47	94.45	0.01
T_2	3	4	62.792	30.428	–62.632	–24.975	159.9	5453.6	94.47	93.93	0.53
L7	4	11	–58.956	–12.124	59.867	14.183	910.9	2059.9	93.93	96.96	3.03
L8	4	10	61.685	18.025	–59.904	–12.790	1781.3	5234.9	93.93	88.08	5.85
L6	5	13	82.402	35.807	–80.421	–27.964	1981.2	7842.5	96.08	90.06	6.01
T_7	5	15	–99.704	–49.680	–99.997	61.999	293.3	12319.0	96.08	103.54	7.46
T_3	6	7	–44.865	–23.418	44.999	28.000	134.4	4581.9	96.50	103.99	7.49
T_4	8	9	–79.761	–29.872	79.999	38.000	238.3	8127.8	94.45	102.13	7.68
T_5	11	12	–59.867	–14.183	60.000	18.711	132.8	4527.8	96.96	104.00	7.04
T_6	13	14	40.422	–2.035	–40.274	6.072	147.8	4036.1	90.06	93.35	3.29

TABLE 6.7

Example 6.5: Load Flow Summary (Constant kVA Loads)

Item	MW	Mvar	MVA	Percentage Power Factor
Swing bus	146.35	30.16	149.43	97.9 lagging
Generators	285.00	146.71	320.55	88.90 lagging
Total demand	431.35	176.875	466.21	92.5 lagging
Total load modeled	422.01 MW, constant kVA	303 Mvar, constant kVA	519.99	81.2 lagging
Power capacitors output		–84.18		
Total losses	9.337	42.742		
System mismatch	0.010	0.003		

- As a constant kVA load model is used, the load demand is the sum of the total load plus the system losses (Table 6.7). It is 431.35 MW and 176.88 Mvar, excluding the Mvar supplied by shunt capacitors.
- There is a total of 100 Mvar of power capacitors in the system. These give an output of 84.14 Mvar. The output is reduced as a square of the ratio of operating voltage to rated voltage. The behavior of shunt power capacitors is discussed in Chapter 8.
- The generators operate at their rated active power output. However, there is little reactive power reserve as the overall operating power factor of the generators is 88.90 lagging. Some reserve spinning power capability should be available to counteract sudden load demand or failure of a parallel line or tie circuit.
- There are two devices in tandem that control the reactive power output of the generators—the voltage-ratio taps on the transformer and the generator voltage regulators. The role of these devices in controlling reactive power flow is discussed in Chapter 8.
- System losses are 9.337 MW and 42 Mvar. An explanation of the active power loss is straightforward; the reactive power loss mainly depends on the series reactance and how far the reactive power has to travel to the load (Chapter 8). Table 6.6 identifies branch flows from-and-to between buses and the losses in each of the circuits. The high-loss circuits can be easily identified.

Example 6.6

The load flow of Example 6.5 is repeated with constant impedance representation of loads. All other data and models remain unchanged.

Table 6.8 shows bus voltages, Table 6.9 branch load flow and losses, and Table 6.10 an overall load summary. The results can be compared with those of the corresponding

TABLE 6.8

Example 6.6: Bus Voltages, Magnitudes, and Phase Angles (Constant Impedance Loads)

Bus#	Voltage	Bus#	Voltage	Bus#	Voltage
1	$1.01 < 0.0°$	6	$0.9750 < -3.5°$	11	$0.9841 < -4.0°$
2	$0.9717 < -4.3°$	7	$1.04 < 0.8°$	12	$1.04 < 0.0°$
3	$0.9659 < -5.0°$	8	$0.9646 < -3.8°$	13	$0.9312 < -10.2°$
4	$0.9692 < -8.2°$	9	$1.04 < 0.9°$	14	$0.9729 < -15.2°$
5	$0.9765 < -5.1°$	10	$0.9375 < -14.7°$	15	$1.04 < 0.2°$

TABLE 6.9

Example 6.6: Active and Reactive Power Flows, Losses, Bus Voltages, and Percentage Voltage Drops (Constant Impedance Type Loads)

	Connected buses		From-to Flow		To-from Flow		Losses		Bus Voltage (%)		
Circuit	From Bus#	To Bus #	MW	Mvar	MW	Mvar	kW	kvar	From	To	Voltage Drop (%)
L1	1	3	51.949	3.668	–51.134	–26.625	815.4	–22957.8	101.00	97.17	3.83
L2	1	2	60.344	6.662	–59.240	–27.892	1103.8	–21230.4	101.00	96.59	4.41
L3	2	3	5.512	–18.254	–5.498	–20.993	14.3	–39247.6	97.17	96.59	0.58
L4	2	6	–16.287	–7.255	16.331	–5.751	44.1	–13005.9	97.17	97.50	0.33
T_1	2	5	12.807	15.307	–12.794	–14.885	12.4	421.6	97.17	97.65	0.47
L5	3	8	–23.729	–2.245	23.826	–10.291	97.6	–12536.2	96.59	96.64	0.05
T_2	3	4	51.148	23.141	–51.049	–19.764	99.0	3376.6	96.59	96.92	0.33
L7	4	11	–59.046	–0.277	59.877	1.785	830.8	1508.3	96.92	98.41	1.49
L8	4	10	53.830	5.994	–52.652	–4.345	1178.1	1648.2	96.92	93.75	3.17
L6	5	13	74.394	25.049	–72.930	–20.070	1463.3	4978.4	97.65	93.12	4.53
T_7	5	15	–99.738	–38.769	–100.00	49.774	262.0	11005.2	97.65	104.00	6.35
T_3	6	7	–44.880	–17.905	45.00	24.301	120.0	4091.6	97.50	104.00	6.50
T_4	8	9	–79.773	–27.016	80.00	21.996	222.7	7592.4	96.64	104.00	7.36
T_5	11	12	–59.878	–1.785	60.00	34.608	122.2	4167.1	98.41	104.00	5.59
T_6	13	14	38.248	–5.943	–38.121	9.396	126.5	3453.3	94.12	97.29	4.17

TABLE 6.10

Example 6.6: Load Flow Summary (Constant Impedance-Type Load)

Item	MW	Mvar	MVA	Percentage Power Factor
Swing bus	112.29	10.39	112.77	99.6 lagging
Generators	285.00	112.33	306.34	93.0 lagging
Total demand	397.293	122.660	415.80	95.5 lagging
Total load demand	390.785	189.40	434.26	89.9 lagging
Total constant impedance-type load modeled	422.01	303 (capacitors)		
Total losses	6.51	66.74		
System mismatch	0.011	0.001		

tables for the constant kVA load model of Example 6.5. The following observations are of interest:

- In Example 6.5, with a constant kVA load, the overall demand (excluding the reactive power supplied by capacitors) is 431.35 MW and 176.88 Mvar. With the constant impedance load model, "off-loading" occurs with voltage reduction, and the load demand reduces to 397.29 MW and 122.66 Mvar. This helps the system to recover and the resulting voltage drops throughout are smaller as compared to those of a constant kVA load.
- The bus voltages improve in response to the reduced load demand. These are 2–5% higher as compared to a constant kVA load model.

- The generators reactive output reduces in response to increased bus voltage. The redistribution of active power changes the phase angles of the voltages.
- The active power loss decreases and the reactive power loss increases. This can be explained in terms of improved bus voltages, which allow a greater transfer of power between buses and hence a larger reactive power loss.

This demonstrates the importance of correct load models in load flow calculations.

6.12.1 Contingency Operation

Consider now that the largest generator of 100 MW on bus 15 is tripped. Assuming that there is no widespread disruption, the operating voltages on buses 13–15 dip by 12%–15%. Also, the voltage on the 230 kV bus 9 is down by approximately 9%. Even if the swing bus has all the spinning reserve and the capability to supply the increased demand due to loss of the 100 MW generator, and none of the system components, transformers, or lines is overloaded, the system has potential voltage problems.

While this load flow is illustrative of what to expect in load flow results, the number of buses in a large industrial distribution system may approach 500 or more and in the utility systems 5000 or more. Three-phase models may be required. The discussions are continued in the chapters to follow.

Problems

(These problems can be solved without modeling on a digital computer.)

6.1 Solve the following equations, using the NR method, to the second iteration. Initial values may be assumed as equal to zero:

$$9x_1 - x_1x_2 - 6 = 0$$
$$x_1 + 6x_2 - x_3^2 - 10 = 0$$
$$x_2x_3^2 - 10x_3 + 4 = 0$$

6.2 Write the power flow equations of the network shown in Figure P6.1, using the NR method in polar form.

6.3 Solve the three-bus system of Figure P6.1, using the NR method, to two iterations.

6.4 Solve Problem 6.3, using the decoupled NR method, to two iterations.

6.5 Construct P–θ and Q–V networks of the system shown in Figure P6.1.

6.6 Solve the network of Figure P6.1, using the fast decoupled method, to two iterations.

6.7 Solve the network of Figure P6.1 using the Ward–Hale method.

6.8 What are the major differences in constant kVA, constant impedance, and constant current loads? How will each of these behave under load flow?

6.9 What are the dimensions of the Y matrix in Example 6.5? What is the percentage of populated elements with respect to the total elements?

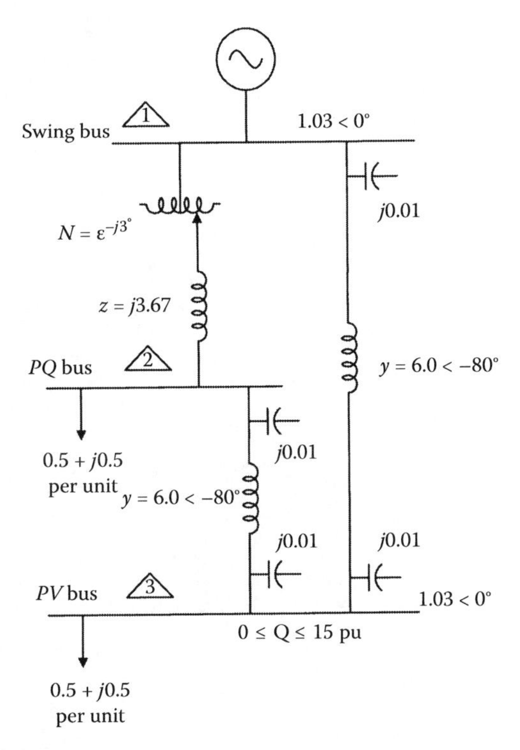

FIGURE P6.1
A three-bus system for Problems 2, 3, 5, 6, and 7.

References

1. RJ Brown, WF Tinney. Digital solution of large power networks. *Trans AIEE Power App Syst*, 76, 347–355, 1957.
2. WF Tinney, CE Hart. Power flow solution by Newton's method. *Trans IEEE Power App Syst*, 86, 1449–1456, 1967.
3. B Stott, O Alsac. Fast decoupled load flow. *Trans IEEE Power App Syst*, 93, 859–869, 1974.
4. NM Peterson, WS Meyer. Automatic adjustment of transformer and phase-shifter taps in the Newton power flow. *IEEE Trans Power App Syst*, 90, 103–108, 1971.
5. MS Sachdev, TKP Medicheria. A second order load flow technique. *IEEE Trans Power App Syst*, PAS-96, 189–195, 1977.
6. MH Kent, WR Schmus, FA McCrackin, LM Wheeler. Dynamic modeling of loads in stability studies. *Trans IEEE Power App Syst*, 88, 139–146, 1969.
7. EPRI. Load Modeling for Power Flow and Transient Stability Computer Studies. Report EL-5003, 1987.

7

AC Motor Starting Studies

7.1 Induction Motor Models

The induction motor models and transformation to d–q axis are discussed in Volume 1 with respect to short-circuit calculations. Here we are interested in load flow. The equivalent circuit of the induction motor, shown in Figure 7.1, is derived in many texts. The power transferred across the air gap is

$$P_g = I_2^2 \frac{r_2}{s} \tag{7.1}$$

Referring to Figure 7.1, r_2 is the rotor resistance, s is the motor slip, and I_2 is the rotor current. Mechanical power developed is the power across the air gap minus copper loss in the rotor, i.e.,

$$(1-s)P_g \tag{7.2}$$

Thus, the motor torque T in Newton meters can be written as

$$T = \frac{1}{\omega_s} I_2^2 \frac{r_2}{s} \approx \frac{1}{\omega_s} \frac{V_1^2(r_2/s)}{(R_1 + r_2/s)^2 + (X_1 + x_2)^2} \tag{7.3}$$

Note that this equation does not contain circuit elements g_m and b_m shown in the equivalent circuit. More accurately, we can write the torque equation as

$$T = \frac{1}{\omega_s} I_2^2 \frac{r_2}{s} \approx \frac{1}{\omega_s} \frac{V_1^2(r_2/s)}{(R_1 + c(r_2/s))^2 + (X_1 + cx_2)^2} \tag{7.4}$$

where c is slightly >1, depending on y_m.

R_1 is the stator resistance, V_1 is the terminal voltage, and ω_s is the synchronous angular velocity = $2\pi f/p$ with p being the number of *pairs* of poles. From Equation 7.3, the motor torque varies approximately as the square of the voltage. Also, if the load torque remains constant and the voltage dips, there has to be an increase in the current.

Figure 7.2 shows the typical torque–speed characteristics of an induction motor at rated voltage and reduced voltage. Note the definitions of locked rotor torque, minimum accelerating torque, and breakdown torque during the starting cycle. Referring to Figure 7.2, there is a cusp or reverse curvature in the accelerating torque curve, which gives the minimum accelerating torque or breakaway torque (= CD). The maximum torque, called the breakdown torque in the curve, occurs at slip s_m and is given by EF. The normal operating

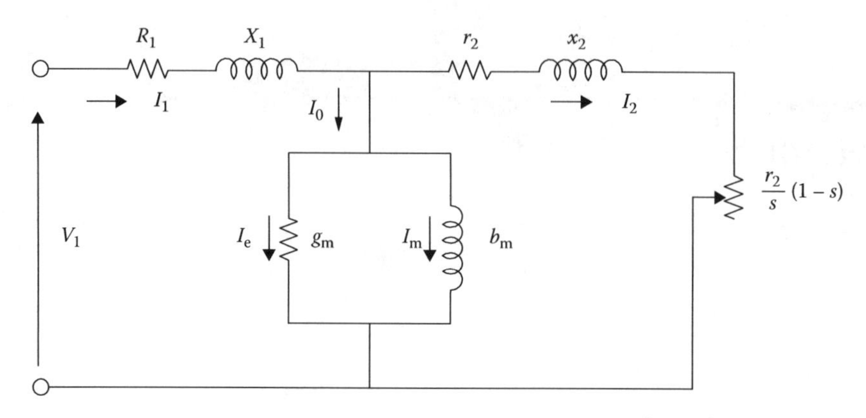

FIGURE 7.1
Equivalent circuit of an induction motor for balanced positive sequence voltages.

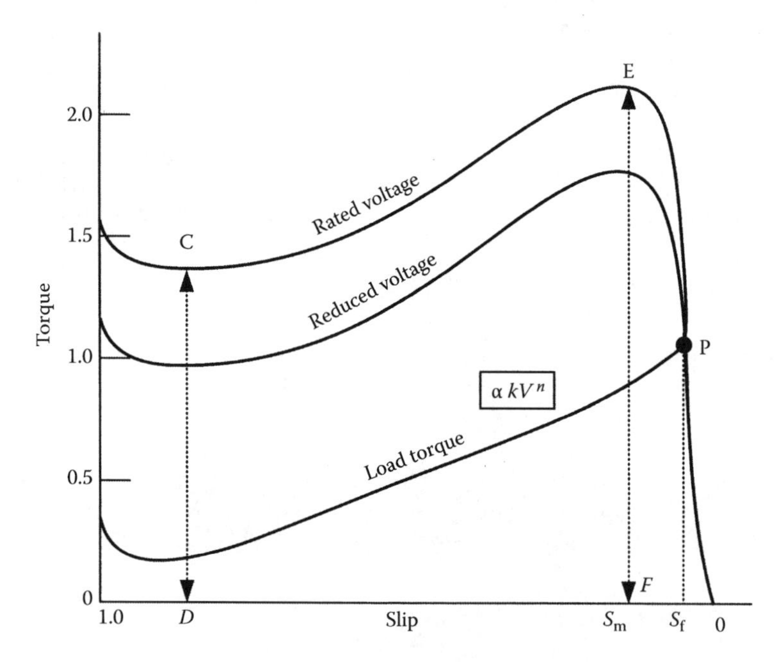

FIGURE 7.2
Torque–speed characteristics of an induction motor at rated voltage and reduced voltage with superimposed load characteristics.

full load point and slip s_f is defined by operating point P. The starting load characteristics shown in this figure are for a fan or blower and varies widely depending on the type of load to be accelerated. Assuming that the load torque remains constant, i.e., a conveyor motor, when a voltage dip occurs, the slip increases and the motor torque will be reduced. It should not fall below the load torque to prevent a stall. Considering a motor breakdown torque of 200%, and full load torque, the maximum voltage dip to prevent stalling is 29.3%. Figure 7.3 shows the torque–speed characteristics for NEMA (National Electrical Manufacturer's Association) design motors A, B, C, D, and F [1].

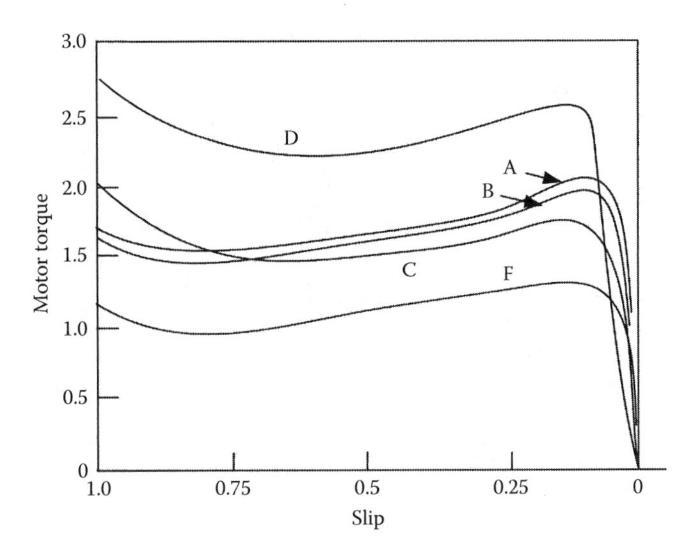

FIGURE 7.3
Torque–speed characteristics of NEMA designs A, B, C, D and E motors.

7.1.1 Double Cage Rotors

The maximum torque is proportional to the square of the applied voltage; it is reduced by stator resistance and leakage reactance, but is independent of rotor resistance. The slip for maximum torque is obtained from the above equation by $dT/ds = 0$, which gives

$$s = \pm \frac{r_r}{\sqrt{r_s^2 + (X_s + X_r)^2}} \approx \frac{r_r}{(X_s + X_r)} \tag{7.5}$$

The maximum torque is obtained by inserting the value of s in Equation 7.4:

$$T_{\mathrm{m}} = \frac{v_s^2}{2\left[\sqrt{\left(r_s^2 + (X_s + X_r)^2\right.} + r_s\right]} \tag{7.6}$$

The slip at the maximum torque is directly proportional to the rotor resistance (Equation 7.5). In many applications, it is desirable to produce higher starting torque at the time of starting. External resistances can be introduced in wound rotor motors which are short-circuited as the motor speeds up. In squirrel cage designs, deep rotor bars or double cage rotors are used. The top bar is of lower cross section, has higher resistance than the bottom bar of larger cross section. The flux patterns make the leakage resistance of the top cage negligible compared to the lower cage. At starting, the slip frequency is equal to the supply system frequency and most of the starting current flows in the top cage, and the motor produces high starting torque. As the motor accelerates, the slip frequency decreases and the lower cage takes more current because of its low leakage reactance compared to the resistance. Figure 7.4a shows the equivalent circuit of a double cage induction motor and Figure 7.4b its torque–speed characteristics.

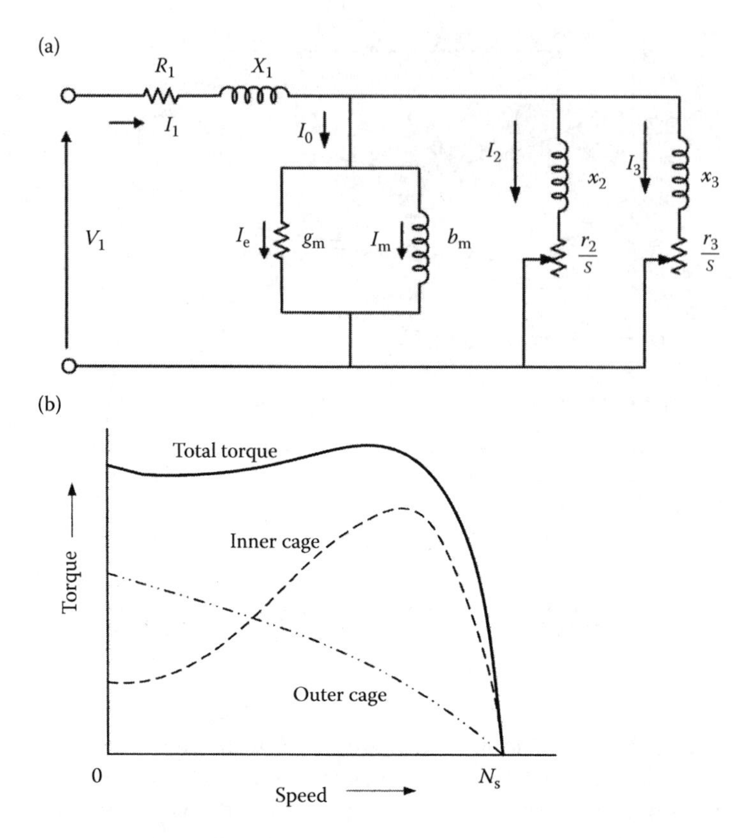

FIGURE 7.4
(a) Equivalent circuit of a double cage induction motor and (b) torque–speed characteristics.

7.1.2 Effects of Variations in Voltage and Frequency

The effects of voltage and frequency variations on induction motor performance are shown in Table 7.1 [2]. We are not much concerned about the effect of frequency variation in load flow analysis, though this becomes important in harmonic analysis. The EMF of a three-phase AC winding is given by

$$E_{ph} = 4.44 K_w f T_{ph} \Phi \text{ volts} \tag{7.7}$$

where E_{ph} is the phase emf, K_w is the winding factor, f is the system frequency, T_{ph} are the turns per phase, and Φ is the flux. Maintaining the voltage constant, a variation in frequency results in an inverse variation in the flux. Thus, a lower frequency results in overfluxing the motor and its consequent derating. In variable-frequency drive systems, V/f is kept constant to maintain a constant flux relation.

We discussed the negative sequence impedance of an induction motor for calculation of an open conductor fault in Volume 1. A further explanation is provided with respect to Figure 7.5, which shows the negative sequence equivalent circuit of an induction motor. When a negative sequence voltage is applied, the MMF wave in the air gap rotates backward at a slip of 2.0 per unit. The slip of the rotor with respect to the backward rotating field is $2-s$. This results in a retarding torque component and the net motor torque reduces to

TABLE 7.1

Effects of Voltage Variations on Induction Motor Performance

Characteristics of Induction Motor	Variation with Voltage	Performance at Rated Voltage (1.0 per Unit) and Other than Rated Voltages				
		0.80	0.95	1.0	1.05	1.10
Torque	$= V^2$	0.64	0.90	1.0	1.10	1.21
Full load slip	$= 1/V^2$	1.56	1.11	1.0	0.91	0.83
Full load current	$\approx 1/V$	1.28	1.04	1.0	0.956	0.935
Full load efficiency		0.88	0.915	0.92	0.925	0.92
Full load power factor		0.90	0.89	0.88	0.87	0.86
Starting current	$= V$	0.80	0.95	1.0	1.05	1.10
No load losses (watts)	$= V^2$	0.016	0.023	0.025	0.028	0.030
No load losses (vars)	$= V^2$	0.16	0.226	0.25	0.276	0.303

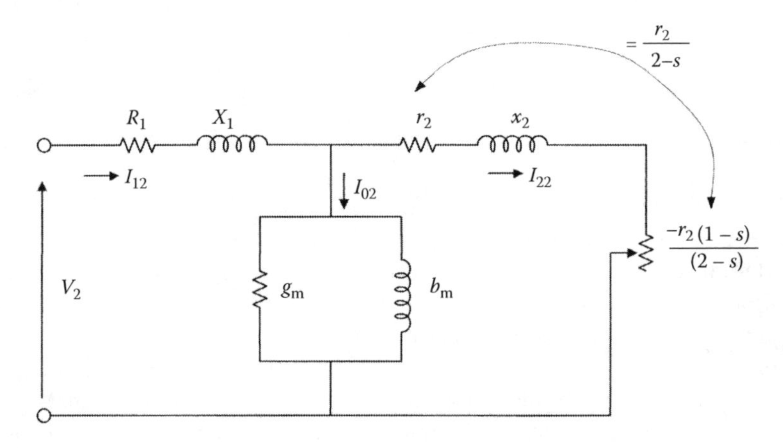

FIGURE 7.5
Equivalent circuit of an induction motor for negative sequence voltage.

$$T = \frac{r_2}{\omega_s}\left(\frac{I_2^2}{s} - \frac{I_{22}^2}{2-s}\right) \tag{7.8}$$

where I_{22} is the current in the negative sequence circuit.

From equivalent circuits of Figures 7.1 and 7.5, we can write the approximate positive and negative sequence impedances of the motor as follows:

$$\begin{aligned} Z_1 &= \left[\left(R_1 + \frac{r_2}{s}\right)^2 + (X_1 + x_2)^2\right]^{1/2} \\ Z_2 &= \left[\left(R_1 + \frac{r_2}{2-s}\right)^2 + (X_1 + x_2)^2\right]^{1/2} \end{aligned} \tag{7.9}$$

Therefore, approximately, the ratio $Z_1/Z_2 = I_s/I_f$, where I_s is the starting current or the locked rotor current of the motor and I_f is the full load current ($s = 1$ at starting). For an induction motor with a locked rotor current of six times the full load current, the negative sequence impedance is one-sixth of the positive sequence impedance. A 5% negative sequence component in the supply system will produce 30% negative sequence current in the motor, which gives rise to additional heating and losses. The rotor resistance will change with respect to high rotor frequency and rotor losses are much higher than the stator losses. A 5% voltage unbalance may give rise to 38% negative sequence current with 50% increase in losses and 40°C higher temperature rise as compared to operation on a balanced voltage with zero negative sequence component. Also, the voltage unbalance is not equivalent to the negative sequence component. The NEMA definition of percentage voltage unbalance is maximum voltage deviation from the average voltage divided by the average voltage as a percentage. Operation above 5% unbalance is not recommended. The zero sequence impedance of motors, whether the windings are connected in wye or delta formation, is infinite. The motor windings are left ungrounded.

From load flow considerations, it is conservative to assume that a balanced reduction in voltage at the motor terminals gives rise to a balanced increase in line current, inversely proportional to the reduced voltage. More accurate models may use one part of the motor load as constant kVA load and the other part as constant impedance load.

7.2 Impact Loads and Motor Starting

Load flow presents a frozen picture of the distribution system at a given instant, depending on the load demand. While no idea of the transients in the system for a sudden change in load application or rejection or loss of a generator or tie-line can be obtained, a steady-state picture is presented for the specified loading conditions. Each of these transient events can be simulated as the initial starting condition, and the load flow study rerun as for the steady-state case. Suppose a generator is suddenly tripped. Assuming that the system is stable after this occurrence, we can calculate the redistribution of loads and bus voltages by running the load flow calculations afresh, with generator omitted. Similarly, the effect of an outage of a tie-line, transformer, or other system component can be studied.

7.2.1 Motor Starting Voltage Dips

One important application of the load flow is to calculate the *initial* voltage dip on starting a large motor in an industrial distribution system. From the equivalent circuit of an induction motor in Figure 7.1 and neglecting the magnetizing and eddy current loss circuit, the starting current or the locked rotor current of the motor is

$$I_{\mathrm{lr}} = \frac{V_1}{(R_1 + r_2) + j(X_1 + x_2)} \tag{7.10}$$

The locked rotor current of squirrel cage induction motors is, generally, six times the full load current on across the line starting, i.e., the full rated voltage applied across the motor terminals with the motor at standstill. Higher or lower values are possible, depending on motor design. Note that the system impedance to the motor, that is, of transformers, cables and source impedance act in series with the motor starting impedance. Thus, the voltage at the motor terminals will change and so the locked rotor current.

Wound rotor motors may be started with an external resistance in the circuit to reduce the starting current and increase the motor starting torque. This is further discussed in a section to follow. Synchronous motors are asynchronously started and their starting current is generally 3–4.5 times the rated full load current on across the line starting. The starting currents are at a low power factor and may give rise to unacceptable voltage drops in the system and at motor terminals. On large voltage dips, the stability of running motors in the same system may be jeopardized, the motors may stall, or the magnetic contactors may drop out. The voltage tolerance limit of solid-state devices is much lower and a voltage dip >10% may precipitate a shutdown. As the system impedances or motor reactance cannot be changed, impedance may be introduced in the motor circuit to reduce the starting voltage at motor terminals. This impedance is short-circuited as soon as the motor has accelerated to approximately 90% of its rated speed. The reduction in motor torque, acceleration of load, and consequent increase in starting time are of consideration. Some methods of reduced voltage starting are as follows: reactor starting, autotransformer starting, wye–delta starting (applicable to motors which are designed to run in delta), capacitor start, and electronic soft start. Other methods for large synchronous motors may be part-winding starting or even variable-frequency starting. The starting impact load varies with the method of starting and needs to be calculated carefully along with the additional impedance of a starting reactor or autotransformer introduced into the starting circuit.

7.2.2 Snapshot Study

This will calculate only the initial voltage drop and no idea of the time-dependent profile of the voltage is available. To calculate the starting impact, the power factor of the starting current is required. If the motor design parameters are known, this can be calculated from Equation 7.10; however, rarely, the resistance and reactance components of the locked rotor circuit will be separately known. The starting power factor can be taken as 20% for motors under 1000 hp (746 kW) and 15% for motors >1000 hp (746 kW). The manufacturer's data should be used when available.

Consider a 10,000 hp (7460 kW), four-pole synchronous motor, rated voltage 13.8 kV, rated power factor 0.8 leading, and full load efficiency 95%. It has a full load current of 410.7 A. The starting current is four times the full load current at a power factor of 15%. Thus, the starting impacts are 5.89 MW and 38.82 Mvar.

During motor starting, generator transient behavior is important. On a simplistic basis, the generators may be represented by a voltage behind a transient reactance, which for motor starting may be taken as the generator transient reactance. Prior to starting the voltage behind this reactance is simply the terminal voltage plus the voltage drop caused by the load through the transient reactance, i.e., $V_t = V + jIX_d$, where I is the load current. For a more detailed solution the machine reactances change from subtransient to transient to synchronous, and open-circuit subtransient, transient, and steady-state time constants should be modeled with excitation system response.

Depending on the relative size of the motor and system requirements, more elaborate motor starting studies may be required. The torque speed and accelerating time of the motor is calculated by step-by-step integration for a certain interval, depending on the accelerating torque, system impedances, and motor and loads' inertia. A transient stability program can be used to evaluate the dynamic response of the system during motor starting [3, 4], also EMTP type programs can be used.

7.3 Motor Starting Methods

Table 7.2 shows a summary of the various starting methods. System network stiffness, acceptable voltage drops, starting reactive power limitation, motor and load parameters need to be considered when deciding upon a starting method. Limitation of stating voltage dips is required for the following:

- The utilities will impose restrictions on the acceptable voltage dips in their systems on starting of motors.
- The starting voltage dips will impact the other running motors in service. The stability of motors in service may be jeopardized on excessive voltage dips.
- The motor starter contactors may drop out on excessive voltage dips. Low-voltage motor starters contactors may drop out in the first cycle of the voltage dip anywhere from 15% to 30%. Medium-voltage DC motor contactors may ride through voltage dips of 30% or more for a couple of cycles depending upon the trapped magnetism, but the auxiliary relays in the motor starting circuit may drop out first.
- The drive and electronic loads will be more sensitive to the voltage dips and may set an upper limit to the acceptable voltage dip.
- On restoration of the voltage, the induction motors will reaccelerate, increasing the current demand from the supply system, which will result in more voltage drops. This phenomenon is cumulative and the resulting large inrush currents on reacceleration may cause a shutdown. This situation can be evaluated with a transient stability type program with dynamic load and generation models.
- A large starting voltage dip may reduce the net accelerating torque of the motor below the load torque during acceleration and result in lockout and unsuccessful starting or synchronizing.

The characteristics of starting methods in Table 7.2 and the associated diagram of starting connections in Figure 7.6 are summarized as follows:

(i) *Full voltage starting* (Figure 7.6a)

This is the simplest method for required starting equipment and controls and is the most economical. It requires a stiff supply system because of mainly reactive high starting currents. The motor terminal voltage during starting and, therefore, the starting torque will be reduced, depending upon the voltage drop in the impedance of the supply system. It is the preferred method, especially when high inertia loads requiring high breakaway torques are required to be accelerated. Conversely, a too fast run-up of low inertia loads can subject the drive shaft and coupling to high mechanical torsional stresses.

(ii) *Reactor* (Figure 7.6b)

A starting reactor tapped at 40%, 65%, and 80% and rated on an intermittent duty basis is generally used. It allows a smooth start with an almost unobservable disturbance in transferring from the reduced to full voltage (when the bypass breaker is closed). The reactor starting equipment and controls are simple, though the torque efficiency is poor.

TABLE 7.2

Starting Method of Motors

Starting Method	Reference Figure No.	*I* Starting	*T* Starting	Cost Ratio	Qualifications
Full voltage starting	a	1	1	1	Simplest starting method giving highest starting efficiency, provided voltage drops due to inrush currents are acceptable
Reactor starting	b	α	α^2	2.5	Simple switching closed transition, torque per kVA is lower compared to auto-transformer starting; a single reactor can be used to start a number of motors
Krondrofer starting	c	α^2	α^2	3.5	Closed-circuit transition requires complex switching, (a) close Y, then S, (b) open Y, (c) close R, (d) open S; applicable to weak electrical systems, where the reduced starting torque of motor can still accelerate the loads
Part-winding starting	d	α	α^2	Varies	Inrush current depends upon design of starting winding; closed-circuit transition by switching the parallel stator winding by closing R; the starting torque cannot be varied and fixed at the design stage; start: close S, run: close R
Shunt capacitor starting	e	Varies	Varies	3	May create harmonic pollution (harmonic filters are required); capacitors switched off by voltage/current control when the speed approaches approximately 95% of the full load speed (bus voltage rises as the motor current falls); start: close S1, S2; run: open S2
Shunt capacitor in conjunction with other reduced voltage starting	f	Varies	Varies	Varies	A reactor starting with shunt capacitors has two-fold reduction in starting current, due to reactor and capacitor; the motor terminal voltage and torque is increased compared to reactor start
Low-frequency starting	g	Varies (150%–200% of load current)	Slightly > load torque	6–7	May create harmonic pollution, not generally used due to high cost; gives a smooth acceleration; the current from supply system can be reduced to 150%–200% of the full load current; one low-frequency starting equipment can be used to start a number of motors in succession by appropriate switching
Pony motor	h	–	–	Varies	Not generally used; the starting motor is high torque intermittent rated machine; applicable when load torque during acceleration is small; synchronizing required at S

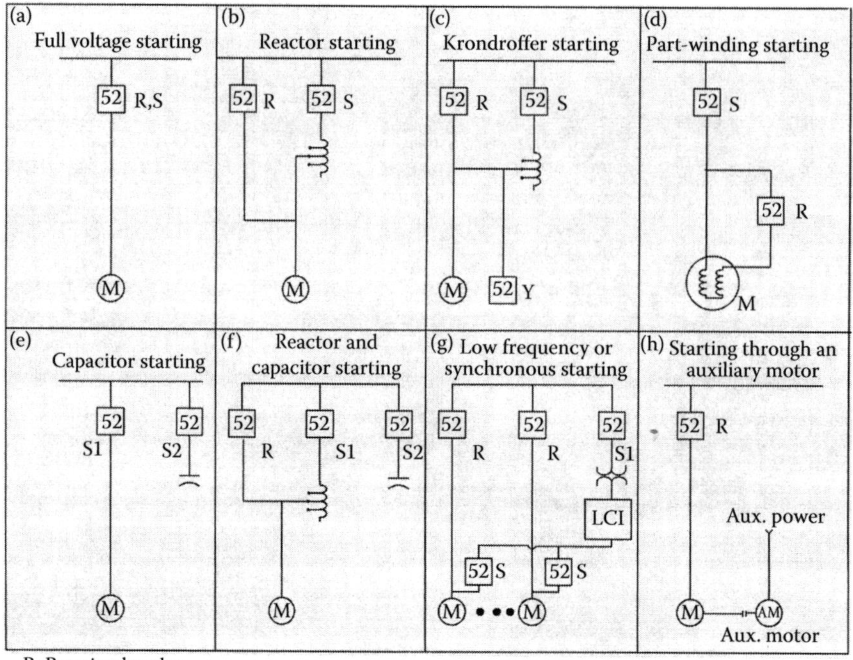

FIGURE 7.6
Starting methods of motors.

Reactance of the reactor for a certain tap can be approximately calculated by the expression $X = (1-R)/R$, where R is the tap ratio or the voltage ratio in per unit of the rated voltage. Thus, the starting current at a particular tap is given by

$$I_s = \frac{I}{Z_s + Z_m + Z_L} \tag{7.11}$$

It is in per unit based upon the motor rated voltage. Z_s, Z_m, and Z_L are the system, motor, and reactor impedances, respectively, reduced to a common base, say motor starting kVA base and are vector quantities, also giving the starting power factor. As these impedances act as a potential divider, the starting voltage at the motor terminals is simply $I_s Z_m$.

The motor starting impedance can be calculated by the formula: $V/\sqrt{3} \times I_s$, where V is the motor rated line to line voltage, and I_s is the starting current at the motor at starting power factor.

(iii) Krondroffer starting (Figure 7.6c)

An advantage of this method of starting is high torque efficiency, at the expense of additional control equipment and three switching devices per starter. Current from the supply system is further reduced by factor α, for the same starting torque obtained with a reactor start. Though close transition reduces the transient current and torque during transition, a quantitative evaluation is seldom attempted. The motor voltage during transition may drop significantly. For weak supply systems, this method of starting at 30%–45% of the voltage can be considered, provided the reduced asynchronous torque is adequate to accelerate the load.

(iv) Part-winding starting (Figure 7.6d)

The method is applicable to large synchronous motors of thousands of horsepower, designed for part-winding starting. These have at least two parallel circuits in the stator winding. This may add 5%–10% to the motor cost. The two windings cannot be exactly symmetrical in fractional slot designs and the motor design becomes specialized regarding winding pitch, number of slots, coil groupings, etc. The starting winding may be designed for a higher temperature rise during starting. Proper sharing of the current between paralleled windings, limiting temperature rises, and avoiding hot spots become a design consideration. Though no external reduced voltage starting devices are required and the controls are inherently simple, the starting characteristics are fixed and cannot be altered. Part-winding starting has been applied to large TMP (Thermo-mechanical pulping) synchronous motors of 10,000 hp and above, yet some failures have known to occur.

(v) Capacitor starting (Figure 7.6e and f)

The power factor of the starting current of even large motors is low, rarely exceeding 0.25. Starting voltage dip is dictated by the flow of starting reactive power over mainly inductive system impedances. Shunt connected power capacitors can be sized to meet a part of the starting reactive kvar, reducing the reactive power demand from the supply system. The voltage at the motor terminals improves and, thus, the available asynchronous torque. The size of the capacitors selected should ensure a certain starting voltage across the motor terminals, considering the starting characteristics and the system impedances. As the motor accelerates and the current starts falling, the voltage will increase and the capacitors are switched off at 100%–104% of the normal voltage, sensed through a voltage relay. A redundant current switching is also provided. For infrequent starting, having shunt capacitors rated at 60% of the motor voltage are acceptable. When harmonic resonance is of concern, shunt capacitor filters can be used.

Capacitor and reactor starting can be used in combination (Figure 7.6f). A reactor reduces the starting inrush current and a capacitor compensates part of the lagging starting kvar requirements. These two effects in combination further reduce the starting voltage dip [5].

(vi) Low-frequency starting or synchronous starting (Figure 7.6g)

Cycloconverters and LCI (Load commutated inverters) have also been used for motor starting [6, 7]. During starting and at low speeds, the motor does not have enough back EMF to commute the inverter thyristors and auxiliary means must be provided. A synchronous motor runs synchronized at low speed, with excitation applied, and accelerates smoothly as LCI frequency increases. The current from the supply system can be reduced to 150%–200% of the full load current, and the starting torque need only be slightly higher than the load torque. The disadvantages are cost, complexity, and large dimensions of the starter. Tuned capacitor filters can be incorporated to control harmonic distortion and possible resonance problems with the load generated harmonics. Large motors require a coordinated starting equipment design.

Figure 7.6g shows that one starter can be switched to start a number of similar rated motors. In fact, any of the reduced voltage systems can be used to start a number of similar motors, though additional switching complexity is involved.

(vii) Starting through an auxiliary motor (Figure 7.6h)

It is an uncommon choice, yet a sound one in some circumstances involving a large machine in relation to power supply system capabilities. No asynchronous torque is required for acceleration and the motor design can be simplified. The starting motor provides an economical solution only when the load torque during acceleration is small. Disadvantages are increased shaft length and the rotational losses of the starting motor, after the main motor is synchronized.

7.3.1 Wye–Delta Starter

We have not discussed wye–delta starter, which is sometimes used for motors of machine tools, pumps, motor–generator sets, etc. A motor must be designed normally to run in delta connection of the windings. Thus, the method is generally applied to low-voltage motors, as the number of turns per phase increase for delta windings and the motor built to run in delta connection will be more expensive for the same rating with wye-connected windings. As the windings are connected in wye during starting, the phase voltage is reduced by a factor of 0.58 of the normal voltage. The starting current per phase is reduced in the same proportion, that is 0.58% of the full voltage starting current. Therefore, the *line current is reduced by* $(0.58)^2 = 0.33$ *times.*

Thus, the stating torque reduces to one-third of the rated voltage starting torque. This method can be employed when the starting torque requirements are low and the starting time is increased.

Any starting method which has open-circuit transition will result in transient torques and currents which may even exceed starting currents.

Where the induction motor is required to run for a considerable period of time on low load, the motor can be run in wye connection during this period, with reduction in magnetizing current and increase in efficiency, see Figure 7.7.

7.3.2 Centrifugal Clutches

Centrifugal clutch is a mechanical device (Figure 7.8). It allows the motor to start on light load, and gradually take up the load as the speed increases. The clutch has two mechanically separate parts; the inner one is attached to the motor shaft and is free to rotate at low speeds within the outer hollow part. Mechanical coupling is afforded within the inner and outer concentric parts by weighted blocks which by centrifugal action come forcibly in contact with the outer part and lock it together. The blocks are coated with friction material or may be metal running in oil, and the centrifugal force will squeeze the oil out. As the motor speeds up, the load is gradually taken up the decreasing clutch slip. By spring loading, the clutch may be prevented from operating till a predetermined speed is obtained.

Figure 7.9 shows the typical current and speed curves with and without the clutch. The total amount of I^2R losses in the motor are very much reduced.

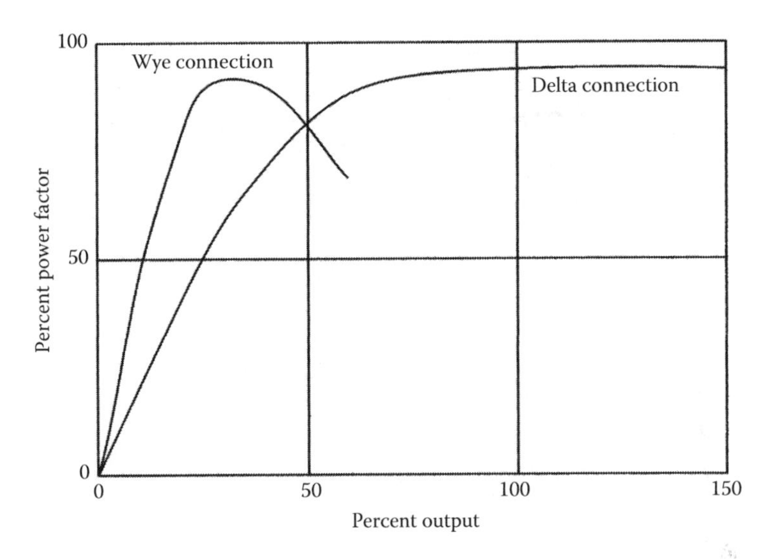

FIGURE 7.7
Power factor of an induction motor running in wye and delta connection.

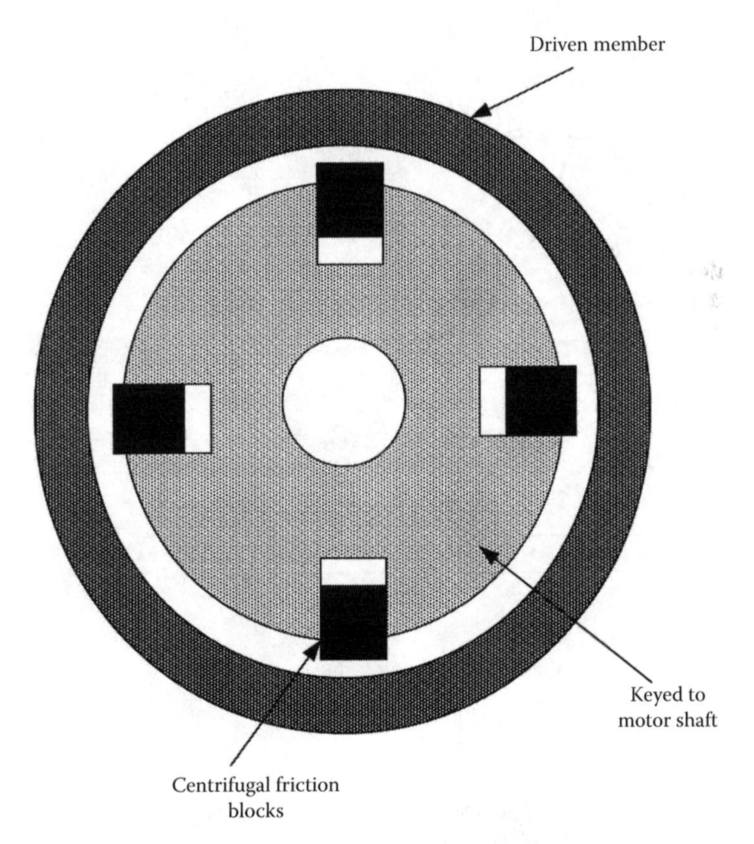

FIGURE 7.8
Centrifugal clutch.

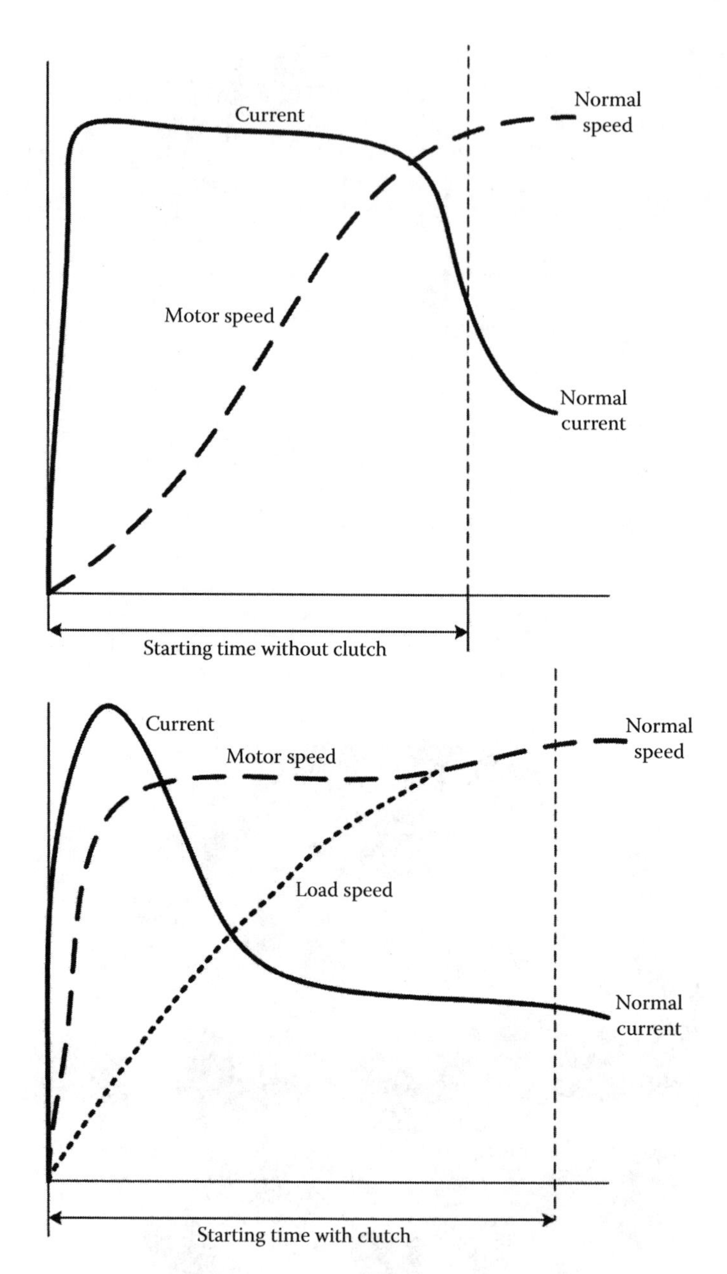

FIGURE 7.9
Starting current, speed profile, starting without and with clutch.

7.3.3 Soft Start

The soft starters using SCRs can be applied for a variety of applications and afford reduced starting currents and smooth starting; however, the cost may be approximately 1.5–2 times the electromechanical starters; for example, for low-voltage applications, these are available for motor horsepower to 800 hp. Typical applications include centrifugal and screw compressors, fans, blowers, pumps, cranes and hoists, rock crushers, HVAC, etc.

- Reduced starting torque stress on mechanical equipment
- Reduction of voltage drop during starting, this may be of much advantage for weak utility systems
- Reduced inrush currents
- Smooth steeples starting

The starting characteristics can be as follows:

7.3.3.1 Voltage Ramp Start

This is the most common method; the initial torque value can be set at 0%–85% of the locked rotor torque (equivalent to starting at full rated voltage), and the ramp time can be adjusted from low values to say 180 s, see Figure 7.10a.

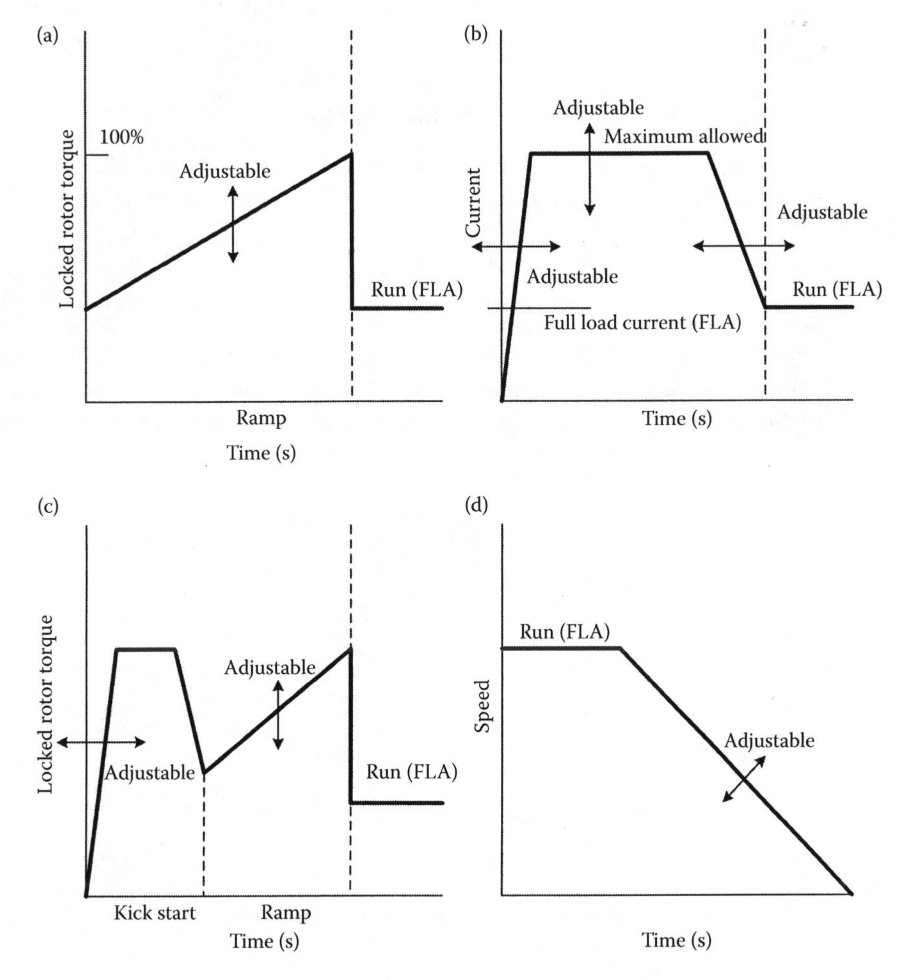

FIGURE 7.10
Various options in soft starting.

7.3.3.2 Current Limit Start

This method limits the maximum per phase current to the motor. This method is used when it is necessary to limit the maximum starting current due to long starting times. The maximum current can be set as a percentage of the locked rotor current. Also the duration of the current limit is adjustable. The ramp time is adjustable from low values to say a maximum of 180 s or more, see Figure 7.10b.

7.3.3.3 Kick Start

It has selectable features in both voltage ramp start and the current limit start modes. It provides a current and torque kick for 0–2.0 s. This provides additional torque to break away from a high friction load, see Figure 7.10c.

7.3.3.4 Soft Stop

It allows controlled stopping of the load. It is applicable when a stop time greater than the coast-to-stop time is desired. It is useful for high friction loads where a quick stop may cause load damage, see Figure 7.10d.

7.3.4 Resistance Starter for Wound Rotor Induction Motors

From Equation 7.3, the maximum torque can be obtained at starting if the rotor resistance including external added resistance is made approximately equal to the combined reactance:

$$(r_2 + R_{ext}) = (X_1 + x_2) \tag{7.12}$$

At $s = 1$, the inclusion of external resistance will reduce the stator current at standstill and also raise the power factor. This is depicted on the induction motor circle diagram in Figure 7.11 [8]. The normal short-circuit point is P_1, and with added resistance it moves to

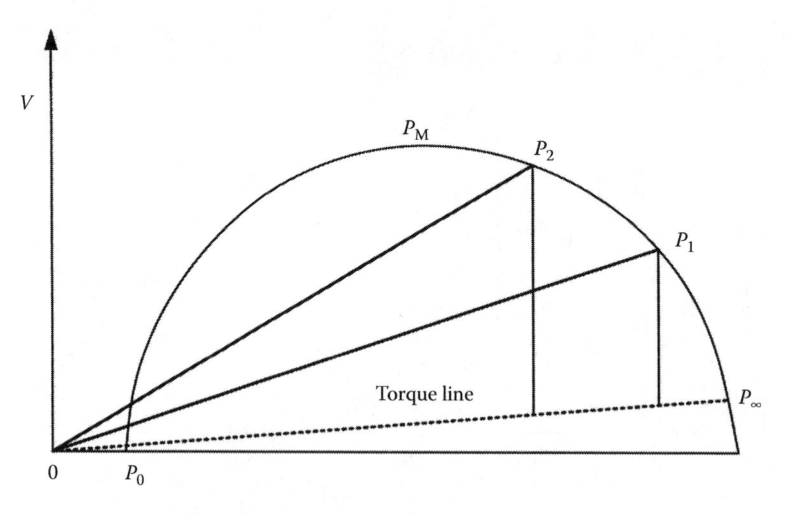

FIGURE 7.11
Circle diagram of an induction motor showing impact of external resistance on starting torque.

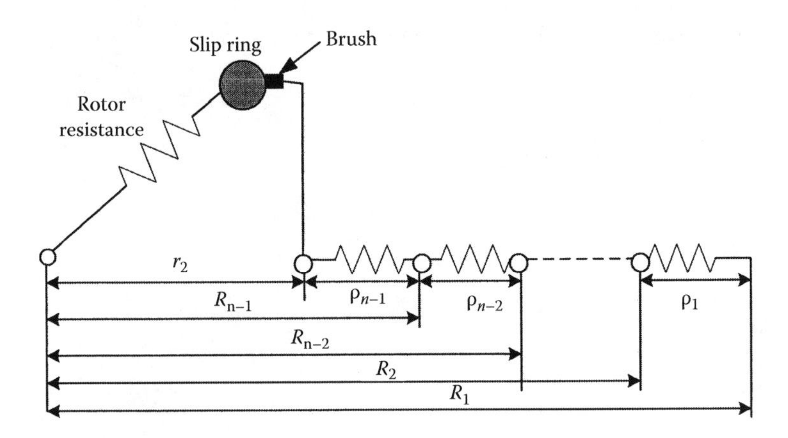

FIGURE 7.12
Configuration for resistance starting of wound rotor induction motors.

P_2. Enough resistance can be added so that maximum torque P_M is obtainable; any additional resistance will decrease the torque.

The configuration for a resistance starter can be depicted in Figure 7.12.

The motor is assumed to start against a constant torque. Let the rotor current fluctuate between the maximum and minimum limits denoted by I_{2max} and I_{2min}, respectively. Let R_1, R_2, ... be the total resistances per phase on the first, second ... steps; which consist of rotor resistance r_2 and the added resistances ρ_1, ρ_2, ρ_3, ..., as shown in Figure 7.12.

At each step, the current is I_{2max}, the resistance in the circuit is R_1 for $s_1 = 1$, R_2 for slip s_2, and so on. The following relation holds:

$$I_{2\max} = \frac{E_2}{\sqrt{\left[(R_1/s_1)^2 + x_2^2\right]}} = \frac{E_2}{\sqrt{\left[(R_2/s_2)^2 + x_2^2\right]}} = \cdots \tag{7.13}$$

where E_2 is the rotor standstill EMF. Then,

$$\frac{R_1}{s_1} = \frac{R_2}{s_2} = \frac{R_3}{s_3} = \cdots = \frac{R_{n-1}}{s_{n-1}} = \frac{r_2}{s_{\max}}. \tag{7.14}$$

On the first step, R_1 remains in the circuit until motor has started and the slip has fallen from unity to s_2, and at the same time the current falls from I_{2max} to I_{2min}. Then,

$$I_{2\min} = \frac{E_2}{\sqrt{\left[(R_1/s_2)^2 + x_2^2\right]}} = \frac{E_2}{\sqrt{\left[(R_2/s_3)^2 + x_2^2\right]}} = \cdots \tag{7.15}$$

so that

$$\frac{R_1}{s_2} = \frac{R_2}{s_3} = \frac{R_3}{s_4} = \cdots = \frac{R_{n-1}}{s_{\max}} \tag{7.16}$$

Therefore, we can write

$$\frac{s_2}{s_1} = \frac{s_3}{s_2} = \frac{s_4}{s_3} = \cdots = \frac{R_2}{R_1} = \frac{R_3}{R_2} = \cdots \frac{r_2}{R_{n-1}} = \gamma \tag{7.17}$$

As $s_1 = 1$, the total resistance in the circuit on the first step is

$$R_1 = \frac{r_2}{s_{max}} \tag{7.18}$$

Thereafter,

$$R_2 = \gamma R_1;\ R_3 = \gamma R_2 = \gamma^2 R_1 \tag{7.19}$$

$$r_2 = \gamma R_{n-1} = \gamma^{n-1} R_1 = \frac{\gamma^{n-1} r_2}{s_{max}} \tag{7.20}$$

Then, it is obvious that

$$\gamma = \sqrt[n-1]{s_{max}} \tag{7.21}$$

The resistance sections can thus be calculated:

$$\begin{aligned}
\rho_1 &= R_1 - R_2 = R_1(1-\gamma) \\
\rho_2 &= R_2 - R_3 = R_1(\gamma-\gamma^2) = \gamma\rho_1 \\
\rho_3 &= \gamma^2\rho_1
\end{aligned} \tag{7.22}$$

If s_{max} is known for maximum starting current I_{2max}, then $n-1$ sections can be found. If s_{min} for starting current I_{2min} is also known or can be estimated, a graphical procedure as shown in Figure 7.13 can be adopted. *The figure is drawn to scale.*

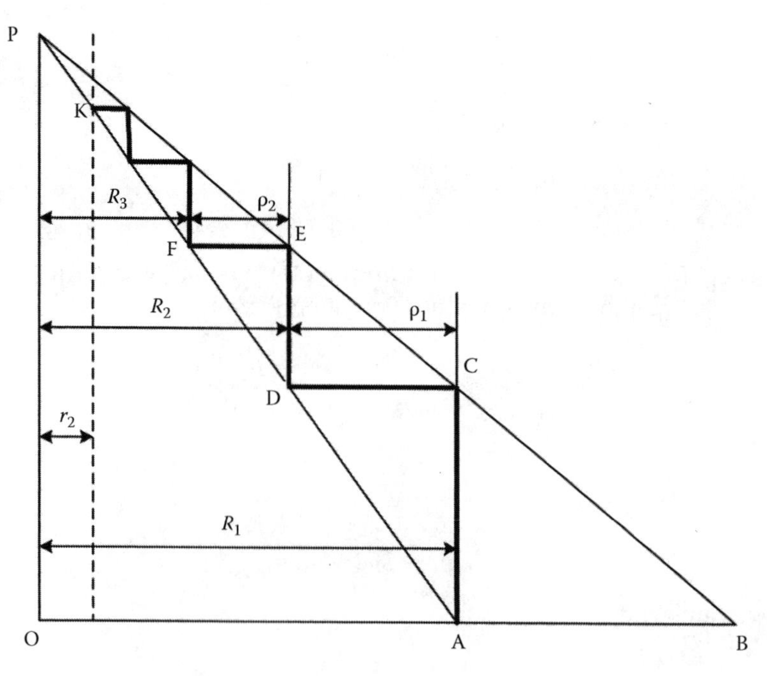

FIGURE 7.13
Graphical construction of resistance starting.

In this figure,

$OA = R_1 = r_2/s_{max}$
$OB = r_2/s_{min}$
$AC/OP = AB/OB = 1 - OA/OB = 1 - (s_{min}/s_{max})$
$= 1 - s_2/s_1 = 1 - s_2$
$CD = OA.AC/OP = R_1(1 - s_2) = \rho_1$

Continuing the construction, the horizontal intercepts between lines AP and BP represent $\rho_1, \rho_2, \rho_3, \ldots$. The intercepts between lines AP and OP represent $R_1, R_2, R_3 \ldots$. The construction is continued till a point K falls on the intercept r_2. The number of required steps is the same as the horizontal lines AB, CD, EF, ….

7.4 Number of Starts and Load Inertia

NEMA [1] specifies two starts in succession with motor at ambient temperature and for the WK^2 of load and the starting method for which motor was designed and one start with motor initially at a temperature not exceeding its rated load temperature. If the motor is required to be subjected to more frequent starts, it should be so designed. The starting time is approximately given by the following expression:

$$t = \frac{2.74 \sum WK^2 N_s^2 10^5}{P_r} \int \frac{dn}{T_a - T_l} \tag{7.23}$$

where t is the accelerating time in seconds, P_r is the rated output, and other terms have been described before. As per Equation 7.23 accelerating time is solely dependent upon the load inertia and inversely proportional to the accelerating torque. The normal inertia loads of the synchronous and induction motors are tabulated in NEMA [1]. For a synchronous machine, it is defined by the following equation:

$$WK^2 = \frac{0.375(\text{hp rating})^{1.15}}{(\text{Speed rpm}/1000)^2} \text{ lb ft}^2 \tag{7.24}$$

The heat produced during starting is given by the following expression:

$$h = 2.74 \sum WK^2 N_s^2 10^{-6} \int \frac{T_a}{T_a - T_l} s \, ds \tag{7.25}$$

where h is heat produced in kW-sec and, again, it depends on load inertia. The importance of load inertia on the starting time and accelerating characteristics of the motor is demonstrated.

The inertia constant H is often used in the dynamic motor starting studies. It is defined as

$$H = \frac{(0.231)(WR^2)(r/\text{min})^2 \times 10^{-6}}{\text{kVA}} s \tag{7.26}$$

where WR^2 is the combined motor and load inertia in lb ft^2. The inertial constant does not vary over large values.

In metric units, H is defined as

$$H = \frac{5.48 \times J \times (r/\text{min})^2 \times 10^{-6}}{\text{kVA}} s$$

where J is in kg m^2. Considering

1 m = 3.281 ft
1 kg = 2.205 lb = 0.0685 slug
1 slug-ft^2 = 1/(0.0685 × 3.281^2) = 1.356 kg m^2

The moment of inertial J in kg m^2 is related to WR^2 as follows:

$$J = \frac{WR^2}{32.32} \times 1.356$$

Reference [9] provides some typical values for synchronous generators. The accelerating torque curve (motor torque–load torque) can be plotted from standstill to motor full load speed. The motor torque–speed curve can be adjusted (by selection and design of the motor) for the expected voltage dip during starting. A severe voltage dip or reduced voltage starting methods may reduce motor torque below the load torque, and the motor cannot be started. The accelerating torque curve thus calculated can be divided into a number of incremental steps and average accelerating torque calculated for each incremental speed (Figure 7.2). Then the accelerating time Δt can be calculated on a step-by-step basis for each speed incremental:

$$\Delta t = \frac{WR^2 \Delta n}{307T} s \tag{7.27}$$

where Wk^2 is in lb ft^2 and T is in lb ft

or

$$\Delta t = \frac{J \Delta n}{19.24T} s$$

where J in kg m^2 and T in kg m.

7.5 Starting of Synchronous Motors

Synchronous motors of the revolving field type can be divided into two categories: salient solid pole synchronous motors (these are provided with concentrated field windings on the pole bodies) and cylindrical rotor synchronous motors (these have distributed phase winding over the periphery of the rotor which is used for starting and excitation).

In the salient pole, synchronous motor can be solid pole type, which has poles in one piece of solid mass, made of cast steel, and forged steel or alloy steel; the starting torque is produced by eddy currents induced in the pole surface. Sometimes, slots are provided in the pole surface to absorb thermal stresses.

The salient laminated pole types have poles of punched sheet steel, laminated, compressed, and formed. The pole head may be equipped with starting windings of the deep-slot squirrel cage type or double squirrel cage type.

Synchronous induction motors can again be of a salient pole or cylindrical rotor type. These motors can be designed to operate as induction motors for a short period when these fall out of step, i.e., due to a sudden reduction in system voltage or excessive load torque and resynchronize automatically on reduction of load torque or restoration of voltage.

The starting of synchronous motors should consider similar factors as discussed for the induction motors, with the addition of synchronization, pulling out of step, and resynchronization.

The synchronous motor design permits lower starting currents for a given starting torque as compared with a squirrel cage induction motor, typically, in ratio of 1:1.75. This may become an important system design consideration, when the starting voltage dips in the power system have to be limited to acceptable levels.

Figure 7.14 depicts the starting characteristics of two basic types of synchronous motors: salient pole and laminated pole. Salient pole motors have higher starting torque (curve a)

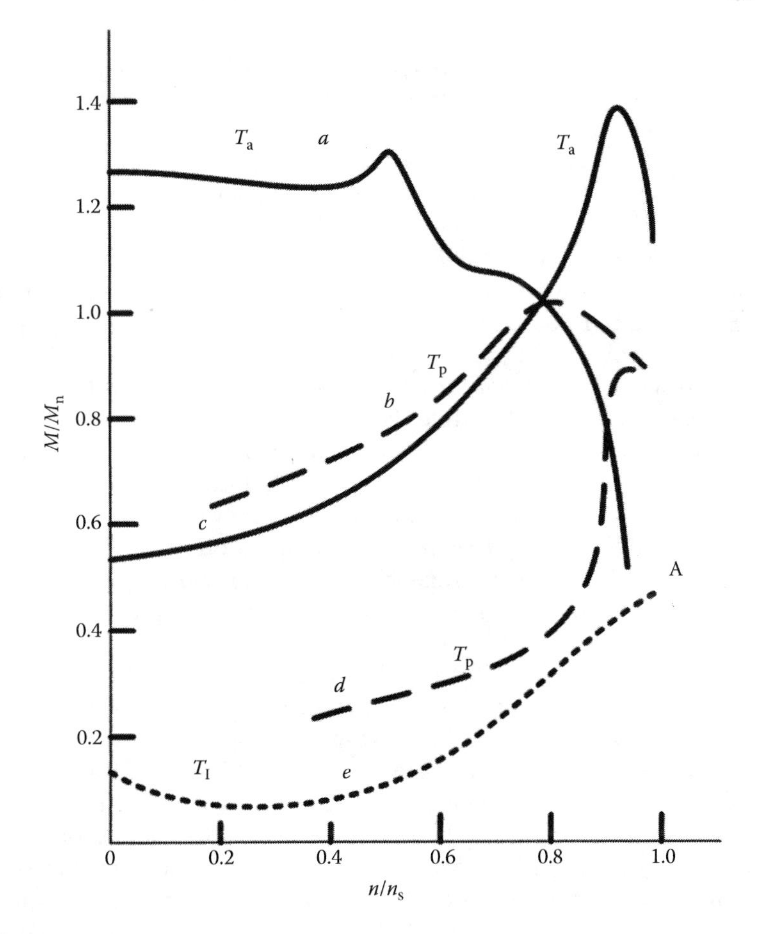

FIGURE 7.14
Starting characteristics of synchronous motors, see the text.

and pulsating or oscillating torque (curve b). The cylindrical rotor machines have starting torque characteristics akin to induction motors (curve c) and smaller oscillating torque (curve d). Salient laminated pole construction with damper windings may produce a characteristic somewhere in between the solid pole and cylindrical rotor designs.

Referring to Figure 7.14, T_a is the asynchronous torque of the motor analogues to the induction motor starting torque, as a synchronous motor is started asynchronously. The rotor saliency causes another torque T_p, which pulsates at a frequency equal to twice the slip frequency. Considering a supply system frequency of f Hz, and a rotor slip of s, the magnetic field in the rotor has a frequency of sf. Due to saliency, it can be divided into two components: (1) a forward revolving field at frequency sf in the same direction as the rotor and (2) a field revolving in reverse direction at sf. Since the rotor revolves at a speed $(1-f)\,s$, the forward field revolves at $(1-s)f + sf$, i.e., at fundamental frequency with respect to stator, in synchronism with the rotating field produced by the stator three-phase windings. The interaction of this field with the stator field produces the torque T_a in a fixed direction. Negative sequence field revolves at $(1-f)s - sf = (1-2f)s$. This is as viewed from the stator and, thus, it has a slip of $f-(1-2s)f = 2sf$ with respect to the field produced by the exciting current. The total torque produced by the synchronous motor is, therefore,

$$T_m = T_a + T_p \cos 2s\omega t \tag{7.28}$$

The currents corresponding to the positive and negative fields are I_a and I_p. The current I_a is at fundamental frequency and I_p is the pulsating current at $(1-2s)f$ Hz. The total current is given by the following expression:

$$I = I_a \cos(\omega t - \varphi) + I_p \cos(1-2s)\omega t \tag{7.29}$$

where φ is the power factor angle at starting.

Due to electrical and magnetic asymmetry, T_a develops a sudden change near 50% speed, resulting in a singular point. This is called the Gorges phenomenon [10]. The shape and size of this dip are governed by machine data and the power system. If resistance is zero, the saddle disappears. Solid poles motor without damper windings and slip-rings will give a more pronounced saddle, while the saddle is a minimum in cylindrical rotor and laminated pole constructions with damper windings. The external resistances during asynchronous starting impact both, T_a, T_p and the saddle. At slip $s = 0.5$, the reverse rotating field comes to a halt as viewed from the stator and torque T_p becomes zero, so also the pulsating current I_p. For slip less than 0.5, the direction of the torque changes; this produces characteristic saddle. The average starting torque and the starting current can be calculated based upon the direct axis and quadrature axis equivalent circuits of the synchronous machine [11].

Figure 7.14 also shows the load torque T_l during starting (curve e). The accelerating torque is $T_a - T_l$ and the motor will accelerate till point A, close to the synchronous speed. At this moment, pull-in torque refers to the maximum load torque under which the motor will pull into synchronism while overcoming inertia of the load and the motor when the excitation is applied. The torque of the motor during acceleration as induction motor at 5% slip at rated voltage and frequency is called nominal pull-in torque. Pull-in and pull-out torques of salient pole synchronous motors for normal values of load inertia are specified in NEMA standards [1]. For motors of 1250 hp (932 kW) and above, speeds 500–1800 rpm, the pull-in torque should not be less than 60% of the full load torque. The maximum slip at pull-in can be approximately determined from the following inequality:

$$s < \frac{242}{N_s}\sqrt{\frac{P_m}{\sum Jf}} \tag{7.30}$$

where N_s is the synchronous speed in rpm, f is the system frequency, P_m is the maximum output of the synchronous motor with excitation applied, when pulling into synchronism, J is the inertia of the motor and driven load in kg m^2.

It is necessary to ascertain the load torque and operating requirements before selecting a synchronous motor application. For example, wood chippers in the paper industry have large pull-out torque requirements of 250%–300% of the full load torque. Banbury mixers require starting, pull-in, and pull-out torques of 125%, 125%, and 250%, respectively. Thus, the load types are carefully considered in applications. Load inertia impacts the heat produced during starting and the starting time. NEMA [1] gives the starting requirements of pumps and compressors and inertia ratios.

Example 7.1

A 3500 hp (2610 kW) induction motor is started in a system configuration as shown in Figure 7.15a. The only load on the transformer secondary is the 3500 hp motor. This is a typical starting arrangement for large motors. While a motor connected to a dedicated transformer may be able to tolerate and accelerate its load, the other loads if connected to the same bus may shutdown or experience unacceptable voltage dips during starting. The motor is 4-pole, rated voltage 4 kV, and has a full load power factor of 92.89%, efficiency = 94.23%, and full load current is 430.4 A. Again the motor rated voltage is slightly lower than the rated bus voltage of 4.16 kV. The locked rotor current is 576% at 19.47% power factor. This gives a starting impact of 3.344 MW, 17.35 Mvar, equivalent to 17.67 MVA. This starting impact is calculated at rated voltage. As the starting impact load is a constant impedance load, if the starting voltage is lower, the starting impact load will reduce proportionally. The transformer is rated 5.00 MVA. Short-term loading of the transformer is acceptable, but if more frequent starts are required, the transformer rating must be carefully considered. The 5 MVA transformer of 6.5% impedance can take a three-phase short-circuit current of 12.3 kA for 2 s according to ANSI/IEEE [12] and also NEMA [1] allows two starts per hour, one with the motor at ambient temperature and the other when the motor has attained its operating temperature.

The motor torque–speed characteristics, power factor, and slip are shown in Figure 7.15b. It has a locked rotor torque of 88.17% and a breakdown torque of 244.5% at rated voltage. The standard load inertial according to NEMA for this size of motor is 8700 lb ft^2 (366.3 kg m^2). The motor drives a boiler ID fan of twice the NEMA inertia = 17,400 lb ft^2 (732.7 kg m^2) (H = 4.536). To this must be added the inertia of the motor and coupling, say the total load inertia is 19,790 lb ft^2 (H = 5.16). The speed torque curve of the load is shown in Figure 7.15b. For a dynamic motor starting study, these data must be obtained from the manufacturer.

The system configuration shown in Figure 7.15a has source impedance (positive sequence) of $0.166 + j\,0.249\ \Omega$. The transformer has an impedance of 6.5%, $X/R = 12.14$, and the motor is connected through two cables in parallel per phase of 350 KCMIL, 400 ft long.

The motor, load, and accelerating torques are shown in Figure 7.16a. The starting current, voltage, and slip are depicted in Figure 17.16b. It is seen that there is a starting voltage dip of 16% and starting current is reduced approximately proportional to the voltage dip. The voltage dip is shown based upon the rated motor voltage of 4 kV. The motor takes approximately 22 s to start. It is not unusual to see a starting time of 40–50 s for boiler ID fan motors. Figure 7.16c shows the active and reactive powers drawn by the motor during starting.

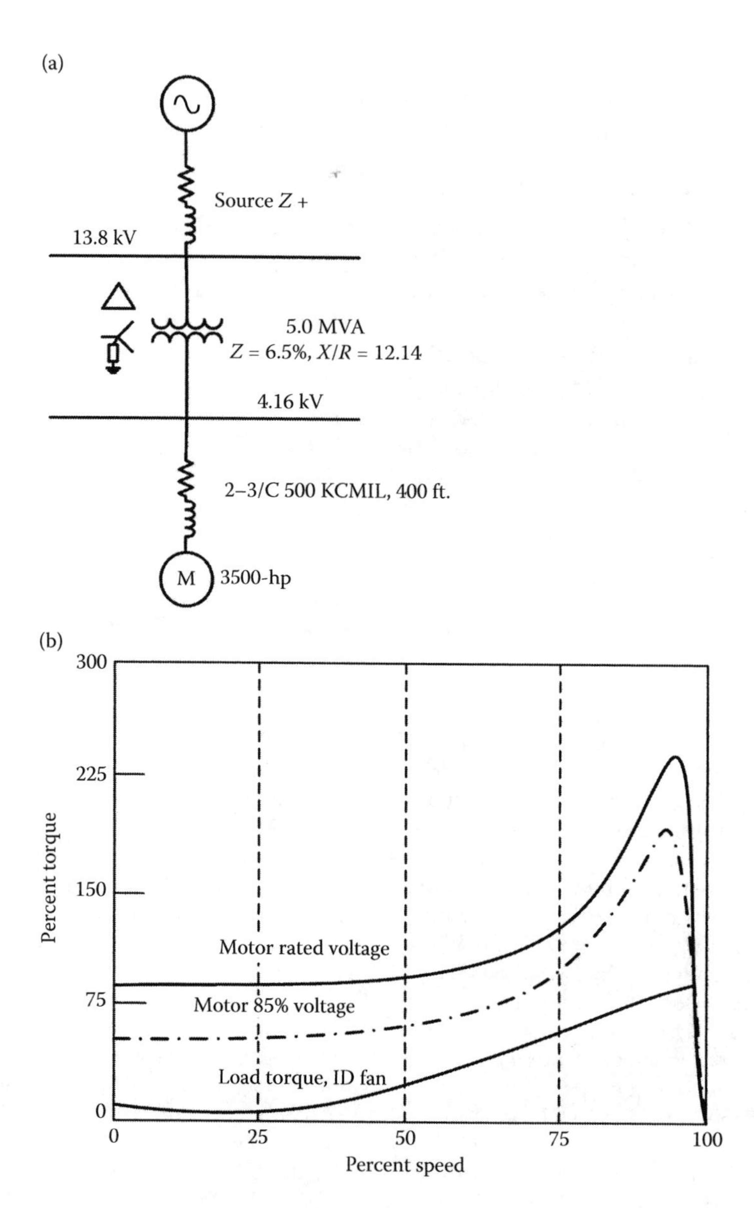

FIGURE 7.15
(a) System configuration and impedances for starting of 3500 hp motor and (b) motor and load torque–speed characteristics.

Example 7.2

Calculate the starting time using the data in Example 7.1 by hand calculations.

As per procedure described above, first calculate the voltage dip at the motor terminals. This is a simple one step load flow problem. Applying a starting impact load of 3.344 MW, 17.35 Mvar to all the impedances shown in series in Figure 7.16b gives an approximate voltage dip of 15%. Assuming that this voltage dip remains practically throughout the starting period, reduce the motor torque–speed characteristics by $(0.85)^2$, that is by 72% (Figure 7.16b). For this example, divide the starting period into four sections and calculate the accelerating torque in each strip. The result is shown in Table 7.3. Calculate the average torque in each period. Then using Equation 7.27, Δt for each incremental Δn is calculated. The summation gives 23.02 s, which is approximately close to the calculation result in Example 7.1. Smaller the steps Δn, more accurate the result.

7.6 Motor Starting Voltage Dips-Stability Considerations

A power system must be designed to ride through the motor starting voltage dip, and many method of reduced voltage starting have been discussed as above.

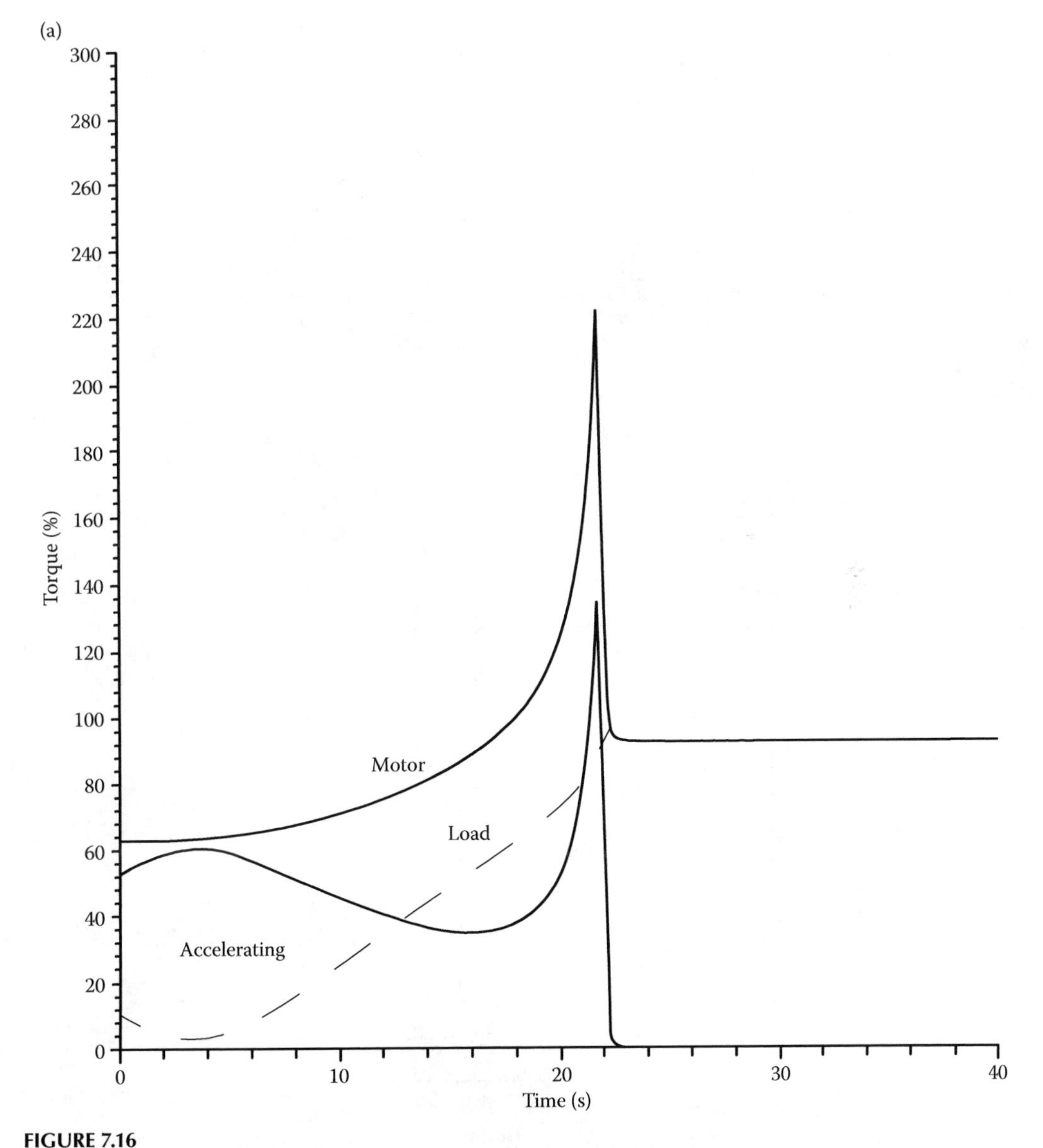

FIGURE 7.16
(a) Motor, load, and accelerating torque, (b) motor terminal current, terminal voltage, and slip, and (c) starting active and reactive power demand.

(Continued)

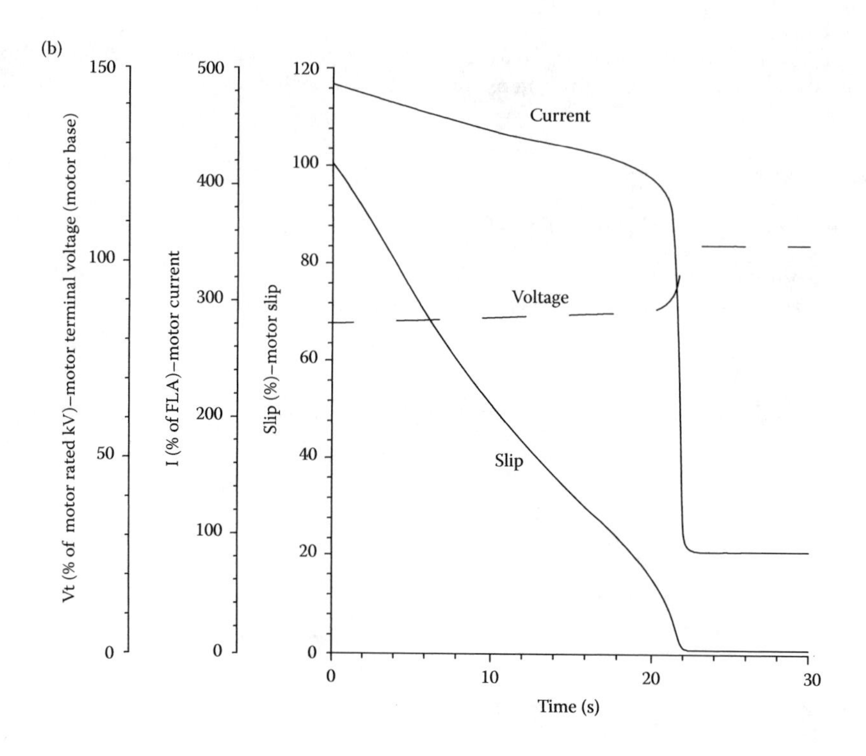

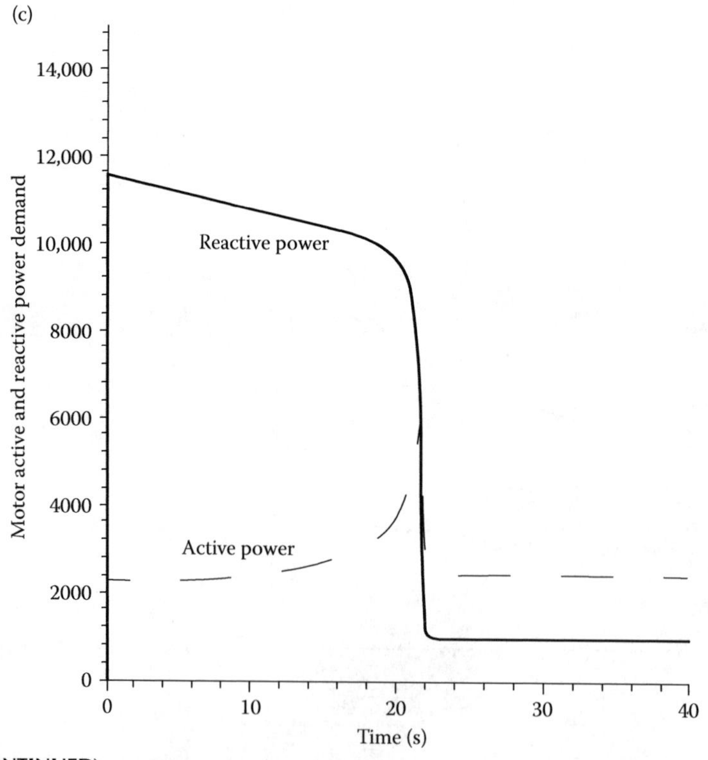

FIGURE 7.16 (CONTINUED)
(a) Motor, load, and accelerating torque, (b) motor terminal current, terminal voltage, and slip, and (c) starting active and reactive power demand.

TABLE 7.3

Calculation of Starting Time of Motor, Example 7.2

No	Speed Range Δn	Average Torque %	Average Torque lb ft²	Time Δt (s)
1	0%–25%	50.3	5274	5.5
2	25%–50%	44.70	4687	6.18
3	50%–75%	39.0	4089	7.09
4	75% to full load	65	6816	4.25

$\Sigma\Delta t = 23.02$.

7.6.1 Induction Motors

If the voltage dip is of sufficient magnitude, the induction motor torque reduces below the load torque and the motor will stall; *provided the voltage is not restored quickly*. It takes finite time for the slip to increase from s_0 to s_u. As torque is proportional to the square of the voltage,

$$T_{\text{reduced}} = T\left(\frac{V_1}{V_0}\right)^2 \tag{7.31}$$

where V_0 is the rated voltage and V is the reduced voltage. The torque T at any slip s is related to the breakdown torque T_p and the critical slip s_p (Figure 7.17) by the expression [3],

$$\frac{T}{T_p} = \frac{2}{(s/s_p)+(s_p/s)} \tag{7.32}$$

Then, we can write

$$\frac{T_{\text{reduced}}}{T_1} = \frac{2k}{(s/s_p)+(s_p/s)} \tag{7.33}$$

where

$$k = \frac{T_p}{T_1}\left(\frac{V_1}{V_0}\right)^2 \tag{7.34}$$

$k = 1$ for the voltage dip that reduces $T_p = T_1$ (see Figure 7.17).

The time t to deaccelerate from full load slip s_0 to a slip s_u is given by

$$t = 2H\int_{s_0}^{s_u} \frac{-\text{d}s}{1-(T_{\text{reduced}}/T_1)} \tag{7.35}$$

This can be written as

$$t = 2s_pH\int_{s_0}^{s_u} \frac{s/s_p + s_p/s}{(s/s_p + s_p/s)-2k}\text{d}(s/s_p) = 2s_pHT \tag{7.36}$$

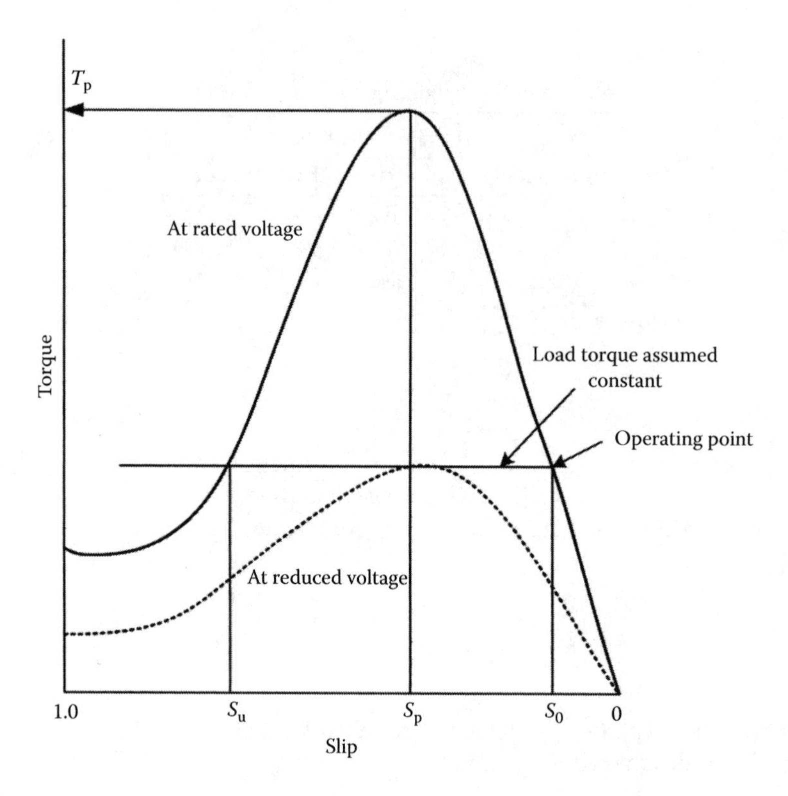

FIGURE 7.17
To illustrate stability on voltage dip of an induction motor.

where

$$T=\left\{s/s_p = k\ln\left[s/s_p\left(s/s_p+s_p/s-2k\right)\right]+\frac{2k^2}{\sqrt{1-k^2}}\tan^{-1}\left(\frac{s/s_p-k}{\sqrt{1-k^2}}\right)\right\}_{s_0}^{s_u} \tag{7.37}$$

The limits of s_0 and s_u are given by

$$s_0, s_u = s_p\left(T_p/T_1 \pm \sqrt{(T_p/T_1)^2-1}\right) \tag{7.38}$$

Example 7.3

We will calculate the capability of 2000 hp (1492 kW) and 1000 hp (746 kW) motors to withstand voltage dips using these equations.

- 2000 hp motor: six pole, 60 Hz, load inertia 16780 lb ft^2, motor rotor inertia = 20,000 lb ft^2, ($H = 3.6$) breakdown torque $T_p = 150\%$, critical slip $s_p = 10\%$, $s_0 = 3.8\%$, and $s_u = 26.2\%$. Repeat with double the inertia.
- 1000 hp motor, 4 pole, $T_p = 200\%$, $H = 3.3$ s, $s_0 = 2.7\%$, $s_u = 37.3\%$, $s_p = 10\%$. Repeat with double the inertia.

The results of the calculations are shown in Figure 7.18.

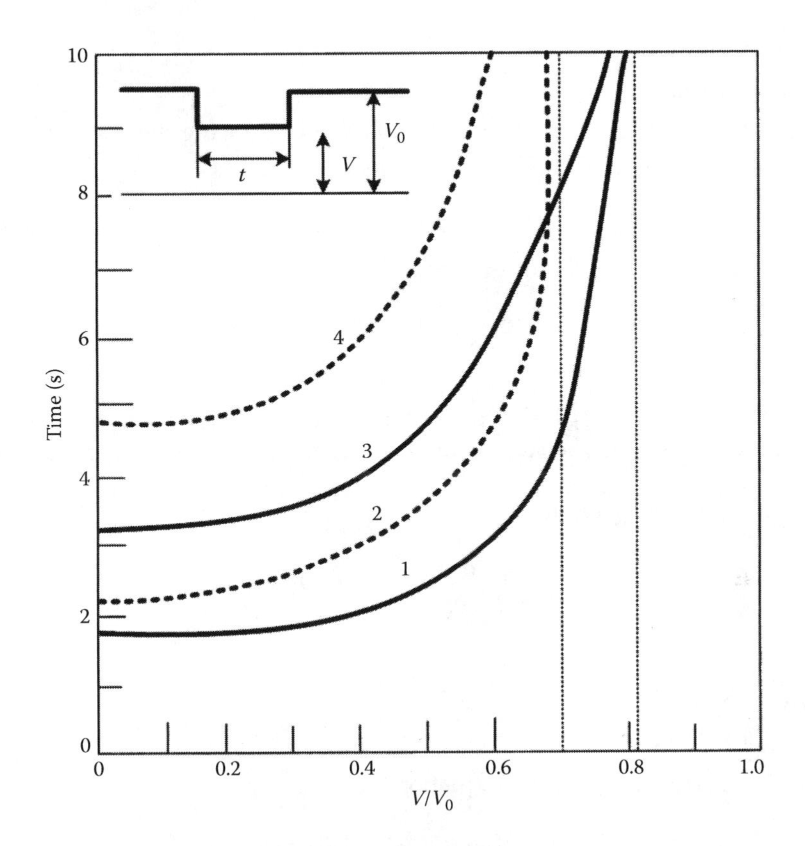

FIGURE 7.18
Calculated ride-through capabilities of 2000 hp and 1000 hp induction motors, parameters as shown in the text.

Curve 1 2000 hp (1491.4 kW) motor, $H = 3.6$
Curve 2 1000 hp (746 kW) motor, $H = 3.3$
Curve 3 2000 hp (1491.4 kW) motor, $H = 7.2$
Curve 4 1000 hp (746 kW) motor, $H = 6.6$

The motors with higher load inertia and higher T_p have better capability to withstand voltage dips.

7.6.2 Synchronous Motors

The analytical treatment of synchronous motors in more complex, see Volumes 1 and 4 for the concepts of equal area criteria of stability. Based on this criterion, a critical clearing angle can be calculated:

$$\delta_c = \cos^{-1}\left[\frac{(P_s/P_m)(\delta_m - \delta_0) + \cos\delta_m - \gamma_1 \cos\delta_0}{1-\gamma_1}\right] \tag{7.39}$$

where

δ_c = critical clearing angle
P_s = suddenly applied load

δ_0 = initial torque angle

P_m = maximum output

γ_1 = a factor by which the maximum output is reduced during the disturbance

δ_m = maximum swing of the torque angle, given by

$$\delta_m = \sin^{-1}(P_s/P_m) \tag{7.40}$$

The output of a synchronous motor is

$$P_1 = \frac{VE}{X_d}\sin\delta + \frac{V^2(X_d - X_q)\sin 2\delta}{2X_dX_q} \tag{7.41}$$

where V is the terminal voltage, E is the excitation voltage, X_d and X_q are the direct and quadrature axis reactances. The second term in Equation 7.41 gives the reluctance torque due to saliency.

The swing equation is given by

$$\frac{GH}{\pi f}\frac{d^2\delta}{dt} = P_a = P_s - P_e \tag{7.42}$$

where

G = MVA rating of the machine

P_a = accelerating power

P_e = electrical gap power given by Equation 7.42

The step-by-step solution is made using the following equation:

$$\Delta_{\delta n} = \Delta_{\delta n-1} + \frac{\pi f P_{a(n-1)}}{GH}(\Delta t)^2 \tag{7.43}$$

Example 7.4

Consider a synchronous motor of 2000 hp (1491.4 kW), 6-pole, 0.8 power factor leading, 175% pullout torque, $H = 3.6$ s, $\delta_0 = 34.85°$, δ_c for a 50% voltage dip = 106.12°.

The calculations are shown in Table 7.4 for a voltage dip of 50%. If the voltage is not restored within 0.35 cycles, the motor falls out of step.

Figure 7.19a shows some similar calculations for complete loss of voltage and with double the inertia.

Curve 1 with complete loss of voltage, $H = 3.6$
Curve 2 with complete loss of voltage, $H = 7.2$
Curve 3 50% voltage dip, $H = 3.6$
Curve 4 50% voltage dip, voltage restored within 0.35 s
Curve 5 50% voltage dip, $H = 7.2$.

Fig. 7.19b shows these characteristics in a more useful form.

7.6.3 Acceptable Voltage Dip on Motor Starting

No guidelines as such can be provided; as demonstrated above, the starting voltage dip is a function of the motor, load, and system parameters. The following considerations apply:

TABLE 7.4

Calculation of Stability for a Voltage Dip of 50%, 2000 hp Motor, Example 7.4

t (s)	P_e	P_a	$\frac{\pi f(\Delta t)^2 P_a}{GH}$	$\Delta\delta_n$	δ_n
0–	1.0	–	–	–	34.85
0+	0.5	0.5	–	–	34.85
0av	–	0.25	1.88	1.88	34.85
0.05	0.52	0.48	3.60	5.48	36.73
0.1	0.59	0.41	3.08	8.56	42.21
0.15	0.68	0.32	2.40	10.96	50.77
0.2	0.77	0.23	1.73	12.69	61.73
0.25	0.84	0.16	1.20	13.89	74.42
0.3	0.87	0.13	0.98	14.87	88.31
0.35	0.85	0.15	1.13	16.0	103.18
0.4	0.76	0.24	1.80	17.80	119.18

- The starting voltage dip should not reduce the torque of the motor below the load torque at any point in the starting cycle. If the accelerating torque becomes negative, the motor will not accelerate to the rated speed and keep crawling at a lower speed, till it is tripped by the protective devices.
- The starting voltage dip should not be of such a magnitude that the stability of the running loads is jeopardized or the running motors are lockout.

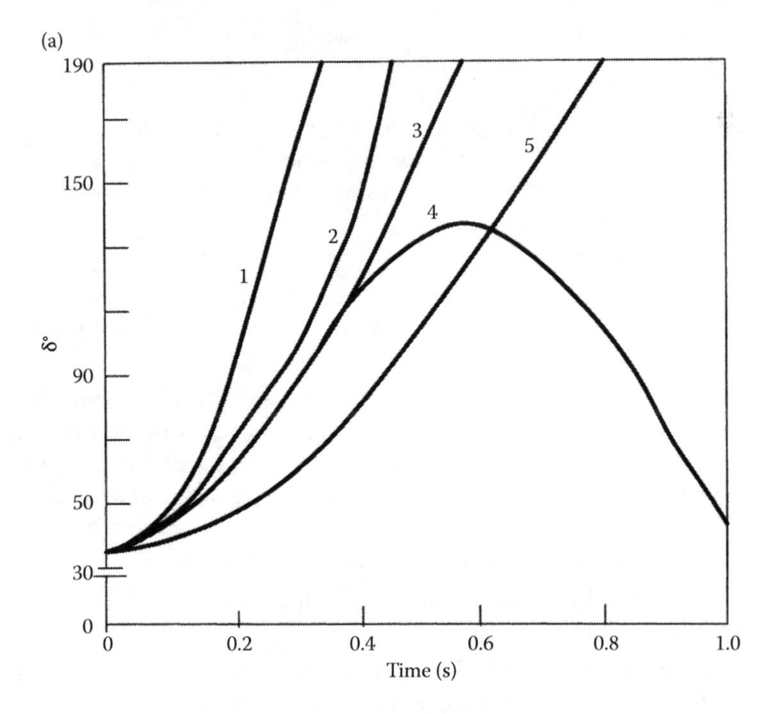

FIGURE 7.19
(a) Step-by step calculated swing curves of 2000 hp synchronous motor, see the text and (b) calculated voltage ride-through capability of 2000 hp synchronous motor, curve 1, $H = 3.6$, curve 2, $H = 7.2$.

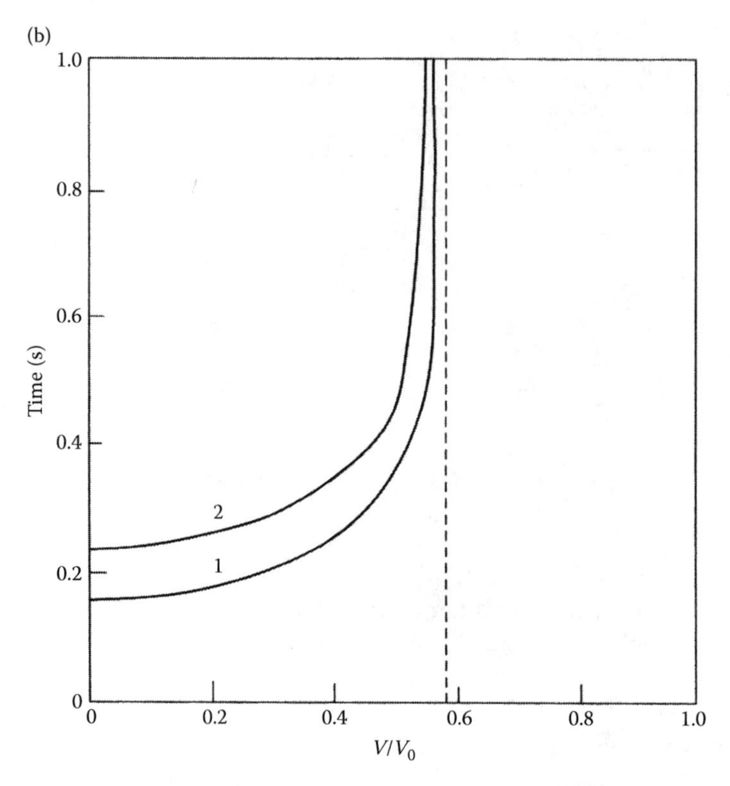

FIGURE 7.19 (CONTINUED)
(a) Step-by step calculated swing curves of 2000 hp synchronous motor, see the text and (b) calculated voltage ride-through capability of 2000 hp synchronous motor, curve 1, $H = 3.6$, curve 2, $H = 7.2$.

- Note that the electronic and drive system loads will be less tolerant to a prolonged voltage dip. When such loads are present, generally, the voltage dip should be limited to no more than 10%; the drive system vendors will specify the tolerable voltage dip and its time duration.
- It should be considered that motors with high inertial loads, like ID fans may take 40–50 s to start. As demonstrated above, the capability to ride-through voltage dips depends not only on the magnitude of the voltage dip, but also the time duration, till it recovers to normal.
- The running loads of induction motors, generally, are constant kVA load. This means that the motor will take a higher current inversely proportional to the voltage. This higher current flowing through the system impedances will create further voltage dips, which will create still higher current: the process is cumulative and can cause voltage instability. Motor starting hand calculations cannot account for this effect, but a proper load flow model of the loads and program will provide accurate results.
- Utilities may impose limits on the starting voltage dips as reflect at the point of supply or coupling with the utility.
- A motor connected to a dedicated transformer can take more voltage dip and the impact on other running loads can be reduced.
- Depending upon the system designs, synchronous motor up to 20,000 hp (14,914 kW) have been started across the line. Synchronous motor design permits

reduction of starting current much better than squirrel cage induction motors. Also at lower operating speeds, the synchronous motors are more efficient.

- A complete system design requires a number of considerations, the fundamental aspects as discussed above; no generalizations can be made.

7.7 EMTP Simulations of Motor Starting Transients

Figure 7.20 shows a simple configuration, there are four motors of 3000 hp (2237 kW) M1, M2, M3, and M4 connected to Bus-A. Motors M1, M2, and M3 are running, while motor M4 is started. The detailed parameters of the modeling in EMTP are not described. The load inertia is adjusted to give a short starting time of approximately 4 seconds; the purpose is to document the real profiles of the transients. The motor is started at 1 s.

- Figure 7.21 shows the starting current of motor M4, which on a transient basis is 8–10 times the motor full load current. This is typical starting current profile curve.
- Figure 7.22 shows the torque curve. Again note that the torque on the instant of switching at 1 s is six times the peak of the full load torque.
- Figure 7.23 depicts the voltage dip during starting, it shows a voltage dip of approximately 22%.
- Figure 7.24 shows the current transient in a running motor M2.
- Figure 7.25 depicts the transients in torque of a running motor. Note that the motor is not fully loaded.
- Figure 7.26 illustrates the torque slip curve of the motor being started and also of a running motor. Note that the slip increases from 0.6% to 1.0% and after the starting period the motor accelerated.

See Reference [13] for further reading.

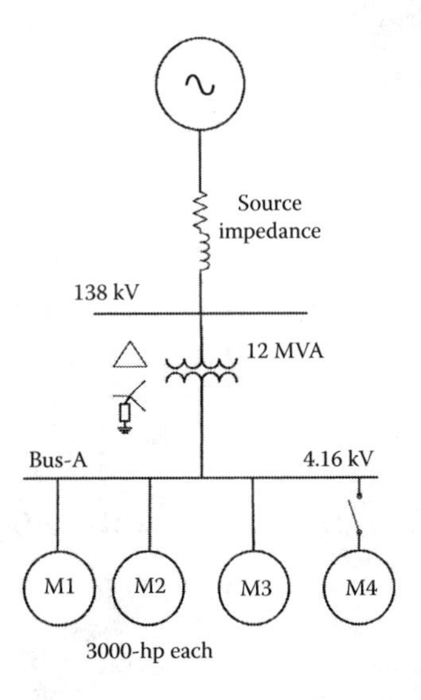

FIGURE 7.20
A configuration for motor starting for EMTP simulation.

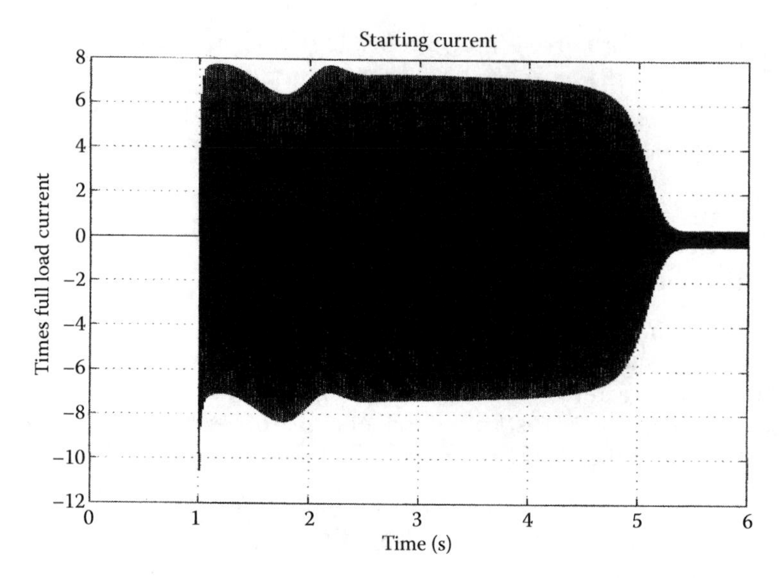

FIGURE 7.21
3000 hp induction motor starting current transient.

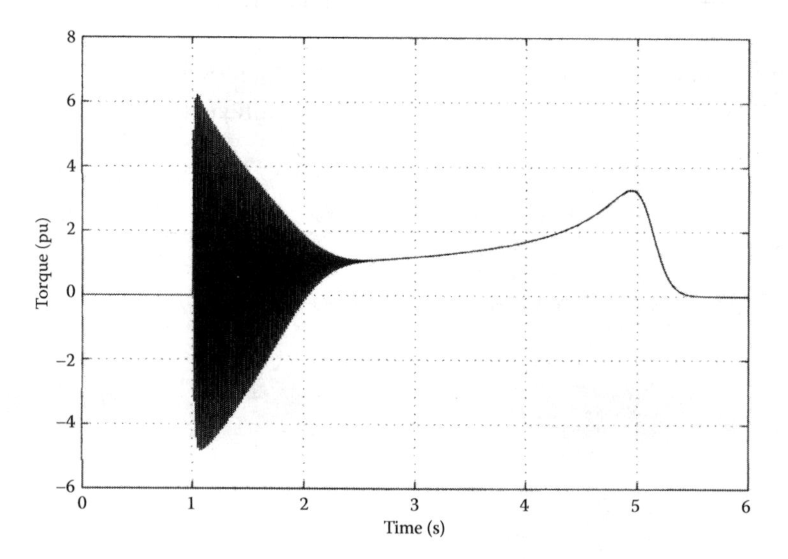

FIGURE 7.22
3000 hp induction motor torque transients during starting.

7.8 Synchronous Motors Driving Reciprocating Compressors

Synchronous motors driving reciprocating compressors give rise to current pulsations in the stator current due to the nature of the reciprocating compressor torque variations. Figure 7.27a shows the crank effort diagrams for one-cylinder, double-acting compressor or two-cylinder, single-acting compressor and Figure 7.27b shows the crank effort diagram for a two-cylinder, double-acting compressor or four-cylinder single-acting compressor.

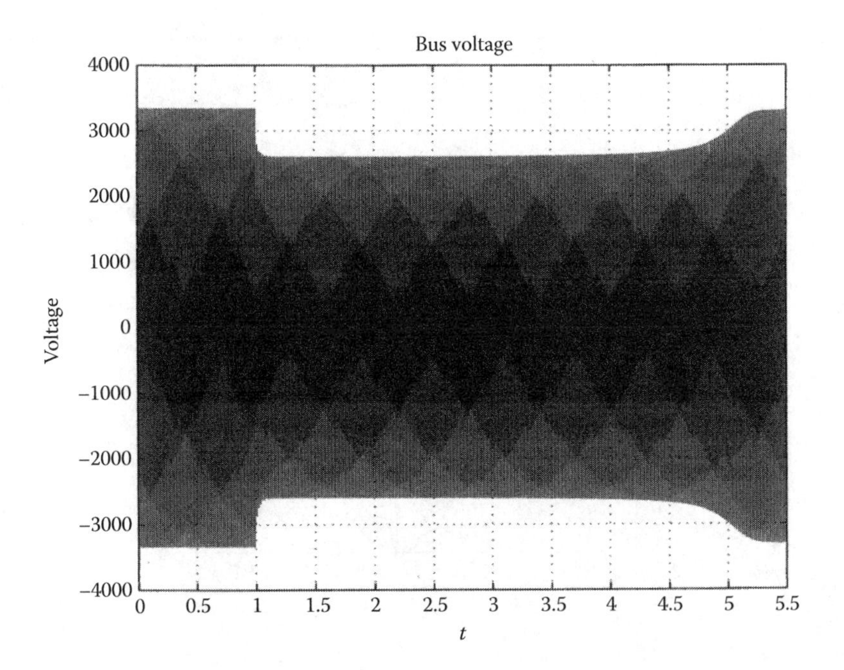

FIGURE 7.23
Bus voltage dip and recovery during starting of 3000 hp induction motor.

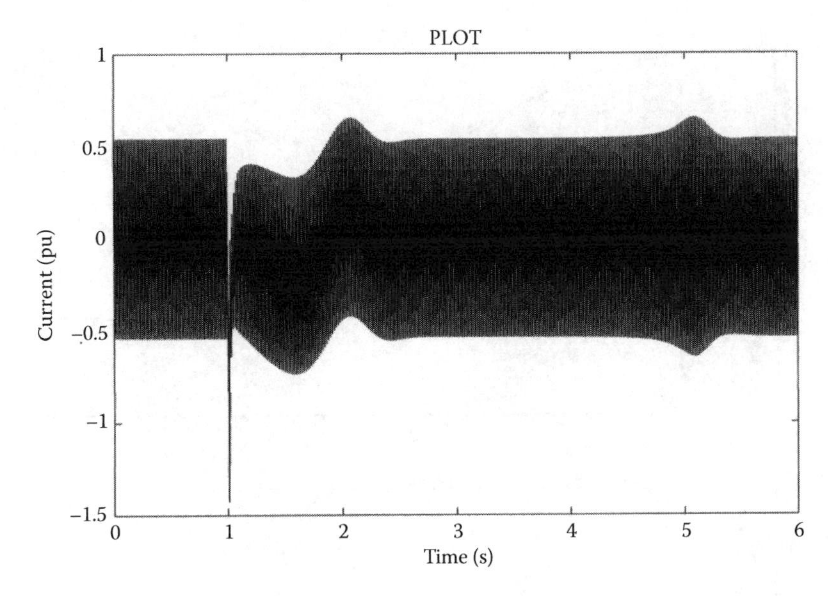

FIGURE 7.24
Impact on the running current of a 3000 hp induction motor running on the same bus.

In some cases, step unloading of compressors is used, which introduces additional irregularity in the crank effort diagram and increases the current pulsation.

NEMA standards limit current pulsation to 66% of rated full load current, corresponding to an angular deviation of approximately 5% from the uniform speed. The

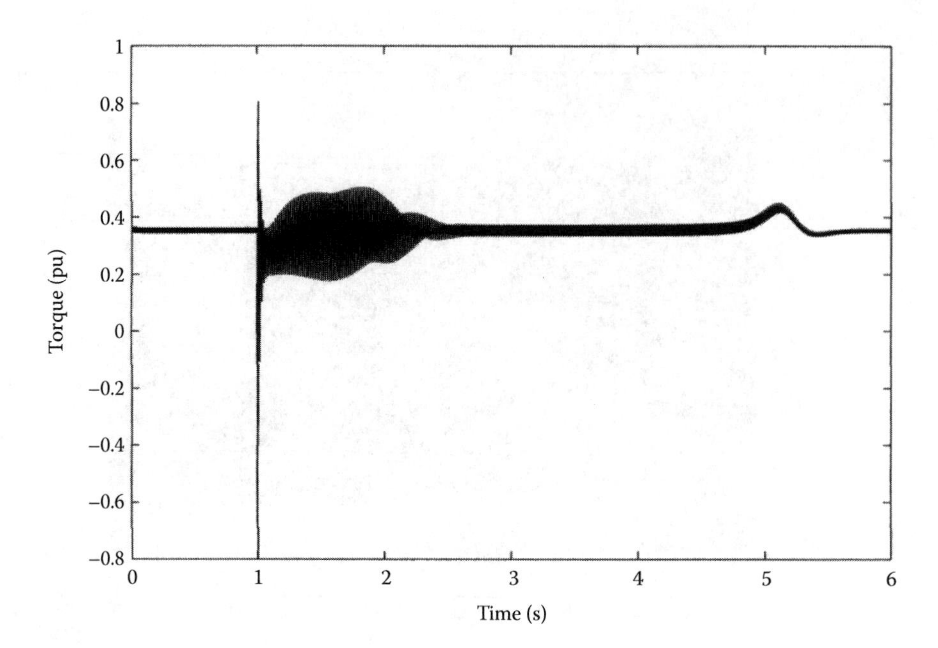

FIGURE 7.25
Impact on the torque of a 3000 hp induction motor running on the same bus.

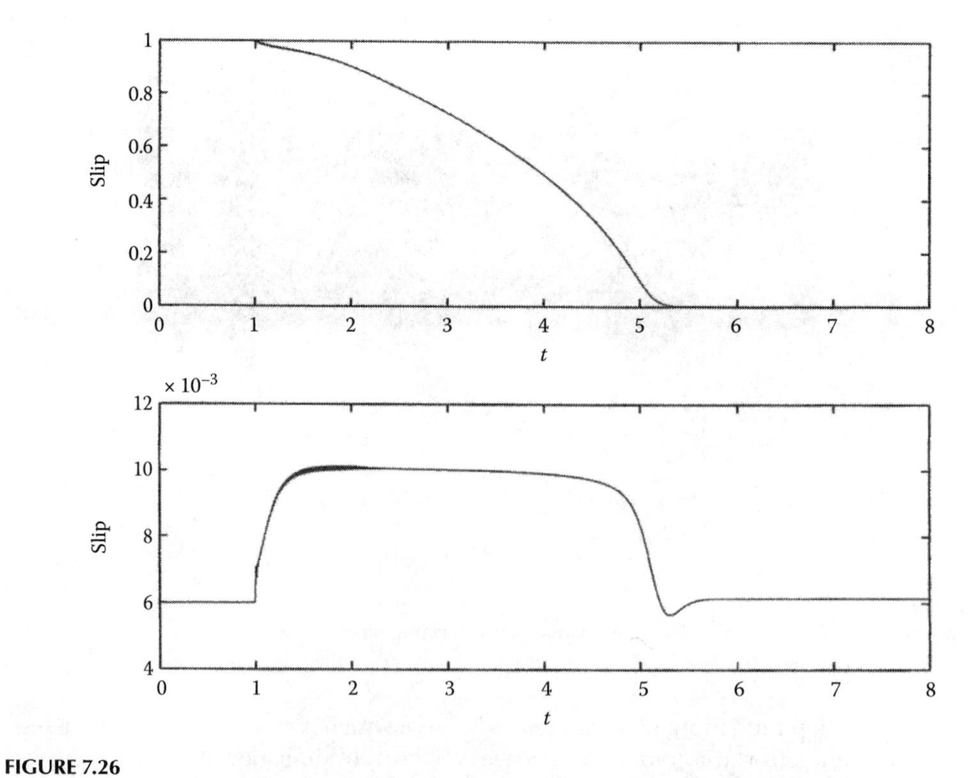

FIGURE 7.26
Slip-time curves of 3000 hp motor during starting and impact on the running motor. The running motor slows down, its slip increases from 0.06% to 0.1%, and the motor accelerates after the starting impact.

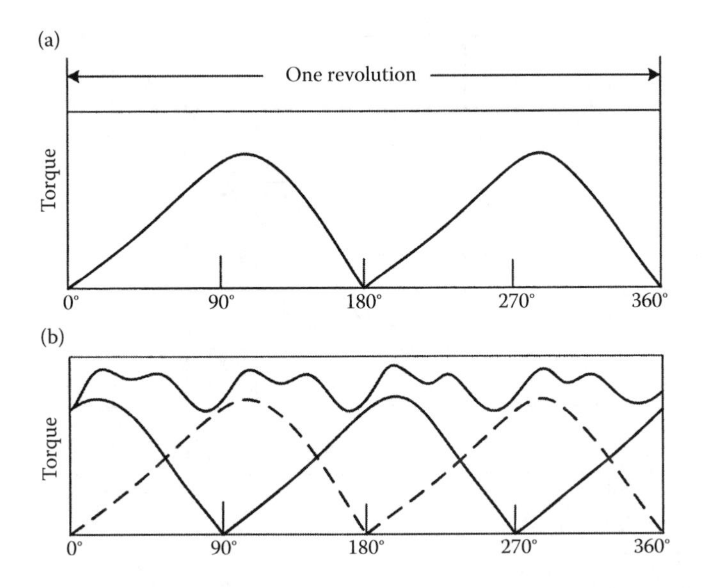

FIGURE 7.27
(a) Crank effort diagram for one-cylinder double-acting compressor or two-cylinder single-acting compressor and (b) crank effort diagram for two-cylinder double-acting compressor or four-cylinder single-acting compressor.

required flywheel effect to limit current pulsation is proportional to compressor factor "X" given by

$$WK^2 = 1.34XfP_r\left(\frac{100}{\text{rpm}}\right)^4 \tag{7.44}$$

where X is the compressor factor, f the frequency, P_r is the synchronizing power, and WK^2 is the total inertia in lb ft^2. The compressor factor curve from NEMA is shown in Figure 7.28.

A system having a mass in equilibrium and a force tending to return this mass to its initial position, if displaced, has a natural frequency of oscillation. For a synchronous motor, it is given by

$$f_n = \frac{35200}{n_s}\sqrt{\frac{P_r f}{WK^2}} \tag{7.45}$$

where f_n is the natural frequency of oscillation, n_s is the synchronous speed. This assumes that there is no damping and the motor is connected to an infinite system. The forcing frequency due to crank effort diagram should differ from the natural frequency by at least 20%.

The output equation is 7.41. A disturbance in relative position of angle δ, results in a synchronizing power flow in the machine. A synchronizing torque is developed which opposes the torque angle change due to the disturbance, i.e., a load change. The final torque angle may be reached only after a series of oscillations, which determine the transient stability of the machine, depending upon whether these converge or diverge, akin to stability of a synchronous generator.

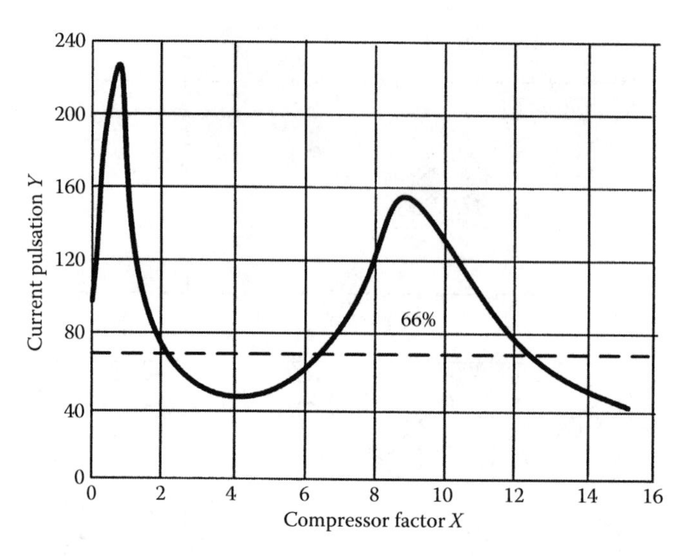

FIGURE 7.28
NEMA compressor factor curve.

The synchronizing power in per unit on machine can be written as

$$P_r = \frac{dP}{d\delta} = \frac{VE_t}{X_d}\cos\delta + \frac{V^2(X_d - X_q)}{X_d X_q}\cos 2\delta \tag{7.46}$$

For any change in P, we can write following the equation:

$$\Delta P = P_r \Delta\delta \tag{7.47}$$

The damping torque occurs due to induced currents in the damper windings, mainly because of quadrature axis flux. The damping power P_{ds} at any slip s can be written as

$$P_{ds} = s\frac{P_0}{s_0} \tag{7.48}$$

where s_0 is the slip, which will produce *asynchronously* the rated power P_0 of the machine.

We can define the slip as the change in the torque angle with time:

$$s = \frac{1}{\omega}\frac{d\delta}{dt} \tag{7.49}$$

Then the acceleration or retardation of the rotor is given by

$$a = \frac{ds}{dt} = \frac{1}{\omega}\frac{d^2\delta}{dt^2} \tag{7.50}$$

From above relations, we can write

$$P_{ds} = \frac{P_0}{s_0\omega}\frac{d\delta}{dt} \tag{7.51}$$

Lastly the power of inertia is

$$P_j = Jg\omega_0^2 a = \frac{T_a P_0}{\omega}\frac{d^2\delta}{dt^2} \tag{7.52}$$

where T_a is the accelerating time constant, J is the moment of inertia, and g is the gravitational constant.

The power balance equation becomes

$$\frac{T_a P_0}{\omega}\frac{d^2\delta}{dt^2} + \frac{P_0}{s_0\omega}\frac{d\delta}{dt} + P_s\delta = P_a \tag{7.53}$$

where P_a is the net power of acceleration or retardation. This differential equation without forcing function is

$$\frac{d^2\delta}{dt^2} + \rho\frac{d\delta}{dt} + \nu^2\delta = 0 \tag{7.54}$$

where the natural frequency is

$$\nu = \sqrt{\frac{\omega P_s}{T_a P_0}} = \sqrt{\frac{\omega P_s}{Jg\omega_0^2}} \tag{7.55}$$

and

$$\rho = \frac{1}{\nu T_a} \tag{7.56}$$

Example 7.5

A machine with $T_a = 10$ s, and synchronizing power two times the rated power has a natural frequency of

$$\nu = \sqrt{\frac{2\pi \times 60 \times 2}{10}} = 8.8 \text{ in } 2\pi\ s = 1.41 \text{ cps}$$

This is fairly low.

The solution of (7.54) is of the form

$$\delta = Ae^{-\frac{\rho}{2}t}\cos(\nu t + B) \tag{7.57}$$

where A and B are the constants of integration.

From Equations 7.49 and 7.50 the rotor slip is given by

$$s = -\frac{\nu}{\omega}Ae^{-\frac{\rho}{2}}\sin(\nu t + B) \tag{7.58}$$

Further discussions are continued in Volume 4. Resonance conditions can be excited.

Problems

7.1 How can the initial starting drop of a motor be calculated using a load flow program? Is a specific algorithm necessary? What is the effect of generator models?

7.2 A 1500 hp (1118 kW), 4 kV, four-pole induction motor has a slip of 1.5%, a locked rotor current six times the full load current, and 94% efficiency. Considering that $X_m = 3$ per unit, and stator reactance = rotor reactance = 0.08 per unit, draw the equivalent positive and negative sequence circuit. If the voltage has a 5% negative sequence component, calculate the positive sequence and negative sequence torque.

7.3 Calculate the starting time, current, voltage, and torque profiles by hand calculations for 2000 hp, 2.4 kV, 4 pole motor, full load slip = 1.75%, full load power factor = 0.945, full load efficiency = 93.5%, starting current = six times the full load current at a power factor of 0.15, total load inertia including that of motor and coupling = 12,000 lb ft^2. Apply the load and motor torque–speed characteristics as shown in Figure 7.8b. The total impedance in the motor circuit at the time of starting = 2.5 pu on 100 MVA base.

7.4 In Problem 7.3, a drive system load can tolerate a maximum 10% voltage dip. Design a motor starting strategy applying at least two starting methods from Table 7.2.

References

1. NEMA. *Large Machines—Induction Motors.* Standard MG1 Part 20, 1993.
2. JR Linders. Effect of power supply variations on AC motor characteristics. *IEEE Trans Ind Appl,* IA-8, 383–400, 1972.
3. JC Das. Effects of momentary voltage dips on the operation of induction and synchronous motors. *Trans IEEE Ind Appl Soc,* 26, 711–718, 1990.
4. JC Das, J Casey. Characteristics and analysis of starting of large synchronous motors. In *Conference Record, IEEE I&CPS Technical Conference,* Sparks, NV, 1999.
5. GS Sangha. Capacitor-reactor start of large synchronous motor on a limited capacity network. *IEEE Trans Ind Appl,* IA-20(5), 1337–1343, 1984.
6. J Langer. Static frequency changer supply system for synchronous motors driving tube mills. *Brown Boveri Rev,* 57, 112–119, 1970.
7. CP LeMone. *Large MV Motor Starting Using AC Inverters.* In *ENTELEC Conference Record,* Houston, TX, 1984.
8. AE Fitzgerlad, C Kingsley, A Kusko. *Electrical Machinery,* 4th ed., McGraw-Hill, New York, 1971.
9. Westinghouse. *Transmission and Distribution Handbook,* 4th ed., East Pittsburg, PA, 1964.
10. J Bredthauer, H Tretzack. HV synchronous motors with cylindrical rotors. *Siemens Power Eng,* V(5), 241–245, 1983.
11. HE Albright. Applications of large high speed synchronous motors. *IEEE Trans Ind Appl,* IA 16(1), 134–143, 1980.
12. IEEE Standard C37.91. *IEEE Guide for Protecting Power Transformers,* 2008.
13. JC Das. *Transients in Electrical Systems,* McGraw-Hill, New York, 2010.

8

Reactive Power Flow and Voltage Control

8.1 Maintaining Acceptable Voltage Profile

Chapters 5 and 6 showed that there is a strong relationship between voltage and reactive power flow, though real power flow on loaded circuits may further escalate the voltage problem. The voltages in a distribution system and to the consumers must be maintained within a certain plus–minus band around the rated equipment voltage, ideally from no-load to full load, and under varying loading conditions. Sudden load impacts (starting of a large motor) or load demands under contingency operating conditions, when one or more tie-line circuits may be out of service, result in short-time or prolonged voltage dips. High voltages may occur under light running load or on sudden load throwing and are of equal considerations, though low voltages occur more frequently. ANSI C84.1 [1] specifies the preferred nominal voltages and operating voltage ranges A and B for utilization and distribution equipment operating from 120 to 34,500 V. For transmission voltages over 34,500 V, only nominal and maximum system voltage is specified. Range B allows limited excursions outside range A limits. As an example, for a 13.8 kV nominal voltage, range A = 14.49 – 12.46 kV and range B = 14.5 – 13.11 kV. Cyclic loads, e.g., arc furnaces giving rise to flicker, must be controlled to an acceptable level, see Table 8.1 (reproduction of Table 3.1 from ANSI standard C84.1). The electrical apparatuses have a certain maximum and minimum operating voltage range in which normal operation is maintained, i.e., induction motors are designed to operate successfully under the following conditions [2]:

1. Plus or minus 10% of rated voltage, with rated frequency.
2. A combined variation in voltage and frequency of 10% (sum of absolute values) provided that the frequency variations do not exceed ±5% of rated frequency.
3. Plus or minus 5% of frequency with rated voltage.

Motor torque, speed, line current, and losses vary with respect to the operating voltage, as shown in Table 7.2. Continuous operation beyond the designed voltage variations is detrimental to the integrity and life of the electrical equipment.

A certain balance between the reactive power consuming and generating apparatuses is required. This must consider losses which may be a considerable percentage of the reactive load demand.

When the reactive power is transported over mainly reactive elements of the power system, the reactive power losses may be considerable and these add to the load demand (Example 6.5). This reduces the active power delivery capability of most electrical equipment rated on a kVA base. As an example, consider the reactive power flow through

TABLE 8.1

Standard Nominal System Voltages and Voltage Ranges

Voltage Class	Nominal System Voltage			Nominal Utilization Voltage	Voltage Range A			Voltage Range B		
					Maximum	Minimum		Maximum	Minimum	
	2-wire	3-wire	4-wire	2-wire 3-wire 4-wire	Utilization and Service Voltage	Service Voltage	Utilization Voltage	Utilization and Service Voltage	Service Voltage	Utilization Voltage
Low voltage	Single-Phase Systems									
	120			115	126	114	108	127	110	104
		120/240		115/230	126/252	114/228	108/216	127/254	110/220	104/208
	Three-Phase Systems									
			208Y/120	200	2I8Y/126	197Y/114	187Y/10S	220Y/127	191Y/110	180Y/104
			240/120	230/115	252/126	228/114	216/108	254/127	220/110	208/104
		240		230	252	228	216	254	220	208
			480Y/277	460Y/266	504Y/291	456Y/263	432Y7249	508Y/293	440Y/254	416Y/240
		480		460	504	456	432	508	440	416
		600		575	630	570	540	635	550	520
Medium voltage		2400			2520	2340	2160	2540	2280	2080
			4160Y/2400		4370/2520	4050Y/2340	3740Y/2160	4400Y/2540	3950Y/2280	3600Y/2080
		4160			4370	4050	3740	4400	3950	3600
		4800			5040	4680	4320	5060	4560	4160
		6900			7240	6730	62)0	7260	6560	5940
			8320Y/4800		8730Y/5040	8110Y/4680		8800Y/5080	7900Y/4560	
			12000Y/6930		12600Y/7270	11700Y/6760		12700Y/7330	11400Y/6580	
			12470Y/7200		13090Y/7560	12160Y/7020		13200Y/7620	11850Y/6840	
			13200Y/7620		13S60Y/8000	12870Y/7430		13970Y/8070	12504Y/7240	
			13800Y/7970		14490Y/8370	13460Y/7770		14520Y/8380	13110Y/7570	

(*Continued*)

TABLE 8.1 (*Continued*)

Standard Nominal System Voltages and Voltage Ranges

	Nominal System Voltage			Nominal Utilization Voltage	Voltage Range A			Voltage Range B		
					Maximum	Minimum		Maximum	Minimum	
Voltage Class	2-wire	3-wire	4-wire	2-wire 3-wire 4-wire	Utilization and Service Voltage	Service Voltage	Utilization Voltage	Utilization and Service Voltage	Service Voltage	Utilization Voltage
Medium voltage		13800			14490	13460	12420	14520	13110	11880
			20780Y/12000		21820Y/12600	20260Y/11700		22000Y/12700	19740Y/11400	
			22860Y/13200		24000Y/13860	22290Y/12870		24200Y/13970	21720Y/12540	
		23000			24150	22430		24340	21850	
			24940Y/14400		26190Y/15120	24320Y/I4040		26400Y/15240	23690Y/13680	
			34500Y/19920		36230Y/20920	33640Y/19420		36510Y/21080	32780Y/18930	
		34500			36230	33640		36510	32780	
					Maximum voltage	See Reference [1] for further details and explanations				
		46000			48300					
		69000			72500					
High voltage		115000			121000					
		138000			145000					
		161000			169000					
		230000			242000					
Extra-high voltage		345000			362000					
		400000			420000					
		500000			550000					
		765000			800000					
Ultra-high voltage		1100000			1200000					

a 0.76 Ω reactor. For a 70 Mvar input the output is 50 and 20 Mvar is lost in the reactor itself. If the load voltage is to be maintained at 1.0 per unit, the source-side voltage should be raised to 1.28 per unit, representing a voltage drop of 28% in the reactor. Figure 8.1 shows the reactive power loss and voltage drops in lumped reactance.

In a loaded transmission line, when power transfer is below the surge impedance loading, the charging current exceeds the reactive line losses and this excess charging current must be absorbed by shunt reactors and generators. Above surge impedance loading, the reactive power must be supplied to the line. At 1.5 times the surge loading, an increase of 150 MW will increase the reactive power losses in a 500 kV transmission line by about

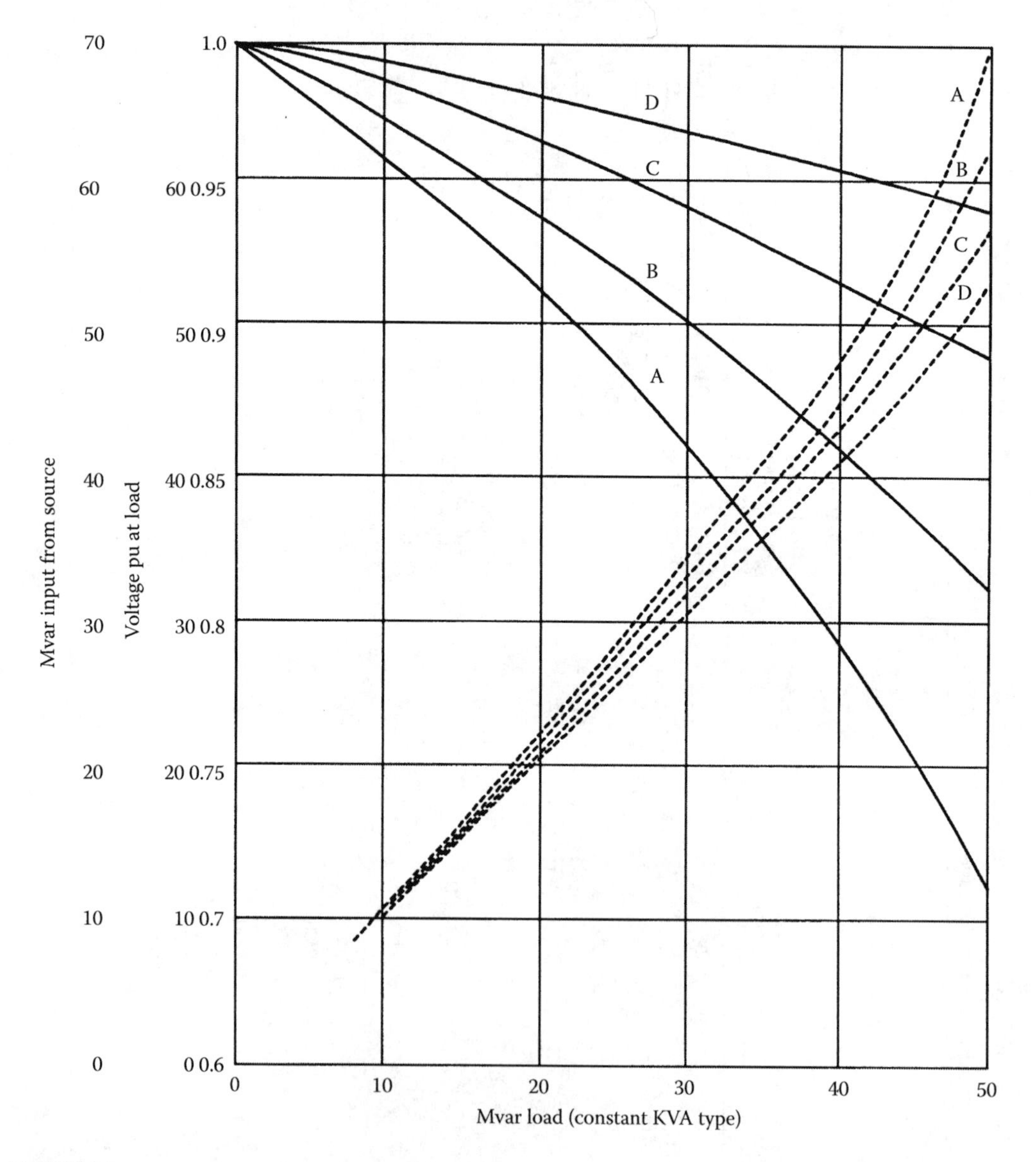

FIGURE 8.1
Voltage drop and reactive power loss in lumped reactance: *A*: 0.76 Ω, *B*: 0.57 Ω, *C*: 0.38 Ω, and *D*: 0.19 Ω.

95 Mvar, or about 50% of the line charging. At twice surge loading, approximately 1800 MW (surge impedance 277 Ω), a 100 MW load increase will increase the reactive losses by 100 Mvar.

A V–Q control may necessitate addition of leading or lagging reactive power sources, which may be passive or dynamic in nature. Shunt reactors and capacitors are examples of passive devices. The SVC (static var controller) is an example of a dynamic device. It should be ensured that all plant generators operate within their reactive power capability limits and remain stable. The on-load tap-changing transformers must be able to maintain an acceptable voltage within their tap setting range.

Assessment of voltage problems in a distribution system under normal and contingency load flow conditions, therefore, requires investigations of the following options:

- Location of reactive power sources, i.e., series and shunt capacitors, synchronous condensers, voltage regulators, overexcited synchronous motors, and SVCs and FACTS devices (Chapter 9) with respect to load.
- Control strategies of these reactive power sources.
- Provisions of on-load tap-changing equipment on tie transformers.
- Undervoltage load shedding.
- Stiffening of the system, i.e., reduction of system reactance which can be achieved by bundle conductors, duplicate feeders, and additional tie-lines.
- Redistribution of loads.
- Induction voltage regulators.
- Step voltage regulators.

8.2 Voltage Instability

Consider the power flow on a mainly inductive tie-line, see Figure 8.2a. The load demand is shown as $P + jQ$, the series admittance $Y_{sr} = g_{sr} + b_{sr}$, and $Z = R_{sr} + jX_{sr}$. A similar circuit is considered in Section 3.2. The power flow equation from the source bus (an infinite bus) is given by

$$\begin{aligned} P + jQ &= V_r e^{-j\theta}\left[(V_s - V_r e^{j\theta})(g_{sr} + jb_{sr})\right] \\ &= \left[(V_s V_r \cos\theta - V_r^2)g_{sr} + V_s V_r b_{sr}\sin\theta\right] \\ &\quad + j\left[(V_s V_r \cos\theta - V_r^2)b_{sr} - V_s V_r g_{sr}\sin\theta\right] \end{aligned} \tag{8.1}$$

If resistance is neglected,

$$P = V_s V_r b_{sr}\cos\theta \tag{8.2}$$

$$Q = \left(V_s V_r \sin\theta - V_r^2\right) b_{sr} \tag{8.3}$$

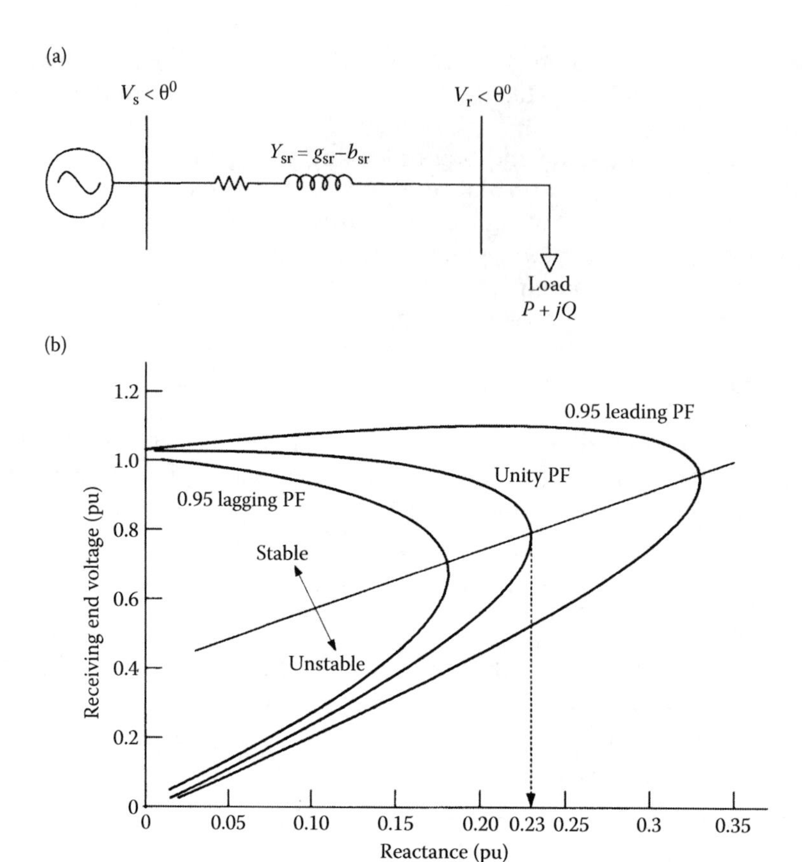

FIGURE 8.2
(a) Active and reactive power flows through a predominantly reactive two-port connector and (b) receiving-end voltage per unit versus system reactance in per unit for different power factors and constant power output. Stable and unstable regions and critical reactance are shown.

These are the same equations as derived in Equations 3.22 and 3.23. If the receiving-end load changes by a factor $\Delta P + \Delta Q$, then

$$\Delta P = (V_s b_{sr} \sin\theta)\Delta V + (V_s V_r b_{sr} \cos\theta)\Delta\theta \tag{8.4}$$

and

$$\Delta Q = (V_s \cos\theta - 2V_r) b_{sr} \Delta V - (V_s V_r b_{sr} \sin\theta)\Delta\theta \tag{8.5}$$

where ΔV_r is the scalar change in voltage V_r and $\Delta\theta$ is the change in angular displacement. If θ is eliminated from Equation 8.1 and resistance is neglected, a dynamic voltage equation of the system is obtained as follows:

$$V_r^4 + V_r^2 (2QX_{sr} - V_s^2) + X_{sr}^2 (P^2 + Q^2) = 0 \tag{8.6}$$

The positive real roots of this equation are

$$V_r = \left[\frac{-2QX_{sr} + V_s^2}{2} \pm \frac{1}{2}\sqrt{\left(2QX_{sr} - V_s^2\right)^2 - 4X_{sr}^2(P^2 + Q^2)} \right]^{1/2} \tag{8.7}$$

This equation shows that, in a lossless line, the receiving-end voltage V_r is a function of the sending-end voltage V_s, series reactance X_{sr}, and receiving-end real and reactive power. Voltage problems are compounded when reactive power flows over heavily loaded active power circuits. If reactive power is considered as zero and the sending-end voltage is 1 per unit, then Equation 8.7 reduces to

$$V_r = \left[\frac{1}{2} \pm \frac{\sqrt{1 - 4X_{sr}^2 P^2}}{2} \right]^{1/2} \tag{8.8}$$

For two equal real value of V_r

$$\left(1 - 4X_{sr}^2 P^2\right)^{1/2} = 0, \text{ i.e., } X_{sr} = 1/2P \tag{8.9}$$

This value of X_{sr} may be called a critical reactance. Figure 8.2b shows the characteristics of receiving-end voltage, against system reactance for constant power flow, at lagging and leading power factors, with the sending-end voltage maintained constant. This figure shows, e.g., that at a power factor of 0.9 lagging, voltage instability occurs for any reactance value exceeding 0.23 per unit. This reactance is the critical reactance. For a system reactance less than the critical reactance, there are two values of voltages: one higher and the other lower. The lower voltage represents unstable operation, requiring large amount of source current. For a system reactance close to the critical reactance, voltage instability can occur for a small positive excursion in the power demand. As the power factor improves, a higher system reactance is permissible for the power transfer. The voltage instability can be defined as the limiting stage beyond which the reactive power injection does not elevate the system voltage to normal. The system voltage can only be adjusted by reactive power injection until the system voltage stability is maintained. From Equation 8.7, the critical system reactance at voltage stability limit is obtained as follows:

$$\left(2QX_{sr} - V_s^2\right) = 4X_{sr}^2\left(P^2 + Q^2\right)$$

$$4X_{sr}^2 P^2 + 4X_{sr} Q V_s^2 - V_s^4 = 0 \tag{8.10}$$

The solution of this quadratic equation gives

$$X_{sr(critical)} = \frac{V_s^2}{2}\left[\frac{Q \pm \sqrt{Q^2 - P^2}}{P^2}\right]$$

$$= \frac{V_s^2}{2P}(-\tan\phi + \sec\phi) \tag{8.11}$$

where ϕ is the power factor angle.

Enhancing the thermal capacity of radial lines by use of shunt capacitors is increasingly common. However, there is a limit to which capacitors can extend the load-carrying capability [3].

In practice, the phenomenon of collapse of voltage is more complex. Constant loads are assumed in the above scenario. At lower voltages, the loads may be reduced, though this is not always true, i.e., an induction motor may not stall until the voltage has dropped more than 25%, even then the magnetic contactors in the motor starter supplying power to the motor may not drop out. This lock out of the motors and loss of some loads may result in voltage recovery, which may start the process of load interruption afresh, as the motors try to reaccelerate on the return voltage.

From Equation 8.5 as θ is normally small,

$$\frac{\Delta Q}{\Delta V} = \frac{V_s - 2V_r}{X_{sr}} \tag{8.12}$$

If the three phases of the line connector are short-circuited at the receiving end, the receiving-end short-circuit current is

$$I_r = \frac{V_s}{X_{sr}} \tag{8.13}$$

This assumes that the resistance is much smaller than the reactance. At no-load $V_r = V_s$; therefore,

$$\frac{\partial Q}{\partial V} = -\frac{V_r}{X_{sr}} = -\frac{V_s}{X_{sr}} \tag{8.14}$$

Thus,

$$\left|\frac{\partial Q}{\partial V}\right| = \text{short-circuit current} \tag{8.15}$$

Alternatively, we could say that

$$\frac{\Delta V_s}{V} \approx \frac{\Delta V_r}{V} = \frac{\Delta Q}{S_{sc}} \tag{8.16}$$

where S_{sc} is the short-circuit level of the system. This means that the voltage regulation is equal to the ratio of the reactive power change to the short-circuit level. This gives the obvious result that the receiving-end voltage falls with the decrease in system short-circuit capacity, or increase in system reactance. A stiffer system tends to uphold the receiving-end voltage. Voltage and reactive power control have significant impact on stability. The problem is that a power system supplies power to a vast number of consumers with different load profiles and there are a number of generating units. As loads vary, the reactive power requirements of the transmission system vary. Since reactive power cannot be transmitted over long distances, voltage control has to be effected by devices dispersed throughout the system.

Example 8.1

Consider the system of Figure 8.3. Bus C has two sources of power, one transformed from the 400 kV bus A and connected through a transmission line and the other from the 230 kV bus B. These sources run in parallel at bus C. The voltages at buses A and B are maintained equal to the rated voltage. A certain load demand at bus C dips the voltage by 10 kV. What is the reactive power compensation required at bus C to bring the voltage to its rated value?

An approximate solution is given by Equation 8.16. Based on the impedance data shown in Figure 8.3, calculate the short-circuit current at bus C. This is equal to 7.28 kA. Therefore,

$$Q_{\mathrm{c}} = \Delta V S_{\mathrm{c}} = (10/230)(230 \times \sqrt{3} \times 7.28) = 126.1 \text{ Mvar}$$

A reactive power injection of 126.1 Mvar is required to compensate the voltage drop of 10 kV. This calculation is approximate; *note that shunt capacitance is ignored.*

The voltage-reactive power stability problem is more involved than portrayed above. Power systems have a hierarchical structure: power generation-transmission lines-subtransmission level-distribution level, and finally to the consumer level. As reactive power does not travel well over long distances, it becomes a local problem. The utility companies operate with an agreed voltage level at intertie connections, and provide for their own reactive power compensation to meet this requirement. It can be said that there are no true hierarchical structures in terms of reactive power flow.

The voltage instability is not a single phenomenon and is closely linked to the electromagnetic stability and shares all aspects of active power stability, though there are differences.

Consider the equations of active and reactive power flows, Equations 3.21 and 3.22, for a short line. Let the voltages be fixed. The angle δ (phasor difference between the sending-end and receiving-end voltage vectors) varies with receiving bus power, as shown in Figure 8.4a. Beyond the maximum power drawn, there is no equilibrium point. Below the maximum power drawn, there are two equilibriums: one in the stable state and the other in the unstable state. A load flow below the maximum point is considered statically stable.

A similar curve can be drawn for reactive power flow, see Figure 8.4b. The angles are fixed, and the bus voltage magnitude changes. The characteristic of this curve is the same as that of the active power flow curve. A reactive power demand above the maximum reactive power results in the nonexistence of a load flow situation. The point ΔV^{u} is statically unstable, corresponding to $\Delta\delta^{u}$. If we define

$$\left|P_{(\max)} - P\right| < \varepsilon \tag{8.17}$$

$$\left|Q_{(\max)} - Q\right| < \varepsilon \tag{8.18}$$

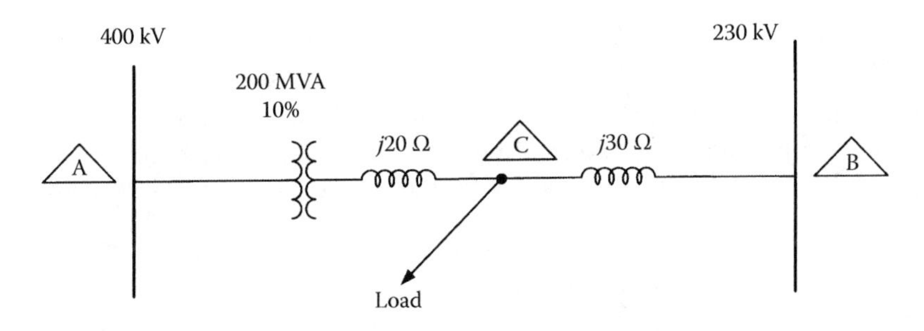

FIGURE 8.3
System for the calculation of reactive power injection on voltage dip (Example 8.1).

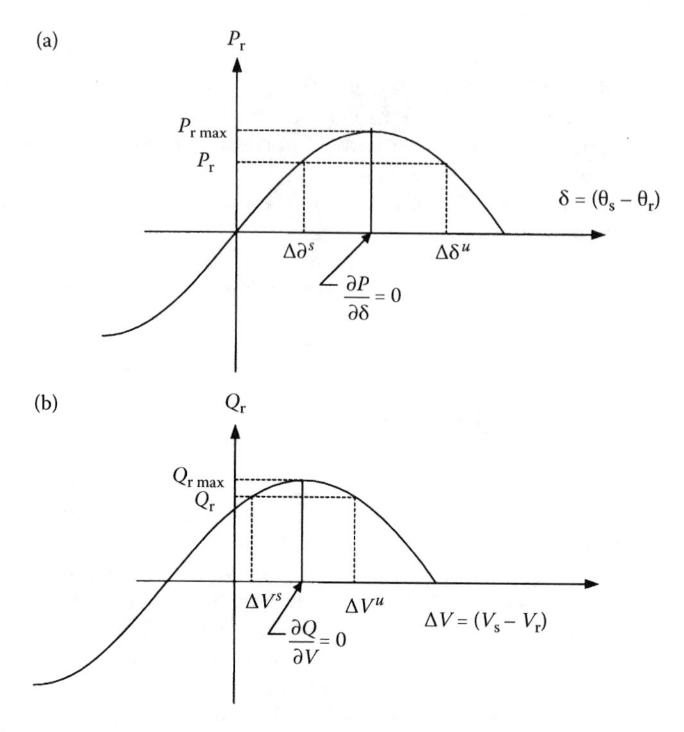

FIGURE 8.4
(a) To illustrate real power steady-state instability and (b) to illustrate reactive power steady-state instability.

then, however, small ε may be, there will always be two equilibrium points. Such an equilibrium point may be called a bifurcation point. The derivatives $\partial P/\partial\delta$ and $\partial Q/\partial V$ are zero at the static stability limit.

8.3 Reactive Power Compensation

The need for reactive power compensation is obvious from the above discussions [4, 5]. An acceptable, if not ideally flat voltage profile, has to be maintained. Concurrently, the stability is improved and the maximum power limit that can be transmitted increases.

8.3.1 Z_0 Compensation

It was shown in Chapter 3 that a flat voltage profile can be achieved with an SIL loading of V^2/Z_0. The surge impedance can be modified with addition of capacitors and reactors so that the desired power transmitted is given by the ratio of the square of the voltage to the modified surge impedance:

$$P_{\text{new}} = V^2/Z_{\text{modified}} \tag{8.19}$$

The load can suddenly change, and, ideally, the compensation should also adjust instantaneously according to Equation 8.19. This is not a practical operating situation and stability becomes a consideration. Passive and active compensators are used to enhance stability. The passive compensators are shunt reactors, shunt capacitors, and series capacitors. The active compensators generate or absorb reactive power at the terminals where these are connected. Their characteristics are described further.

8.3.2 Line Length Compensation

The line length of a wavelength of 3100 mi arrived at in Chapter 3 for a 60 Hz power transmission is too long. Even under ideal conditions, the natural load (SIL) cannot be transmitted $>\lambda/4$ (= 775 mi). Practical limits are much lower. As $\beta=\omega\sqrt{LC}$, the inductance can be reduced by series capacitors, thereby reducing β. The phase-shift angle between the sending-end and receiving-end voltages is also reduced and the stability limit is, therefore, improved.

8.3.3 Compensation by Sectionalization of Line

The line can be sectionalized, so that each section is independent of the others, i.e., meets its own requirements of flat voltage profile and load demand. This is compensation by sectionalizing, achieved by connecting constant-voltage compensators along the line. These are active compensators, i.e., thyristor-switched capacitors (TSCs), thyristor-controlled reactors (TCRs), and synchronous condensers. All three types of compensating strategies may be used in a single line.

Consider that a *distributed* shunt inductance L_{shcomp} is introduced. This changes the effective value of the shunt capacitance as follows:

$$j\omega C_{\text{comp}} = j\omega C + \frac{1}{j\omega L_{\text{shcomp}}} \tag{8.20}$$

where

$$K_{\text{sh}} = \frac{1}{\omega^2 L_{\text{shcomp}} C} = \frac{X_{\text{sh}}}{X_{\text{shcomp}}} = \frac{b_{\text{shcomp}}}{b_{\text{sh}}} \tag{8.21}$$

in which K_{sh} is the degree of shunt compensation. It is negative for a shunt capacitance addition.

Similarly, let a distributed *series* capacitance C_{srcomp} be added. The degree of series compensation is given by K_{sc}:

$$K_{\text{sc}} = \frac{X_{\text{srcomp}}}{X_{\text{sr}}} = \frac{b_{\text{sr}}}{b_{\text{srcomp}}} \tag{8.22}$$

The series or shunt elements added are distinguished by subscript "comp" in the above equations. Combining the effects of series and shunt compensations

$$\begin{aligned} Z_{0\text{comp}} &= Z_0\sqrt{\frac{1-K_{\text{sc}}}{1-K_{\text{sh}}}} \\ P_{0\text{comp}} &= P_0\sqrt{\frac{1-K_{\text{sh}}}{1-K_{\text{sc}}}} \end{aligned} \tag{8.23}$$

Also

$$\beta_{comp} = \beta\sqrt{(1-K_{sh})(1-K_{sc})} \tag{8.24}$$

The effects are summarized as follows:

- Capacitive shunt compensation increases β and power transmitted and reduces surge impedance. Inductive shunt compensation has the opposite effect, reduces β and power transmitted, and increases surge impedance. A 100% inductive shunt capacitance will theoretically increase the surge impedance to infinity. Thus, at no-load, shunt reactors can be used to cancel the Ferranti effect.
- Series capacitive compensation decreases surge impedance and β and increases power transfer capacity. Series compensation is applied more from the steady-state and transient stability considerations rather than from power factor improvement. It provides better load division between parallel circuits, reduced voltage drops, and better adjustment of line loadings. It has practically no effect on the load power factor. Shunt compensation, on the other hand, directly improves the load power factor. Both types of compensations improve the voltages and, thus, affect the power transfer capability. The series compensation reduces the large shift in voltage that occurs between the sending and receiving ends of a system and improves the stability limit.

The performance of a symmetrical line was examined in Section 3.7. At no-load, the midpoint voltage of a symmetrical compensated line is given by

$$V_m = \frac{V_s}{\cos(\beta l/2)} \tag{8.25}$$

Therefore, series capacitive and shunt inductive compensation reduce the Ferranti effect, while shunt capacitive compensation increases it.

The reactive power at the sending end and receiving end of a symmetrical line was calculated in Equation 3.86 and is reproduced as follows:

$$Q_s = -Q_r = \frac{\sin\theta}{2}\left[Z_0 I_m^2 - \frac{V_m^2}{Z_0}\right]$$

This equation can be manipulated to give the following equation in terms of natural load of the line and $P_m = V_m I_m$ and $P_0 = V_0^2/Z_0$:

$$Q_s = -Q_r = P_0\frac{\sin\theta}{2}\left[\left(\frac{PV_0}{P_0 V_m}\right)^2 - \left(\frac{V_m}{V_0}\right)^2\right] \tag{8.26}$$

For $P = P_0$ (natural loading) and $V_m = 1.0$ per unit, $Q_s = Q_r = 0$.

If the terminal voltages are adjusted so that $V_m = V_0 = 1$ per unit:

$$Q_s = -Q_r = P_0 \frac{\sin\theta}{2}\left[\left(\frac{P}{P_0}\right)^2 - 1\right] \tag{8.27}$$

At no-load

$$Q_s = -Q_r = -P_0 \tan\frac{\theta}{2} \approx -P_0 \frac{\theta}{2} \tag{8.28}$$

If the terminal voltages are adjusted so that for a certain power transfer, $V_m = 1$ per unit, then the sending-end voltage is

$$V_s = V_m\left(1 - \sin^2\frac{\theta}{2}\left[1 - \left(\frac{P}{P_0}\right)^2\right]\right)^{1/2} = -V_r \tag{8.29}$$

When series and shunt compensation are used, the reactive power requirement at *no-load* is approximately given by

$$Q_s = -P\frac{\beta l}{2}(1 - K_{sh}) = -Q_r \tag{8.30}$$

If K_{sh} is zero, the reactive power requirement of a series compensated line is approximately the same as that of an uncompensated line, and the reactive power handling capability of terminal synchronous machines becomes a limitation. Series compensation schemes, thus, require SVCs or synchronous condensers/shunt reactors.

8.3.4 Effect on Maximum Power Transfer

The power transfer for an uncompensated lossless line under load, phase angle δ is

$$P = \frac{V_s V_r}{Z_0 \sin\beta l}\sin\delta = \frac{V_s V_r}{Z_0 \sin\theta}\sin\delta \tag{8.31}$$

This can be put in more familiar form by assuming that $\sin\theta = \theta$:

$$\begin{aligned} \beta l &= \theta = \omega l\sqrt{LC} \\ Z_0\theta &= \omega l\sqrt{LC}\left(\sqrt{L/C}\right) = \omega l L = X_{sc} \end{aligned} \tag{8.32}$$

$$P = \frac{V_s V_r}{Z_0 \sin\theta}\sin\delta \approx \frac{P_0}{\sin\theta}\sin\delta \approx \frac{P_0}{\theta}\delta \tag{8.33}$$

We know that $\delta = 90°$ for a theoretical steady-state limit. A small excursion or change of power transmitted or switching operations will bring about instability, see Figure 8.4a. Practically, an uncompensated line is not operated at >30° load angle.

With compensation, the ratios of powers become

$$\frac{P_{\text{comp}}}{P} = \frac{1}{\sqrt{\frac{1-K_{sc}}{1-K_{sh}}}\sin\left[\beta x\sqrt{(1-K_{sc})(1-K_{sh})}\right]} \tag{8.34}$$

Series compensation has a more pronounced effect on P_{max}. Higher values of K_{sc} can give rise to resonance problems and in practice $K_{sc} \not> 0.8$. Series compensation can be used with a line of any length and power can be transmitted over a larger distance than is otherwise possible, also see Refs. [4,5].

Example 8.2

A 650 mi (1046 km) line has $\beta = 0.116°/\text{mi}$ (0.0721°/km).

1. Calculate P_{max} as a function of P_0 for an uncompensated line.
2. What are the limitations of operating under these conditions?
3. Considering a load angle of 30°, calculate the sending-end voltage if the midpoint voltage is held at 1 per unit. Also, calculate the sending-end reactive power of the uncompensated line.
4. Recalculate these parameters with 80% series compensation.
   - The electrical length is $\theta = \beta l = 0.116 \times 650 = 75.4°$. Therefore, the maximum power transfer as a function of natural load is $1/\sin 75.4° = 1.033P_0$.
   - An uncompensated transmission line is not operated close to the steady-state limit.
   - If the midpoint voltage is maintained at 1 per unit, by adjustment of the sending-end and receiving-end voltages, the sending-end voltage from Equation 8.29 is

$$\left[1-(\sin 37.25)^2(1-0.5^2)\right]^{1/2} = 0.852 \text{ pu}$$

The ratio $P/P_0 = 0.5$ for load angle $\delta = 30°$. Thus, the voltages are too low. The reactive power input required from Equation 8.27 is 0.36 kvar per kW of load transferred at each end, i.e., a total of 0.72 kvar per kW transmitted.

   - With 80% series compensation, the maximum power transfer from Equation 8.34 is

$$P_{\text{comp}}/P = \frac{1}{\sqrt{1-0.8}\,\sin\left[75.4\sqrt{1-0.8}\right]} = 4.0278P_0$$

From Equation 8.23, the surge impedance with compensation is $0.447Z_0$. Therefore, the natural loading = $2.237P_0$. If the line is operated at this load, a flat voltage profile is obtained. The new electrical length of the line is

$$\theta' = \sin^{-1}\left(\frac{P_0'}{P_{max}'}\right) = 33.7°$$

8.3.5 Compensation with Lumped Elements

The distributed shunt or series compensation derived above is impractical and a line is compensated by lumped series and shunt elements at the midpoint or in sections, using the same methodology. The problem usually involves steady-state as well as dynamic and transient stability considerations.

Figure 8.5a shows the midpoint compensation [4]. Each half-section of the line is shown as an equivalent Π model. The circuit of Figure 8.5a can be redrawn as shown in Figure 8.5b. The phasor diagram is shown in Figure 8.5c. The degree of compensation for the central half, k_m, is

$$\kappa_m = \frac{b_{shcomp}}{0.5b_{sh}} \tag{8.35}$$

For equal sending- and receiving-end voltages

$$P = \frac{V^2}{X_{sr}(1-s)}\sin\delta \tag{8.36}$$

where

$$s = \frac{X_{sr}}{2}\frac{b_{sh}}{4}(1-\kappa_m).1 \tag{8.37}$$

The midpoint voltage can be expressed as follows:

$$V_m = \frac{V\cos(\delta/2)}{1-s} \tag{8.38}$$

The equivalent circuit of the line with compensation is represented in Figure 8.5d. For $s < 1$, the midpoint compensation increases the midpoint voltage, which tends to offset the series voltage drop in the line. If the midpoint voltage is controlled by variation of the susceptance of the compensating device, then using the relationship in Equation 8.38, Equation 8.36 can be written as follows:

$$P = \frac{V^2}{X_{sr}(1-s)}\sin\delta = \frac{V_m V}{X_{sr}\cos(\delta/2)}\sin\delta = 2\frac{V_m V}{X_{sr}}\sin\frac{\delta}{2} \tag{8.39}$$

If the midpoint voltage is equal to the sending-end voltage and equal to the receiving-end voltage

$$P = \frac{2V^2}{X_{sr}}\sin\frac{\delta}{2} \tag{8.40}$$

This is shown in Figure 8.5e. Thus, the power transmission characteristics are a sinusoid whose amplitude varies as s varies. Each sinusoid promises ever higher power transfer.

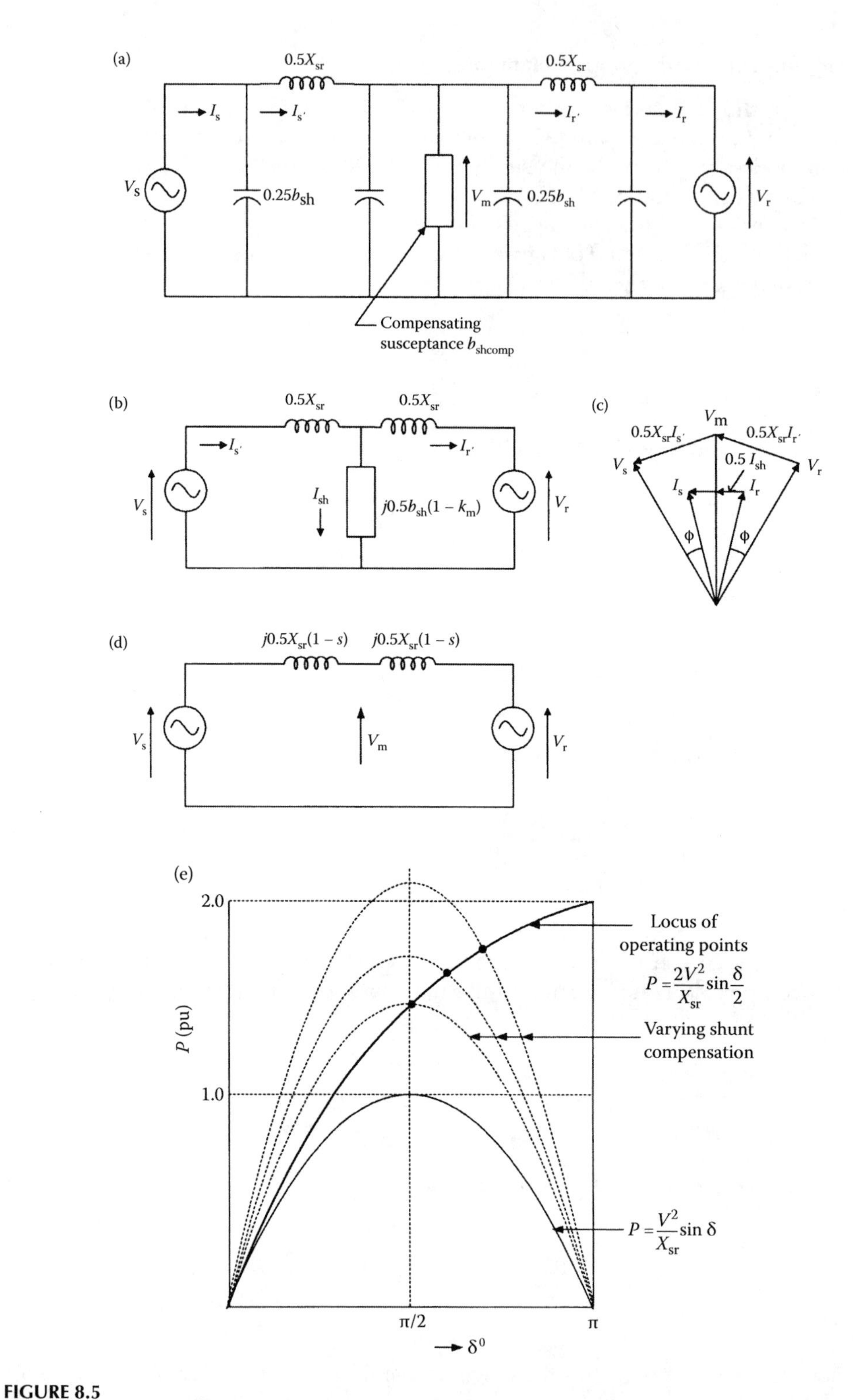

FIGURE 8.5
(a) Midpoint compensation of a transmission line, (b) circuit of (a) redrawn, (c) phasor diagram, (d) equivalent circuit, and (e) P–δ characteristics and dynamic response operation of a midpoint compensator.

For $P > \sqrt{2}P_{max}$, angle $\delta > \pi/2$. When the transmission angle increases, the compensator responds by changing the susceptance to satisfy Equation 8.40. The economic limit of a compensator to put effective capacitive susceptance may be much lower than the maximum power transfer characteristics given by Equation 8.40. When the compensator limit is reached, it does not maintain a constant voltage and acts like a fixed capacitor.

The shunt compensation to satisfy Equation 8.38 can be calculated from the following equation:

$$b_{\mathrm{shcomp}} = -\frac{4}{X_{\mathrm{sr}}}\left[1 - \frac{V}{V_{\mathrm{m}}}\cos\frac{\delta}{2}\right] + \frac{b_{\mathrm{sh}}}{2} \tag{8.41}$$

In Chapter 6, we have seen that reactive power cannot be transferred over long distances. Due to the inherent problem of flow of reactive power connected with reduction of voltage and usable voltage band, voltage controlled (PV) buses must be scattered throughout the power system.

8.4 Reactive Power Control Devices

Some of the devices and strategies available for reactive power compensation and control are as follows:

- Synchronous generators
- Synchronous condensers
- Synchronous motors (overexcited)
- Shunt capacitors and reactors
- Static var controllers
- Series capacitors with power system stabilizers
- Line dropping
- Undervoltage load shedding
- Voltage reduction
- Under load tap-changing transformers
- Setting lower transfer limits

8.4.1 Synchronous Generators

Synchronous generators are primary voltage-control devices and primary sources of spinning reactive power reserve. Figure 8.6 shows the reactive capability curve of a generator. To meet the reactive power demand, generators can be rated to operate at 0.8 power factor, at a premium cost. Due to thermal time constants associated with the generator exciter, rotor, and stator, some short-time overload capability is available and can be usefully utilized. On a continuous basis, a generator will operate successfully at its treated voltage and with a power factor at a voltage not more than 5% above or below the rated voltage, but not necessarily in accordance with the standards established for operation at the rated voltage [6]. The generator capability curves at voltages other than the rated voltage may differ.

Referring to Figure 8.6, the portion SPQ is limited by the generator megawatt output, the portion QN by the stator current limit, the portion NM by the excitation current limit, and the portion ST by the end-iron heating limit. Q is the rated load and power factor operating point. In the leading reactive power region, the minimum excitation limit (MEL) and the normal manufacturer's underexcited reactive ampère (URA) limit are imposed. MEL plays an important role in voltage control. The MEL curve rises slightly above the pull-out curve also called classical static stability curve A, providing an extra margin of stability and preventing stator end-turn heating from low excitation. Where cables or a large shunt capacitors' bank can remain in service after a large disturbance, the capability of the generators to absorb capacitive current comes into play and appropriate settings of MEL are required. Curve B shows the improvement in the stability achieved by high-response excitation systems. The dotted curve shows the effect of operating at a higher than rated voltage, and curve C shows the short-time overload capability in terms of reactive power output.

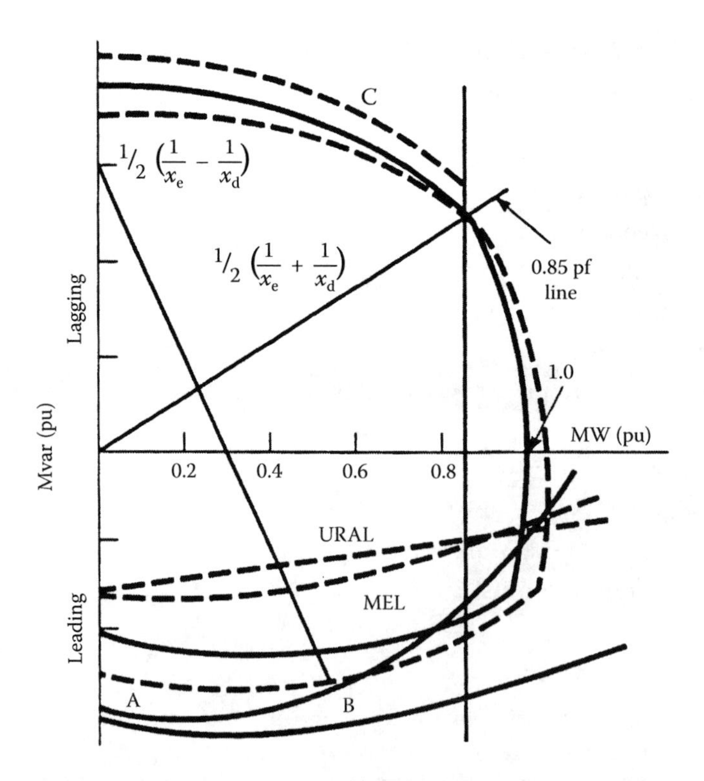

FIGURE 8.6
Reactive capability curve of a 0.85 power factor rated synchronous generator with stability limit curves. Curve A: classical stability limit, curve B: enhanced stability limit with high-response excitation system, and curve C: 20 min overload reactive capability, dotted reactive capability curve shows operation at higher than rated voltage.

The reactive output of a generator decreases if the system voltage increases and conversely it increases if the system voltage dips. This has a *stabilizing* effect on the voltage. The increased or decreased output acts in a way to counteract the voltage dip or voltage rise.

8.4.2 Synchronous Condensers

A synchronous condenser is a dynamic compensator and is characterized by its large synchronous reactance and heavy field windings. It develops a zero power factor leading current with least expenditure in active power losses. Figure 8.7 shows a *V* curve for a constant output voltage. The rated reactive power output for a 0.8 leading power factor synchronous motor is also shown for comparison. In a synchronous condenser at 100% excitation, full load leading kvars are obtained. At about 10% of the excitation current, the leading kvar output falls to a minimum corresponding to the losses. The rated field current is defined as the current for rated machine output at rated voltage. The lagging kvar is usually limited to approximately one-third of the maximum leading kvar to prevent loss of synchronism on a disturbance. The full load power factor may be as low as 0.02. The current drawn by a synchronous condenser can be varied from leading to lagging by change of its excitation, i.e., machine internal voltage behind its synchronous reactance (steady state) and transient reactance (for transient stability) characteristics.

A synchronous condenser provides stepless control of reactive power in both underexcited and overexcited regions. Synchronous condensers have been used for normal voltage control, high-voltage DC (HVDC) applications, and voltage control under upset conditions with voltage regulators. The transient open-circuit time constant is high (of the order of a few seconds) even under field forcing conditions, yet synchronous condensers have been applied to improve the transient stability limits on voltage swings. At present, these are being replaced with static var compensators, which have faster responses.

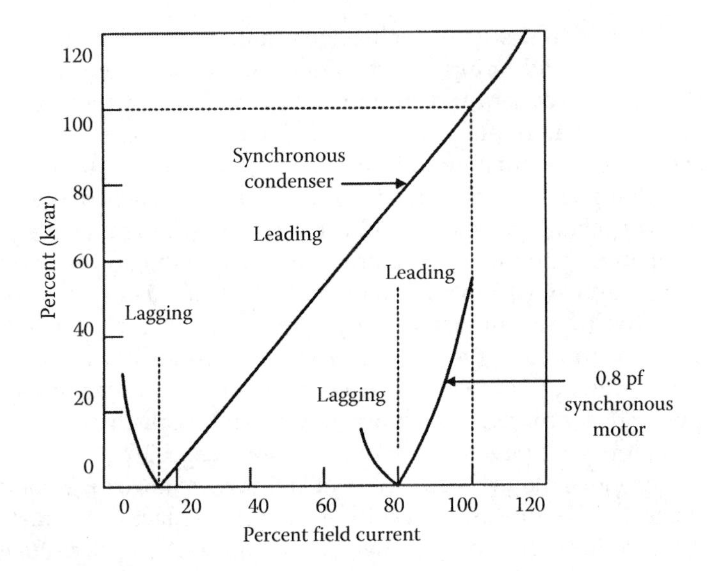

FIGURE 8.7
V curves for a synchronous condenser and 0.8 leading power factor synchronous motor.

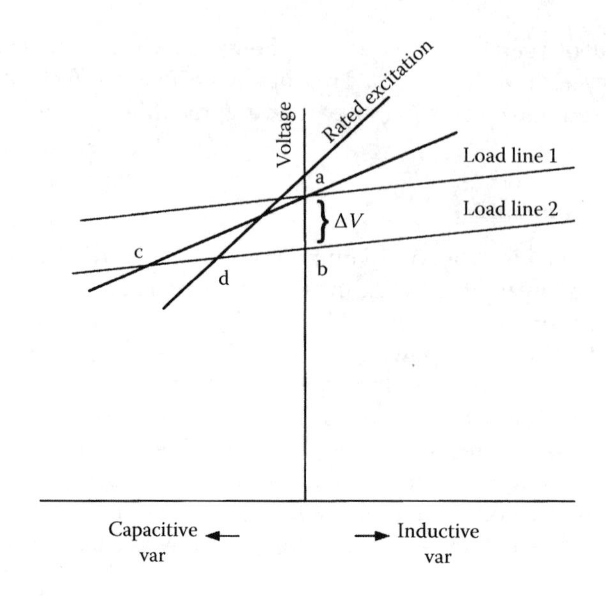

FIGURE 8.8
Operation of a synchronous condenser on sudden voltage dip.

Figure 8.8 shows the response characteristics of a synchronous condenser. Consider that the condenser is initially operating at point *a*, neither providing inductive nor capacitive reactive power. A sudden drop in voltage (ΔV) forces the condenser to swing to operating point *c* along the load line 2. As it is beyond the rated operating point *d* at rated excitation, the operation at *c* is limited to a short time, and the steady-state operating point will be restored at *d*.

8.4.3 Synchronous Motors

In an industrial distribution system, synchronous motors can be economically selected, depending on their speed and rating. Synchronous motors are more suitable for driving certain types of loads, e.g., refiners and chippers in the pulp and paper industry, Banbury mixers, screw-type plasticators and rubber mill line shafts, compressors, and vacuum pumps. The power factor of operation of induction motors deteriorates at low speeds due to considerable overhang leakage reactance of the windings. Synchronous motors will be more efficient in low-speed applications, and can provide leading reactive power. Higher initial costs as compared to an induction motor are offset by the power savings. A careful selection of induction and synchronous motors in an industrial environment can obviate the problems of reactive power compensation and maintain an appropriate voltage profile at the load centers. Synchronous motors for such applications are normally rated to operate at 0.8 power factor leading at full load and need a higher rated excitation system as compared to unity power factor motors. An evaluation generally calls for synchronous motors versus induction motors with power factor improvement capacitors.

The type of synchronous motor excitation control impacts its voltage-control characteristics. The controllers can be constant current, constant power factor, or constant kvar type, though many uncontrolled excitation systems are still in use. A constant-current controller compensates for a decrease in motor field current due to field heating; the reactive power output from the motor increases at no-load and consequently the voltage rises at no-load

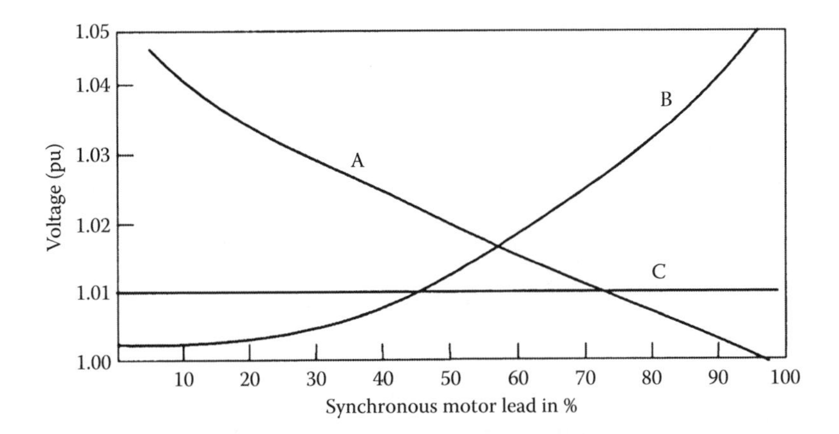

FIGURE 8.9
Effect of synchronous motor controller type on voltage regulation. (a) Constant-current or unregulated controller, (b) constant power-factor controller, and (c) constant kvar controller.

and falls as the motor is loaded. A constant power factor controller increases the reactive power output with load, and the voltage rises with the load. A cyclic load variation will result in cyclic variation of reactive power and thus the voltage. A constant reactive power controller will maintain a constant-voltage profile, irrespective of load, but may give rise to instability when a loaded motor is subjected to a sudden impact load. Constant var and power factor controllers are common. Figure 8.9 shows the voltage-control characteristics of these three types of controllers.

8.4.4 Shunt Power Capacitors

Shunt capacitors are extensively used in industrial and utility systems at all voltage levels. As we will see, the capacitors are the major elements of flexible ac transmission systems (FACTS). Much effort is being directed in developing higher power density, lower cost improved capacitors, and an increase in energy density by a factor of 100 is possible. These present a constant impedance type of load, and the capacitive power output varies with the square of the voltage:

$$\text{kvar}_{V_2} = \text{kvar}_{V_1}\left|\frac{V_2}{V_1}\right|^2 \tag{8.42}$$

where kvar_{V_1} is the output at voltage V_1 and kvar_{v2} is the output at voltage V_2. As the voltage reduces, so does the reactive power output, when it is required the most. This is called the *destabilizing* effect of the power capacitors. (Compare to the stabilizing effect of a generator excitation system.)

Capacitors can be switched in certain discrete steps and do not provide a stepless control. Figure 8.10 shows a two-step sequential reactive power switching control to maintain voltage within a certain band. As the reactive power demand increases and the voltage falls (shown linearly in Figure 8.10 for simplicity), the first bank is switched at A, which compensates the reactive power and suddenly raises the voltage. The second step switching is, similarly, implemented at B. On reducing demand, the banks are taken out from service

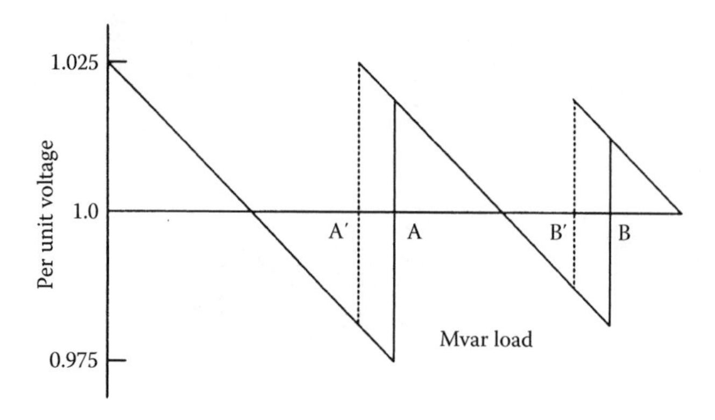

FIGURE 8.10
Sequential switching of capacitors in two steps with reactive power flow control to maintain voltage within a certain acceptable band. The no-load voltage is assumed to be 1.025 per unit.

at A′ and B′. A time delay is associated with switching in either direction to override transients. Voltage or power factor dependent switching controls are also implemented. Power capacitors have a wide range of applications:

- Power factor improvement capacitors switched with motors at low- and medium-voltage levels or in multistep power factor improvement controls in industrial systems.
- Single or multiple banks in industrial distribution systems at low- and medium-voltage substations.
- As harmonic filters in industrial distribution systems, arc furnaces, steel mills, and HVDC transmission.
- Series or shunt devices for reactive power compensation in transmission systems.
- Essential elements of SVC and FACTS controllers.

Some of the problems associated with the shunt power capacitors are as follows:

- Switching inrush currents at higher frequencies and switching overvoltages.
- Harmonic resonance problems.
- Limited overvoltage withstand capability.
- Limitations of harmonic current loadings.
- Possibility of self-excitation of motors when improperly applied as power factor improvement capacitors switched with motors.
- Prolonging the decay of residual motor voltage on disconnection, and trapped charge on disconnection, which can increase the inrush current on reswitching and lead to restrikes
- Requirements of *definite purpose* switching devices.

These are offset by low capital and maintenance costs, modular designs varying from very small to large units and fast switching response. See Volume 3 for the application of capacitors.

8.4.5 Static Var Controllers

The var requirements in transmission lines swing from lagging to leading, depending on the load. Shunt compensation by capacitors and reactors is one way. However, it is slow, and power circuit breakers have to be derated for frequent switching duties. The IEEE and CIGRE definition of a static var generator (SVG) embraces different semiconductor power circuits with their internal control enabling them to produce var output proportional to an input reference. An SVG becomes an SVC when equipped with external or system controls which derive its reference from power system operating requirements and variables. SVCs can be classified into the following categories [7]:

Thyristor-controlled reactor (TCR), see Figure 8.11a

Thyristor-switched capacitor (TSC), see Figure 8.11b

Fixed capacitor and TCR (FC-TCR), see Figure 8.11c

TSC and TCR, see Figure 8.11d

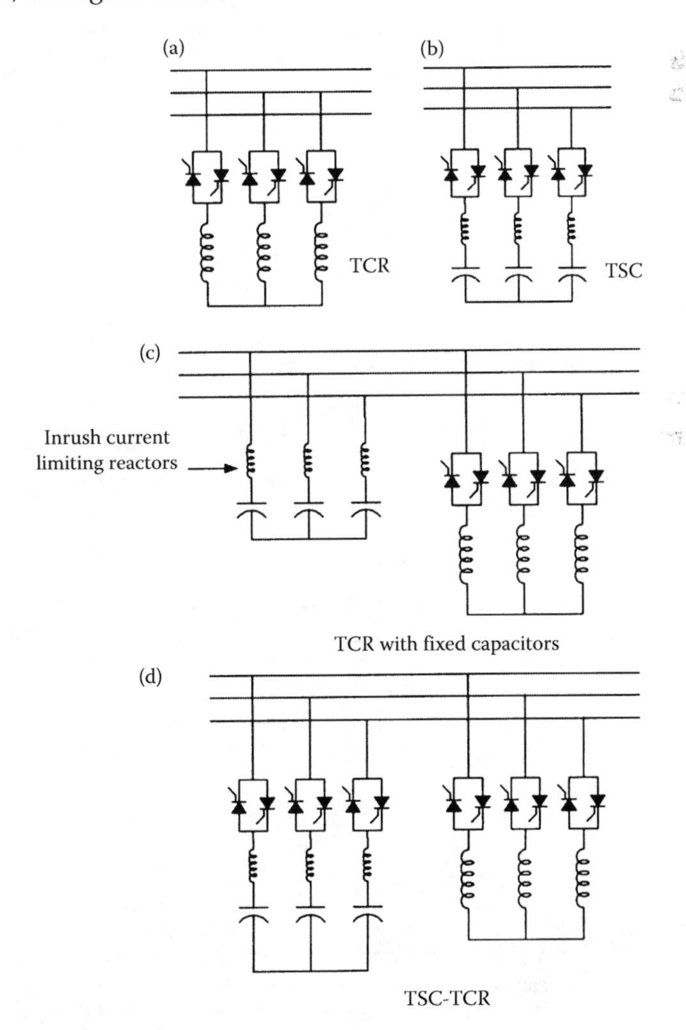

FIGURE 8.11
SVC controllers: (a) TCR, (b) TSC, (c) FC-TCR, and (d) TSC-TCR.

Figure 8.12 shows the circuit diagram of an FC-TCR. The arrangement provides discrete leading vars from the capacitors and continuously lagging vars from TCRs. The capacitors are used as tuned filters, as considerable harmonics are generated by thyristor control, see Volume 3.

The steady-state characteristics of an FC-TCR are shown in Figure 8.13. The control range is AB with a positive slope, determined by the firing angle control:

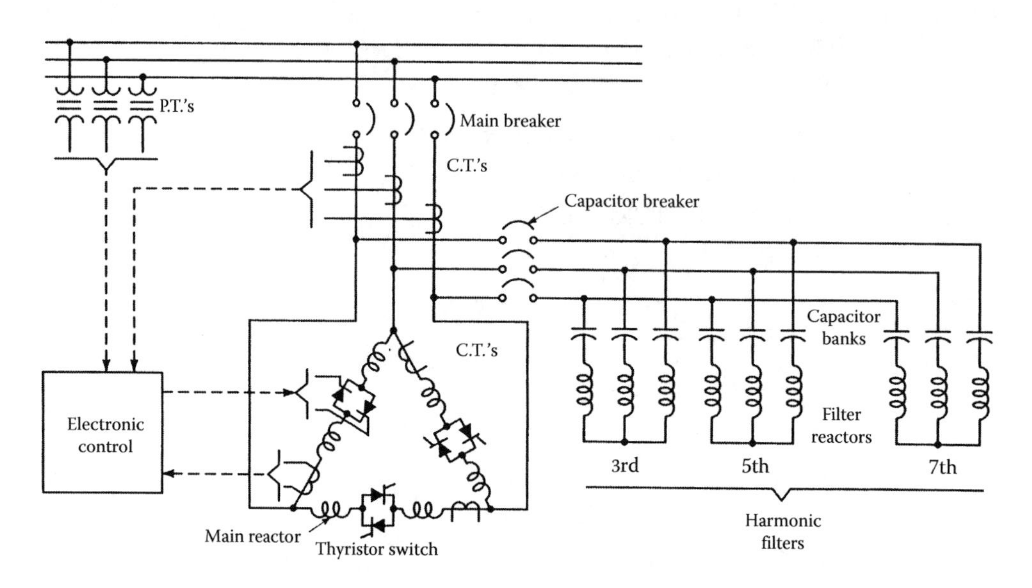

FIGURE 8.12
Circuit diagram of an FC-TCR, with switched capacitor filters.

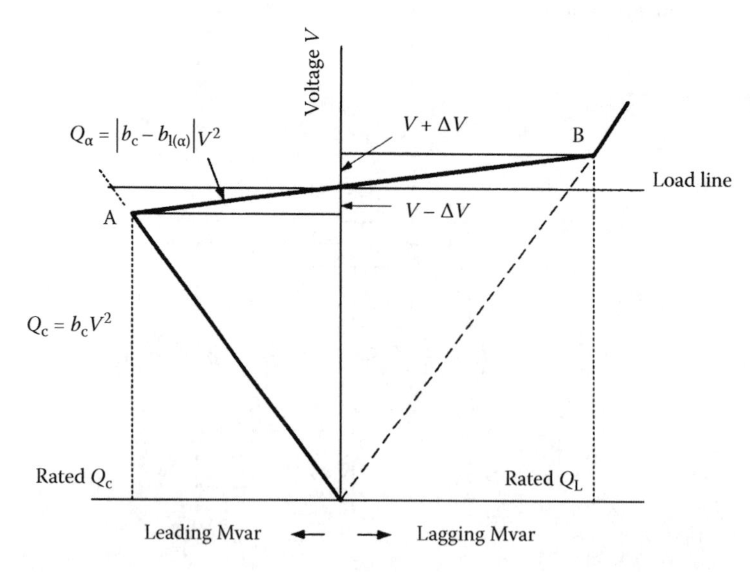

FIGURE 8.13
Steady-state *V–Q* characteristics of an FC-TCR.

$$Q_{\alpha} = \left| b_c - b_{1(\alpha)} \right| V^2 \tag{8.43}$$

where b_c is the susceptance of the capacitor and $b_1(\alpha)$ is the susceptance of the inductor at firing angle *a*. As the inductance is varied, the susceptance varies over a large range. The voltage varies within limits $V \pm \Delta V$. Outside the control interval AB, the FC-TCR acts like an inductor in the high-voltage range and like a capacitor in the low-voltage range. The response time is of the order of 1 or 2 cycles. The compensator is designed to provide emergency reactive and capacitive loading beyond its continuous steady-state rating.

A TSC-TCR provides thyristor control for the reactive power control elements, capacitors, and reactors. Improved performance under large system disturbances and lower power loss are obtained. Figure 8.14 shows the *V–I* characteristics. A certain short-time overload capability is provided both in the maximum inductive and capacitive regions (shown for the inductive region in Figure 8.14).

Voltage regulation with a given slope can be achieved in the normal operating range. The maximum capacitive current decreases linearly with the system voltage, and the SVC becomes a fixed capacitor when the maximum capacitive output is reached. The voltage support capability decreases with decrease in system voltage.

SVCs are ideally suited to control the varying reactive power demand of large fluctuating loads (i.e., rolling mills and arc furnaces) and dynamic overvoltages due to load rejection and are used in HVDC converter stations for fast control of reactive power flow.

Compensation by sectionalizing is based on a midpoint dynamic shunt compensator. With a dynamic compensator at the midpoint, the line tends to behave like a symmetrical line. The power transfer equation for equal sending- and receiving-end voltages is Equation 8.40. The advantage of static compensators is apparent. The midpoint voltage will vary with the load, and an adjustable midpoint susceptance is required to maintain

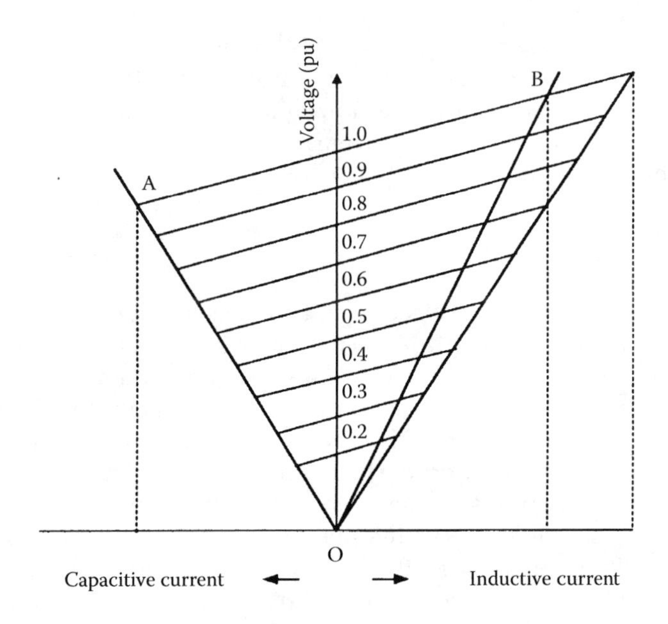

FIGURE 8.14
V–I characteristics of an SVC (TSC-TCR) showing reduced output at lower system voltages.

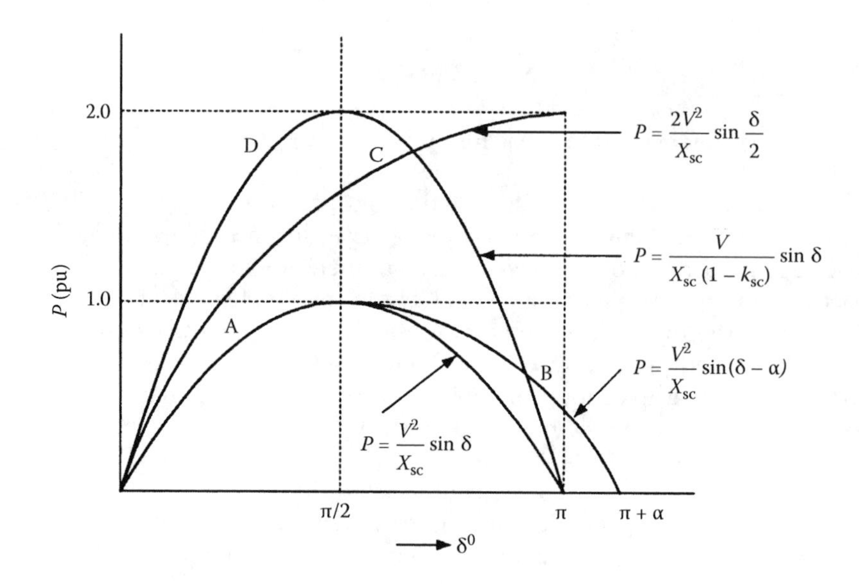

FIGURE 8.15
P–δ characteristics. (a) Uncompensated line, (b) with phase-angle regulator, (c) with midpoint shunt compensation, and (d) with series compensation.

constant-voltage magnitude. With rapidly varying loads, it should be possible for the reactive power demand to be rapidly corrected, with least overshoot and voltage rise. Figure 8.15 shows the transient power angle curves for an uncompensated line, with phase angle shift, shunt compensation, and series compensation. With the midpoint voltage held constant, the angles between the two systems can each approach 90°, for a total static stability limit angle of 180°.

The power system oscillation damping can be obtained by rapidly changing the output of the SVC from capacitive to inductive so as to counteract the acceleration and deceleration of interconnected machines. The transmitted electrical power can be increased by capacitive vars when the machines accelerate and it can be decreased by reactive vars when the machines decelerate.

8.4.6 Series Compensation of HV Lines

An implementation schematic of the series capacitor installation is shown in Figure 8.16. The performance under normal and fault conditions should be considered. Under fault conditions, the voltage across the capacitor rises, and unlike a shunt capacitor, a series capacitor experiences many times its rated voltage due to fault currents. A zinc oxide varistor in parallel with the capacitor may be adequate to limit this voltage. Thus, in some applications, the varistor will reinsert the bank immediately on termination of a fault. For locations with high fault currents, a parallel fast acting triggered gap is introduced which operates for more severe faults. When the spark gap triggers, it is followed by closure of the bypass breaker. Immediately after the fault is cleared, to realize the beneficial effect of series capacitor on stability, it should be reinserted quickly, and the main gap is made self-extinguishing. A high-speed reinsertion scheme can reinsert the series capacitors in a few

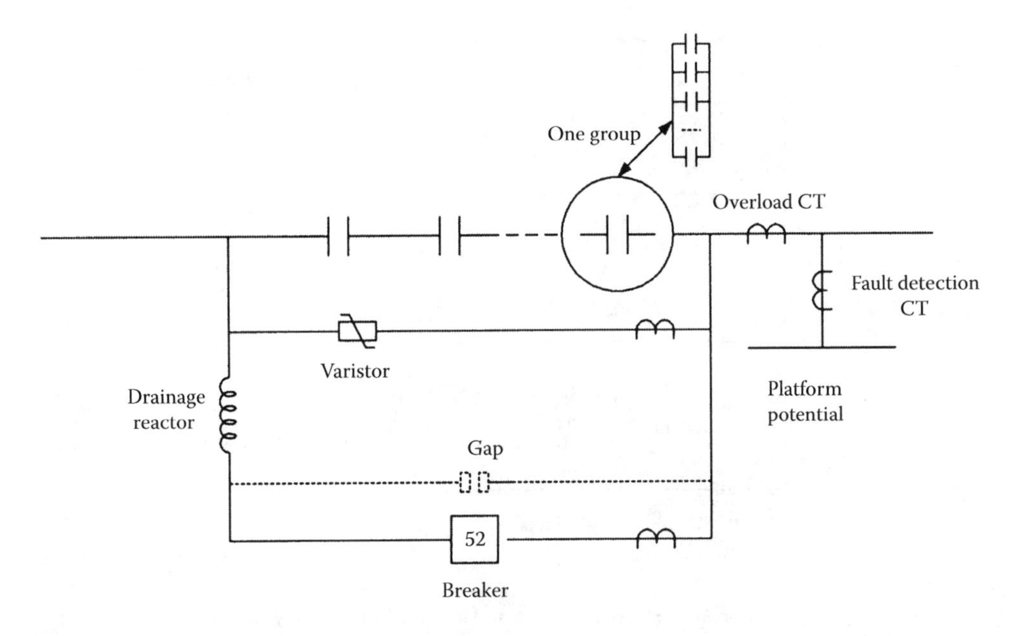

FIGURE 8.16
Schematic diagram of a series capacitor installation.

cycles. The bypass switch must close at voltages in excess of nominal, but not at levels too low to initiate main gap sparkover.

The discharge reactor limits the magnitude and frequency of the current through the capacitor when the gap sparks over. This prevents damage to the capacitors and fuses. A series capacitor must be capable of carrying the full line current. Its reactive power rating is

$$I^2X_c \text{ per phase} \tag{8.44}$$

and, thus, the reactive power output varies with the load current.

Figure 8.17 shows the impact of series versus shunt compensation at the midpoint of a transmission line. Both systems are, say, designed to maintain 95% midpoint voltage. The midpoint voltage does not vary much when an SVC is applied, but with series compensation, it varies with load. However, for a transfer of power higher than the SVC control limit, the voltage falls rapidly as the SVC hits its ceiling limit, while series compensation holds the midpoint voltage better.

A series capacitor has a natural resonant frequency given by

$$f_n = \frac{1}{2\pi\sqrt{LC}} \tag{8.45}$$

where f_n is usually less than the power system frequency. At this frequency, the electrical system may reinforce one of the frequencies of the mechanical resonance, causing *subsynchronous resonance* (SSR). If f_r is the SSR frequency of the compensated line, then at resonance

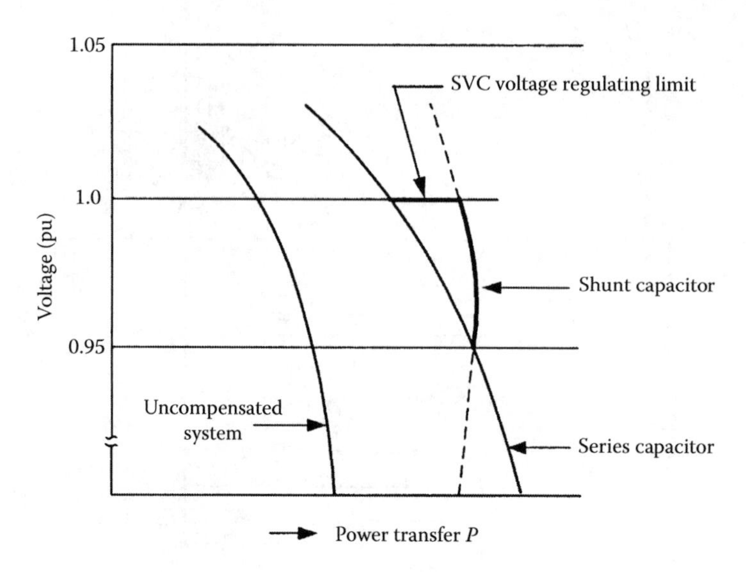

FIGURE 8.17
Series versus shunt compensation; power transfer characteristics for a mid-compensated line.

TABLE 8.2
Characteristics of Reactive Power Sources

Equipment	Characteristics
Generators	Primary source of reactive power reserve, continuously adjustable reactive power in inductive and capacitive regions limited by generator reactive power-reactive capability curve. Response time depending on excitation system, fast response excitation systems of 0.1 s possible. Limited capacitive reactive power on underexcitation. In this mode the generator acts like a shunt reactor absorbing reactive power from the system. The industrial generators are used in PV mode, to maintain the bus voltage within their reactive power capability
Synchronous condensers	Continuously adjustable reactive power in inductive and capacitive regions, slow response (1 s). Limited inductive reactive power
SVC	Same as synchronous condensers, wider range of control, fast response of 1–2 cycles, better range of reactive power capability. Harmonic pollution and SSR of consideration
Power capacitors	Switchable in certain steps, only capacitive reactive power, response dependent on control system. On a voltage dip, the reactive power output decreases as square of the voltages
Shunt reactors	Single unit per line, inductive reactive power only, switching response dependent on control system, of the order of a couple of cycles with power circuit breaker controls. Not applied for industrial systems
Load and line dropping	Emergency measures. Load dropping also reduces active power
ULTC transformers	In certain steps only, slow response, does not generate reactive power, only reroutes it
Step voltage regulators	Discussed in Appendix C, Volume 1
FACTS controllers	Discussed in Chapter 9

$$2\pi f_r L = \frac{1}{2\pi f_r C}$$
$$f_r = f\sqrt{K_{sc}} \tag{8.46}$$

This shows that the SSR occurs at frequency f_r, which is equal to normal frequency multiplied by the square root of the degree of compensation. The transient currents at subharmonic frequency are superimposed upon the power frequency component and may be damped out within a few cycles by the resistance of the line. Under certain conditions, subharmonic currents can have a destabilizing effect on rotating machines. A dramatic voltage rise can occur if the generator quadrature axis reactance and the system capacitive reactance are in resonance. There is no field winding or voltage regulator to control quadrature axis flux in a generator. Magnetic circuits of transformers can be driven to saturation and surge arresters can fail. The inherent dominant subsynchronous frequency characteristics of the series capacitor can be modified by a parallel connected TCR. For further reading, see Volume 4.

Table 8.2 shows the comparative characteristics and applications of reactive power compensating devices.

8.4.7 Shunt Reactors

We discussed the Ferranti effect in Chapter 3. Shunt reactors are used to limit the voltage on open circuit or light load. These are usually required for line lengths exceeding 200 km. A shunt reactor may be permanently connected to a line to limit fundamental frequency overvoltages to about 1.5 pu for a duration of less than 1 s. Figure 8.18a shows the connections of fixed and switched shunt reactors. Depending on the stiffness of the system, permanently connected shunt reactors may not be needed. Figure 8.18b depicts a tapped shunt reactor connected to the tertiary winding of a transformer.

See Volume 1 for the switching transients and specifications of shunt reactors.

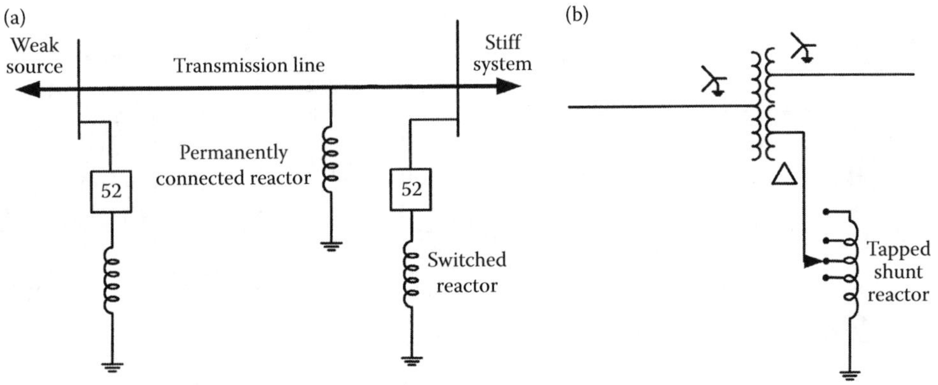

FIGURE 8.18
(a) Switched and permanently connected shunt reactors and (b) tapped reactor connected to tertiary delta winding of a transformer.

8.4.8 Induction Voltage Regulators

Induction voltage regulators can be applied for voltage control of feeders (so also step voltage regulators, discussed in Appendix C of Volume 1). Single-phase or three-phase units can be used. For long feeders, the regulators and shunt capacitors can be used in combination.

In a wound rotor three-phase induction motor, the secondary (rotor) EMF magnitude depends upon the stator/rotor windings turn ratio and its phase angle relative to stator supply system voltage depends upon the relative position of rotor and stator windings. If the EMF E_2 in the rotor circuit is injected into supply system voltage V_1, it can be raised or lowered.

The windings on the stator and rotor are reversed, Figure 8.19a, primary windings on the rotor. As the flux linkage depends upon the relative positions of the windings on stator and rotor, when the windings are at physical 90° angle, maximum linkage occurs. In phasor diagram in Figure 8.19b, the flux linkage is maximum, and E_2 directly adds to V_1. In phasor diagram in Figure 8.19c, a 90° rotation of the rotor from the maximum flux linkage will result in zero flux linkages and $E_2 = 0$, while a further rotation by 90° reverses the flux linkage and E_2 is now reverse of that in Figure 8.19d The short-circuited compensating winding is in the quadrature axis with the primary. Without this compensating

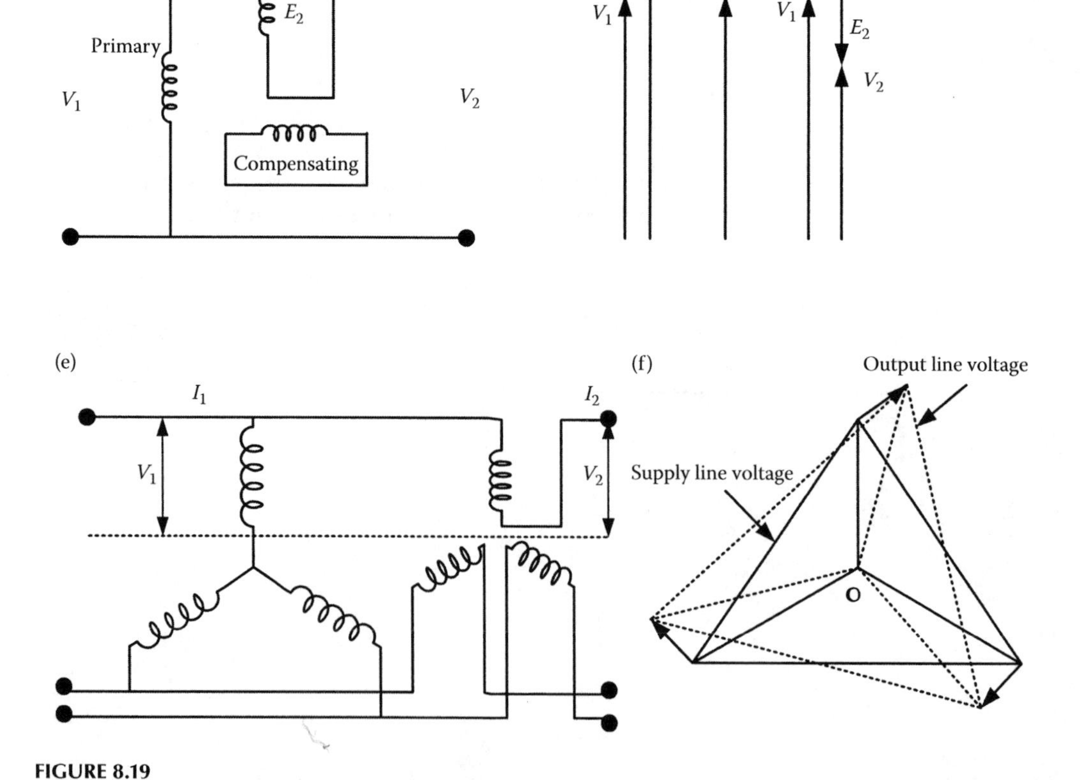

FIGURE 8.19
(a) through (d) Single-phase induction regulator and phasor diagrams and (e, f) three-phase induction voltage regulator.

winding, the leakage reactance of the stator (secondary) winding will be large when the rotor is in position of zero boost.

The operation of a three-phase induction voltage regulator is similar, see Figure 8.19e. As the movement of the rotor is limited, slip rings are generally not necessary. Variation in output voltage is obtained by altering the angular position of the rotor, causing a phase shift but not a change in the magnitude of the induced EMF E_2, see Figure 8.19f.

8.5 Some Examples of Reactive Power Compensation

Three examples of the reactive power flow are considered to illustrate reactive power/voltage problems in transmission [8], generating and industrial systems.

Example 8.3

Figure 8.20a shows three sections of a 230 kV line, with a series impedance of 0.113 + j0.80 Ω/mi and a shunt capacitance of 0.2 M Ω/mi. Each section is 150 mi long. A voltage-regulating transformer is provided at each bus. The reactive power flow is considered under the following conditions:

- The rated voltage is applied at slack bus 1 at no-load.
- A load of 40 Mvar is applied at furthermost right bus 4. No tap adjustment on transformers and no reactive power injection are provided.
- The transformer's tap adjustment raises the voltage at each bus to approximately 230 kV.
- A 30 Mvar capacitive injection is provided at bus 3.
- Transformer tap adjustment and reactive power injection of 5 Mvar each are provided at buses 2 and 3.

1. At no-load, the voltage rises at the receiving end due to charging capacitance current. Each line section is simulated by a Π network. The voltage at bus 4 is 252.6 kV. The charging current of each section flows cumulatively to the source, and 25.2 Mvar capacitive power must be supplied from the slack bus.
2. The line sections in series behave like a long line and the voltage will rise or fall, depending on the loading, unless a section is terminated in its characteristic impedance. On a lossless line, the voltage profile is flat and the reactive power is limited. On a practical lossy line and termination in characteristic impedance, the voltage will fall at a moderate rate, and when not terminated in characteristic impedance, the voltage will drop heavily, depending on the reactive power flow. Figure 8.20b shows the progressive fall of voltage in sections, and the voltage at bus 4 is 188.06 kV, i.e., 81.7% of nominal voltage. At this reduced voltage, only 26 Mvar of load can be supplied.
3. Each tap changer raises the voltage to the uniform level of the starting section voltage. Note that the charging kvar of the line, modeled as shunt admittance, acts like a reactive power injection at the buses, augmenting the reactive power flow from the upstream bus. The voltages on the primary of the transformer, i.e., at the termination of each line section, are shown in Figure 8.20c.
4. A reactive power injection of 30 Mvar at bus 3 brings the voltages of buses 2 and 3 close to the rated voltage of 230 kV. However, the source still supplies capacitive charging power of the transmission line system and the reactive power to the load flows from the nearest injection point. Thus, reactive power

injections in themselves are not an effective means to transmit reactive power over long distances, see Figure 8.20d.

5. This shows the effect of tap changers as well as reactive power injection. Reactive power injection at each bus with tap-changing transformers can be an effective way to transmit reactive power over long distance, but the injections at each bus may soon become greater than the load requirement. This shows that the compensation is best provided as close to the load as practicable, see Figure 8.20e.

Figure 8.20f shows the voltage profile in each of the above cases.

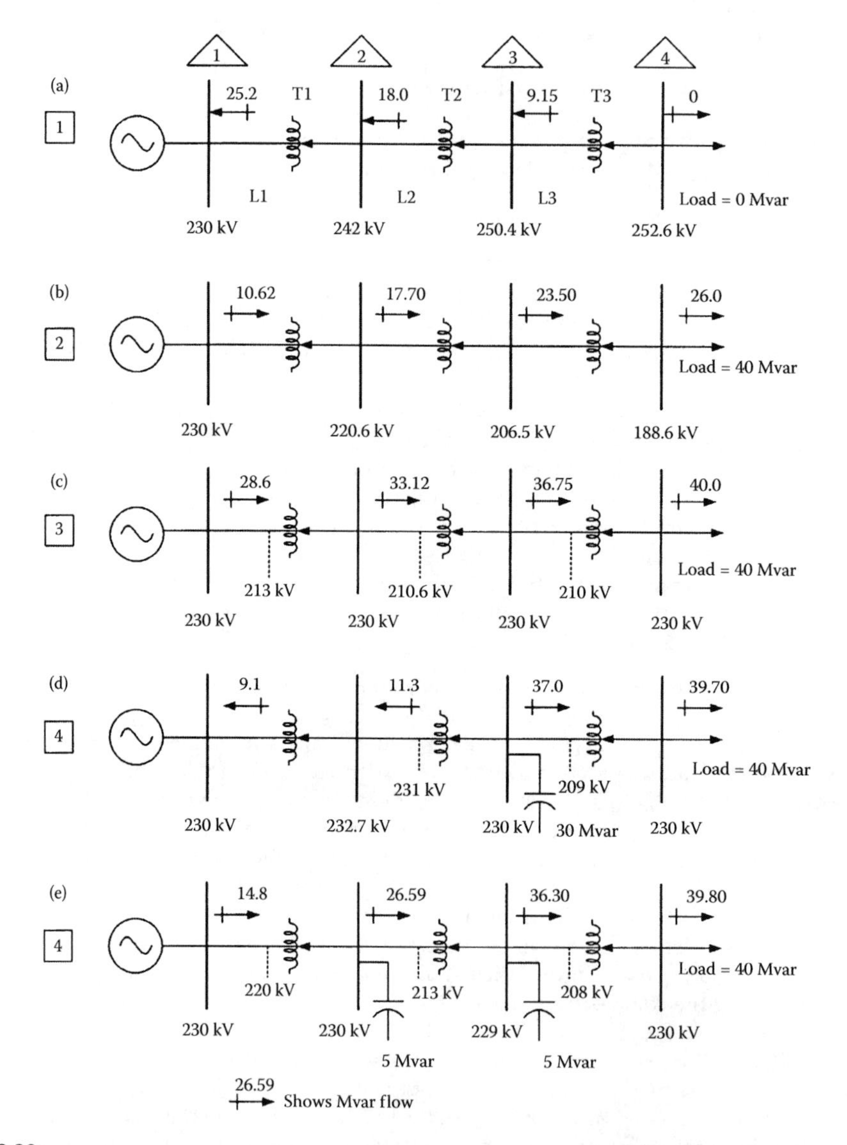

FIGURE 8.20
(a) A three-section 230 kV line at no-load, (b) with 40 Mvar load at bus 4, no tap adjustments, and no reactive power injection, (c) with transformer tap adjustment to compensate for the line voltage drop, (d) with reactive power injection of 30 Mvar at bus 3, (e) with reactive power injection of 5 Mvar at buses 2 and 3 and transformer tap adjustments, and (f) voltage profile along the line in all five cases of study.

(Continued)

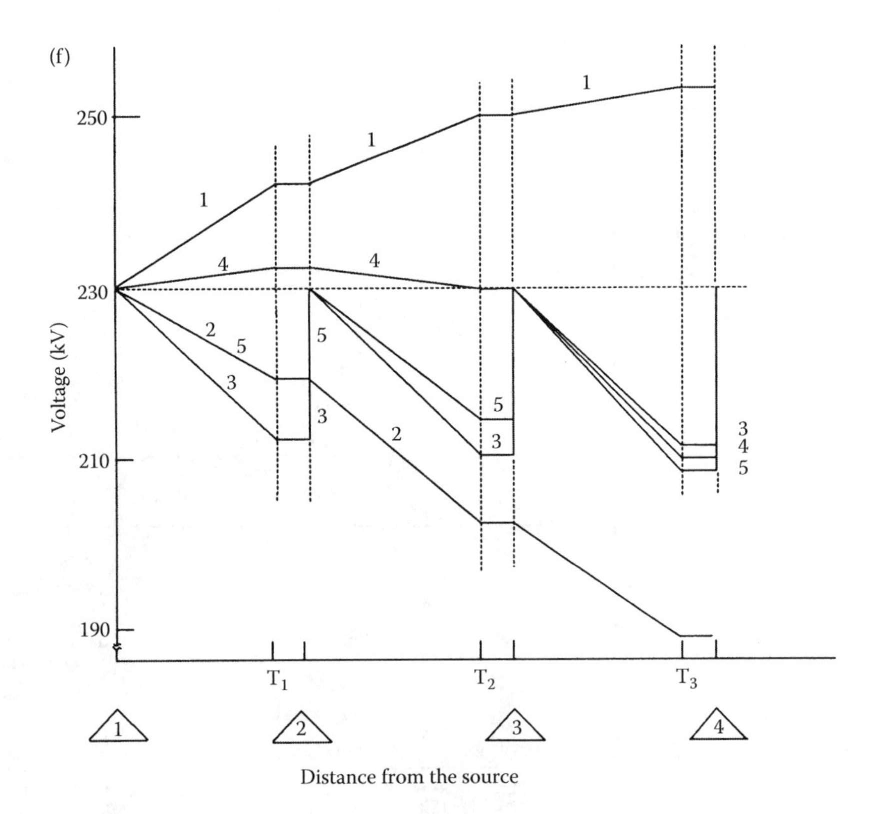

FIGURE 8.20 (CONTINUED)
(a) A three-section 230 kV line at no-load, (b) with 40 Mvar load at bus 4, no tap adjustments, and no reactive power injection, (c) with transformer tap adjustment to compensate for the line voltage drop, (d) with reactive power injection of 30 Mvar at bus 3, (e) with reactive power injection of 5 Mvar at buses 2 and 3 and transformer tap adjustments, and (f) voltage profile along the line in all five cases of study.

Example 8.4

This example illustrates that relative location of power capacitors and the available tap adjustments on the transformers profoundly affect the reactive power flow.

Figure 8.21 illustrates the effects of off-load and on-load tap changing with shunt capacitors located at the load or at the primary side of the transformer. Figure 8.21a through c shows the load flow when there are no power capacitors and only a tap-changing transformer. Figure 18.21d through f shows the capacitors located on the load bus. Figure 8.21g through i shows the capacitors located at the primary side of the transformer. In each of these cases, the objective is to maintain load voltage equal to rated voltage for a primary voltage dip of 10%. In each case, a constant power load of 30 MW and 20 Mvar is connected to the transformer secondary. Only reactive power flows are shown in these figures for clarity.

Figure 8.21a shows that with the transformer primary voltage at rated level, taps on the transformer provide a 5% secondary voltage boost to maintain the load voltage at 1.00 per unit. If the primary voltage dips by 10%, the load voltage will be 0.89 per unit and the reactive power input to transformer increases by 1.0 Mvar. An underload adjustment of taps must provide a 15% voltage boost to maintain the load voltage equal to the rated voltage, see Figure 8.21c.

For load flow in Figure 8.21d, the transformer secondary voltage is at rated level, as capacitors directly supply 20 Mvar of load demand and no secondary voltage boost is

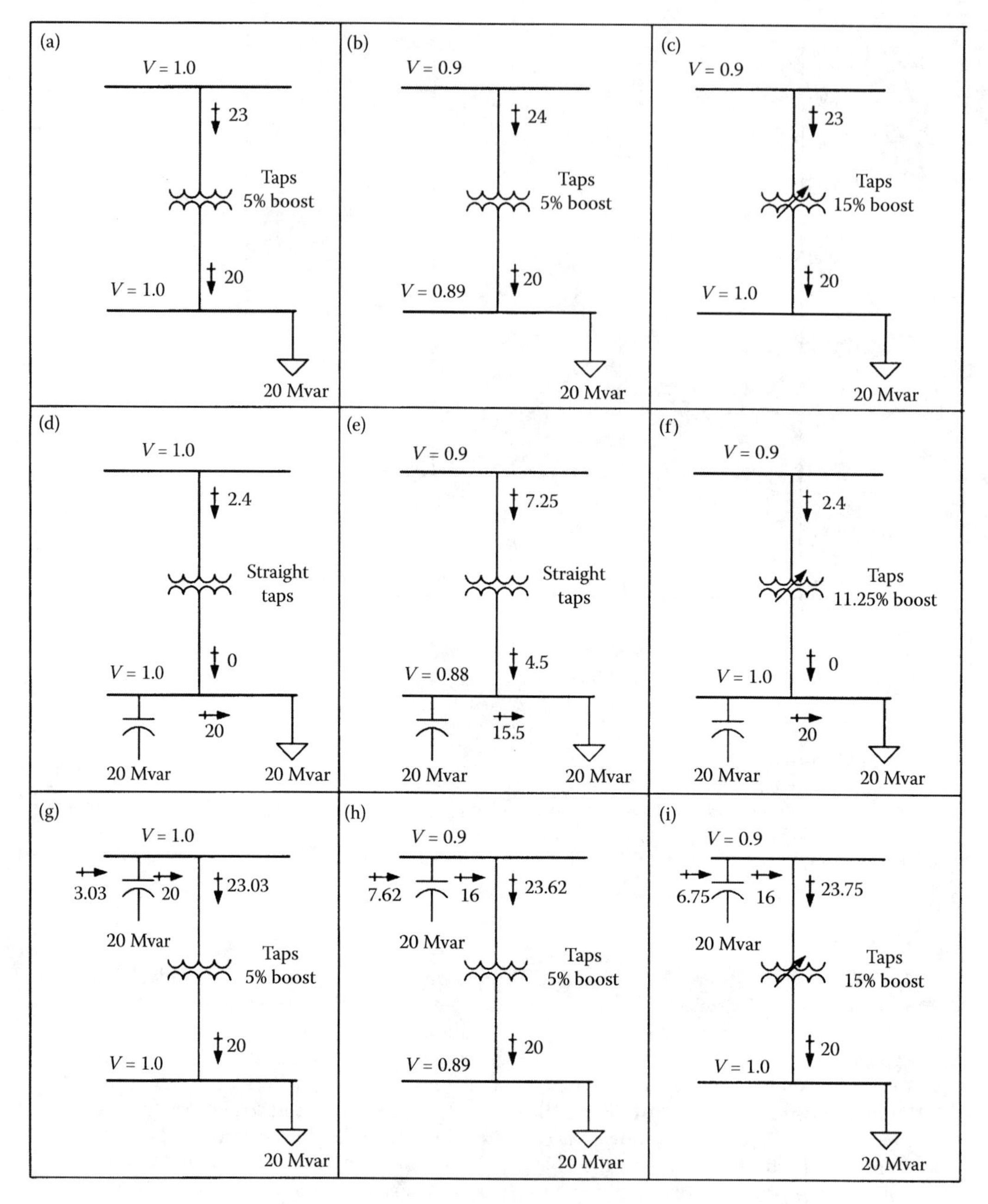

FIGURE 8.21
Reactive power flow with power capacitors and transformer tap changing.

required from the transformer tap settings. Yet, 2.4 Mvar is drawn from the source, which represents reactive power losses in the transformer itself. A 10% dip in the primary voltage results in a load voltage of 0.88 per unit and the reactive power requirement from the source side increases to 7.25 Mvar. Thus, load capacitors have not helped the voltage as compared to the scenario in Figure 8.21b. The combination of power capacitor and on-load tap changing to maintain the rated secondary voltage is shown in Figure 8.21f.

Figure 8.21g through i shows the voltages, tap settings, and reactive power flows when the power capacitors of the same kvar rating are located on the primary side of the transformer. The source has to supply higher reactive power as compared to the capacitors located on the load bus.

Example 8.5

Figure 8.22 shows a distribution system with a 71 MVA generator, a utility tie transformer of 40/63 MVA, and loads lumped on the 13.8 kV bus. A 0.38 Ω reactor in series with the generator limits the short-circuit currents on the 13.8 kV system to acceptable levels. However, it raises the generator operating voltage and directly reduces its reactive power capability due to losses. The total plant load including system losses is 39 MW and 31 Mvar. The excess generator power is required to be supplied into the utility's system and also it should be possible to run the plant on an outage of the generator. The utility transformer is sized adequately to meet this contingency.

In order to maintain voltages at acceptable levels in the plant distribution, it is required to hold the 13.8 kV bus voltage within ±2% of the rated voltage. Also, the initial voltage dip on sudden outage of the generator should be limited to 12% to prevent a process loss.

Figure 8.23 shows the reactive power sharing of load demand from the utility source and generator versus generator operating voltage. To meet the load reactive power demand of 31 Mvar, the generator must operate at 1.06 per unit voltage. It delivers its rated Mvar output of 37.4. Still, approximately 4 Mvar of reactive power should be supplied by the utility source. A loss of 10.4 Mvar occurs in the generator reactor and transformer. To limit the generator operating voltage to 1.05, still higher reactive power must be supplied from the utility's source.

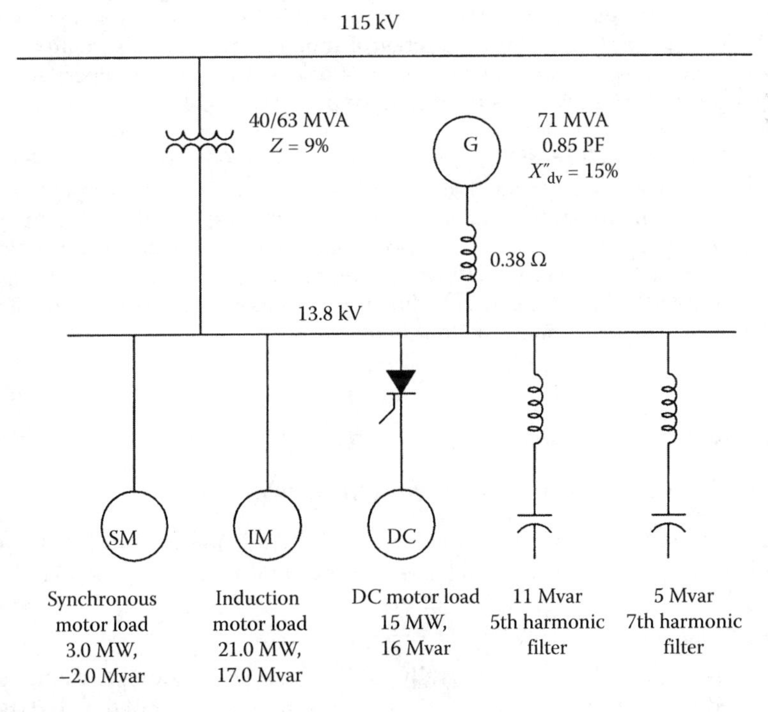

FIGURE 8.22
A 13.8 kV industrial cogeneration facility (Example 8.5).

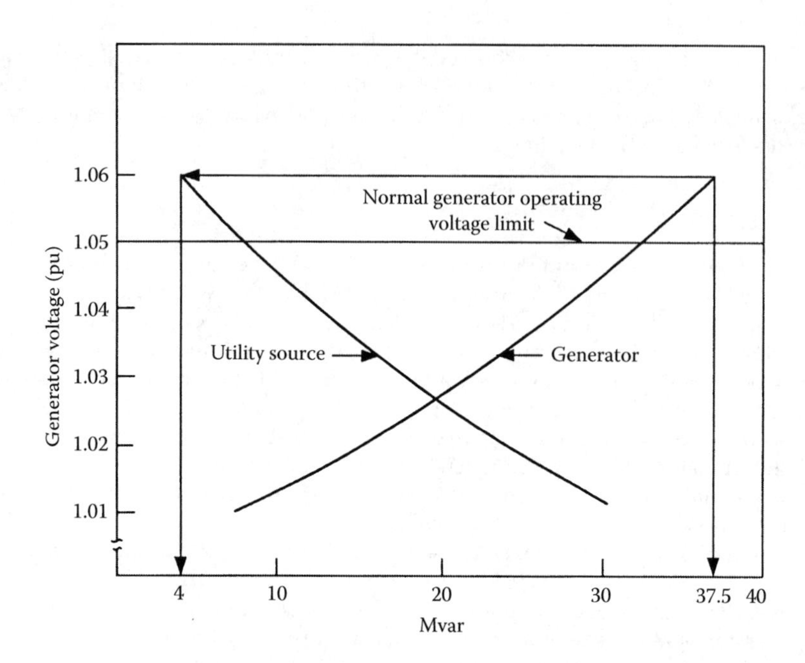

FIGURE 8.23
Reactive power sharing between generator and utility system versus voltage of generation for Example 8.5.

If reactive power compensation is provided in the form of shunt harmonic filters, the performance is much improved. With an 11 Mvar filter alone, the generator can be operated at 1.05 per unit voltage. Its reactive power output is 28.25 Mvar, the source contribution is 1 Mvar, and the loss is 9 Mvar. Addition of another 5 Mvar of the seventh harmonic filter results in a flow of 6 Mvar into the utility's system; the generator operates at 30.25 Mvar, approximately at 81% of its rated reactive power capability at rated load, and the remaining reactive power capability of the generator cannot be normally utilized with its maximum operating voltage of 1.05 per unit. It serves as a reserve in case of a voltage dip.

On the sudden loss of a generator, initial voltage dips of 17.5%, 5.5%, and 4%, respectively, occur with no harmonic filter, with only the fifth harmonic filter, and with both fifth and seventh harmonic filters in service. This is another important reason to provide reactive power compensation. It will prevent a process loss, and to bring the bus voltage to its rated level, some load shedding is required. Alternatively, the transformer should be provided with under load tap changing.

8.6 Reactive Power Compensation-Transmission Line

The load flow and reactive power compensation along a transmission line are examined, Reference [8]. Figure 8.24a shows a 350 kV transmission line, which is 480 km long.

Conductors

589 mm^2, 54 strands, Al, Hurdles1, Prelli. GMR = 0.01276 m, outside diameter = 3.15 cm. The configuration in Figure 8.25 gives the following impedance data, based on a computer subroutine:

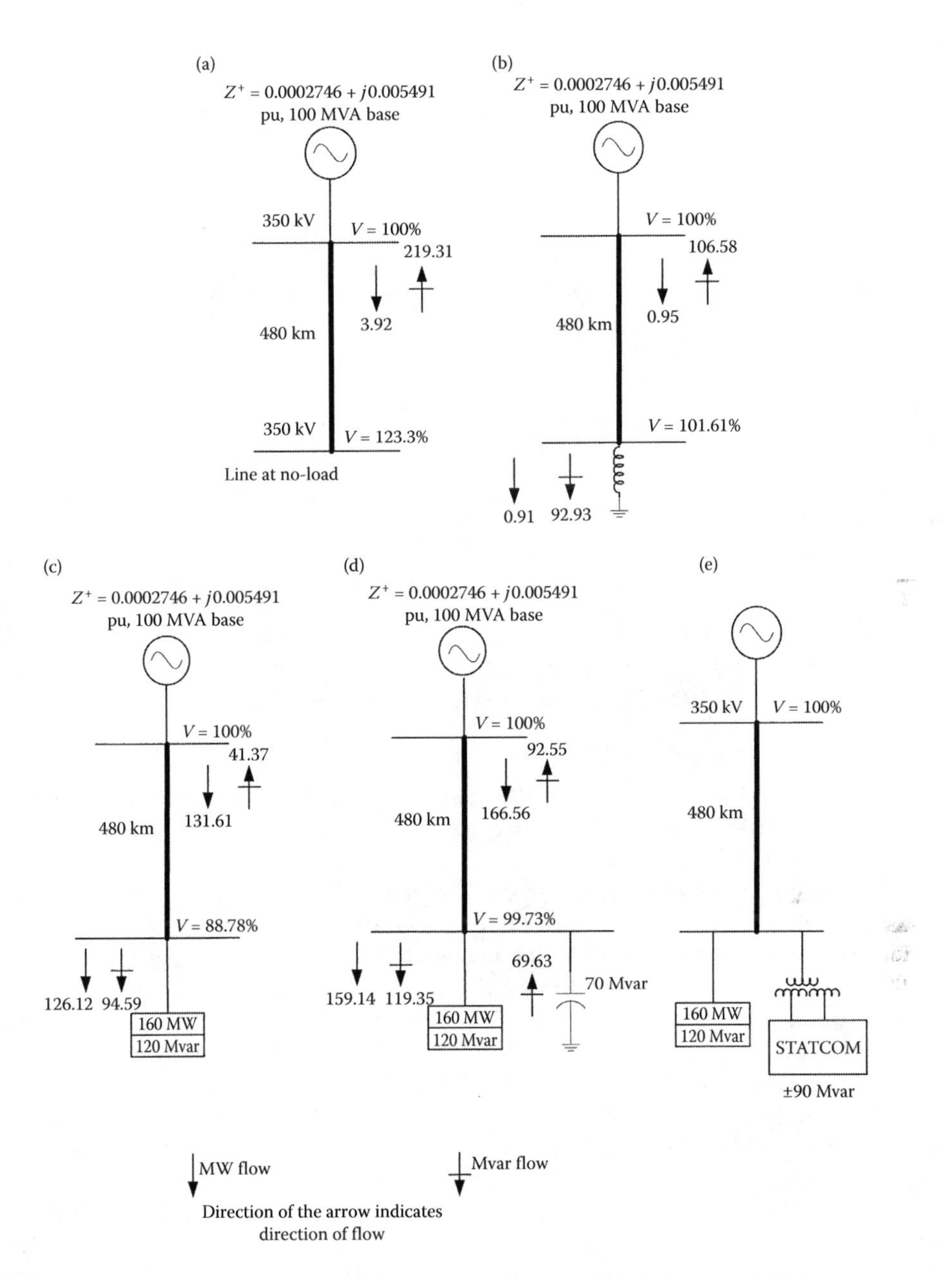

FIGURE 8.24
Power flow over a 350 kV 480 km long line: (a)–(d) show various study cases and (e) shows application of a STATCOM.

$$Z^{+} = Z^{-} = 0.05621 + j0.49286\ \Omega/\text{km}$$

$$Z^{0} = 0.33819 + j1.1565\ \Omega/\text{km}$$

$$Y^{+} = Y^{-} = 3.34161 \times 10^{-6}\ \text{S/km}$$

$$Y^{0} = 3.19538 \times 10^{-6}\ \text{S/km}$$

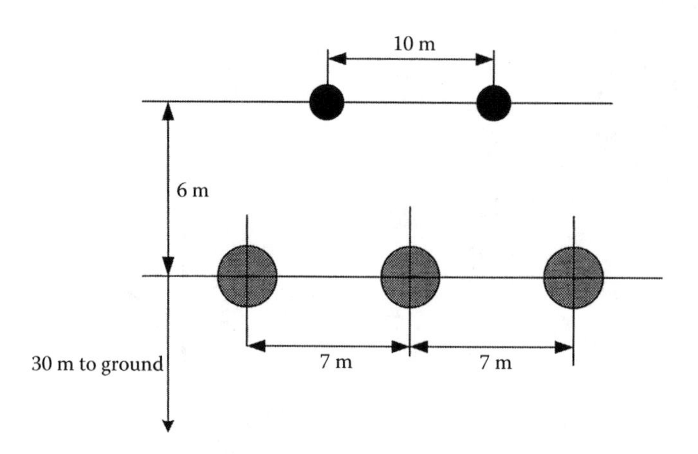

FIGURE 8.25
Configuration of 350 kV line conductors and spacing.

These parameters are calculated at 50 Hz, and the soil resistivity is 100 Ω-m.

Ground Wires

Two ground wires, spacing and configuration as shown in Figure 8.25, Prelli, Chlorine, GMR = 0.00272 m, and outside diameter = 0.75 cm.

Case 1: The line at no-load

The receiving-end voltage rises to 123.34%. The entire 3.92 MW of active power is lost in the line, and 219.31 Mvar is supplied into the utility system. The sending-end power factor is 1.97% leading, see Figure 8.24a.

Case 2: Reactive power compensation at the receiving end

Consider that a 90 Mvar lagging reactive power, in the form of a shunt reactor, is provided at the receiving end. The situation is depicted in Figure 8.24b. The source now supplies 0.019 MW and 106.58 Mvar of capacitive reactive power, power factor = 0.02 leading. 199.504 Mvar capacitive is the loss in the line. The receiving-end voltage is now 101.61%.

Case 3: The line supplies a load of 200 MVA at 0.8 PF at the receiving end

The voltage at the receiving end is now 88.78%. The load flow is depicted in Figure 8.24c. Still 41.37 capacitive reactive power is supplied into the utility source. The power factor at the utility source is 87.41 leading. The load modeled is a constant impedance load, and due to reduced voltage at the bus only 126.12 MW out of 160 MW modeled is supplied.

Case 4: As in Case 3, but a 70 Mvar capacitor bank is added at the receiving end

The load flow is depicted in Figure 8.24d. The receiving-end voltage is now 99.79%. 131.61 MW is supplied from the source; the power factor of the load supplied from the source is 95.40 leading.

A summary of the load flow solutions in these four cases is shown in Table 8.3. It may be concluded that 90 Mvar reactive power lagging to 70 Mvar reactive power leading is required at the receiving end. A fixed or switched compensation device will be difficult to control as the load varies from low load to full expected load. A STATCOM or TCR will be an appropriate choice, see Figure 8.24e. Note that TCR gives rise to harmonic pollution, which is a consideration when harmonic loads are present—which is invariably the case, see Volume 3. Also see Chapter 9.

The utilities rightly penalize for poor power factor; the reactive power compensation increases with poor power factor.

8.7 Reactive Power Compensation in an Industrial Distribution System

Figure 8.26a shows a large distribution system, which serves a total load of 118 MVA at 0.85 PF from three 13.8 kV buses, shown as buses 1, 2, and 3. Eighty percent of the load is considered a constant kVA load and rest 20% as constant impedance load, see Chapter 6. There are a number of low-voltage and medium-voltage substations connected to three 13.8 kV buses: consider that the total load reflected on each bus is the running load plus system losses. There are two generators of 40 MW each operating power factor of 0.85 and these are operated in PV mode; the maximum and minimum reactive power limits are set at 24 and 1 Mvar for each of the generators. There is one utility tie through 60/100 MVA transformer of 138–13.8 kV. The load flow data are shown in Table 8.4. Table 8.5 shows the voltages on 13.8 kV buses for each case of the study and Table 8.6 summarizes the overall load flow picture. The step-by-step load flow and reactive power compensation strategy is described in the following study cases.

Case 1: Base case with no adjustments

The load flows in major sections of the distribution system and bus voltages are shown in Figure 8.26a. The voltage on 13.8 kV bus 3 is 97.17% and on bus 2 it is 98.18%. Note that, both generators G1 and G2 operate with their maximum reactive power output of 24 Mvar, yet they are not able to maintain 100% voltages on buses 1 and 2. Approximately 15.2 Mvar is supplied from the 138 kV utility source.

Case 2: Set utility tie transformer taps at –5%

The utility tie transformer is provided with off-load taps only on the 138 kV windings; selecting a tap of –5% (signifying a 5% voltage boost at 13.8 kV secondary voltage at no load) reduces the reactive power output of the generator G1 to 18.23 and that of generator G2 to 22.0 Mvar, and the voltages at buses 1 and 2 are now 100%. The reactive power from the utility source is now 23.93 Mvar, i.e., an increase of 8.73 Mvar compared to case 1. Bus 3 voltage is 101%. The power factor of the power supply from the utility degrades from 79% to 65%, see Table 8.6 and Figure 8.26b.

Case 3: Generator 2 is out of service

It is required that full load plant operation should be sustained when either of the generators G1 or G2 goes out of service. The load flow with G2 out of service is shown in

TABLE 8.3

Load Flow over 480 km, 350 kV Line (50 Cycles Operation)

Case No.	Load Bus Voltage (%)	Load Supplied from Source	Load at Receiving End	System Losses
1	123.3	3.921 MW, –219.31 Mvar, 219.34 MVA at 1.79 PF leading	0	3.921 MW, –219.31 Mvar
2	101.61	0.019 MW, –106.58 Mvar, 106.58 MVA at 0.02 PF leading	0.92 MW, 92.93 Mvar	0.92 MW, –199.5 Mvar
3	88.78	131.61 MW, –41.372 Mvar, 137.96 MVA at 95.4 PF leading	126. 12 MW, 94.59 Mvar, 157.64 MVA at 0.8 PF lagging	5.491 MW, –135.958 Mvar
4	99.73	166.56 MW, –92.55 Mvar, 190% Mvar at 87.41 PF leading	159 MW, 49.73 Mvar, 166.75 MVA at 95.45 PF lagging	7.421 MW, –142.285 Mvar

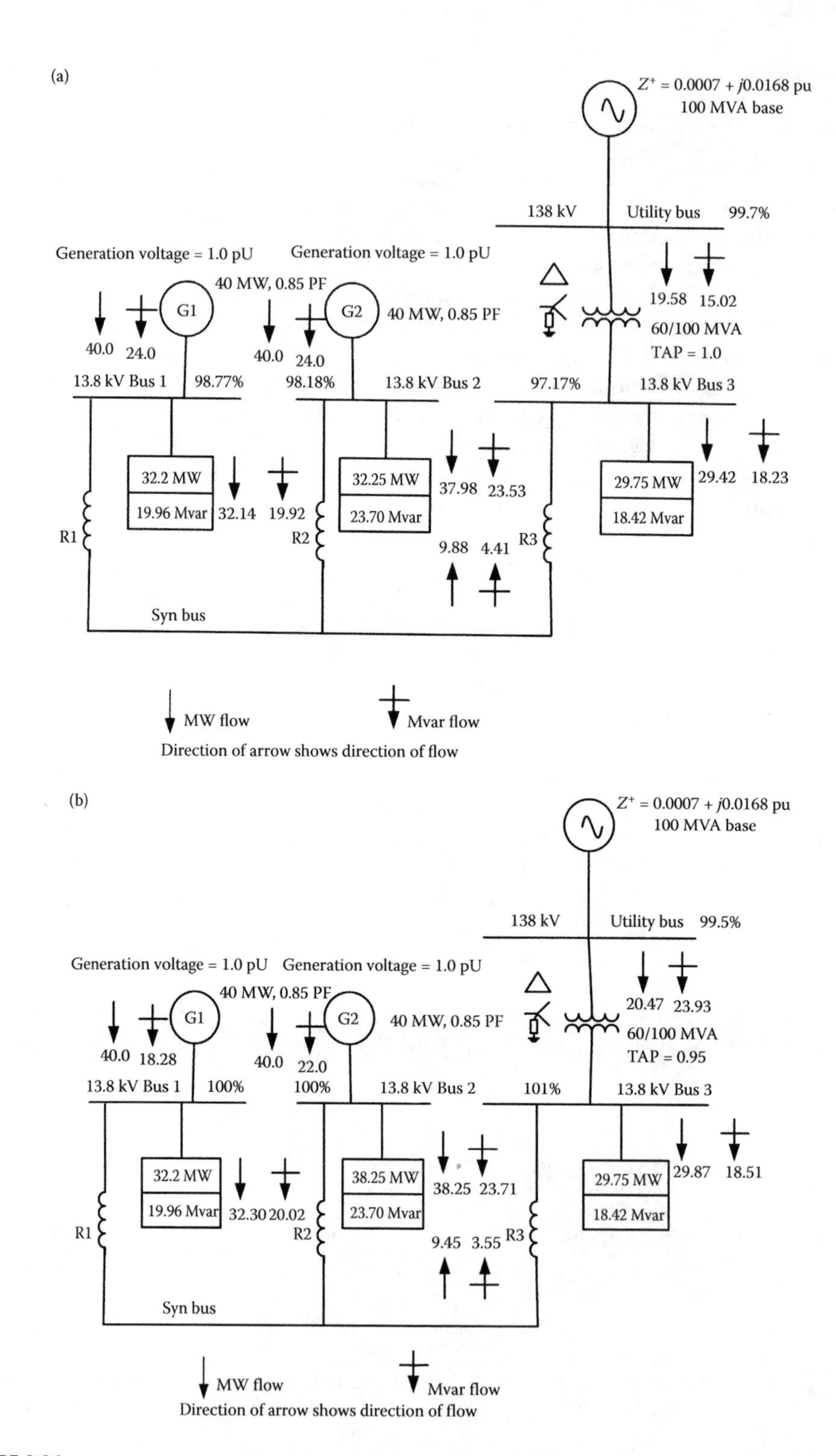

FIGURE 8.26
(a)–(j) Load flow and reactive power compensation from study cases (a) through (f).

(Continued)

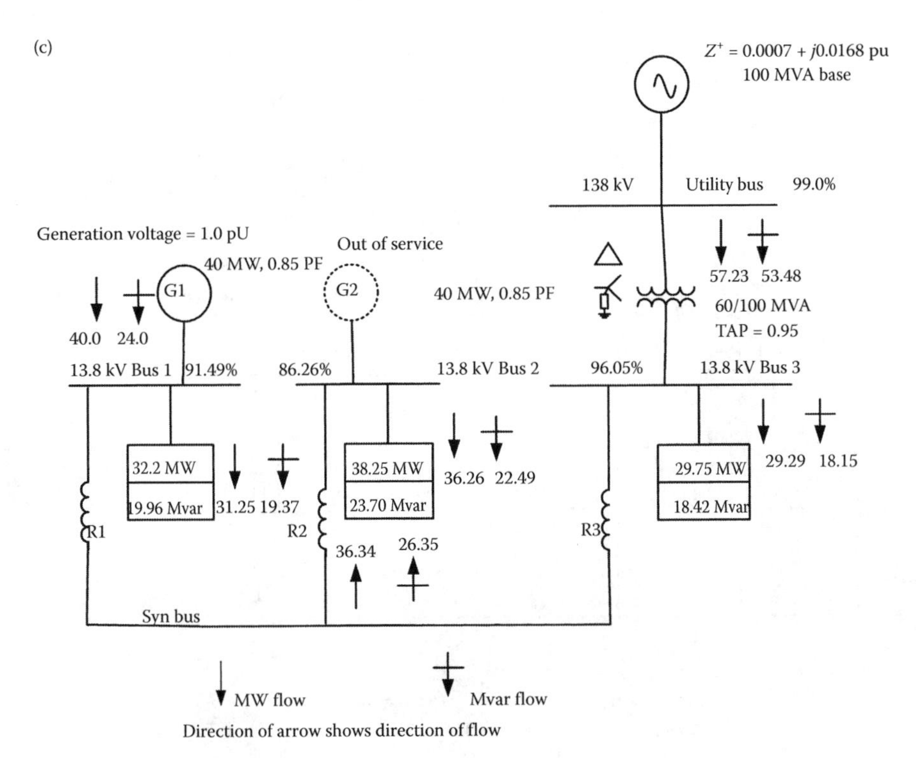

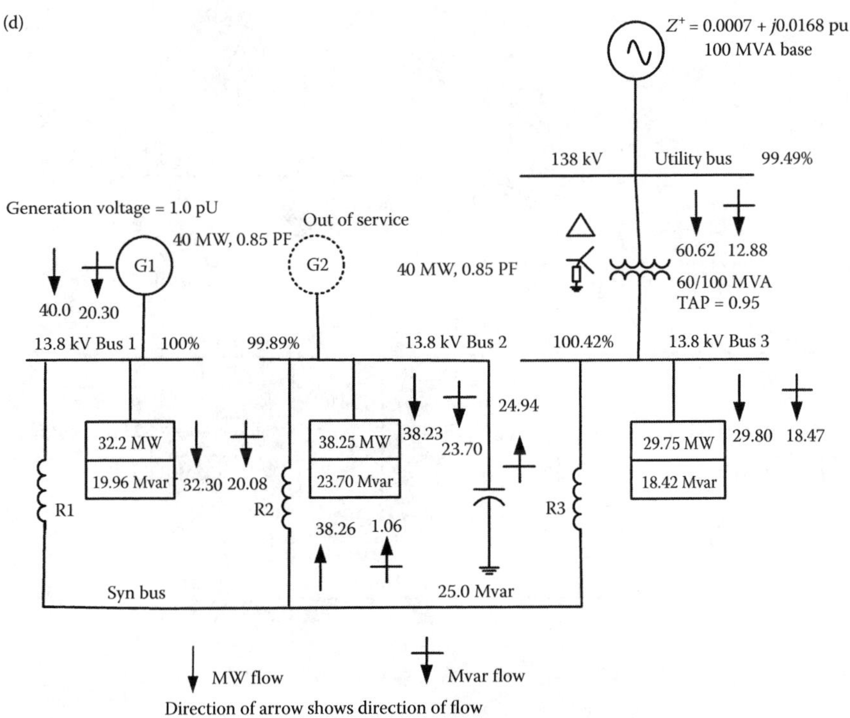

FIGURE 8.26 (CONTINUED)
(a)–(j) Load flow and reactive power compensation from study cases (a) through (f).

(*Continued*)

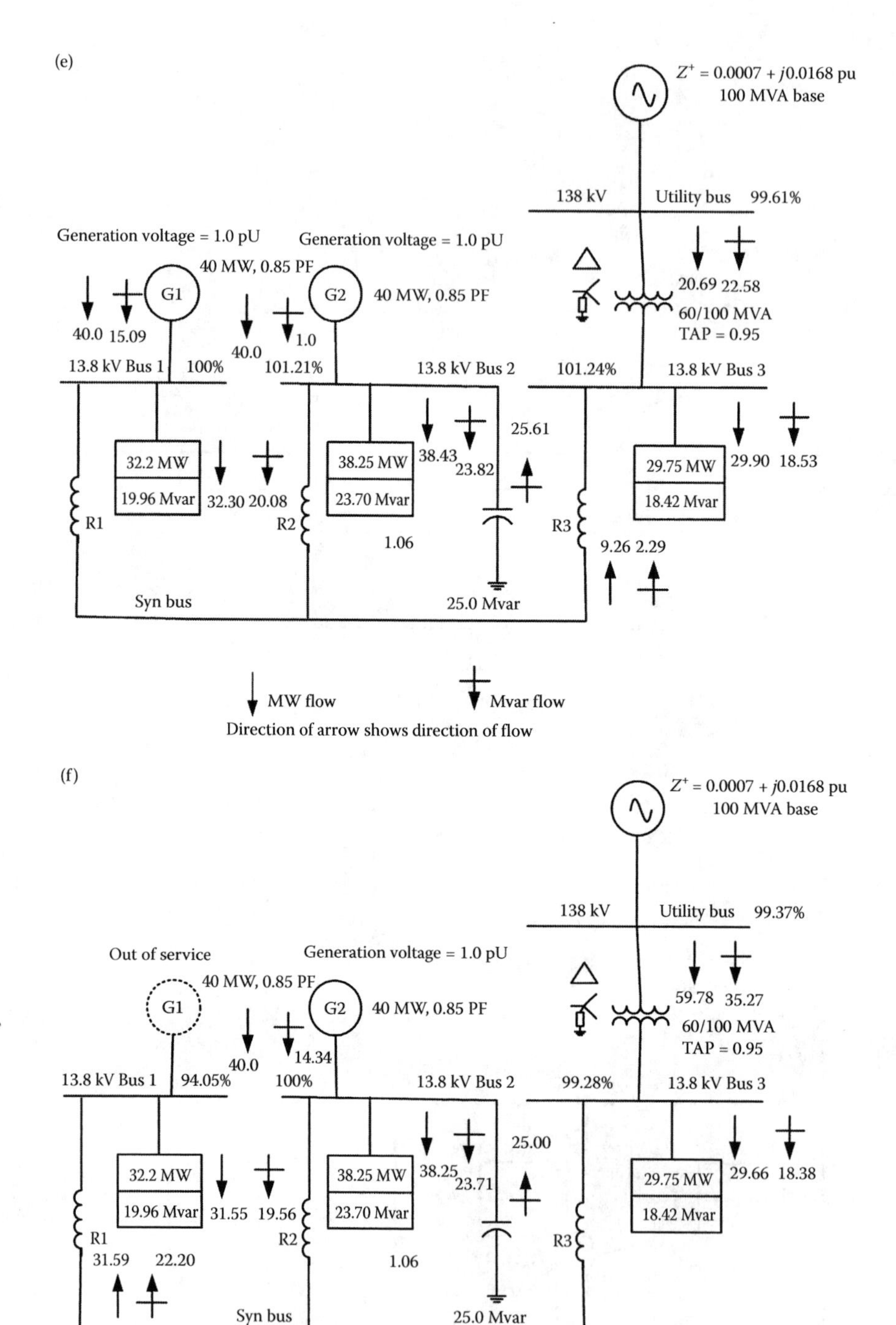

FIGURE 8.26 (CONTINUED)
(a)–(j) Load flow and reactive power compensation from study cases (a) through (f).

(Continued)

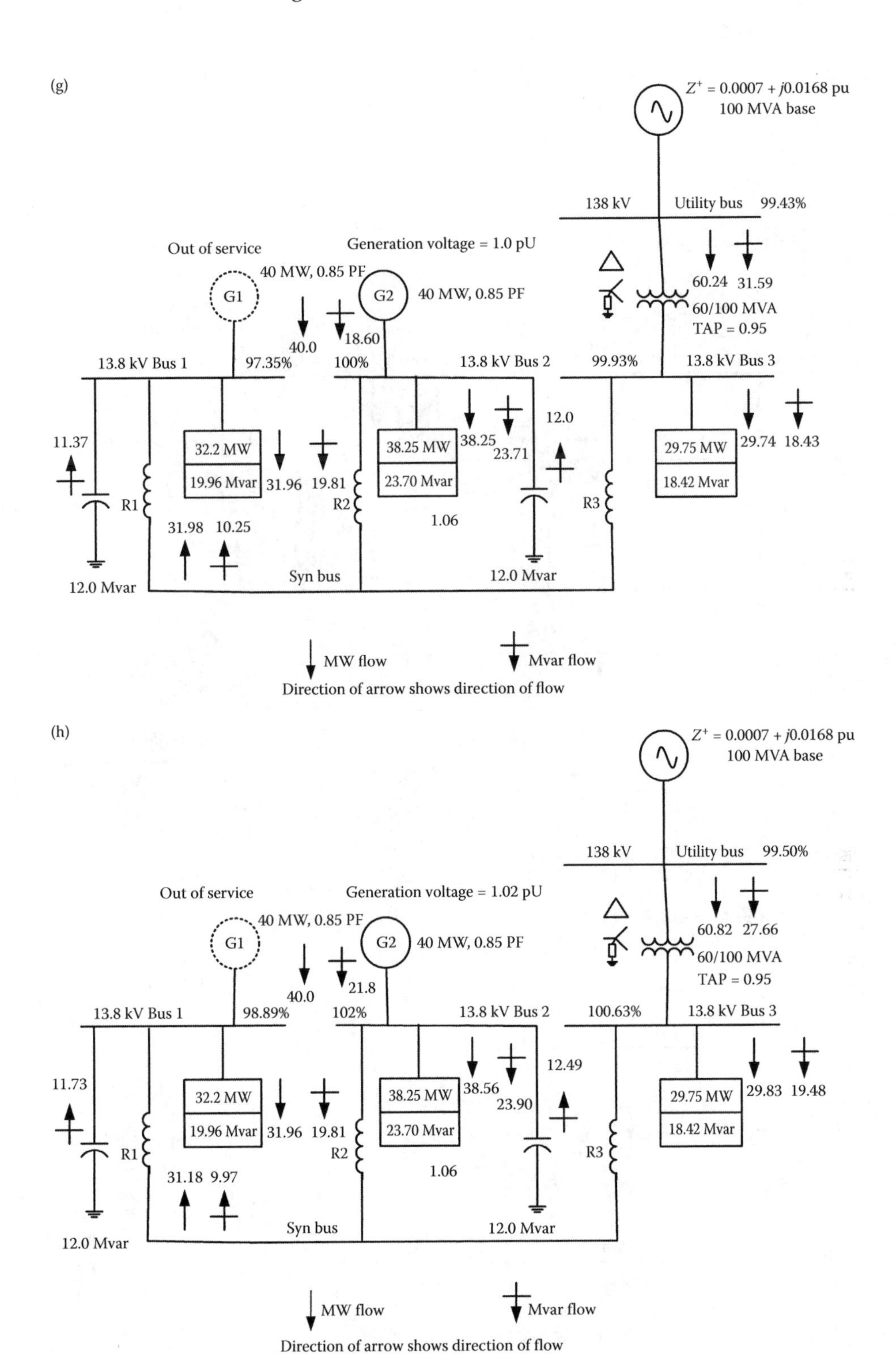

FIGURE 8.26 (CONTINUED)
(a)–(j) Load flow and reactive power compensation from study cases (a) through (f).

(Continued)

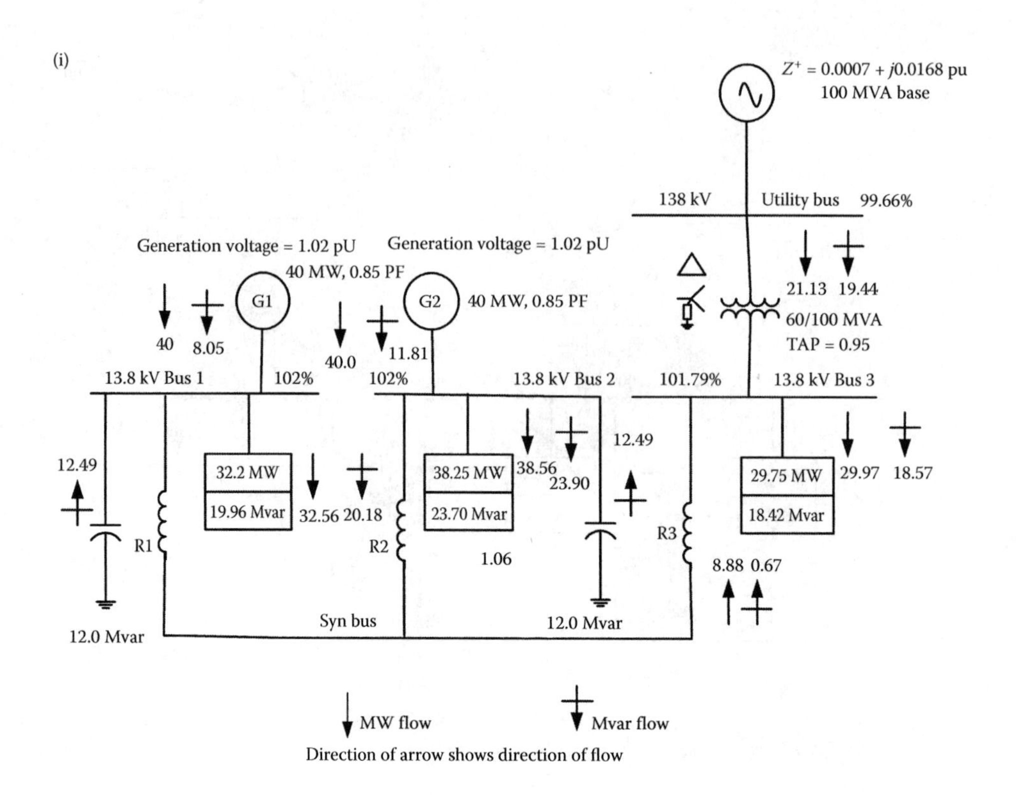

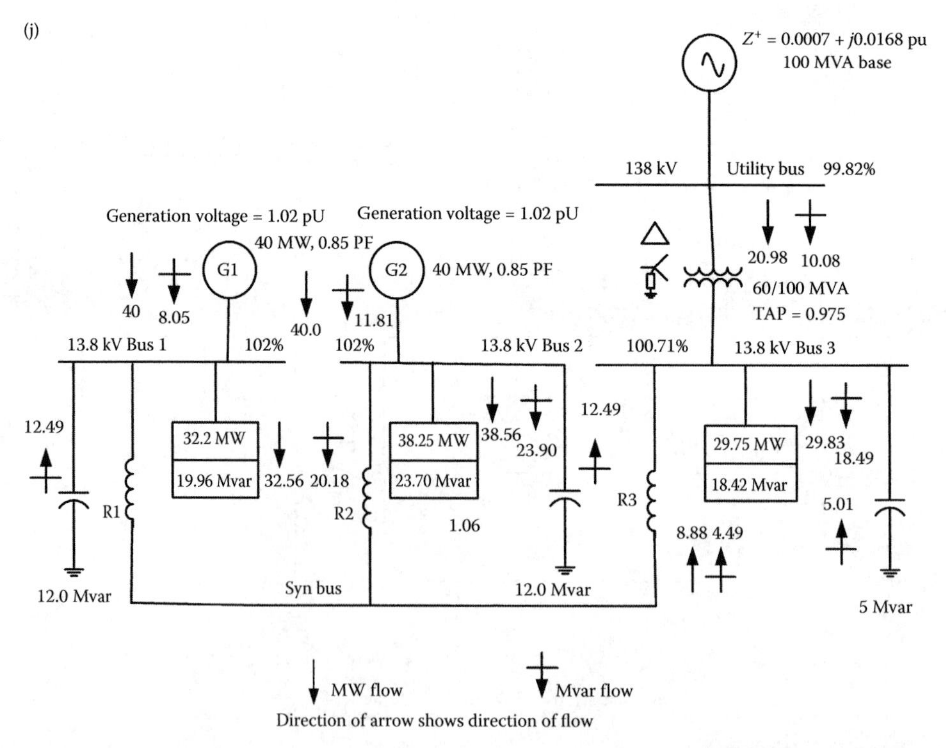

FIGURE 8.26 (CONTINUED)
(a)–(j) Load flow and reactive power compensation from study cases (a) through (f).

TABLE 8.4

Load Flow Data

Description	System Data
Utility source 138 kV	0.0007 + *j*0.0168 pu source impedance
Utility tie transformer, delta–wye connected 60/100 MVA, Z = 10%	0.0046 + *j*0.1583 pu transformer impedance
Reactor R1, 0.3 Ω	0.002 + *j*0.1575 pu reactor impedance
Reactor R2, 0.3 Ω	0.002 + *j*0.1575 pu reactor impedance
Reactor R4, 0.4 Ω	0.0026 + *j*0.2100 pu reactor impedance
Load bus 1, 13.8 kV	36 MVA at 0.85 PF, 80% constant KVA, and the rest 20% constant impedance
Load bus 2, 13.8 kV	45 MVA at 0.85 PF, 80% constant KVA, and the rest 20% constant impedance
Load bus 3, 13.8 kV	35 MVA at 0.85 PF, 80% constant KVA, and the rest 20% constant impedance
Generator G1	40MW, 0.85 PF, reactive power limits 24 Mvar maximum, 1 Mvar minimum
Generator G2	40MW, 0.85 PF, 24 Mvar maximum, 1 Mvar minimum

Impedances on a 100 MVA base.

TABLE 8.5

Calculated Bus Voltages in Percentage

Case No.	Brief Description, See the Text	Bus 1	Bus 2	Bus 3
1	Base	98.77	98.19	97.17
2	Utility transformer tap at –5%	100	100	101
3	G2 out of service	91.49	86.26	96.05
4	G2 out of service and 25 Mvar capacitor bank at bus 2	100	99.89	100.42
5	G1 and G2 in service and 25 Mvar capacitor bank at bus 2	100	101.21	101.24
6	G1 out of service and 25 Mvar capacitor bank at bus 2	94.05	100	99.28
7	G1 out of service and 12.0 Mvar capacitor banks at buses 1 and 2	97.35	100	99.93
8	G1 out of service and 12.0 Mvar capacitor banks at buses 1 and 2 and G2 operated at 2% higher voltage	98.89	102	100.63
9	G1 and G2 in service and 12.0 Mvar capacitor banks at buses 1 and 2 and G1 and G2 operated at 2% higher voltage	102	102	101.79
10	G1 and G2 in service and 12.0 Mvar capacitor banks at buses 1 and 2, 5 Mvar capacitor bank at bus 3, utility tie transformer taps at 0.975, and G1 and G2 operated at 2% higher voltage	102	102	100.71

Figure 8.26c. The voltage of bus 2 is now 86.26% (a voltage dip of 13.74%), whereas the voltages of buses 1 and 3 are 91.24% and 96.05%, respectively. Generator 1 again hits its reactive power output of 24 Mvar and the utility source supplies 53.48 Mvar. Reactive power loss is 17.47 in the system as the reactive power has to flow mainly reactive power system components. Also note the change in load supplied and total maximum demand from case to case. This case shows that reactive power compensation should be provided.

Case 4: 25 Mvar shunt capacitor bank at bus 2

With a 25 Mvar capacitor bank provided on bus 2, the load flow is depicted in Figure 8.26d. The bus voltages are acceptable. A question arises whether the 25 Mvar bank should be switched on bus 2 as soon as generator G2 goes out of service or it can remain switched in at all times. Switching of a capacitor bank on loss of a source is fraught out with two problems: (1) switching transients and (2) switching delay. It may take 10–15 cycles before

TABLE 8.6

Load Flow Summary

Case No.	G1 Output		G2 Output		Load Supplied from Utility Source		Total Load Demand (Considers Installed Capacitor Banks and Losses)		System Losses	
	MW	Mvar	MW	Mvar	MVA	PF	MVA	PF	MW	Mvar
1	40	24	49	24	24.79	78.98Lag	117.94	84.43 Lag	0.039	1.516
2	40	18.28	40	22	51.49	65.01 Lag	119.24	84.26 Lag	0.054	1.978
3	40	24	–	–	78.33	73.07 Lag	124.32	78.21 Lag	0.402	17.47
4	40	20.3	–	–	67.13	90.30 Lag	111.973	89.17 Lag	0.284	11.90
5	40	15.09	40	1.0	30.62	67.55 Lag	107.86	93.35 Lag	0.052	1.90
6	–	–	40	14.34	69.40	86.13 Lag	111.43	89.54 Lag	0.310	12.96
7	–	–	40	18.6	68.02	88.56 Lag	112.10	89.42 Lag	0.284	11.62
8	–	–	40	21.8	66.81	91.03 Lag	112.30	89.78 Lag	0.260	11.37
9	40	8.05	40	11.81	28.72	73.60 Lag	108.50	93.21 Lag	0.045	1.630
10	40	9.97	40	13.72	23.283	90.14 Lag	106.49	94.84 Lag	0.033	1.254

the standby capacitor can be brought in line and this time delay is not acceptable for critical loads to ride through the voltage dips.

Case 5: Load flow with 25 Mvar capacitor bank at bus 2 continuously in service

Figure 8.26e shows the load flow in this situation. It is noted that the capacitor bank of 25 Mvar can be continuously in service. The bus voltages are acceptable. The generators back out on their reactive power output: it has a desirable impact on the field system as less heat will be produced in the field windings.

Case 6: Operating conditions as in Case 5, but G1 is out of service

The voltage at bus 1 is 94.05% (a voltage dip of approximately 6%), see Figure 8.26f. This is unacceptable.

Case 7: Reactive power compensation of 12 Mvar at buses 1 and 2

Therefore, the reactive power compensation of 25.0 Mvar is divided approximately 50%–50% on buses 1 and 2. *Furthermore, the maximum size of capacitor bank at 13.8 kV should be limited to 10–12 Mvar.* The switching concerns are described in Volume 1. The load flow is shown in Figure 8.26g. The voltage at bus 1 is improved from 94.05% in case 6 to 97.35%.

Case 8: Raise the operating voltages of generators G1 and G2 to 1.02 pu

As stated earlier, as per ANSI/IEEE standards, a generator can be operated up to 5% higher voltage subject to some qualifications. All other conditions remaining the same, with generators operating at 1.02 pu, the load flow is shown in Figure 8.26h. Now, the voltage of bus 1 is 98.89%.

Case 9: Load flow with all sources and reactive power compensation in service

This is now the load flow under normal conditions, with both generators and 12.0 Mvar capacitor banks at buses 1 and 2 in service, as shown in Figure 8.26i. Buses 1 and 2 operate at 102% voltage. It is desirable to keep operating voltages slightly above normal operating voltages; therefore, it is acceptable. However, the power factor of the utility source is 73.6%. The utilities levy penalties for low power factor and it should be improved.

Case 10: Final configuration

It is desired that the utility power factor under normal operation is no less than 90%. To achieve this objective, the utility transformer tap is adjusted to 0.975 (2.5% 13.8 kV voltage boost at no-load) and another 5 Mvar capacitor bank is added at bus 3. The load flow is shown in Figure 8.26j.

Thus, overall

- Set utility tie transformer taps to 0.975, to provide 2.5% 13.8 kV voltage boost at no-load.
- Provide shunt capacitor banks of 12 Mvar each at 13.8 kV buses 1 and 2, and 5 Mvar at bus 3. In case harmonic loads are present, these capacitor banks should be turned into harmonic filters, see Volume 3.
- Operate generators at 1.02 pu rated voltage.

This will meet the requirements of normal load flow, a power factor of 0.9 lagging from the utility source, and full load operation on loss of either of generators 1 or 2, one at a time.

This demonstrates that there are a number of constraints in load flow and a certain objective is achieved with variations of the parameters of operation. Optimal load flow is discussed in Chapters 11 and 12.

Problems

8.1 Mathematically derive Equations 8.6 and 8.37.

8.2 The voltage under load flow at a certain bus in an interconnected system dips by 10%, while the voltages on adjacent buses are held constant. The available short-circuit current at this bus is 21 kA, voltage 230 kV. Find the reactive power injection to restore the voltage to its rated value of 230 kV.

8.3 In Example 8.1, the line is compensated by a shunt compensation of $K_{sh} = 0.3$. Calculate all the parameters of Example 8.2. Repeat for series and shunt compensation of $K_{sh} = K_{sc} = 0.3$.

8.4 A 100 MVA load at 0.8 power factor and 13.8 kV is supplied through a short transmission line of 0.1 per unit reactance (100 MVA base). Calculate the reactive power loss and the load voltage. Size a capacitor bank at load terminals to limit the voltage drop to 2% at full load. What are the capacitor sizes for three-step switching to maintain the load voltage no more than 2% below the rated voltage as the load varies from zero to full load?

8.5 A 200 MW, 18 kV, 0.85 power factor generator is connected through a 200 MVA step-up transformer of 18–500 kV and of 10% reactance to a 500 kV system. The transformer primary windings are provided with a total of five taps, two below the rated voltage of 2.5% and 5% and two above the rated voltage of 2.5% and 5%. Assuming that initially the taps are set at a rated voltage of 18–500 kV, what is the generation voltage to take full rated reactive power output from the generator? If the operating voltage of the generator is to be limited to rated voltage, find the tap setting on the transformer. Find the reactive power loss through the transformer in each case. Neglect resistance.

8.6 A 100 mi (160 km) long line has $R = 0.3\ \Omega$, $L = 3$ mH, and $C = 0.015\ \mu$F. It delivers a load of 100 MVA at 0.85 power factor at 138 kV. If the sending-end voltage is maintained at 145 kV, find the Mvar rating of the synchronous condenser at the receiving end at no-load and full load.

8.7 Derive an equation for the load line of the synchronous condenser shown in Figure 8.8.

8.8 Plot P–δ curves of: (1) 230 kV uncompensated line, (2) with midpoint shunt compensation of 0.6, and (3) with series compensation of 0.6. The line is 200 mi long, has a series reactance of 0.8 Ω/mi, and a susceptance $y = 5.4 \times 10^{-6}$ S/mi.

References

1. ANSI Standard C84.1. Voltage rating for electrical power systems and equipments *(60 Hz)*, 1988.
2. NEMA. Large machines-induction motors. Standard MG1-Part 20, 1993.
3. P Kessel, R Glavitsch. Estimating voltage stability of a power system. *Trans IEEE Power Delivery*, 1, 346–352, 1986.
4. TJE Miller. *Reactive Power Control in Electrical Power Systems*, Wiley, New York, 1982.
5. IEEE Committee Report. Var management-problem and control. *Trans IEEE Power App Syst*, 103, 2108–2116, 1984.
6. ANSI Standard C50.13. *American National Standard Requirements for Cylindrical Rotor Synchronous Generators*, 1977.
7. CIGRE WG 30-01, Task Force No. 2. In: IA Erinmez, ed. *Static Var Compensators*. CIGRE, Paris, 1986.
8. TM Kay, PW Sauer, RD Shultz, RA Smith. EHV and UHV line loadability dependent on var supply capability. *Trans IEEE Power App Syst*, 101, 3586–3575, 1982.

9

FACTS—Flexible AC Transmission Systems

9.1 Power Quality Problems

Power quality problems include voltage sags and swells, high-frequency line-to-line surges, steep wave fronts, or spikes caused by switching of loads, harmonic distortions, and outright interruptions, which may extend over prolonged periods. An impulse is a unidirectional pulse of 10 ms in duration. Voltage sag is a reduction in nominal voltage for more than 0.01 s and less than 2.5 s. A swell is an increase in voltage for more than 0.01 s and less than 2.5 s. Low and high voltages are reduction and increase in voltage for more than 2.5 s. All these events are observable. A problem that is not easily detected is the common-mode noise that occurs on all conductors of an electrical circuit at the same time. The tolerance of processes to these power quality problems is being investigated more thoroughly; however, the new processes and electronic controls are more sensitive to power quality problems. It is estimated that the power quality problems cost billions of dollars per year to American industry. Though this book is not about mitigating the power quality problems and detailed discussions of this subject, two figures of interest are Figures 9.1 and 9.2. Figure 9.1 categorizes the various power quality problems and the technologies that can be applied to mitigate these problems [1]. Figure 9.2 is the new ITI (Information Technology Industry Council, earlier CBEMA) curve, which replaces the earlier CBEMA (Computer and Business Equipment Manufacturer's Association) [2] curve for acceptable power quality of computers and electronic equipment. The voltage dips and swells and their time duration become of importance, which are a subject of load flow and contingency load flow. Volume 3 addresses harmonic pollution and Volume 4 has a chapter on surge protection.

9.2 FACTS

The concept of flexible AC transmission systems (FACTS) was developed by EPRI and many FACTS operating systems are already implemented [3–24]. The world's first thyristor-controlled series capacitor was put in service on the Bonneville Power Authority's 500 kV line in 1993 [4].

We have examined the problem of power flow over transmission lines and to change the impedance of the line or its transmission angle, the role of SVCs is described in Chapter 8. Advances in recent years in power electronics, software, microcomputers, and fiber-optic transmitters that permit signals to be sent to and fro from high-voltage levels make

Power quality condition		Power conditioning technology								
		A	B	C	D	E	F	G	H	I
Transient voltage surge	Common mode									
Transient voltage surge	Normal mode									
Noise	Common mode									
Noise	Normal mode									
Notches										
Voltage distortion										
Sag										
Swell										
Undervoltage										
Overvoltage										
Momentary interruption										
Long term interruption										
Frequency variations										

It is reasonable to expect that the indicated condition will be corrected

The indicated condition may or may not be conditioning product performance corrected, due to significant variations in power

The indicated condition is not corrected

A = TVSS
B = EMI/RFI filter
C = Isolation transformer
D = Electronic voltage regulator
E = Ferroresonance voltage regulator
F = Motor generator
G = Standby power system
H = Uninterruptible power supply (UPS)
I = Standby engine generator

FIGURE 9.1
Power quality problems and their mitigation technologies.

possible the design and use of fast FACTS controllers. For example, the use of thyristors makes it possible to switch capacitors' orders of magnitude faster than those with circuit breakers and mechanical devices.

The use of electronic devices in processes, industry, and the home has created an entirely new demand for power quality that the energy providers must meet.

The FACTS use a voltage source bridge, rather than current source configuration, which is not discussed in this volume, and is covered in Volume 3. The requirements of FACTS controllers are as follows:

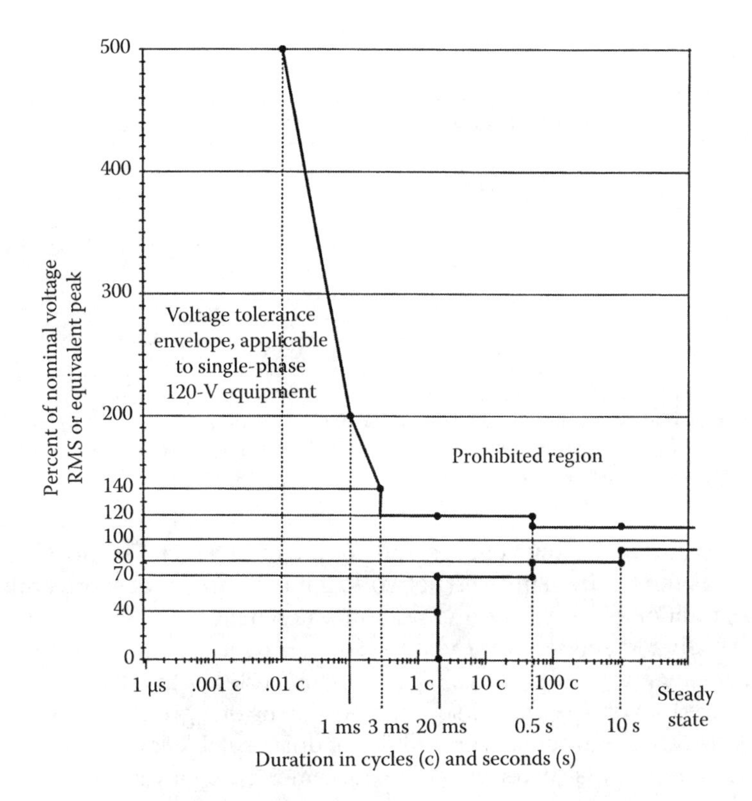

FIGURE 9.2
ITI curve, 2000.

- The converter should be able to act as an inverter or rectifier with leading or lagging reactive power, i.e., four-quadrant operation is required, compared to current source line commutated converter that has two-quadrant operations [6].
- The active and reactive powers should be independently controllable with control of phase angle.

The basic electronic devices giving rise to this thrust in the power quality are listed in Table 9.1. FACTS devices control the flow of AC power by changing the impedance of the transmission line or the phase angle between the ends of a specific line. Thyristor controllers can provide the required fast control, increasing or decreasing the power flow on a specific line and responding almost instantaneously to stability problems. FACTS devices can be used to dampen the subsynchronous oscillations which can be damaging to rotating equipment, i.e., generators. These devices and their capabilities are only briefly discussed in this chapter.

9.3 Synchronous Voltage Source

A solid-state synchronous voltage source (SS) can be described as analogous to a synchronous machine. A rotating condenser has a number of desirable characteristics, i.e., high

TABLE 9.1

FACTS Devices

Application	STATCOM or STATCON	SPFC (SSSC)	UPFC	NGH-SSR Damper
Voltage control	X	X	X	
Var compensation	X		X	
Series impedance control		X	X	X
MW flow control		X	X	
Transient stability	X	X	X	X
System isolation		X	X	
Damping of oscillations	X	X	X	X

Each of these devices has varying degrees of control characteristics broadly listed above.
STATCOM, STATCON, static synchronous compensator or static condenser; *SPFC, SSSC*, series power flow controller, static series synchronous compensator; *UPFC*, unified power flow controller; *NGH-SSR*, Narain G. Hingorani subsynchronous resonance damper [6].

capacitive output current at low voltages and source impedance that does not cause harmonic resonance with the transmission network. It has a number of shortcomings too, e.g., slow response, rotational instability, and high maintenance.

An SS can be implemented with a voltage source inverter using GTOs (gate turn-off thyristors). An elementary six-pulse voltage source inverter with a DC voltage source can produce a balanced set of three quasi-square waveforms of a given frequency. The output of the six-pulse inverter will contain harmonics of unacceptable level for transmission line application, and a multipulse inverter can be implemented by a variety of circuit arrangements (see Volume 3).

The reactive power exchange between the inverter and AC system can be controlled by varying the amplitude of the three-phase voltage produced by the SS. Similarly, the real power exchange between the inverter and AC system can be controlled by phase shifting the output voltage of the inverter with respect to the AC system. Figure 9.3a shows the coupling transformer, the inverter, and an energy source which can be a DC capacitor, battery, or superconducting magnet.

The reactive and real power generated by the SS can be controlled independently and any combination of real power generation/absorption with var generation and absorption is possible, as shown in Figure 9.3b. The real power supplied/absorbed must be supplied by the storage device, while the reactive power exchanged is internally generated in the SS.

The reactive power exchange is controlled by varying the amplitude of three-phase voltage. For a voltage greater than the system voltage, the current flows through the reactance from the inverter into the system, i.e., the capacitive power is generated. For an inverter voltage less than the system voltage, the inverter absorbs reactive power. The reactive power is, thus, exchanged between the system and the inverter, and the real power input from the DC source is zero. In other words, the inverter simply interconnects the output terminals in such a way that the reactive power currents can freely flow through them.

The real power exchange is controlled by phase shifting the output voltage of the inverter with respect to the system voltage. If the inverter voltage leads the system voltage, it provides real power to the system from its storage battery. This results in a real component of the current through tie reactance, which is in phase opposition to the AC voltage. Conversely, the inverter will absorb real power from the system to its storage device if the voltage is made to lag the system voltage.

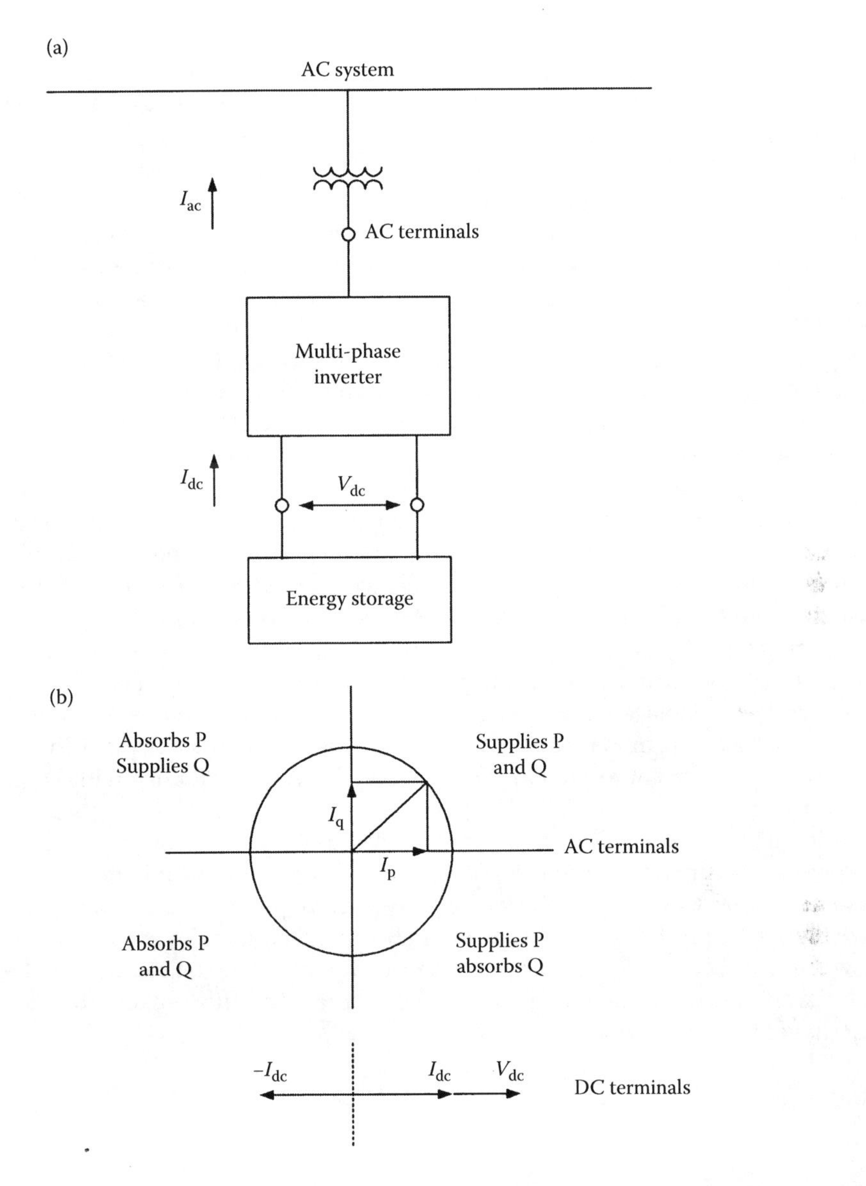

FIGURE 9.3
(a) A shunt-connected synchronous voltage source and (b) possible modes of operation for real and reactive power generation.

This bidirectional power exchange capability of the SS makes complete temporary support of the AC system possible.

9.3.1 Static Synchronous Compensator (STATCOM)

The static var compensators discussed in Chapter 8 are a controllable source of reactive power by synchronously switching reactors or capacitors. For example, a thyristor-controlled reactor (TCR) consists of a continuously variable reactor and shunt capacitors. By varying the reactive power output of the reactor, the capacitive reactive power output can be varied. It can be made zero, and then inductive by proper choice of controllable reactor and fixed capacitor bank. Thus, variable reactive shunt impedance is created which can meet the compensation requirement of a transmission network [10, 17–20, 22, 23].

The concept of generating reactive power without storage elements like capacitors and reactors dates back to 1976. These converters are operated as voltage or current sources and produce reactive power without reactive power storage elements by circulating alternating current among the phases of an AC system. Functionally, their operation is equivalent to that of a synchronous machine, whose reactive power is varied by an excitation controller. Analogously, these are termed static synchronous compensators (STATCOM).

Figure 9.3a is also the schematic of an inverter-based shunt STATCOM, sometimes called a static condenser (STATCON). It is a shunt reactive power compensating device. It can be considered as an SS with a storage device as DC capacitor. A GTO-based power converter produces an AC voltage in phase with the transmission line voltage. When the voltage source is greater than the line voltage ($V_L < V_0$), leading vars are drawn from the line and the equipment appears as a capacitor; when voltage source is less than the line voltage ($V_L > V_0$), a lagging reactive current is drawn. The basic building block is a voltage-source inverter which converts DC voltage at its terminals into three-phase AC voltage. A STATCON may use many six-pulse inverters, output phase shifted and combined magnetically to give a pulse number of 24 or 48 for the transmission systems. Using the principle of harmonic neutralization, the output of n basic six-pulse inverters, with relative phase displacements, can be combined to produce an overall multiphase system. The output waveform is nearly sinusoidal and the harmonics present in the output voltage and input current are small. This ensures waveform quality without passive filters, see Volume 3.

Note that while a STATCOM can operate in all the four-quadrants, see Figure 9.3b, a synchronous machine with excitation controller will generally supply reactive power into the system and has limited capability to absorb reactive power from the system. It will supply active power into the system but cannot absorb or reverse the active power. The reactive power current I drawn by the synchronous compensator is determined by the system voltage V, and internal voltage E, dictated by excitation and the total circuit reactance including that of the transformer:

$$I = \frac{V - E}{X} \tag{9.1}$$

The corresponding reactive power exchanged is

$$Q = \frac{1 - \dfrac{E}{V}}{X} V^2 \tag{9.2}$$

Increasing E above V, i.e., overexcitation, results in a leading current, i.e., the machine is seen as an AC capacitor by the system. Conversely, decreasing E below V produces a lagging current and the machine is seen as a reactor by the system.

Figure 9.4 shows a circuit diagram of a STATCOM. It shows a three-level, 12-pulse bridge, see Volume 3. The other configurations are H-bridges, or three-phase, two-level, 6-pulse bridges. A number of GTOs may be connected with reverse parallel diodes. Each

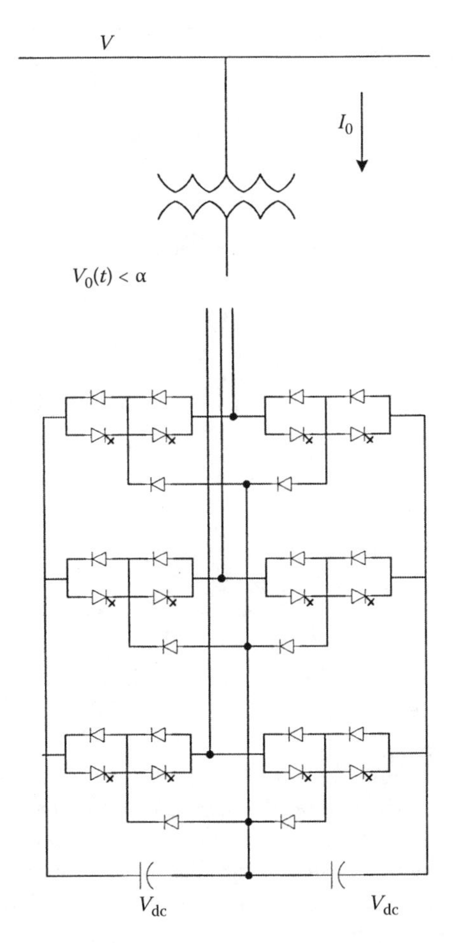

FIGURE 9.4
Schematic of a STATCOM or STATCON inverter-based static shunt compensator.

converter produces a square, quasi-square, or pulse width modulated phase-shifted waveform, Volume 3, and the final voltage waveform can be made to approximate a sine wave, so that no filtering is required, see Figure 9.5, which depicts a 48-pulse output waveform. According to instantaneous power theory [25], Volume 3, the net instantaneous power at AC terminals is equal to the net instantaneous power at the DC terminals (neglecting losses). As the converter supplies only reactive power, the real input power supplied by the DC source is zero. As the reactive power at zero frequency supplied by the DC capacitor is zero, the capacitor plays no part in the reactive power generation. *The converter simply interconnects the ac terminals so that reactive power can flow freely between them, with zero instantaneous power exchange. The converter only draws a fluctuating ripple current from the DC storage capacitor that provides a constant voltage.*

Practically, the GTOs are not lossless, and the energy stores in the capacitor are used up by internal losses. These losses can be supplied from the AC system by making output voltages slightly lag the AC system voltages. In this way, the converter absorbs a small amount of real power. The reactive power generation or absorption can be controlled by capacitor voltage control, by increasing or decreasing it. If the converter is equipped with a

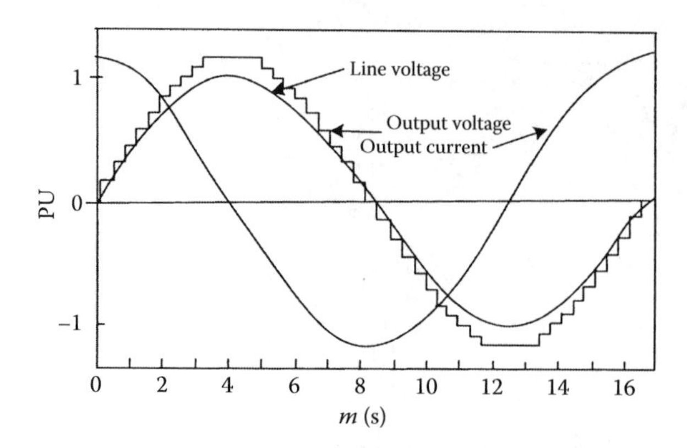

FIGURE 9.5
A waveform with 48-pulse operation.

large DC source, like a large battery or superconducting magnet (Volume 1), the converter can control the real and reactive power exchange.

9.4 Fundamentals of Control

The gating signals for the GTOs are generated by the *internal converter control* in response to the demand for reactive power and or real power reference signals. The reference signals are provided by the *external or system control*, from operator intervention and system variables, which determine the functional operation of STATCOM. If it is operated as an SVC, the reference input to the internal control is the reactive current. The magnitude of the AC voltage is directly proportional to the DC capacitor voltage. Thus, the internal control should establish the required DC voltage on the capacitor.

The internal controls must establish the capability to produce a synchronous output voltage waveform that forces the real and reactive power exchange with the system. As one option, the reactive output current can be controlled *indirectly* by controlling the DC capacitor voltage (as magnitude of AC voltage is directly proportional to the DC capacitor voltage), which is controlled by the angle of the output voltage. Alternatively, it can be *directly* controlled by internal voltage control mechanism, i.e., PWM converter, in which case the DC voltage is kept constant by control of the angle.

The control circuit in Figure 9.6 shows the simplified block circuit diagram of internal control. The inputs to internal control are V, the system voltage (obtained through potential transformers), output current of the converter I_0, which is broken into active and reactive components. These components are compared with an external reactive current reference, determined from the compensation requirements, and the internal real current reference, derived from DC voltage regulation loop. Voltage V operates a phase-locked loop that produces the synchronizing signal angle θ. The reactive current error amplifier produces control angle α. Angle $\Theta + \alpha$ operates the gate pattern logic. After amplification, the real and reactive current signals are converted into magnitude and angle of the required converter output voltage. The DC voltage reference determines the real power the converter must absorb from the system to supply losses.

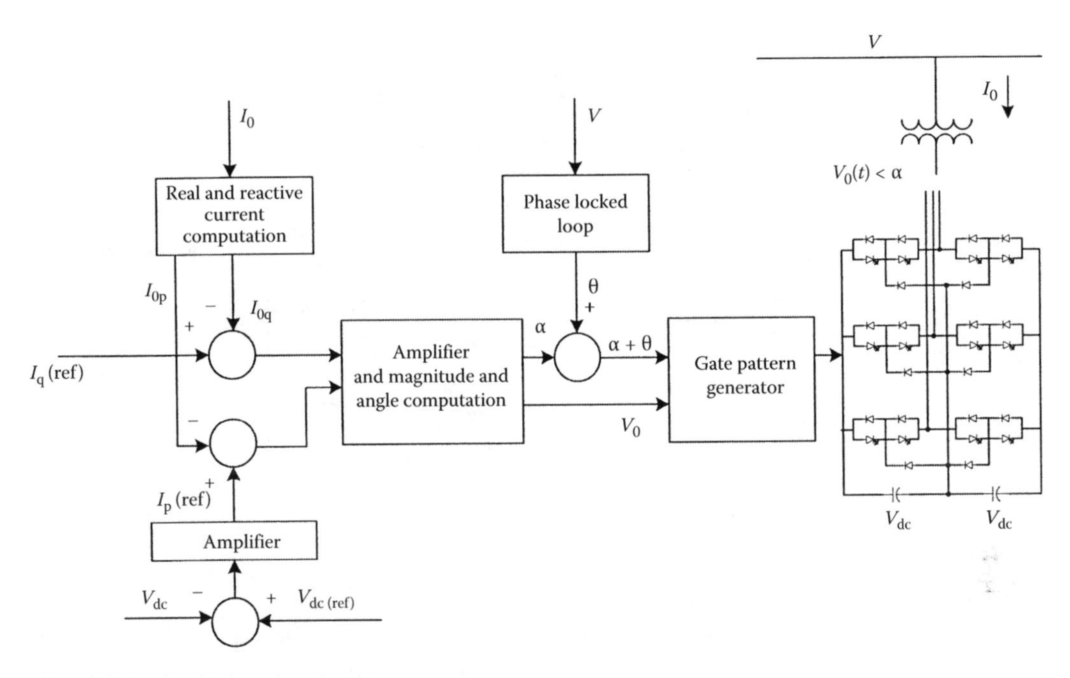

FIGURE 9.6
Fundamental control circuit of a STATCOM, see the text.

In summary, the following advantages are obtained:

- Interface real power sources
- Higher response to system changes
- Mitigation of harmonics compared to a fixed capacitor-thyristor-controlled reactor (FC-TCR)
- Superior low-voltage performance.

9.4.1 *V–I* Characteristics

The *V–I* characteristics are shown in Figure 9.7. Compared to the characteristics of an SVC, the STATCON is able to provide rated reactive current under reduced voltage conditions. It also has transient overload capacity in both the inductive and capacitive region, the limit being set by the junction temperature of the semiconductors. By contrast, an SVC can only supply diminishing output current with decreasing system voltage.

The ability to produce full capacitive current at low voltages makes it ideally suitable for improving the first swing (transient) stability. The dynamic performance capability far exceeds that of a conventional SVC. It has been shown that the current system can transition from full-rated capacitive to full-rated inductive vars in approximately a *quarter cycle*. The dynamic response is much faster than attainable with variable impedance (TCR, FC-TCR, TSC-TCR—see Chapter 8) compensators.

STATCON, just like an SVC, behaves like an ideal midpoint compensator, see Figure 8.5, until the maximum capacitive output current is reached. The reactive power output of a STATCOM varies linearly with the system voltage, while that of the SVC varies with the square of the voltage. In an SVC, TCRs produce high harmonic content, as the current

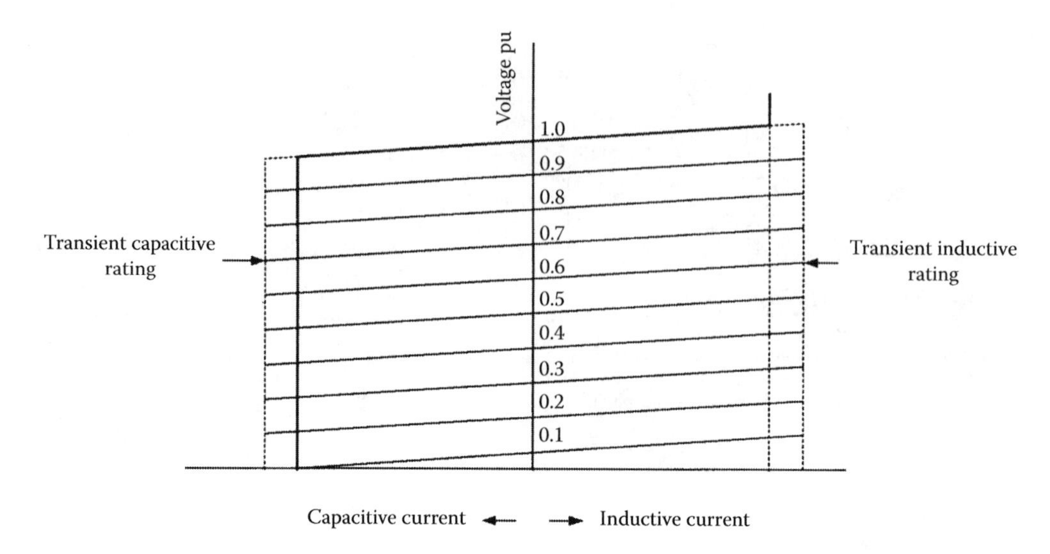

FIGURE 9.7
V–I characteristics of a STATCON.

waveform is chopped off in the phase-controlled rectifiers, and passive filters are required. A STATCON uses phase multiplication or pulse-width modulation and the harmonic generation is a minimum.

The Sullivan STATCOM of TVA is rated ±100 Mvar to provide day-to-day regulation of voltage and another contingency operation under a major disturbance to the power system. The solid-state power circuit combines eight three-phase inverters, each with a nominal rating of 12.5 MVA, into 48-pulse operation. The inverter is coupled to 161 kV line by a single step-down transformer having a wye and delta secondary. The nominal secondary voltage is 5.1 kV line-to-line. The inverter is operated from a DC storage capacitor charged nominally to 6.6 kV.

The loss versus reactive power output current characteristics of a 48-pulse, 100 Mvar voltage source STATCOM is shown in Figure 9.8; the losses include coupling transformer. The losses are switching and snubber losses, which are dependent on the characteristics of semiconducting devices; this figure is valid for GTOs.

9.4.2 Regulation

The STATCOM is not used as an ideal voltage regulator and the terminal voltage is allowed to vary in proportion to the regulating current. The linear operating range can be extended if a regulation slope is allowed. This also enables load sharing between other compensating devices that may be present on the transmission system. Under steady state, the compensator terminal voltage V_T change can be written as

$$\Delta V_T = \Delta V \frac{1}{1 + X/\kappa} \tag{9.3}$$

where X is the system reactance and k is the slope. Figure 9.9a shows the *V–I* characteristics in the operating range of a STATCOM versus Figure 9.9b which shows the same characteristics for an SVC. In contrast to STATCOM, an SVC becomes a fixed capacitor at full output,

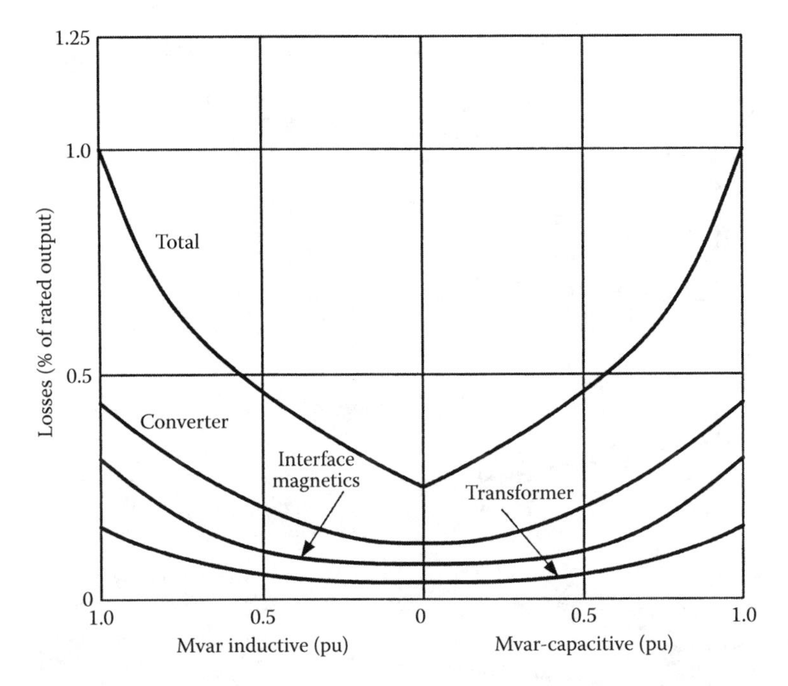

FIGURE 9.8
Loss curves with respect to output.

and its compensating current decreases with AC system voltage, and maximum var output decreases as square of the voltage. The STATCOM is, thus, superior in providing voltage support under large system voltage excursions. As demonstrated in Figure 9.9a, the STATCOM has increased transient rating in both the inductive and capacitive operating regions.

9.4.2.1 Hybrid Connections of STATCOM

The STATCOM can be applied with a fixed capacitor, a fixed reactor, TCR, TSC, or TCR. This is depicted in Figure 9.10. The *V–I* characteristics will vary.

9.5 Static Series Synchronous Compensator

A static series synchronous compensator may also be called a series power flow controller (SPFC) [21, 22, 24]. The basic circuit is that of an SS in *series* with the transmission line (Figure 9.11). We have observed that conventional series compensation can be considered as reactive impedance in series with the line, and the voltage across it is proportional to the line current. A series capacitor increases the transmitted power by a certain percentage, depending on the series compensation for a given δ. In contrast, an SSSC injects a compensating voltage, V_q, in series with the line *irrespective of the line current*. The SSSC increases the maximum power transfer by a fraction of the power transmitted, nearly independent of δ:

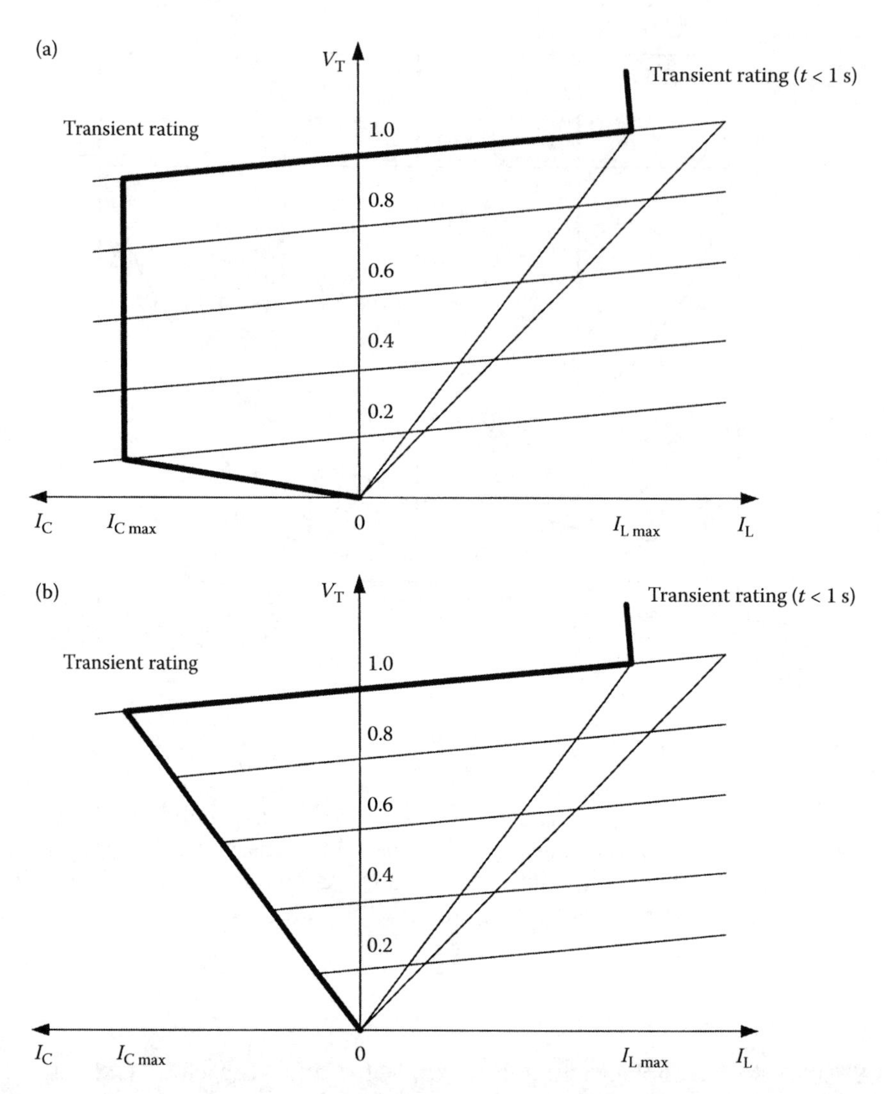

FIGURE 9.9
(a) V–I characteristics of a STATCOM and (b) of an SVC.

$$P_q = \frac{V^2}{X_{sc}} \sin\delta + \frac{V}{X_{sc}} V_q \cos\left(\frac{\delta}{2}\right) \tag{9.4}$$

While a capacitor can only increase the transmitted power, the SSSC can decrease it by simply reversing the polarity of the injected voltage. The reversed voltage adds directly to the reactive power drop in the line and the reactive line impedance is increased. If this reversed polarity voltage is larger than the voltage impressed across the line by sending and receiving end systems, the power flow will reverse:

$$|V_q| = |V_s - V_r + IX_L| \tag{9.5}$$

Thus, stable operation of the system is possible with positive and negative power flows, and due to the response time being less than one cycle, the transition from positive to

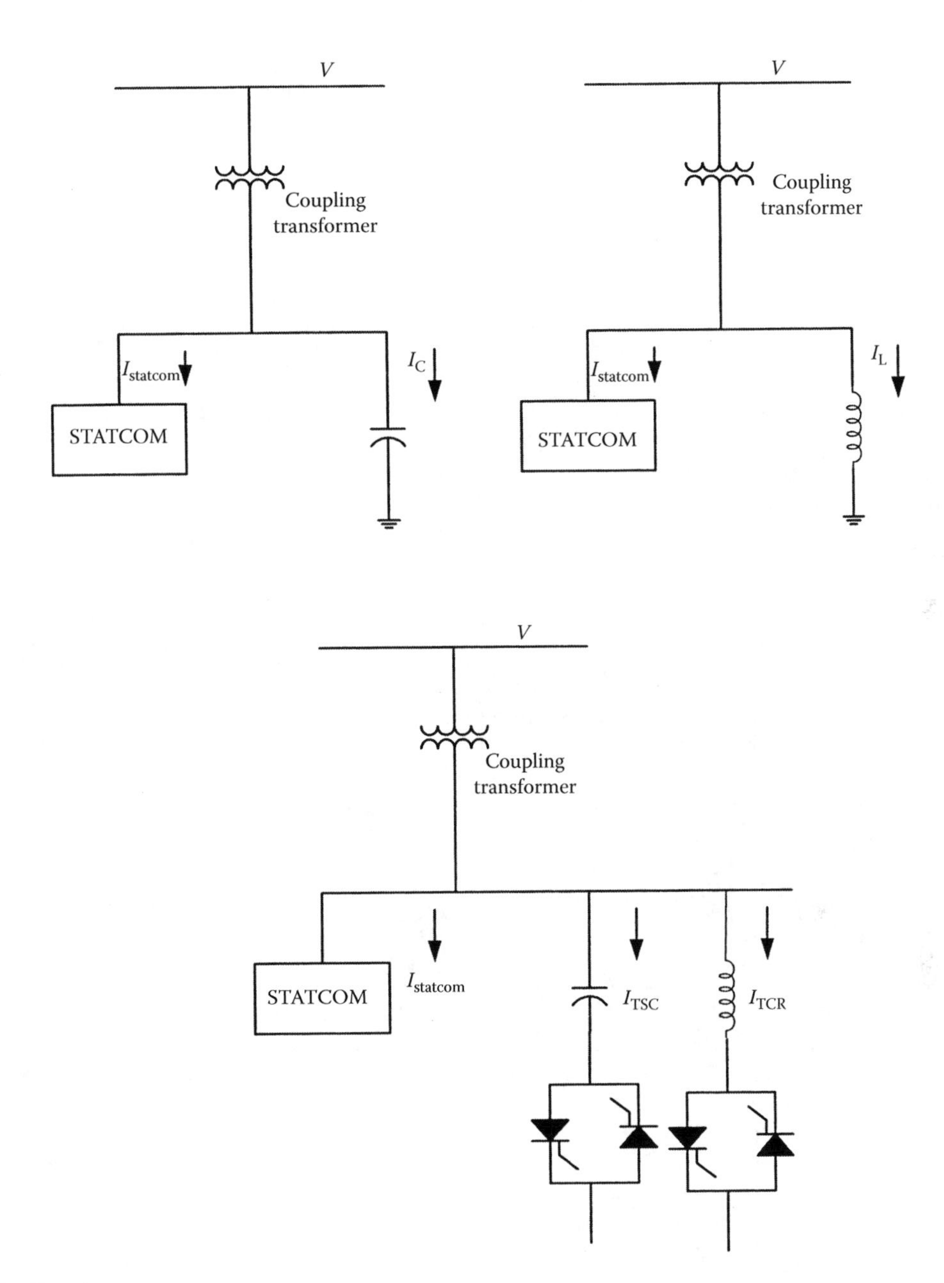

FIGURE 9.10
Hybrid connections of a STATCOM, with shunt capacitors, with shunt reactor, with TCR and TSC.

negative power flow is smooth and continuous. Figure 9.12a shows the P–δ curves of a series capacitor and Figure 9.12b that of an SSSC.

The SSSC can negotiate both reactive and active powers with an AC system, simply by controlling the angular position of the injected voltage with respect to the line current. One important application is simultaneous compensation of both reactive and resistive elements of the line impedance. By applying series compensation, the X/R ratio decreases. As R remains constant, the ratio is $(X_L - X_c)/R$. As a result, the reactive component of the

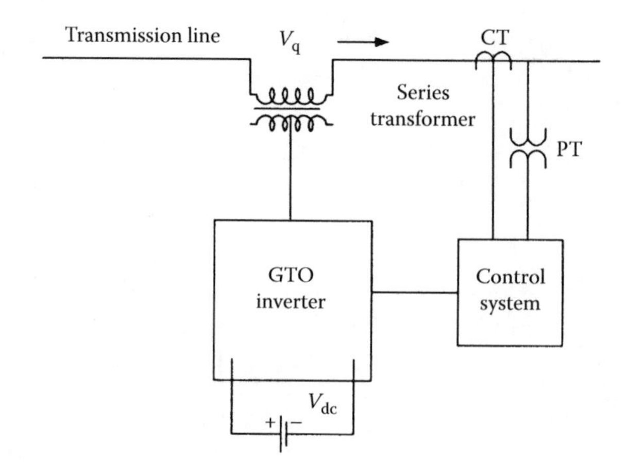

FIGURE 9.11
Series-connected synchronous voltage source, series power flow controller (SPFC), or solid-state series compensator (SSSC).

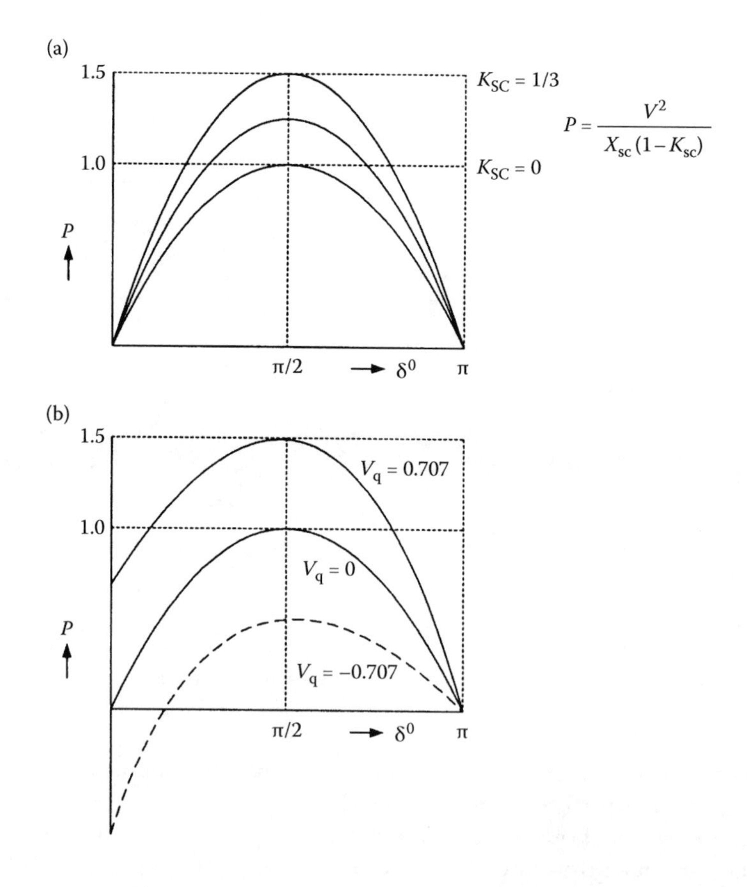

FIGURE 9.12
(a) P–δ characteristics of conventional series capacitive compensation and (b) P–δ characteristics of an SSSC as a function of the compensating voltage V_q.

current supplied by the receiving end progressively increases, while the real component of the current transmitted to the receiving end progressively decreases. An SSSC can inject a component of voltage in antiphase to that developed by the line resistance drop to counteract the effect of resistive voltage drop on the power transmission.

The dynamic stability can be improved, as the reactive line compensation with simultaneous active power exchange can damp power system oscillations. During periods of angular acceleration (increase of torque angle of the machines), an SSSC with suitable energy storage can provide maximum capacitive line compensation to increase the active power transfer and also absorb active power, acting like a damping resistor in series with the line.

The problems of SSR stated for a series capacitor in Section 8.4.6 are avoided. The SSSC is essentially an AC voltage source which operates only at the fundamental frequency and its output impedance at other frequencies can be theoretically zero, though SSSC does have a small inductive impedance of the coupling transformer. An SSSC does not form a series resonant circuit with the line inductance, rather it can damp out the subsynchronous oscillations that may be present due to existing series capacitor installations.

Figure 9.13 illustrates the characteristics of an SSSC. The VA rating is simply the product of maximum line current and the maximum series compensating voltage. Beyond the maximum rated current, the voltage falls to zero. The maximum current rating is practically the maximum steady-state line current. In many practical applications, only capacitive series line compensation is required and an SSSC can be combined with a fixed series capacitor.

If the device is connected to a short line with infinite buses, unity voltages, and constant phase angle difference, the characteristic can be represented by a circle in the P–Q plane with

$$\text{Center} = S_0 Z^*/2R \tag{9.6}$$

$$\text{Radius} = |S_0 Z^*/2R| \tag{9.7}$$

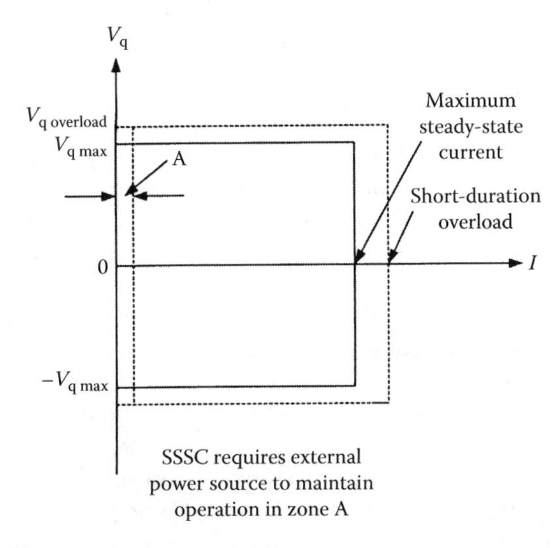

FIGURE 9.13
Range of SSSC voltage versus current, with overload capacity shown in dotted lines.

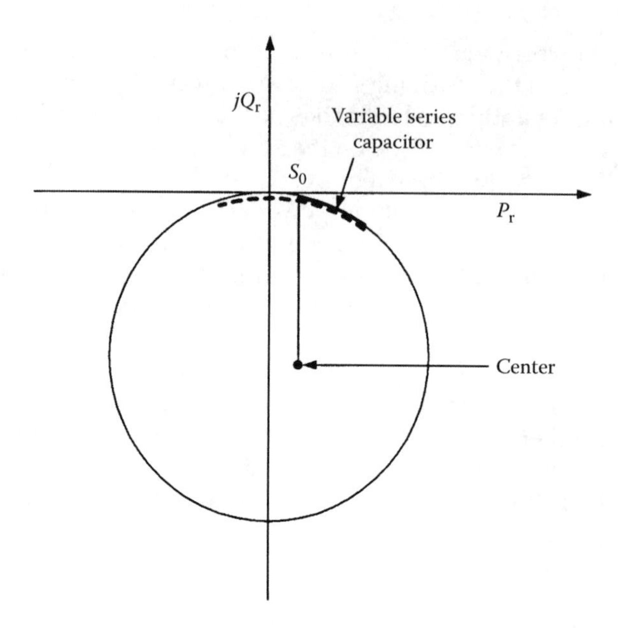

FIGURE 9.14
Operating characteristics of an SSSC in the P–Q plane, showing comparison with variable series capacitor.

where $S_0 = P_0 + jQ_0$ = uncompensated power flow, Z is the series impedance of the line = $R_{sc} + jX_{sc}$, and Z^* is the complex conjugate of Z. The operating characteristics are defined by the *edge* of the circle only. Figure 9.14 shows the power transfer capability of a variable series capacitor with an SPFC controller. The portion of the characteristics to the left of the origin shows power reversal capability.

9.6 Unified Power Flow Controller

A unified power controller consists of two voltage source switching converters, a series and a shunt converter, and a DC link capacitor (Figure 9.15). The arrangement functions as an ideal AC-to-AC power converter in which real power can flow in either direction between AC terminals of the two converters, and each inverter can independently generate or absorb reactive power at its own terminals. Inverter 2 injects an AC voltage V_q with controllable magnitude and phase angle (0–360°) at the power frequency in series with the line. This injected voltage can be considered as an SS.

The real power exchanged at the AC terminal is converted by inverter into DC power which appears as DC link voltage. The reactive power exchanged is generated internally by the inverter.

The basic function of inverter 1 is to absorb the real power demanded by inverter 2 at the DC link. This DC link power is converted back into AC and coupled with the transmission line through a shunt transformer. Inverter 1 can also generate or absorb controllable reactive power. The power transfer characteristics of a short transmission line with a unified

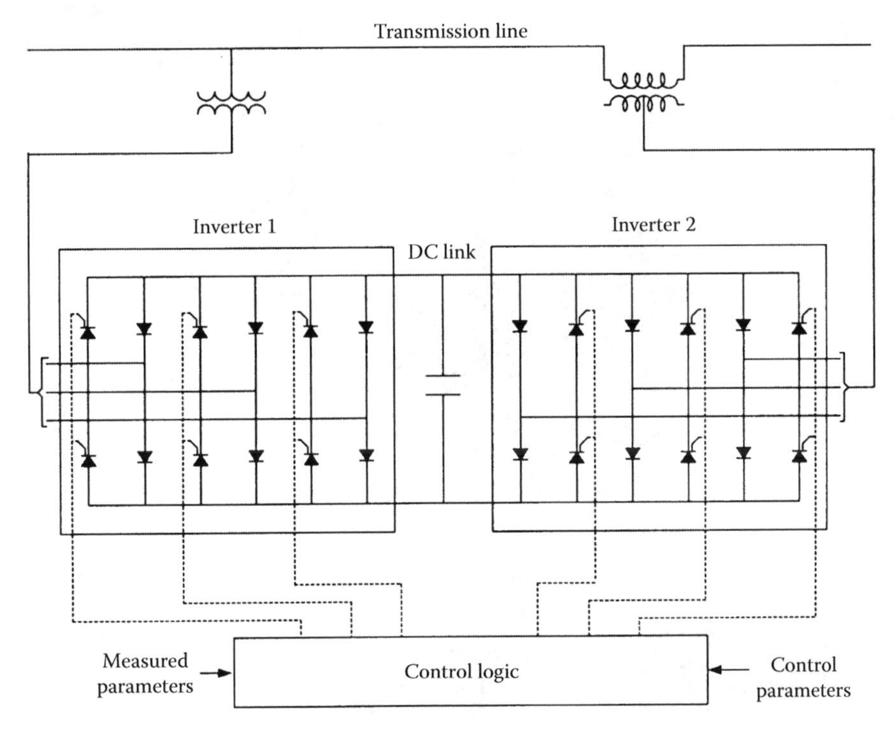

FIGURE 9.15
Schematic diagram of a unified power flow controller (UPFC).

power flow controller (UPFC) connecting two infinite buses, unity voltage, and a constant phase angle, can be represented by a circle on the P–Q plane:

$$(P_R - P_0)^2 + (Q_R - Q_0)^2 = \left(\frac{V_i}{Z}\right)^2 \tag{9.8}$$

where P_0 and Q_0 are the line uncompensated real and reactive power, V_i is the magnitude of the injected voltage, and Z is the line series impedance. The center is at the uncompensated power level S_0 and the radius is V_i/Z. Consider a UPFC with 0.25 per unit (pu) voltage limit. Let the series reactance of the line be 1.0 pu and the uncompensated receiving end power be $1 + j0$ pu. The UPFC can then control the receiving end power within a circle of 2.5 per unit. With its center being at (1, 0), power transfer could be controlled between +3.5 and −1.5 pu.

The allowable operating range with the UPFC is anywhere inside the circle, while the SPFC operating range is the circle itself. The portion of the UPFC circle inside the SPFC circle represents operation of the UPFC with real power transfer from the transmission system to the series inverter to the shunt inverter, while the portion of the UPFC circle outside the SPFC circle represents the transfer of real power from the shunt inverter to the series inverter to the transmission system (Figure 9.16). A comparison with phase angle regulator and series capacitors is superimposed in this figure. These devices provide one-dimensional control, while the UPFC provides simultaneous and independent P and Q control over a wide range.

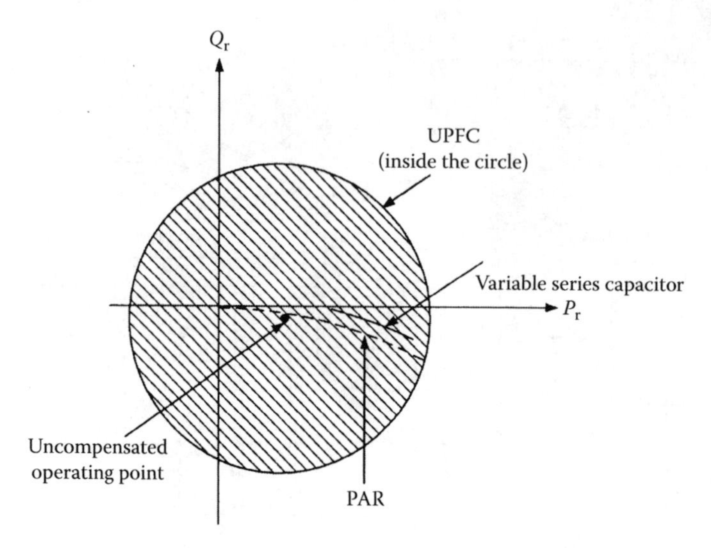

FIGURE 9.16
Comparison of characteristics of UPFC, SPFC, and series capacitor in the *P*–*Q* plane.

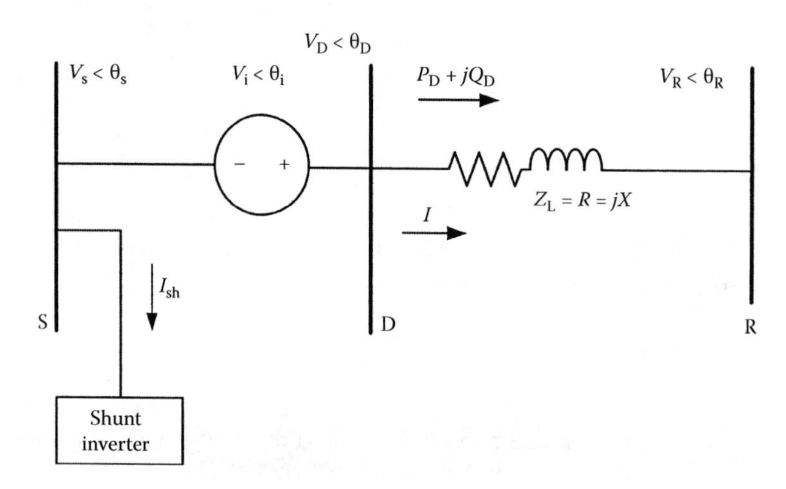

FIGURE 9.17
Equivalent circuit of application of a UPFC to a transmission line.

Figure 9.17 shows a two-bus system with a UPFC controller, represented as a shunt inverter and series voltage source. Referring to the symbols shown in this figure, consider that we want to maximize PD, subject to some constraints:

$$
\begin{aligned}
& V_i \leq L_{Vi} \\
& I \leq L_I \\
& I_{sh} \leq L_{Ish} \\
& LV_{D\min} \leq V_D \leq L_{VD\max} \\
& |P_{dc}| \leq L_{Pdc}
\end{aligned} \tag{9.9}
$$

The letter "L" denotes the specified limits, P_{dc} is the real power transfer between series and shunt inverters, and I_{sh} is the magnitude of the reactive current of the shunt inverter. By eliminating θ_i, the line flow obeys the equation of an ellipse:

$$A(P_D - P_c)^2 + B(P_D - P_c)(Q_D - Q_C) + C(Q_D - Q_C)^2 = D^2 \tag{9.10}$$

where

$$P_C = P_0 + \frac{V_i^2 R}{Z_L^2}, \quad Q_C = Q_0 + \frac{V_i^2 X}{Z_L^2} \tag{9.11}$$

P_0 and Q_0 are given by

$$S_0 = P_0 + jQ_0 = V_s\left(\frac{V_s - V_L}{Z_L}\right)^* \tag{9.12}$$

$$A = 4X^2\left[V_s^2 - V_s V_R \cos(\theta_s - \theta_R)\right] - 4RXV_s V_R \sin(\theta_s - \theta_r) + V_R^2 Z_L^2 \tag{9.13}$$

$$B = -8XR\left[V_s^2 - V_s V_R \cos(\theta_s - \theta_R)\right] + 4V_s V_R \sin(\theta_s - \theta_r)(R^2 - X^2) \tag{9.14}$$

$$C = 4R^2\left[V_s^2 - V_s V_R \cos(\theta_s - \theta_R)\right] + 4RXV_s V_R \sin(\theta_s - \theta_r) + V_R^2 Z_L^2 \tag{9.15}$$

$$D = V_i\left[2V_s V_R \cos(\theta_s - \theta_R) - V_R^2\right] \tag{9.16}$$

These equations will give unconstrained power flow of line with series injected voltage.

Applying Limits

For the limit $I \le L_I$

$$\cos\left(\theta_i - \phi_{Z_L} - \phi_{I_0}\right) \le \frac{1}{2I_0(V_i / Z_L)}\left(L_I^2 - I_0^2 - \frac{V_i^2}{Z_L^2}\right) \tag{9.17}$$

where

$$I_0 = \frac{V_s - V_R}{Z_L} \quad I = \frac{V_s + V_i - V_R}{Z_L} \tag{9.18}$$

ϕ_{Z_L}, ϕ_{I_0} are the phase angles of Z_L and I_0, respectively.

The real power transfer between series and shunt inverter is

$$\begin{aligned} P_i &= \mathrm{Re}\left(V_i I^*\right) \\ &= \frac{1}{Z_L^2}\left(RV_i^2 + EV_i \cos\theta_i + FV_i \sin\theta_i\right) \end{aligned} \tag{9.19}$$

where

$$E = RV_s \cos\theta_s - RV_R \cos\theta_R + XV_s \sin\theta_S - XV_R \sin\theta_R \tag{9.20}$$

$$F = RV_s \sin\theta_s - RV_R \sin\theta_R - XV_s \cos\theta_S + XV_R \cos\theta_R \tag{9.21}$$

The constraint $|P_{dc}| \le L_{PDC}$ is equivalent to the following inequalities:

$$\sin(\theta_i - \alpha) \le \frac{1}{V_i\sqrt{E^2+F^2}}\left[L_{PDC}(R^2+X^2) - RV_i^2\right] \tag{9.22}$$

$$\sin(\theta_i - \alpha) \ge \frac{1}{V_i\sqrt{E^2+F^2}}\left[-L_{PDC}(R^2+X^2) - RV_i^2\right] \tag{9.23}$$

where

$$\sin\alpha = \frac{E}{\sqrt{E^2+F^2}}, \quad \cos\alpha = \frac{F}{\sqrt{E^2+F^2}} \tag{9.24}$$

The constraint $L_{VD\min} \le V_D \le V_{D\max}$ is equivalent to

$$\begin{aligned} \cos(\theta_i - \theta_s) &\le \frac{1}{2V_iV_s}\left(L_{VD\max}^2 - V_i^2 - V_s^2\right) \\ \cos(\theta_i - \theta_s) &\ge \frac{1}{2V_iV_s}\left(L_{VD\min}^2 - V_i^2 - V_s^2\right) \end{aligned} \tag{9.25}$$

Figure 9.18 shows the line flow with constraints; for further reading see Reference [9].

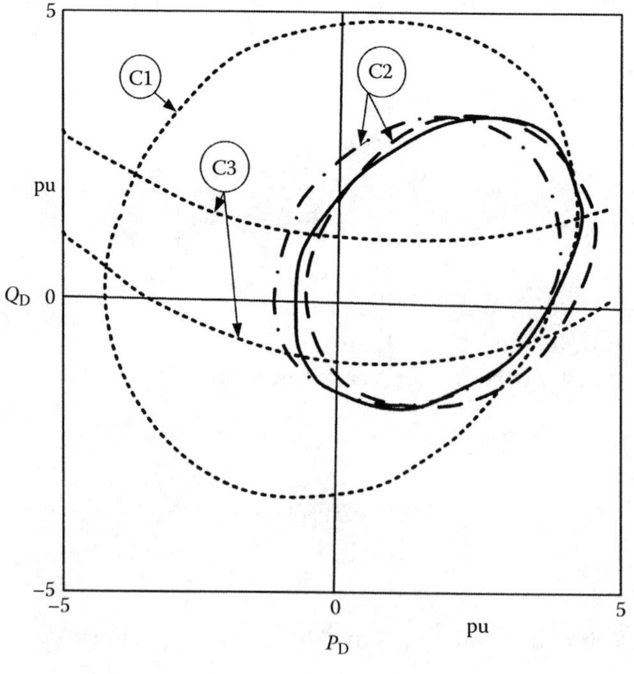

FIGURE 9.18
UPFC power flow characteristics with constraints.

9.7 FACTS for Distribution Systems

When applied to distribution systems, STATCOM is called D-STATCOM. Another analogues term is dynamic voltage restorer (DVR). The DVR is a series connecting device but it should not be confused with UPFC or SSSC described above.

A solid-state circuit breaker and current limiter are described in Volume 1.

A fault on a feeder serving an industrial facility can give rise to voltage sags. The DVR is a custom power device for series connection. It can control the voltage supplied to the load by injecting a voltage of arbitrary magnitude in the line. This allows the correction of the voltage seen by the load to the desired magnitude. The DVR is capable of supplying and absorbing both real and reactive power.

Figure 9.19a shows the connection of a DVR in a line. The injection transformers are actually three single-phase transformers and their primary windings are rated for the full line current. The primary voltage rating is the maximum voltage the DVR can inject into the line. The DVR rating per phase is the voltage multiplied by the line current.

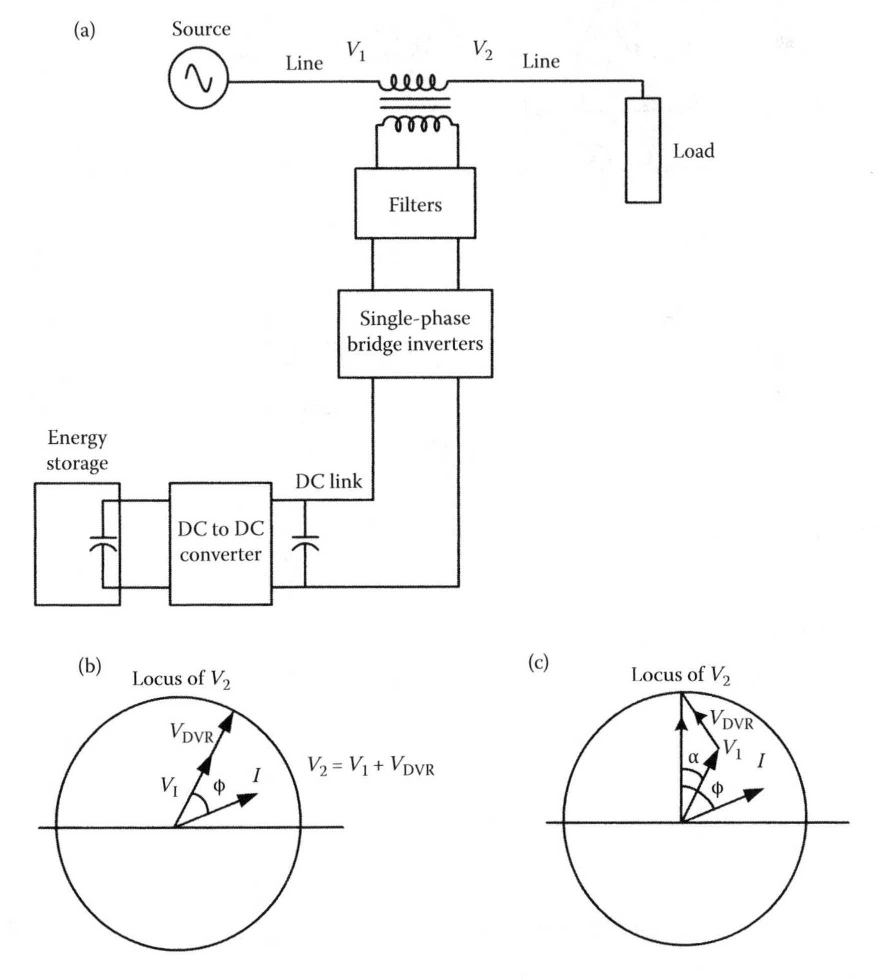

FIGURE 9.19
(a) Connection circuit of a DVR, (b) phasor diagram of voltage correction, and (c) with power factor limits.

Voltage on the secondary of the transformers is controlled by single-phase bridge inverters connected to a common DC link. The bridges are independently controlled to allow their phases to be compensated individually. The output voltage waveform is generated with pulse-width modulated switching, see Volume 3.

For compensating the large-voltage sags, it is necessary to exchange real power with the line. This requires an energy storage device like a capacitor, the size of which is a function of external sages and their duration.

Figure 9.19b shows the phasor diagram of a voltage correction of 0.2 per unit. The power supplied by DVR is

$$P_{DVR} = V_{DVR} I \cos\phi \tag{9.26}$$

where V_{DVR} is the injected voltage, I is the load current, and ϕ is the load power factor angle. This power must come from the stored energy device. This energy could possibly be reduced to almost zero, by using reactive power to compensate the source voltage. It is, generally, not necessary that the compensated voltage is in phase with the source voltage. The locus of the load voltage is a circle, as shown in Figure 9.19c. The angle α for zero real power can be found as follows:

$$\begin{aligned} P_{in} &= V_{1A}\cos(\phi-\alpha)+V_{1B}\cos(\phi-\alpha)+V_{1C}\cos(\phi-\alpha) \\ P_{out} &= V_{2A}\cos(\phi)+V_{2B}\cos(\phi)+V_{2C}\cos(\phi) \end{aligned} \tag{9.27}$$

where V_1 is the source-side phase voltage and V_2 is the load-side phase voltage.

For zero real power

$$\begin{aligned} P_{in} &= P_{out} \\ \alpha &= \phi - \arccos\left(\frac{3V_2\cos(\phi)}{V_{1A}+V_{1B}+V_{1C}}\right) \end{aligned} \tag{9.28}$$

For a balanced system $V_{1A} = V_{1B} = V_{1C}$ and the maximum voltage that can be compensated with zero energy is

$$V_1 = V_2 \cos\phi \tag{9.29}$$

For single-phase sag on phase A

$$V_{1B} = V_{1C} = V_1 \tag{9.30}$$

and the correction limit is

$$V_{1A} = 3V_2\cos\phi - 2V_1 \tag{9.31}$$

For a power factor of 0.8, the limits are 0.4 pu for a single-phase event and 0.8 for a three-phase event, see Figure 9.19c.

9.8 Application of a STATCOM to an Industrial Distribution System

A study of the application of a STATCOM in an industrial distribution system, compressed configuration for load flow shown in Figure 9.20, is conducted. This shows three plant

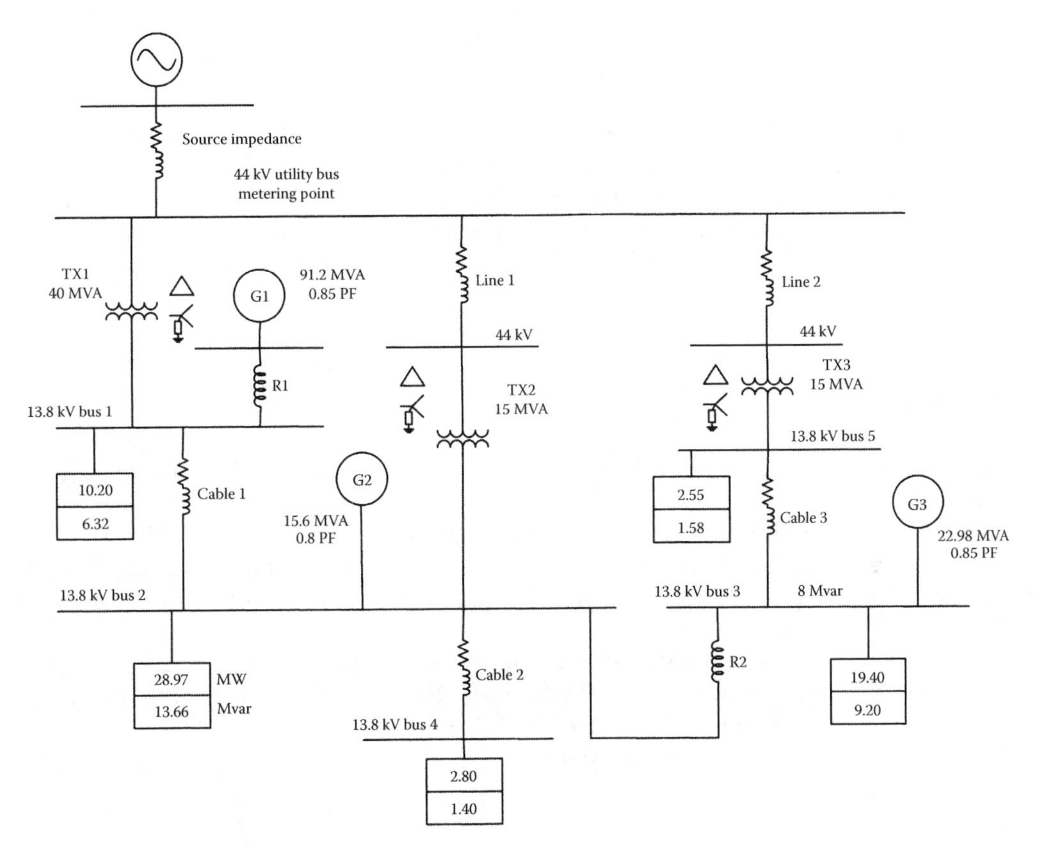

FIGURE 9.20
An industrial distribution system for the study of voltage dips and reactive power compensation.

generators, ratings as shown, three utility tie transformers, and the sum total of loads on each of the buses. Referring to this figure, the following operating conditions are applicable:

- The utility requires that the power factor as seen at the point of interconnection at 44 kV utility bus should remain preferably unity or within a variation of 0.95 lagging to 0.95 leading.
- The utility three-phase short-circuit level at 44 kV is only 151.7 MVA, $X/R = 3.75$. This represents a weak utility system. In fact, the power is received over a considerable length of 44 kV line. The utility source cannot be represented as a swing bus; considerable voltage dip will occur on load flow and contingency load flow.
- All loads are composite loads, 80% constant KVA and balance 20% constant impedance type.
- The contingency load flow conditions must be studied. These include loss of one generator at a time or loss of any two generators at the same time.
- It is required that preferably no load shedding should be resorted to in case of loss of *any two generators* simultaneously, which is a stringent condition.
- Under normal operation, with all generators in service, the excess generated power is supplied into the utility system; i.e., the system operates in cogeneration mode.

The impedance data for the load flow are shown in Table 9.2.

A number of trial load flow runs are made, which are not documented. The following settings are applied:

Generator G1 is constrained to operate at a voltage of 1.02 pu of its rated voltage, while generators G2 and G3 are constrained to operate at their rated voltage of 13.8 kV.

All generators are operated in PV mode, and their reactive power output is constrained within the maximum rated reactive power output and a minimum of 1 Mvar output.

The transformer TX1, TX2, and TX3 off-load taps at 13.8 kV are adjusted as follows:

TX1: Rated taps

TX2: 0.975 (13455 V)

TX3: 0.975 (13455 V).

As we have seen, the generator operating voltages and tap settings impact the reactive power flow.

Case 1: Normal load flow with all generators in service

The summary results for all load flow cases are shown in Tables 9.3 and 9.4. Table 9.3 shows the voltage profiles on major 13.8 kV buses and Table 9.4 shows the power balance.

When all generators are in operation, the voltages on 13.8 kV buses are a maximum 1.46% below rated operating voltage and 31.03 MW and 7.0 Mvar is supplied into the utility system, the power factor of the utility source is 0.97 leading.

Case 2: Add shunt capacitors in the system

TABLE 9.2

Load Flow Impedance Data, Figure 9.20

	Description	Load Flow Data
1	Utility short-circuit level	151.7 MVA, $X/R = 3.75$
2	Model utility short-circuit level As an impedance connected to infinite bus (swing bus)	$3.283 + j12.31\ \Omega$
3	Reactor R1, 2000 A	$0.4\ \Omega$, $X/R = 100$
4	Reactor R2, 1200 A	$0.8\ \Omega$, $X/R = 60$
5	Transformer TX1, 40/67 MVA, 44–13.8 kV, delta–wye connected	$Z = 9\%$, $X/R = 23.70$
6	Transformer TX2, 40/67 MVA, 44–13.8 kV, delta–wye connected	$Z = 7.66\%$, $X/R = 18.60$
7	Transformer TX3, 40/67 MVA, 44–13.8 kV, delta–wye connected	$Z = 7.89\%$, $X/R = 18.60$
8	Cable 1, 500 KCMIL, 133% insulation, MV90, three-per phase, 1000 ft (305 m)	$0.026706 + j0.035052\ \Omega$ per 1000 ft, *per conductor*
9	Cable 2, 500 KCMIL, 133% insulation, MV90, one-per phase, 2000 ft (610 m)	$0.026706 + j0.035052\ \Omega$ per 1000 ft, *per conductor*
10	Cable 3, 500 KCMIL, 133% insulation, MV90, two-per phase, 615 ft (186 m)	$0.026706 + j0.035052\ \Omega$ per 1000 ft, *per conductor*
11	Line 1, 336 KCMIL, ACSR with ground conductor, 1320 ft (402 m)	$0.062463 + j0.100392\ \Omega$ per 1000 ft, $Y = 0.0000014$ S
12	Line 1, 336 KCMIL, ACSR with ground conductor, 1320 ft (402 m)	$0.062463 + j0.100392\ \Omega$ per 1000 ft, $Y = 0.0000014$ S

TABLE 9.3

Load Flow in a Large Distribution System, Voltage Profiles

	Utility Source Metering Point		Bus 1		Bus 2		Bus 3		Bus 4	
Case #	Voltage	VD ±	Voltage	VD ±	Voltage	VD ±	Voltage	VD ±	Voltage	VD ±
1	98.54	−1.46	99.08	−0.92	99.88	−0.12	98.57	−1.43	98.72	−1.28
2	100.49	+0.49	101.59	+1.59	101.13	+1.13	101.23	+1.23	101.23	+1.23
3	97.34	−2.76	99.86	−0.14	99.83	−0.17	100	0.00	99.67	−0.33
4	100.27	+0.27	101.36	+1.36	101.13	+1.13	100.96	+0.96	100.96	+0.96
5	99.96	−0.04	101.13	+1.13	100.13	+0.13	100.23	+0.23	100.74	+0.74
6	99.45	−0.55	100.71	+0.71	100.44	+0.44	99.71	−0.29	100.28	+0.28
7	79.59	−20.41	79.41	−20.59	79.29	−20.71	81.11	−19.89	79.10	−19.99
8	76.26	−23.73	76.47	−23.53	76.38	−23.62	74.52	−25.48	74.19	−24.81
9	95.27	−4.73	99.10	−0.90	99.90	−0.10	96.75	−3.25	99.83	−0.17

Note: The negative sign shows voltage dip and the positive sign shows voltage rise. The voltage at utility source is inconsequential in plant operation.

TABLE 9.4

Generation, Power from or to Utility Source, and System Losses

	G1		G2		G3		Utility Source			System Losses	
Case #	MW	Mvar	MW	Mvar	MW	Mvar	MW	Mvar	PF%	MW	Mvar
1	70	18.59	12	9.40	13.0	9.8	−31.03	−7.0	97.55 Lead	2.04	19.36
2	70	6.26	12	1.0	13	1.0	−30.47	−4.16	99.08 Lead	1.85	17.95
3	0	0	12	9.40	13	6.32	39.40	−1.66	99.93 Lead	3.12	13.44
4	70	7.39	0	0	13	2.0	−18.55	−3.32	98.43 Lead	0.85	13.41
5	70	8.54	12.00	1.00	0	0	−17.62	−3.66	97.91 Lead	0.83	14.52
6	70	10.57	0	0	0	0	−5.74	−2.27	92.99 Lead	0.35	12.15
7	0	0	0	0	13.00	9.80	51.51	14.23	96.3 Lag	6.10	26.27
8	0	0	12.00	9.40	0	0	52.34	16.34	95.45 Lag	6.82	29.20
9	0	0	12.00	9.40	0	0	56.67	1.98	99.93 Lag	5.83	25.43

Note: The losses include the losses in the utility source impedance. The negative sign shows active and reactive powers supplied into the utility system.

The trial runs with one and two generators out of service show that the voltages dip severely, and processes cannot be maintained. Note that load shedding may take 10–15 cycles. The load management system must summate the loads and generation, compare it with set maximum power input from utility source, and shed designated loads by opening the desired circuit breakers. However, it is too long a time for the systems to ride through the voltage dips. The unstabilized motor contactors can haphazardly drop out with first cycle of the voltage dip causing a process shutdown. Similarly, ULTC (under-load tap changing) on transformers is not fast enough to prevent a shutdown.

Therefore, add maximum ratings of shunt capacitor banks at various buses by trial load flow, so that under normal operation, with all the generators and all the shunt capacitors in service, there are no overvoltages in the distribution system.

When this is done, the shunt capacitors supply the reactive power, and compensate for the loss of reactive power due to outage of a generator. As we have seen, the voltages are mainly dependent on the reactive power flow in the distribution system. The following capacitor banks are added:

Bus 1: 15 Mvar

Bus 2: 10 Mvar

Bus 3: 6 Mvar

Thus, a total of 33 Mvar of shunt capacitor banks are added, see Figure 9.21. The load flow with all generators in service and all capacitor banks in service show that the voltage profiles on the buses are a maximum of 1.6% above the rated voltages. It is desirable to operate the system with slightly higher voltages than the rated voltages. *This will be the normal operating mode.* In this case, 30.5 MW is supplied into the utility system at a power factor of 0.99 leading, close to unity. This meets the requirements of utility's stipulation of a power factor close to unity.

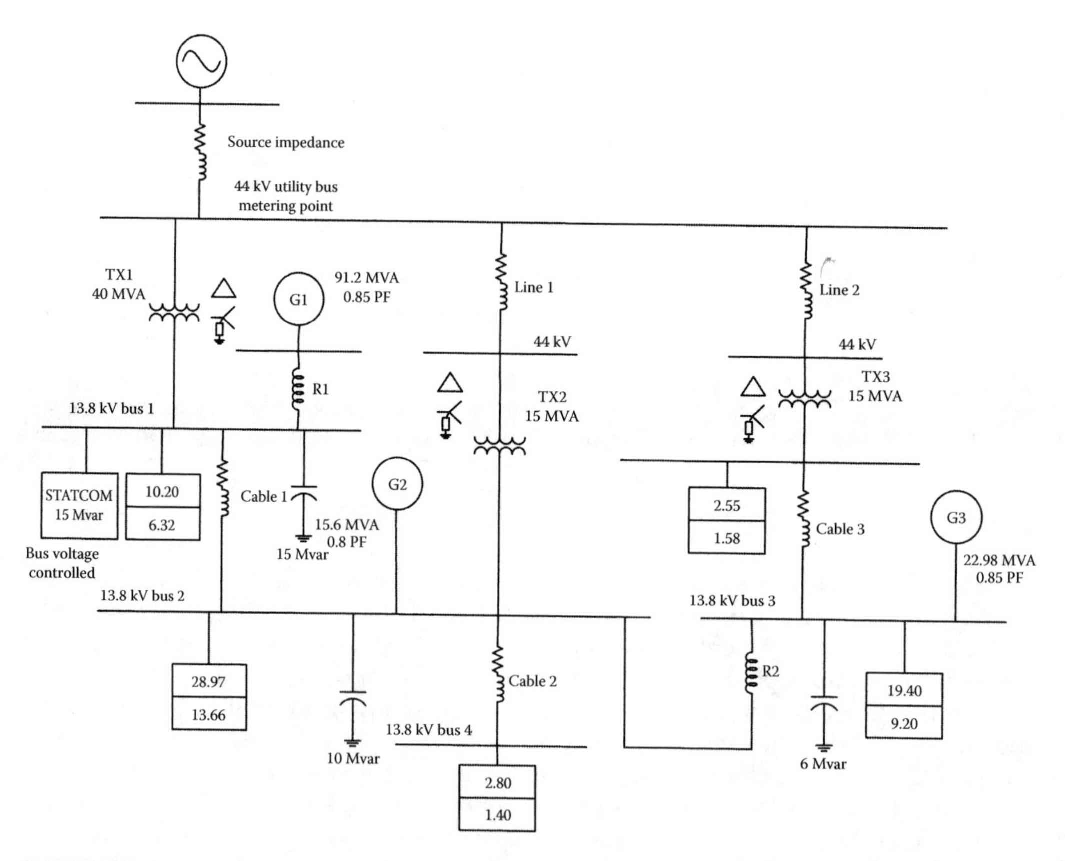

FIGURE 9.21
The distribution system as in Figure 9.20 with additions of shunt capacitor banks and STATCOM.

Case 3: Generator G1 is out of service; generators 2 and 3 are in service and also all shunt capacitor banks

Case 4: Generator G2 is out of service; generators 2 and 3 are in service and also all shunt capacitor banks

Case 5: Generator G3 is out of service; generators 1 and 2 are in service and also all shunt capacitor banks

Cases 3, 4, and 5 simulate outage of one generator at a time. The results of the calculations are shown in Tables 9.3 and 9.4. It is seen that almost rated voltages are maintained at 13.8 kV buses, and the power factor of the supply from the utility varies from 92.99 to 99.99 leading. Thus, the operation with outage of one generator at a time is acceptable, and no loads need to be shed.

Case 6: Generators G2 and G3 are out of service; generator 1 is in service and also all shunt capacitor banks

The voltages on 13.8 kV buses are close to the rated voltages. The power factor of the power supplied into the utility system is 93% leading.

Case 7: Generators G1 and G2 are out of service; generator 3 is in service and also all shunt capacitor banks

Voltage dip of the order of 21% occurs on 13.8 kV buses; the operation cannot be sustained.

Case 8: Generators G1 and G3 are out of service; generator 2 is in service and also all shunt capacitor banks

Voltage dip of the order of 25% occurs on 13.8 kV buses; the operation cannot be sustained.

Thus, the maximum voltage dip occurs when generators G1 and G3 simultaneously go out of service. The voltage dip of 25% will usher a widespread process shutdown. Though this may be a rare occurrence that two in-plant generators will go out of service simultaneously, these machines are old and need additional maintenance. An outage of a running unit can occur, while the other is under maintenance. The alternatives of additional reactive power compensation are discussed in Chapter 8 and summarized in Table 8.1.

A STATCOM is a better choice because of faster response compared to SVCs and much less harmonic pollution. The harmonic pollution can be a major concern when the system has harmonic producing loads, like ASDs. Though a STATCOM can supply both capacitive and inductive power, in this application, the capability to supply inductive reactive power or real power exchange is not required. Only capacitive reactive power needs to be supplied.

The results show that the voltages on all buses are close to the rated voltage, except bus 3, which experiences a voltage dip of 3.25%. The power factor of the supply from the utility source is 99.3 lagging.

9.8.1 Dynamic Simulation

Figure 9.22 shows the dynamic simulation response of the STATCOM, for the maximum voltage dip, when generators G1 and G3 simultaneously go out of service. These show that the reactive power output of the STATCOM floating on bus 1 increases to approximately 15 Mvar in about 1/2 cycle, bringing up the bus voltage. The control circuit of the STATCOM and dynamic model of generator G2, which remain in service, are used for this simulation. See Reference [26].

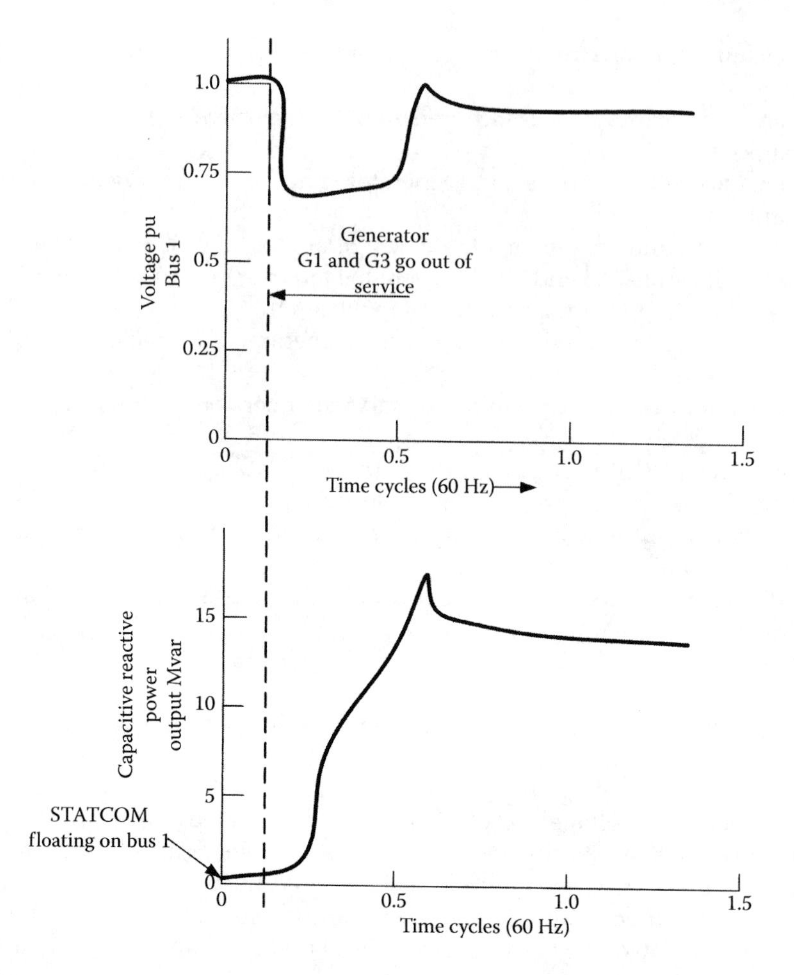

FIGURE 9.22
Dynamic response characteristics of STATCOM on simultaneous loss of generators 1 and 3, voltage, and reactive power output profile.

Problems

9.1 State 5 points of superior performance of STATCOM compared to a TCR.

9.2 A generator and utility source are connected to the same bus and operate in synchronism. The generator has a direct connection with the bus, while the utility source is connected through a transformer. What impact will the transformer tap settings have on the reactive power output of the generator?

References

1. JC Das. *Industrial Power Systems: In Encyclopedia of Electrical and Electronic Engineering*, vol. 10, John Wiley, New York, 2002.

2. IEEE Standard 446. *Recommended Practice for Emergency and Standby Power Systems for Industrial and Commercial Applications*, New York, 1995.
3. NG Hingorani. Flexible AC transmission. *IEEE Spectrum*, 30, 40–45, 1993.
4. C Schauder, M Gernhard, E Stacey, T Lemak, L Gyugi, TW Cease, A Edris. Development of a 100 Mvar static condenser for voltage control of transmission systems. *IEEE Trans Power Delivery*, 10, 1486–1993, 1995.
5. RJ Nelson, J Bain, SL Williams. Transmission series power flow control. *IEEE Trans Power Delivery*, 10(1), 504, 1995.
6. NG Hingorani, L Gyugyi. *Understanding FACTS*, IEEE Press, New York, 2000.
7. IA Erinmez, Ed. Static var compensators, CIGRE Working Group 38-01, Task Force No. 2, 1986.
8. KK Sen. STATCOM: Theory modeling and applications. *Proceedings of IEEE/PES Winter Meeting* (Paper: 99WM706), New York, 1999.
9. MA Kamarposshti, M Alinezhad. Comparison of SVC and STATCOM in static voltage stability margin enhancement. *World Acad Sci Eng Technol*, 3, 722–727, 2009.
10. R Doncker. High power semiconductor development for FACTS and custom power applications. *EPRI Conference on Future Power Delivery*, Washington, DC, April 1996.
11. P Kessel, R Glavitsch. Estimating the voltage stability of a power system. *IEEE Trans*, PWRD-1(3), 346–354, 1986.
12. MH Baker. An assessment of FACTS controllers for transmission system enhancement. *CIGRE-SC14, International Colloquium on HVDC and FACTS*, Montréal, September 1995.
13. TM Kay, PW Sauer, RD Shultz, RA Smith. EHV and UHV line loadability dependence on var supply capability. *IEEE Trans*, PAS 101(9), 3568–3575, 1982.
14. IEEE Committee Report. *VAR* management—Problem recognition and control. *IEEE Trans Power App Syst*, PAS-103, 2108–2116, 1984.
15. CA Canizares. Analysis of SVC and TCSC in voltage collapse. *Conference Record, IEEE PES 1998 Summer Meeting*, San Diego, CA, July 1998.
16. JB Ekanayake, N Jenkins. A three level advanced static var compensator. *IEEE Trans Power Delivery*, 10(2), 1996.
17. CJ Hatziadoniu, FE Chalkiadakis. A 12-pulse static synchronous compensator for the distribution system employing the three level GTO inverter. *Conference Record, IEEE 1997 PES Meeting*, Paper No. PE-542-PWRD-0-01, 1997.
18. L Gyugyi. Principles and applications of static, TC shunt compensators. *IEEE Trans Power App Syst*, PAS-97(5), 1935–1945, 1978.
19. H Akagi, H Fujita, S Yonetani, Y Kondo. A 6.6 kV transformer-less STATCOM based on a five level diode-clamped PWM converter: System design and experimentation of 200-V, 10 kVA laboratory model. *IEEE Trans Ind Appl*, 44(2), 672–680, 2008.
20. BL Agarwal et al. Advanced series compensation, steady state, transient stability, and subsynchronous resonance studies. *Proceedings of Flexible AC Transmission Systems Conference*, Boston, MA, May 1992.
21. L Gyugyi et al. Static synchronous series compensator: A solid state approach to the series compensation of the transmission lines. *IEEE Trans Power Delivery*, 12(1), 406–417, 1997.
22. S Mori, K Matsuno. Development of large static var generator using self commutated inverters for improving power system stability. *IEEE Trans Plasma Sci*, 8(1), 371–377, 1993.
23. RJ Nelson, J Bain, SL Williams. Transmission series power flow control. *IEEE Trans Power Delivery*, 10(1), 504–510, 1995.
24. H Akagi, Y Kanazawa, A Nabae. Generalized theory of instantaneous reactive power in three-phase circuits. *Proceedings of the International Power Electronics Conference*, Tokyo, pp. 1375–1386.
25. H Akagi, EH Watanbe, M Aredes. *Instantaneous Power Theory and Applications to Power Conditioning*, Wiley Intersciences, Piscataway, NJ, 2007.
26. JC Das. Application of a STATCOM to an industrial distribution system connected to a weak utility system. *IEEE Trans. Industry Application*, 5(2), 5345–5354, 2016.

10

Three-Phase and Distribution System Load Flow

Normally, three-phase systems can be considered as balanced. Though some unbalance may exist, due to asymmetry in transmission lines, machine impedances, and system voltages, yet these are often small and may be neglected. A single-phase positive sequence network of a three-phase system is adequate for balanced systems. The unbalances cannot be ignored in every case, i.e., a distribution system may serve considerable single-phase loads. In such cases, three-phase models are required. A three-phase network can be represented in both impedance and admittance forms. The matrix methods for network solution, the primitive network, formation of loop and bus impedance and admittance matrices, and transformations are discussed in Volume 1.

For a three-phase network, the power flow equations can be written as follows:

$$\begin{vmatrix} V_{pq}^{a} \\ V_{pq}^{b} \\ V_{pq}^{c} \end{vmatrix} + \begin{vmatrix} e_{pq}^{a} \\ e_{pq}^{b} \\ e_{pq}^{c} \end{vmatrix} = \begin{vmatrix} Z_{pq}^{aa} & Z_{pq}^{ab} & Z_{pq}^{ac} \\ Z_{pq}^{ba} & Z_{pq}^{bb} & Z_{pq}^{bc} \\ Z_{pq}^{ca} & Z_{pq}^{cb} & Z_{pq}^{cc} \end{vmatrix} \begin{vmatrix} i_{pq}^{a} \\ i_{pq}^{b} \\ i_{pq}^{c} \end{vmatrix} \tag{10.1}$$

The equivalent three-phase circuit is shown in Figure 10.1a, and its single line representation in Figure 10.1b. In the condensed form, Equation 10.1 is

$$V_{pq}^{abc} + e_{pq}^{abc} = Z_{pq}^{abc} i_{pq}^{abc} \tag{10.2}$$

Similarly, for a three-phase system, in the admittance form

$$\begin{vmatrix} i_{pq}^{a} \\ i_{pq}^{b} \\ i_{pq}^{c} \end{vmatrix} + \begin{vmatrix} j_{pq}^{a} \\ j_{pq}^{b} \\ j_{pq}^{c} \end{vmatrix} = \begin{vmatrix} y_{pq}^{aa} & y_{pq}^{ab} & y_{pq}^{ac} \\ y_{pq}^{ba} & y_{pq}^{bb} & y_{pq}^{bc} \\ y_{pq}^{ca} & y_{pq}^{cb} & y_{pq}^{cc} \end{vmatrix} \begin{vmatrix} V_{pq}^{a} \\ V_{pq}^{b} \\ V_{pq}^{c} \end{vmatrix} \tag{10.3}$$

In the condensed form, we can write

$$i_{pq}^{abc} + j_{pq}^{abc} = y_{pq}^{abc} V_{pq}^{abc} \tag{10.4}$$

A three-phase load flow study is handled much like a single-phase load flow. Each voltage, current, and power becomes a three-element vector and each single-phase admittance element is replaced by a 3×3 admittance matrix.

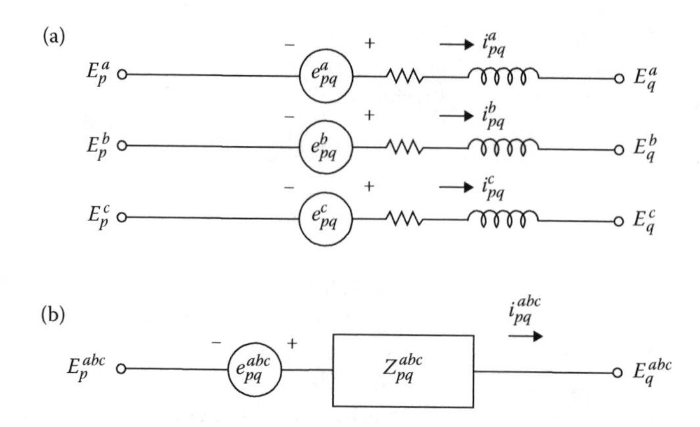

FIGURE 10.1
(a) Three-phase network representation, primitive impedance matrix and (b) single line representation of a three-phase network.

10.1 Phase Coordinate Method

The concepts of symmetrical components are discussed in Volume 1. A symmetrical system, when transformed by symmetrical component transformation, is decoupled and there is an advantage in arriving at a solution using this method. The assumptions of a symmetrical system are not valid when the system is unbalanced. Untransposed transmission lines, large single-phase traction loads, and bundled conductors are some examples. Unbalanced currents and voltages can give rise to serious problems in the power system, i.e., negative sequence currents have a derating effect on generators and motors and ground currents can increase the coupling between transmission line conductors. Where the systems are initially coupled (Example A.1, Volume 1), then even after symmetrical component transformation, the equations remain coupled. The method of symmetrical components does not provide an advantage in arriving at a solution. By representing the system in phase coordinates, i.e., phase voltages, currents, impedances, or admittances, the initial physical identity of the system is maintained. Using the system in the phase frame of reference, a generalized analysis of the power system network can be developed for unbalance, i.e., short-circuit or load flow conditions [1–3]. The method uses a nodal Y admittance matrix and, due to its sparsity, optimal ordering techniques are possible. Series and shunt faults and multiple unbalanced faults can be analyzed. The disadvantage is that it takes more iterations to arrive at a solution.

Transmission lines, synchronous machines, induction motors, and transformers are represented in greater detail. The solution technique can be described in the following steps:

- The system is represented in phase frame of reference.
- The nodal admittance matrix is assembled and modified for any changes in the system, see Volume 1 for the formation of nodal impedance method.
- The nodal equations are formed for the solution.

The nodal admittance equation is the same as for a single-phase system:

$$\bar{Y}\bar{V} = \bar{I} \tag{10.5}$$

Each node is replaced by three equivalent separate nodes. Each voltage and current is replaced by phase-to-ground voltages and three-phase currents; I and V are the column vectors of nodal phase currents and voltages, respectively. Each element of $\bar{Y}$ is replaced with a 3×3 nodal admittance submatrix. Active sources such as synchronous machines can be modeled with a voltage source in series with passive elements. Similarly, transformers, transmission lines, and loads are represented on a three-phase basis. The system base Y matrix is modified for the conditions under study, e.g., a series fault on opening a conductor can be simulated by Y-matrix modification. The shunt faults, i.e., single phase-to-ground, three phase-to-ground, two phase-to-ground, and their combinations, can be analyzed by the principle of superimposition.

Consider a phase-to-ground fault at node k in a power system. It is equivalent to setting up a voltage V_f at k equal in magnitude but opposite in sign to the prefault voltage of the node k. The only change in the power system that occurs due to fault may be visualized as the application of a fault voltage V_f at k and the point of zero potential. If the effect of V_f is superimposed upon the prefault state, the fault state can be analyzed. To account for the effect of V_f, all emf sources are replaced by their internal admittances and converted into equivalent admittance based on the prefault nodal voltage. Then, from Equation 10.5

$$\begin{aligned} &I_i = 0, \quad \text{and} \quad V_k = \text{prefault voltage} \\ &i = 1,2,\ldots,N, \quad i \neq k \end{aligned} \tag{10.6}$$

The fault current is

$$i_k = \sum_{i=1}^{i=N} Y_{ki} E_i \tag{10.7}$$

where E_i is the net *postfault* voltage.

For two single line-to-ground faults occurring at two different nodes p and q:

$$\begin{aligned} &I_i = 0, \quad i = 1,2,\ldots,N \\ &i \neq p,q \end{aligned} \tag{10.8}$$

where V_p and V_q are equal to prefault values. Nodes p and q may represent any phase at any busbar. The currents I_p and I_q are calculated from

$$I_k = \sum_{i=1}^{N} Y_{ki} E_i \quad k = p,q \tag{10.9}$$

Thus, calculation of multiple unbalanced faults is as easy as a single line-to-ground fault, which is not the case with the symmetrical component method.

10.2 Three-Phase Models

Three-phase models of cables and conductors are examined in Appendix A. Y matrices of three-phase models are examined in this chapter, mainly for use in the factored Gauss–Y admittance method of load flow.

10.2.1 Conductors

A three-phase conductor with mutual coupling between phases and ground wires has an equivalent representation shown in Figure 10.2a and b, and the following equations are then written for a line segment:

$$\begin{vmatrix} V_a - V_a' \\ V_b - V_b' \\ V_c - V_c' \end{vmatrix} = \begin{vmatrix} Z_{aa'-g} & Z_{ab'-g} & Z_{ac'-g} \\ Z_{ba'-g} & Z_{bb'-g} & Z_{bc'-g} \\ Z_{ca'-g} & Z_{cb'-g} & Z_{cc'-g} \end{vmatrix} \begin{vmatrix} I_a \\ I_b \\ I_c \end{vmatrix} \tag{10.10}$$

See also Equation A.22 in Appendix A, Volume 1.

In the admittance form, Equation 10.10 can be written as follows:

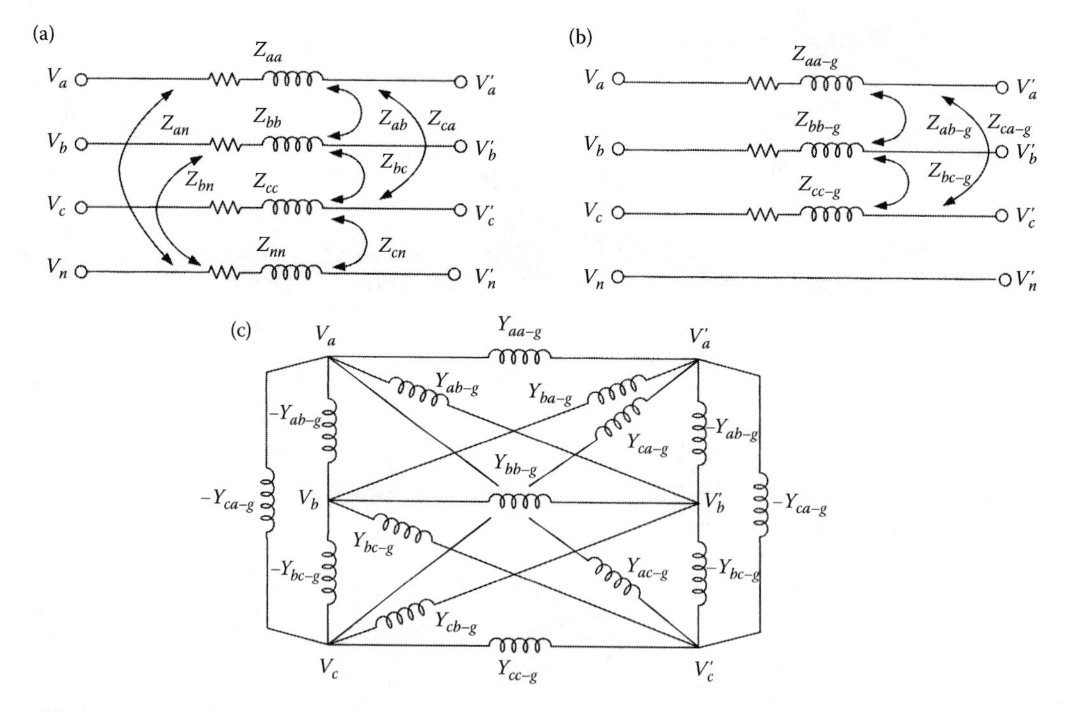

FIGURE 10.2
(a) Mutual couplings between a line section with ground wire in the impedance form, (b) transformed network in the impedance form, and (c) equivalent admittance network of a series line section.

$$\begin{vmatrix} I_a \\ I_b \\ I_c \end{vmatrix} = \begin{vmatrix} Y_{aa-g} & Y_{ab-g} & Y_{ac-g} \\ Y_{ba-g} & Y_{bb-g} & Y_{bc-g} \\ Y_{ca-g} & Y_{cb-g} & Y_{cc-g} \end{vmatrix} \begin{vmatrix} V_a - V_a' \\ V_b - V_b' \\ V_c - V_c' \end{vmatrix} \tag{10.11}$$

Equation 10.11 can be rearranged as follows:

$$\begin{aligned} I_a &= Y_{aa-g}(V_a - V_a') + Y_{ab-g}(V_b - V_b') + Y_{ac-g}(V_c - V_c') \\ &= Y_{aa-g}(V_a - V_a') + Y_{ab-g}(V_a - V_b') + Y_{ac-g}(V_a - V_c') \\ &\quad - Y_{ab-g}(V_a - V_b) + Y_{ac-g}(V_a - V_c) \\ I_b &= Y_{bb-g}(V_b - V_b') + Y_{ba-g}(V_b - V_a') + Y_{bc-g}(V_b - V_c') \\ &\quad - Y_{ba-g}(V_b - V_a) + Y_{bc-g}(V_b - V_c) \\ I_c &= Y_{cc-g}(V_c - V_c') + Y_{cb-g}(V_c - V_b') + Y_{ca-g}(V_c - V_a') \\ &\quad - Y_{cb-g}(V_c - V_b) + Y_{ca-g}(V_c - V_a) \end{aligned} \tag{10.12}$$

The same procedure can be applied to nodes V_a', V_b', and V_c'. This gives the equivalent series circuit of the line section as shown in Figure 10.2c. The effect of coupling is included in this diagram. Therefore, in the nodal frame, we can write the three-phase Π model of a line as follows:

$$\begin{vmatrix} I_a \\ I_b \\ I_c \\ I_a' \\ I_b' \\ I_c' \end{vmatrix} = \begin{vmatrix} Y^{abc} + \frac{1}{2}Y_{sh} & -Y^{abc} \\ -Y^{abc} & Y^{abc} + \frac{1}{2}Y_{sh} \end{vmatrix} \begin{vmatrix} V_a \\ V_b \\ V_c \\ V_a' \\ V_b' \\ V_c' \end{vmatrix} \tag{10.13}$$

where

$$\bar{Y}^{abc} = \bar{Z}^{-1,abc} \tag{10.14}$$

There is a similarity between the three-phase and single-phase admittance matrix, each element being replaced by a 3×3 matrix.

The shunt capacitance (line charging) can also be represented by current injection. Figure 10.3a shows the capacitances of a feeder circuit and Figure 10.3b shows the current injection. The charging currents are

$$\begin{aligned} I_a &= -\frac{1}{2}[Y_{ab} + Y_{ac} + Y_{an}]V_a + \frac{Y_{ab}}{2}V_b + \frac{Y_{ac}}{2}V_c \\ I_b &= -\frac{1}{2}[Y_{ab} + Y_{ac} + Y_{an}]V_b + \frac{Y_{ab}}{2}V_a + \frac{Y_{bc}}{2}V_c \\ I_c &= -\frac{1}{2}[Y_{ab} + Y_{ac} + Y_{an}]V_c + \frac{Y_{ac}}{2}V_a + \frac{Y_{bc}}{2}V_b \end{aligned} \tag{10.15}$$

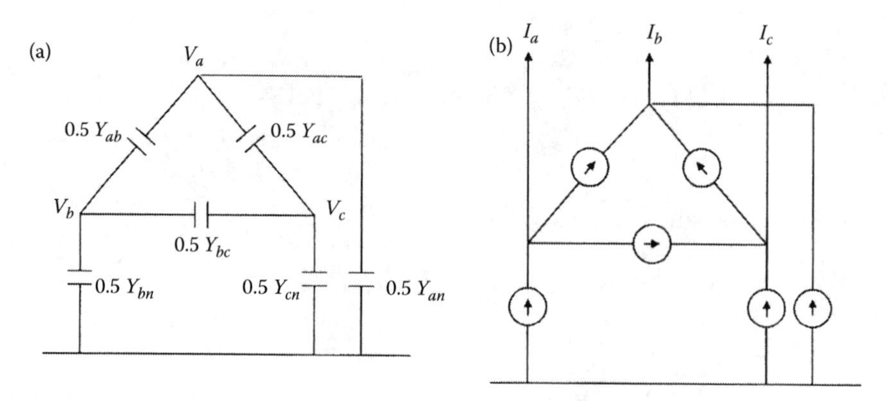

FIGURE 10.3
(a) Capacitances in a three-phase circuit and (b) equivalent current injections.

10.2.2 Generators

The generators can be modeled by an internal voltage behind the generator transient reactance. This model is different from the power flow model of a generator, which is specified with a power output and bus voltage magnitude.

The positive, negative, and zero sequence admittances of a generator are well identified. The zero sequence admittance is

$$Y_0 = \frac{1}{R_0 + jX_0 + 3(R_g + jX_g)} \tag{10.16}$$

where R_0 and X_0 are the generator zero sequence resistance and reactance and R_g and X_g are the resistance and reactance added in the neutral grounding circuit; R_g and X_g are zero for a solidly grounded generator. Similarly,

$$Y_1 = \frac{1}{X'_d} \tag{10.17}$$

$$Y_2 = \frac{1}{X_2} \tag{10.18}$$

where X'_d is the generator direct axis transient reactance and X_2 is the generator negative sequence reactance (resistances ignored). These sequence quantities can be related to the phase quantities as follows:

$$\bar{Y}^{abc} = \begin{vmatrix} Y_{11} & Y_{12} & Y_{13} \\ Y_{21} & Y_{22} & Y_{23} \\ Y_{31} & Y_{32} & Y_{33} \end{vmatrix}$$

$$= \frac{1}{3}\bar{T}_s \bar{Y}^{012} \bar{T}_s^t = \begin{vmatrix} Y_0 + Y_1 + Y_2 & Y_0 + aY_1 + a^2Y_2 & Y_0 + a^2Y_1 + aY_2 \\ Y_0 + a^2Y_1 + aY_2 & Y_0 + Y_1 + Y_2 & Y_0 + aY_1 + a^2Y_2 \\ Y_0 + aY_1 + a^2Y_2 & Y_0 + a^2Y_1 + aY_2 & Y_0 + Y_1 + Y_2 \end{vmatrix} \tag{10.19}$$

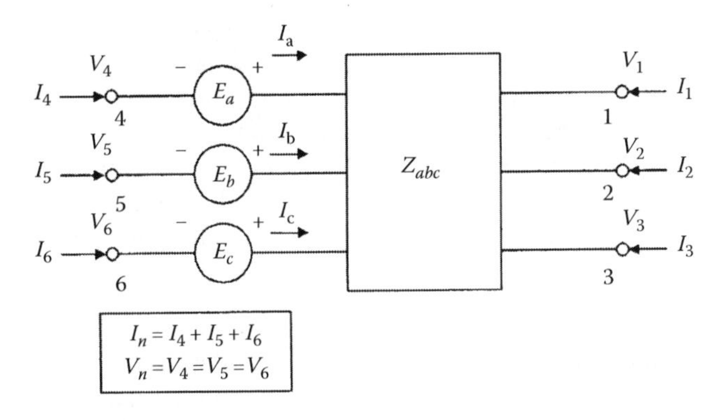

FIGURE 10.4
Norton equivalent circuit of a generator.

where a is the vector operator $1<120°$.

The machine model suitable for unbalance loading and neutral current flow is written as follows:

$$\begin{vmatrix} I_1 \\ I_2 \\ I_3 \\ S^*/E_1^* \\ I_n \end{vmatrix} = \begin{vmatrix} Y_{11} & Y_{12} & Y_{13} & -Y_1 & -Y_0 \\ Y_{21} & Y_{22} & Y_{23} & -a^2Y_1 & Y_0 \\ Y_{31} & Y_{32} & Y_{33} & -aY_1 & -Y_0 \\ -Y_1 & -aY_1 & -a^2Y_1 & 3Y_1 & 0 \\ -Y_0 & -Y_0 & -Y_0 & 0 & 3Y_0 \end{vmatrix} \begin{vmatrix} V_1 \\ V_2 \\ V_3 \\ E_1 \\ V_n \end{vmatrix} \tag{10.20}$$

Referring to Figure 10.4, we can write

$$I_a = S_1^*/E_1^* \quad I_b = S_2^*/E_2^* \quad I_c = S_c^*/E_3^* \tag{10.21}$$

$$I_a = S_1^*/E_1^* \quad I_b = S_2^*/aE_1^* \quad I_c = S_c^*/a^2E_1^* \tag{10.22}$$

$$I_a + I_b + I_c = \frac{S^*}{E_1^*} = \frac{S_1^* + S_2^* + S_3^*}{E_1^*} \tag{10.23}$$

where S_1, S_2, and S_3 are the individual phase powers, S is the total power, and E_1 is the positive sequence voltage behind the transient reactance. For a solidly grounded system, the neutral voltage is zero. The internal machine voltages E_1, E_2, and E_3 are balanced; however, the terminal voltages V_1, V_2, and V_3 depend on internal machine impedances and unbalance in machine currents, I_a, I_b, and I_c. Because of unbalance each phase power is not equal to one-third of the total power. I_1, I_2, and I_3 are the injected currents and I_n is the neutral current. Equation 10.20 can model unbalances in the machine inductances and external circuit [2].

10.2.3 Generator Model for Cogeneration

The cogenerators in distribution system load flow are not modeled as PV type machines, i.e., to control the bus voltage. They are controlled to maintain a constant power and power factor, and power factor controllers may be required. Thus, for load flow, the synchronous generators can be modeled as constant complex power devices. The induction generators require reactive power that will vary with terminal voltage. Assuming a voltage close to rated voltage these can also be modeled as *P–Q* devices. Figure 10.5a shows the Norton equivalent of the generator and Figure 10.5b shows the load flow calculation procedure [4]. The generator is represented by three injected currents. For the short-circuit calculations, the generator model is the same, except that I_1 is kept constant, as the internal voltage of the generator can be assumed not to change immediately after the fault.

10.2.4 Three-Phase Transformer Models

Three-phase transformer models considering winding connections, and turns ratio, are described in this section [5]. Consider a 12-terminal coupled network, consisting of three primary windings and three secondary windings mutually coupled through the transformer core (Figure 10.6). The short-circuit primitive matrix for this network is

$$\begin{vmatrix} i_1 \\ i_2 \\ i_3 \\ i_4 \\ i_5 \\ i_6 \end{vmatrix} = \begin{vmatrix} y_{11} & y_{12} & y_{13} & y_{14} & y_{15} & y_{16} \\ y_{21} & y_{22} & y_{23} & y_{24} & y_{25} & y_{26} \\ y_{31} & y_{32} & y_{33} & y_{34} & y_{35} & y_{36} \\ y_{41} & y_{42} & y_{43} & y_{44} & y_{45} & y_{46} \\ y_{51} & y_{52} & y_{53} & y_{54} & y_{55} & y_{56} \\ y_{61} & y_{62} & y_{63} & y_{64} & y_{65} & y_{66} \end{vmatrix} \begin{vmatrix} v_1 \\ v_2 \\ v_3 \\ v_4 \\ v_5 \\ v_6 \end{vmatrix} \tag{10.24}$$

This ignores tertiary windings. It becomes a formidable problem for calculation if all the *Y* elements are distinct. Making use of the symmetry, the *Y* matrix can be reduced to

$$\begin{vmatrix} y_p & -y_m & y'_m & y''_m & y'_m & y''_m \\ -y_m & y_s & y''_m & y''_m & y''_m & y''_m \\ y'_m & y''_m & y_p & -y_m & y'_m & y''_m \\ y''_m & y''_m & -y_m & y_s & y''_m & y''_m \\ y'_m & y''_m & y'_m & y''_m & y_p & -y_m \\ y''_m & y''_m & y''_m & y''_m & -y_m & y_s \end{vmatrix} \tag{10.25}$$

This considers that windings 1, 3, and 5 are the primary windings and windings 2, 4, and 6 are the secondary windings with appropriate signs for the admittances. The primed elements are all zero if there are no mutual couplings, e.g., in the case of three single-phase transformers:

$$\begin{vmatrix} i_1 \\ i_2 \\ i_3 \\ i_4 \\ i_5 \\ i_6 \end{vmatrix} = \begin{vmatrix} y_p & -y_m & 0 & 0 & 0 & 0 \\ -y_m & y_s & 0 & 0 & 0 & 0 \\ 0 & 0 & y_p & -y_m & 0 & 0 \\ 0 & 0 & -y_m & y_s & 0 & 0 \\ 0 & 0 & 0 & 0 & y_p & -y_m \\ 0 & 0 & 0 & 0 & -y_m & y_s \end{vmatrix} \begin{vmatrix} v_1 \\ v_2 \\ v_3 \\ v_4 \\ v_5 \\ v_6 \end{vmatrix} \tag{10.26}$$

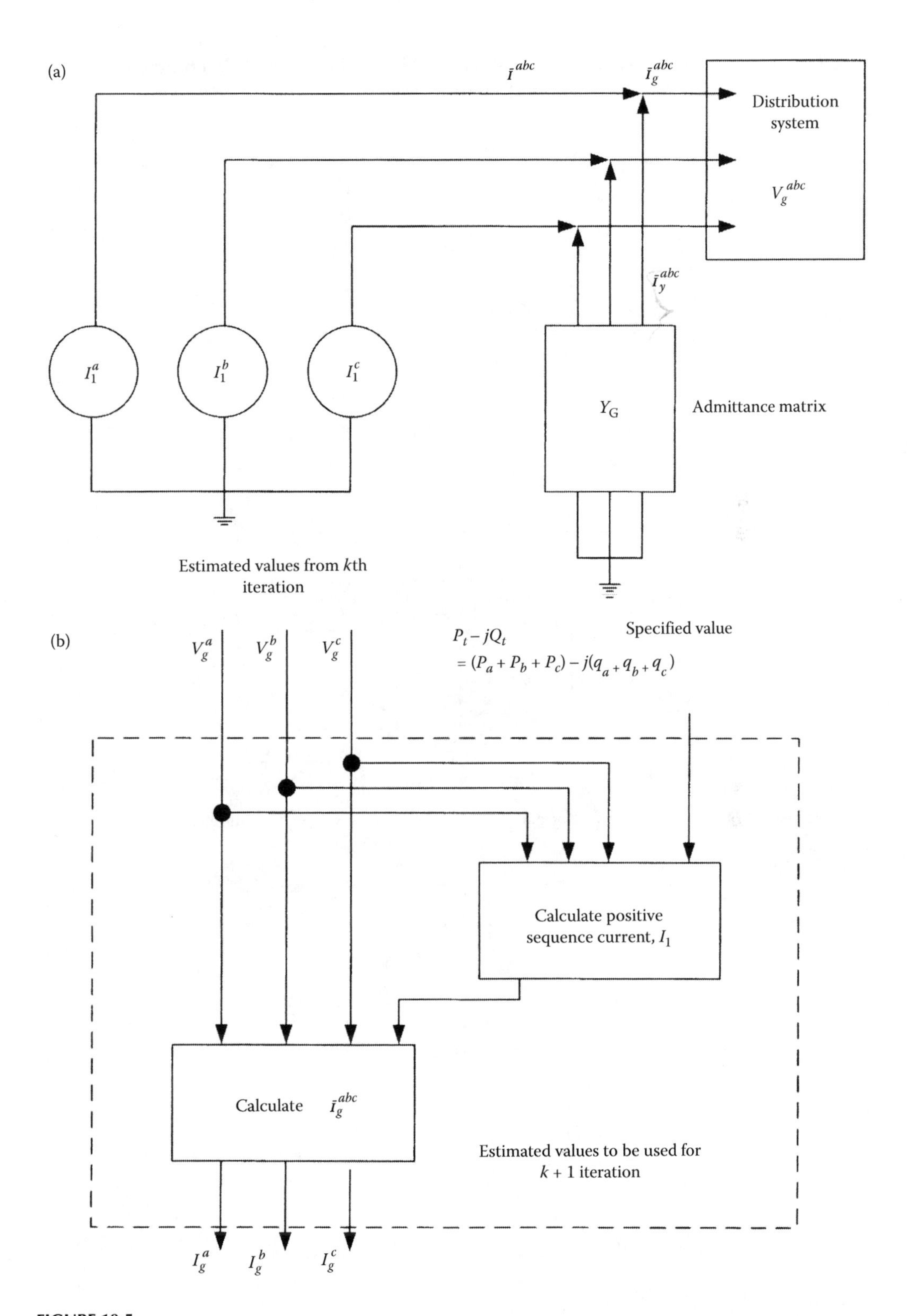

FIGURE 10.5
(a) Norton equivalent circuit of a generator for distribution systems and (b) circuit for calculations of load flow.

Consider a three-phase wye–delta transformer (Figure 10.7). The branch and node voltages in this figure are related by the following connection matrix:

$$\begin{vmatrix} v_1 \\ v_2 \\ v_3 \\ v_4 \\ v_5 \\ v_6 \end{vmatrix} = \begin{vmatrix} 1 & 0 & 0 & 0 & 0 & 0 \\ 0 & 0 & 0 & 1 & -1 & 0 \\ 0 & 1 & 0 & 0 & 0 & 0 \\ 0 & 0 & 0 & 0 & 1 & -1 \\ 0 & 0 & 1 & 0 & 0 & 0 \\ 0 & 0 & 0 & -1 & 0 & 1 \end{vmatrix} \begin{vmatrix} V_a \\ V_b \\ V_c \\ V_A \\ V_B \\ V_C \end{vmatrix} \tag{10.27}$$

or we can write

$$\bar{\upsilon}_{\text{branch}} = \bar{N}\bar{V}_{\text{node}} \tag{10.28}$$

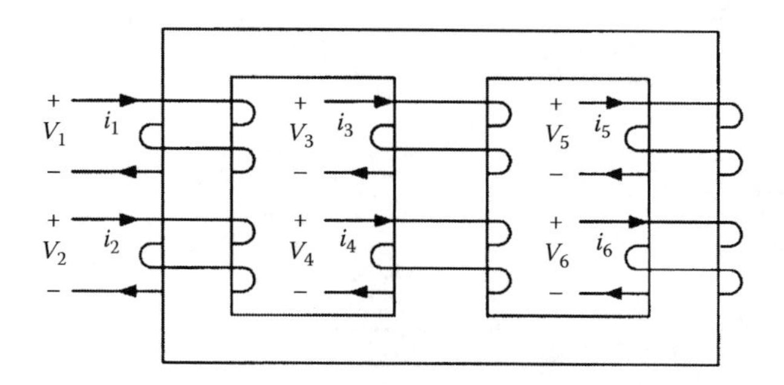

FIGURE 10.6
Elementary circuit of a three-phase transformer showing a 12-terminal coupled primitive network.

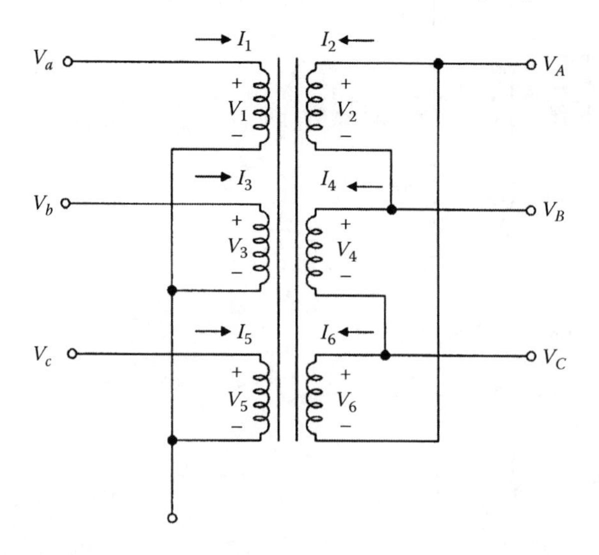

FIGURE 10.7
Circuit of a grounded wye–delta transformer with voltage and current relations for the derivation of connection matrix.

where $\bar{N}$ is the connection matrix.

Kron's transformation [6] is applied to the connection matrix $\bar{N}$ to obtain the node admittance matrix:

$$\bar{Y}_{\text{node}} = \bar{N}^t \bar{Y}_{\text{prim}} \bar{N} \tag{10.29}$$

The node admittance matrix is obtained in phase quantities as follows

$$\bar{Y}_{\text{node}} = \begin{vmatrix} y_s & y'_m & y'_m & -(y_m + y''_m) & (y'_m + y''_m) & 0 \\ y'_m & y_s & y'_m & 0 & -(y_m + y''_m) & (y_m + y''_m) \\ y'_m & y'_m & y_s & (y_m + y''_m) & 0 & -(y_m + y''_m) \\ -(y_m + y''_m) & 0 & (y_m + y''_m) & 2(y_s - y'''_m) & -(y_s - y'''_m) & -(y_s - y'''_m) \\ (y_m + y''_m) & -(y_m + y''_m) & 0 & -(y_s - y'''_m) & 2(y_s - y'''_m) & -(y_s - y'''_m) \\ 0 & (y_m + y''_m) & -(y_m + y''_m) & -(y_s - y'''_m) & -(y_s - y'''_m) & 2(y_s - y'''_m) \end{vmatrix} \tag{10.30}$$

The primed y_m vanish when the primitive admittance matrix of three-phase bank is substituted in Equation 10.29. The primitive admittances are considered on a per unit basis, and both primary and secondary voltages are 1.0 per unit. But a wye–delta transformer so obtained must consider a turn's ratio of $\sqrt{3}$, so that wye and delta node voltages are still 1.0 per unit. The node admittance matrix can be divided into submatrices as follows:

$$\bar{Y}_{\text{node}} = \begin{vmatrix} \bar{Y}_{\text{I}} & \bar{Y}_{\text{II}} \\ Y^t_{\text{II}} & \bar{Y}_{\text{III}} \end{vmatrix} \tag{10.31}$$

where each 3×3 submatrix depends on the winding connections, as shown in Table 10.1. The submatrices in this table are defined as follows:

$$\bar{Y}_{\text{I}} = \begin{vmatrix} y_t & 0 & 0 \\ 0 & y_t & 0 \\ 0 & 0 & y_t \end{vmatrix} \quad \bar{Y}_{\text{II}} = \frac{1}{3}\begin{vmatrix} 2y_t & -y_t & -y_t \\ -y_t & 2y_t & -y_t \\ -y_t & -y_t & 2Y_t \end{vmatrix} \quad \bar{Y}_{\text{III}} = \frac{1}{\sqrt{3}}\begin{vmatrix} -y_t & y_t & 0 \\ 0 & -y_t & y_t \\ y_t & 0 & -y_t \end{vmatrix} \tag{10.32}$$

Here, y_t is the leakage admittance per phase in per unit. Note that all primed y_m are dropped. Normally, these primed values are much smaller than the unprimed values. These are considerably smaller in magnitude than the unprimed values and the numerical values of y_s; y_p and y_m are equal to the leakage impedance y_t obtained by the short-circuit test, see Appendix C vol. 1. It is assumed that all three transformer banks are identical.

Table 10.1 can be used as a simplified approach to the modeling of common core-type three-phase transformer in an unbalanced system. With more complete information, the model in Equation 10.30 can be used for benefits of accuracy.

TABLE 10.1

Submatrices of Three-Phase Transformer Connections

Winding Connections		Self-Admittance		Mutual Admittance	
Primary	**Secondary**	**Primary**	**Secondary**	**Primary**	**Secondary**
Wye–G	Wye–G	$\bar{Y}_{\mathrm{I}}$	$\bar{Y}_{\mathrm{I}}$	$-\bar{Y}_{\mathrm{I}}$	$-\bar{Y}_{\mathrm{I}}$
Wye–G	Wye	$\bar{Y}_{\mathrm{II}}$	$\bar{Y}_{\mathrm{II}}$	$-\bar{Y}_{\mathrm{II}}$	$-\bar{Y}_{\mathrm{II}}$
Wye	Wye–G				
Wye	Wye				
Wye–G	Delta	$\bar{Y}_{\mathrm{I}}$	$\bar{Y}_{11}$	$\bar{Y}_{\mathrm{III}}$	$\bar{Y}^{t}_{\mathrm{III}}$
Wye	Delta	$\bar{Y}_{\mathrm{II}}$	$\bar{Y}_{11}$	$\bar{Y}_{\mathrm{III}}$	$\bar{Y}^{t}_{\mathrm{III}}$
Delta	Wye	$\bar{Y}_{\mathrm{II}}$	$\bar{Y}_{\mathrm{III}}$	$\bar{Y}^{t}_{\mathrm{III}}$	$\bar{Y}_{\mathrm{III}}$
Delta	Wye–G	$\bar{Y}_{\mathrm{II}}$	$\bar{Y}_{11}$	$\bar{Y}^{t}_{\mathrm{III}}$	$\bar{Y}_{\mathrm{III}}$
Delta	Delta	$\bar{Y}_{\mathrm{II}}$	$\bar{Y}_{\mathrm{II}}$	$-\bar{Y}_{\mathrm{II}}$	$-\bar{Y}_{\mathrm{II}}$

Y^{t}_{III} is the transpose of Y_{III}.

10.2.4.1 Symmetrical Components of Three-Phase Transformers

As stated in earlier chapters, in most cases, it is sufficient to assume that the system is balanced. Then, symmetrical components models can be arrived at by using symmetrical component transformations. Continuing with wye–ground–delta transformer of Figure 10.7 first consider the self-admittance matrix. The transformation is

$$\bar{Y}^{pp}_{012} = \bar{T}_s^{-1}\begin{vmatrix} y_p & y'_m & y'_m \\ y'_m & y_p & y'_m \\ y'_m & y'_m & y_p \end{vmatrix}\bar{T}_s = \begin{vmatrix} y_p + 2y'_m & 0 & 0 \\ 0 & y_p - y'_m & 0 \\ 0 & 0 & y_p - y'_m \end{vmatrix} \tag{10.33}$$

Note that zero sequence admittance is different from the positive and negative sequence impedances. If y'_m is neglected, all three are equal.

Similarly,

$$\bar{Y}^{ss}_{012} = \frac{1}{3}\bar{T}_s^{-1}\begin{vmatrix} 2(y_s - y'''_m) & -(y_s - y'''_m) & -(y_s - y'''_m) \\ -(y_s - y'''_m) & 2(y_s - y'''_m) & -(y_s - y'''_m) \\ -(y_s - y'''_m) & -(y_s - y'''_m) & 2(y_s - y'''_m) \end{vmatrix}\bar{T}_s = \begin{vmatrix} 0 & 0 & 0 \\ 0 & (y_s - y'''_m) & 0 \\ 0 & 0 & (y_s - y'''_m) \end{vmatrix} \tag{10.34}$$

As expected, there is no zero sequence self-admittance in the delta winding.

Finally, the mutual admittance matrix gives

$$\bar{Y}_{012}^{ps} = \frac{1}{\sqrt{3}}\bar{T}_s^{-1}\begin{vmatrix} -(y_m + y_m'') & 0 & (y_m + y_m'') \\ (y_m + y_m'') & -(y_m + y_m'') & 0 \\ 0 & (y_m + y_m'') & -(y_m + y_m'') \end{vmatrix}\bar{T}_s$$

$$= \begin{vmatrix} 0 & 0 & 0 \\ 0 & -(y_m + y_m'') < 30° & 0 \\ 0 & 0 & -(y_m + y_m'') < -30° \end{vmatrix} \quad (10.35)$$

A phase shift of 30° occurs in the positive sequence network and a negative phase shift of −30° occurs in the negative sequence. Similar results were arrived at in Chapter 2, Section 2.5.3. No zero sequence currents can occur between wye and delta side of a balanced wye–delta transformer. Based upon above equations, the positive, negative, and zero sequence admittances of the transformer are shown in Figure 10.8.

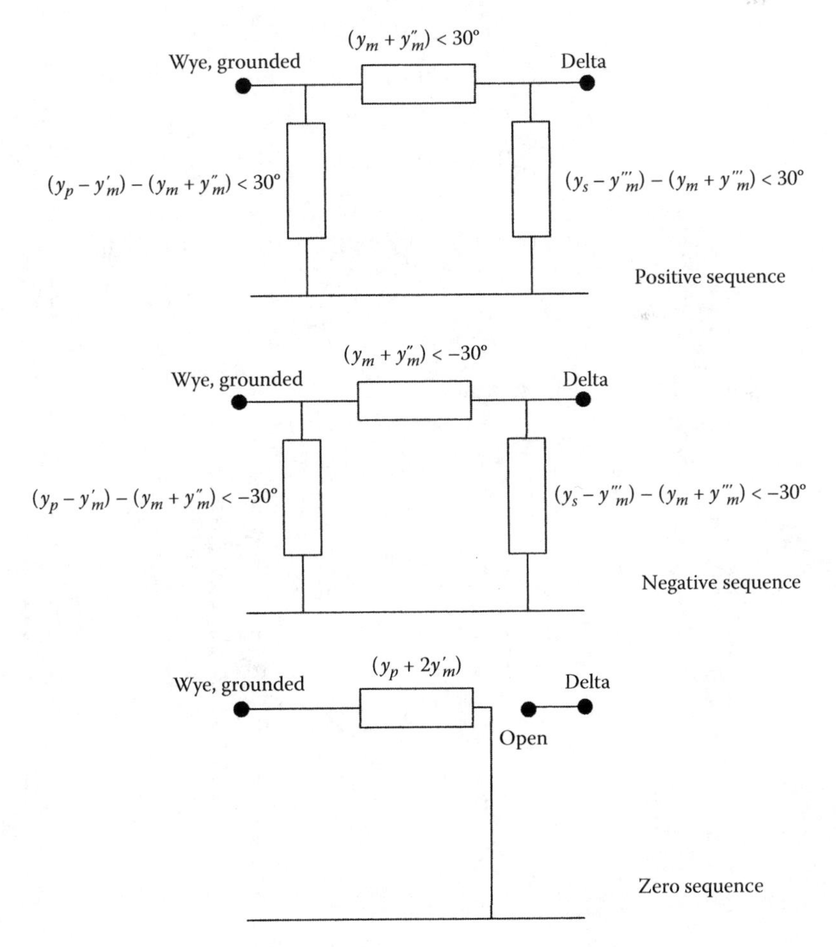

FIGURE 10.8
Sequence impedance circuits of a wye-grounded delta two-winding transformer.

If the phase shift is ignored,

- $y'_m = y''_m = y'''_m$, which are zero in a three-phase bank and
- $y_p - y_m$ is small admittance, as y_p is only slightly $> y_m$ and $y_p - y_m$ and $y_s - y_m$ as open-circuit admittances.

Then, the simplified model returns to that shown in Table 3.1.

If the off-nominal tap ratio between primary and secondary windings is α:β, where α and β are the taps on the primary and secondary sides, respectively, in per unit, then the submatrices are modified as follows:

- Divide self-admittance of the primary matrix by α^2
- Divide self-admittance of the secondary matrix by β^2
- Divide mutual admittance matrixes by αβ

Consider a wye-grounded transformer. Then, from Table 10.1

$$\bar{Y}^{abc} = \begin{vmatrix} \bar{Y}_I & -\bar{Y}_I \\ -\bar{Y}_I & \bar{Y}_I \end{vmatrix} = \begin{vmatrix} y_t & 0 & 0 & -y_t & 0 & 0 \\ 0 & y_t & 0 & 0 & -y_t & 0 \\ 0 & 0 & y_t & 0 & 0 & -y_t \\ -y_t & 0 & 0 & y_t & 0 & 0 \\ 0 & -y_t & 0 & 0 & y_t & 0 \\ 0 & 0 & -y_t & 0 & 0 & y_t \end{vmatrix} \tag{10.36}$$

For off-nominal taps, the matrix is modified as

$$\bar{Y}^{abc} = \begin{vmatrix} \frac{y_t}{\alpha^2} & 0 & 0 & \frac{y_t}{\alpha\beta} & 0 & 0 \\ 0 & \frac{y_t}{\alpha^2} & 0 & 0 & \frac{y_t}{\alpha\beta} & 0 \\ 0 & 0 & \frac{y_t}{\alpha^2} & 0 & 0 & \frac{y_t}{\alpha\beta} \\ \frac{y_t}{\alpha\beta} & 0 & 0 & \frac{y_t}{\beta^2} & 0 & 0 \\ 0 & \frac{y_t}{\alpha\beta} & 0 & 0 & \frac{y_t}{\beta^2} & 0 \\ 0 & 0 & \frac{y_t}{\alpha\beta} & 0 & 0 & \frac{y_t}{\beta^2} \end{vmatrix} = \left|\bar{Y}_{yg-y}\right| \tag{10.37}$$

A three-phase transformer winding connection of delta primary and grounded-wye secondary is commonly used. From Table 10.1, its matrix equation is

$$\bar{Y}^{abc} = \begin{vmatrix} \frac{2}{3}y_t & -\frac{1}{3}y_t & -\frac{1}{3}y_t & -\frac{y_t}{\sqrt{3}} & \frac{y_t}{\sqrt{3}} & 0 \\ -\frac{1}{3}y_t & \frac{2}{3}y_t & -\frac{1}{3}y_t & 0 & -\frac{y_t}{\sqrt{3}} & \frac{y_t}{\sqrt{3}} \\ -\frac{1}{3}y_t & -\frac{1}{3}y_t & \frac{2}{3}y_t & \frac{y_t}{\sqrt{3}} & 0 & -\frac{y_t}{\sqrt{3}} \\ -\frac{y_t}{\sqrt{3}} & 0 & \frac{y_t}{\sqrt{3}} & y_t & 0 & 0 \\ \frac{y_t}{\sqrt{3}} & -\frac{y_t}{\sqrt{3}} & 0 & 0 & y_t & 0 \\ 0 & \frac{y_t}{\sqrt{3}} & -\frac{y_t}{\sqrt{3}} & 0 & 0 & y_t \end{vmatrix} \quad (10.38)$$

where y_t is the leakage reactance of the transformer.

For an off-nominal transformer, the Y matrix is modified as shown in the following:

$$\bar{Y}^{abc} = \begin{vmatrix} \frac{2}{3}\frac{y_t}{\alpha^2} & \frac{1}{3}\frac{y_t}{\alpha^2} & -\frac{1}{3}\frac{y_t}{\alpha^2} & -\frac{y_t}{\sqrt{3}\alpha\beta} & \frac{y_t}{\sqrt{3}\alpha\beta} & 0 \\ -\frac{1}{3}\frac{y_t}{\alpha^2} & \frac{2}{3}\frac{y_t}{\alpha^2} & -\frac{1}{3}\frac{y_t}{\alpha^2} & 0 & -\frac{y_t}{\sqrt{3}\alpha\beta} & \frac{y_t}{\sqrt{3}\alpha\beta} \\ -\frac{1}{3}\frac{y_t}{\alpha^2} & -\frac{1}{3}\frac{y_t}{\alpha^2} & \frac{2}{3}\frac{y_t}{\alpha^2} & \frac{y_t}{\sqrt{3}\alpha\beta} & 0 & -\frac{y_t}{\sqrt{3}\alpha\beta} \\ -\frac{y_t}{\sqrt{3}\alpha\beta} & 0 & \frac{y_t}{\sqrt{3}\alpha\beta} & \frac{y_t}{\beta^2} & 0 & 0 \\ \frac{y_t}{\sqrt{3}\alpha\beta} & -\frac{y_t}{\sqrt{3}\alpha\beta} & 0 & 0 & \frac{y_t}{\beta^2} & 0 \\ 0 & \frac{y_t}{\sqrt{3}\alpha\beta} & -\frac{y_t}{\sqrt{3}\alpha\beta} & 0 & 0 & \frac{y_t}{\beta^2} \end{vmatrix} \quad (10.39)$$

where a and β are the taps on the primary and secondary sides in per unit.

In a load flow analysis, the equation of a wye-grounded delta transformer and $a = \beta = 1$ can be written as follows:

$$\begin{vmatrix} I_A \\ I_B \\ I_C \\ I_a \\ I_b \\ I_c \end{vmatrix} = \begin{vmatrix} y_t & 0 & 0 & -\frac{1}{\sqrt{3}}y_t & \frac{1}{\sqrt{3}}y_t & 0 \\ 0 & y_t & 0 & 0 & -\frac{1}{\sqrt{3}}y_t & \frac{1}{\sqrt{3}}y_t \\ 0 & 0 & y_t & \frac{1}{\sqrt{3}}y_t & 0 & -\frac{1}{\sqrt{3}}y_t \\ -\frac{1}{\sqrt{3}}y_t & 0 & \frac{1}{\sqrt{3}}y_t & \frac{2}{3}y_t & -\frac{1}{3}y_t & -\frac{1}{3}y_t \\ \frac{1}{\sqrt{3}}y_t & -\frac{1}{\sqrt{3}}y_t & 0 & -\frac{1}{3}y_t & \frac{2}{3}y_t & -\frac{1}{3}y_t \\ 0 & \frac{1}{\sqrt{3}}y_t & -\frac{1}{\sqrt{3}}y_t & -\frac{1}{3}y_t & -\frac{1}{3}y_t & \frac{2}{3}y_t \end{vmatrix} \begin{vmatrix} V_A \\ V_B \\ V_C \\ V_a \\ V_b \\ V_c \end{vmatrix} \tag{10.40}$$

Here, the currents and voltages with capital subscripts relate to primary and those with lower case subscripts relate to secondary. In the condensed form, we will write it as follows:

$$\bar{I}_{ps} = \bar{Y}_{Y-\Delta}\bar{V}_{ps} \tag{10.41}$$

Using symmetrical component transformation

$$\begin{vmatrix} \bar{I}_P^{012} \\ \bar{I}_s^{012} \end{vmatrix} = \begin{vmatrix} \bar{T}_s & 0 \\ 0 & \bar{T}_s \end{vmatrix}^{-1} \bar{Y}_{y-\Delta} \begin{vmatrix} \bar{T}_s & 0 \\ 0 & \bar{T}_s \end{vmatrix} \begin{vmatrix} \bar{V}_P^{012} \\ \bar{V}_s^{012} \end{vmatrix} \tag{10.42}$$

Expanding

$$\begin{vmatrix} \bar{I}_P^{012} \\ \bar{I}_s^{012} \end{vmatrix} = \begin{vmatrix} y_t & 0 & 0 & 0 & 0 & 0 \\ 0 & y_t & 0 & 0 & y_t < -30^\circ & 0 \\ 0 & 0 & y_t & 0 & 0 & y_t < 30^\circ \\ 0 & 0 & 0 & 0 & 0 & 0 \\ 0 & y_t < 30^\circ & 0 & 0 & y_t & 0 \\ 0 & 0 & y_t < -30^\circ & 0 & 0 & y_t \end{vmatrix} \begin{vmatrix} \bar{V}_P^{012} \\ \bar{V}_s^{012} \end{vmatrix} \tag{10.43}$$

The positive sequence equations are

$$\begin{aligned} I_{p1} &= y_t V_{p1} - y_t < -30^\circ V_{s1} \\ I_{s1} &= y_t V_{s1} - y_t < 30^\circ V_{p1} \end{aligned} \tag{10.44}$$

The negative sequence equations are

$$\begin{aligned} I_{p2} &= y_t V_{p2} - y_t < 30°V_{s2} \\ I_{s2} &= y_t V_{s2} - y_t < -30°V_2 \end{aligned} \tag{10.45}$$

The zero sequence equations are

$$\begin{aligned} I_{p0} &= y_t V_{p0} \\ I_{s0} &= 0 \end{aligned} \tag{10.46}$$

For a balanced system, only the positive sequence component needs to be considered. The power flow on the primary side is

$$\begin{aligned} S_{ij} &= V_i I_{ij}^* = V_i \left(y_t^* V_i^* - y_t^* < 30°V_j^* \right) \\ &= \left[y_t V_i^2 \cos\theta_{yt} - y_t \left|V_i V_j\right| \cos\left(\theta_i - \theta_{yt} - (\theta_j + 30°)\right)\right] \\ &\quad + j\left[-y_t V_i^2 \sin\theta_{yt} - y_t \left|V_i V_j\right| \sin\left(\theta_i - \theta_{yt} - (\theta_j + 30°)\right)\right] \end{aligned} \tag{10.47}$$

and on the secondary side

$$\begin{aligned} S_{ji} &= V_i I_{ji}^* = V_j \left(y_t^* V_j^* - y_t^* < -30°V_i^* \right) \\ &= \left[y_t V_j^2 \cos\theta_{yt} - y_t \left|V_j V_i\right| \cos\left(\theta_j - \theta_{yt} - (\theta_i - 30°)\right)\right] \\ &\quad + j\left[-y_t V_j^2 \sin\theta_{yt} - y_t \left|V_j V_i\right| \sin\left(\theta_j - \theta_{yt} - (\theta_i - 30°)\right)\right] \end{aligned} \tag{10.48}$$

10.2.5 Load Models

For a distribution system, the load window concept is discussed in Chapter 6. Based on test data, a detailed load model can be derived, and the voltage/current characteristics of the models are considered. The models are not entirely three-phase balanced types and single-phase loads give rise to unbalances. A load window can be first constructed and percent of each load type allocated, see Figure 10.9. By testing, the power/voltage

Incandescent lighting	Fluorescent lighting	Space heating	Dryer	Refrig. freezer	Elect. range	TV	Others	Total = 100%

FIGURE 10.9
A load window showing composition of various load types.

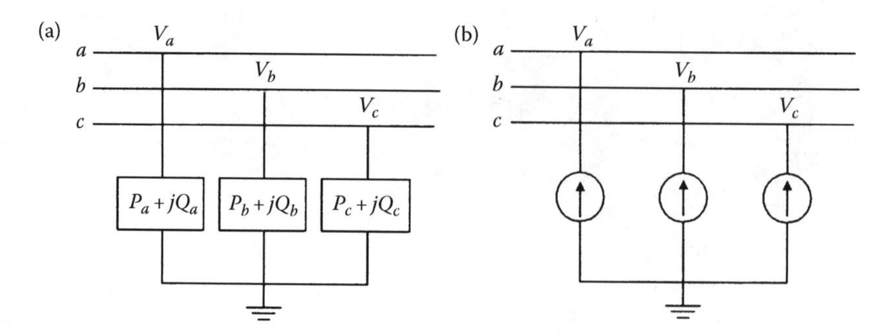

FIGURE 10.10
(a) Three-phase load representation and (b) equivalent current injection.

characteristics of most load types are known. As an example fluorescent lighting, the power requirement reduces when the voltage dips and increases as the voltage is restored to the operating voltage. Conversely for air conditioning loads, the power requirement will increase as the voltage rises and also as the voltage dips below rated, giving a U-shaped curve [7]. A typical three-phase load is shown in Figure 10.10. The unbalance is allowed by load current injections.

10.3 Distribution System Load Flow

The distribution system analysis requirements are as follows:

- It should be capable of modeling 4–44 kV primary distribution feeders and networks, 120/208 V secondary networks, and isolated 277/480 V systems simultaneously. These systems become very large and may consist of 4000 secondary distribution buses (three-phase models = 12,000 buses) and 2000 primary distribution buses. The system must be, therefore, capable of handling 20,000 buses. As an example, Southern California Edison serves 4.2 million customers over a territory of 50,000 mi^2 and there are 600 distribution substations, 3800 distribution circuits over 38,000 circuit mi, 61,000 switches, 800 automatic reclosers, and 7600 switched capacitor banks. Data generation and modeling become very time consuming.
- The system is inherently unbalanced. In the planning stage, it may be adequate to model the system on a single-phase basis, but the operation requires three-phase modeling. The capability of modeling three-phase systems, line segments, and transformers with phase shifts is a must. The core and copper losses need to be considered. A nonlinear model of the core losses is appropriate.
- The cogenerators should be capable of being modeled on primary and secondary systems.
- Contingency analysis is required to study the effect of outage of feeders.
- Short-circuit calculations are performed using the same database. The load currents may not be small compared to short-circuit currents and cannot be neglected. Prefault voltages and current injections obtained in load flow are inputs into the

short-circuit calculations. Contributions of the cogenerators to faults must be included. Figure 10.11 shows a flowchart.

- A feeder has several interfeeder switches to link with other feeders. Under heavy loading or contingency conditions, normally open switches are closed to prevent system overloads. Optimal interfeeder switching decisions are required to be made in a distribution system. As any switching operation must be carried out to prevent overloads, the problem of optimal switching translates into that of dispatching currents through alternate routes to meet the load demand to prevent overloads and achieve phase balance. The radial nature of the network is preserved and the number of switching operations is minimized.

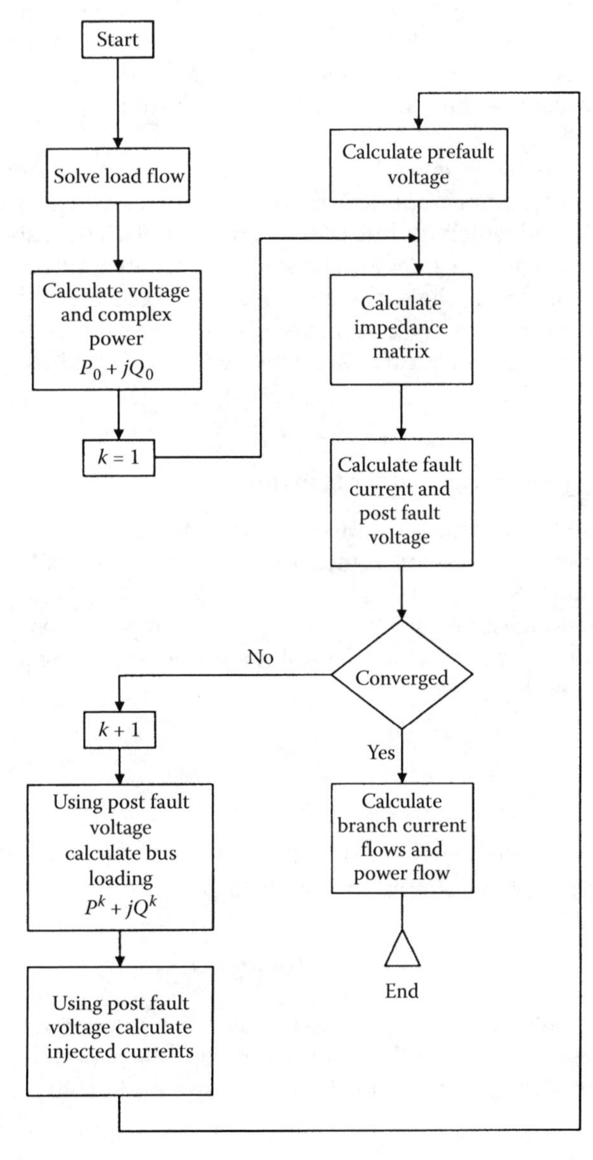

FIGURE 10.11
Flowchart for short-circuit calculations considering prefault voltages and currents.

- Optimal capacitor location on distribution feeders for voltage control and energy loss reduction is required.

In today's environment, distribution networks are being influenced by energy savings and improvements in power quality. Regenerative methods of generation and storage, e.g., fuel cells, are receiving impetus. Optimization of alternative sources of energy and maximization of network utilization without overloading a section are required [8].

10.3.1 Methodology

The Gauss and Newton–Raphson (NR) methods are applicable. In the NR approach, because of the low X/R ratio of conductors associated with distribution system line segments compared to transmission systems, the Jacobian cannot be decoupled. A Gauss method using a sparse bifactored bus admittance matrix flowchart is shown in Figure 10.12. As the voltage-specified bus in the system is only the swing bus, the rate of convergence is comparable to that of the NR method.

The voltage at each bus can be considered as having contributions from two sources, and the theorem of superimposition is applied. These two sources are the voltage-specified station bus and the load and generator buses, see Figure 10.13a. The loads and cogenerators are modeled as current injection sources. The shunt capacitance currents are also included in the current injections. Using the superimposition principle, only one source is active at a time. When the swing bus voltage is activated, all current sources are disconnected, see Figure 10.13b. When the current sources are activated, the swing bus is short circuited to ground, see Figure 10.13c.

10.3.2 Distribution System as a Ladder Network

A distribution system forms a ladder network, with loads teed-off in a radial fashion, see Figure 10.14a. It is a nonlinear system, as most loads are of constant kW and kvar. However, linearization can be applied.

For a linear network, assume that the line and load impedances are known and the source voltage is known. Starting from the last node, and assuming a node voltage of V_n, the load current is given by

$$I_n = \frac{V_n}{Z_{1n}} \tag{10.49}$$

Note that I_n is also the line current, as this is the last node. Therefore, the voltage at node $n-1$ can be obtained simply by subtracting the voltage drop:

$$V_{n-1} = V_n - I_{n-1,n} Z_{n-1,n} \tag{10.50}$$

This process can be carried out until the sending-end node is reached. The calculated value of the sending-end voltage will be different from the actual applied voltage. Since the network is linear, all the line and load currents and node voltages can be multiplied by the ratio:

$$V_{\text{actual}} / V_{\text{calculated}} = V_s / V_{\text{calculated}} \tag{10.51}$$

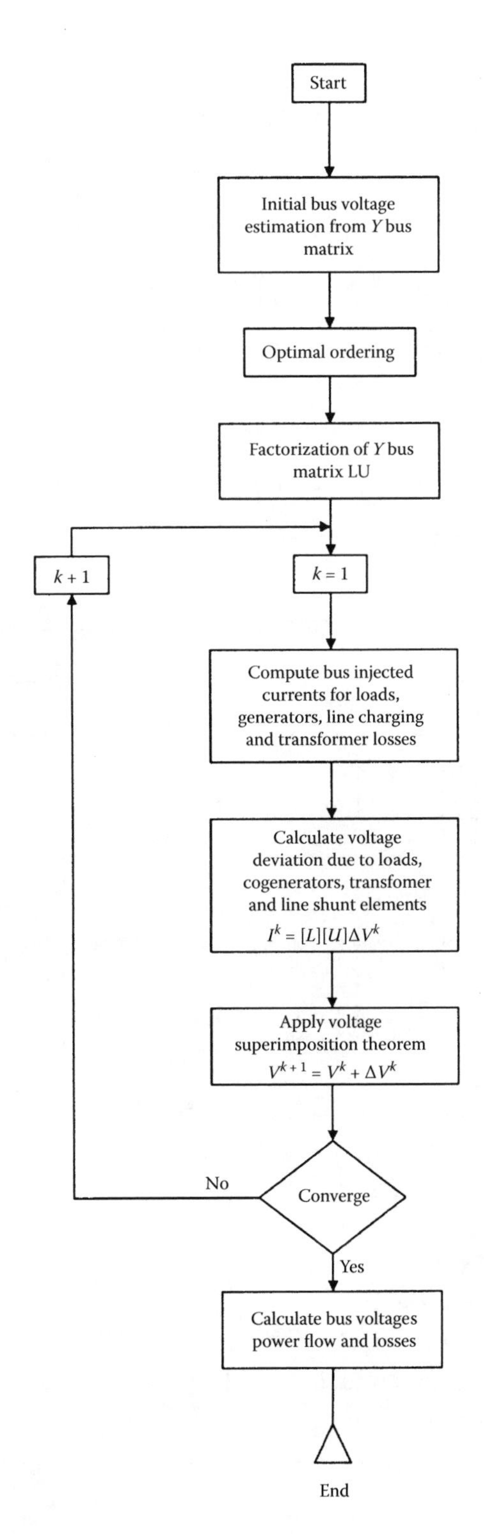

FIGURE 10.12
Flowchart for distribution system load flow using Gauss factored Y matrix.

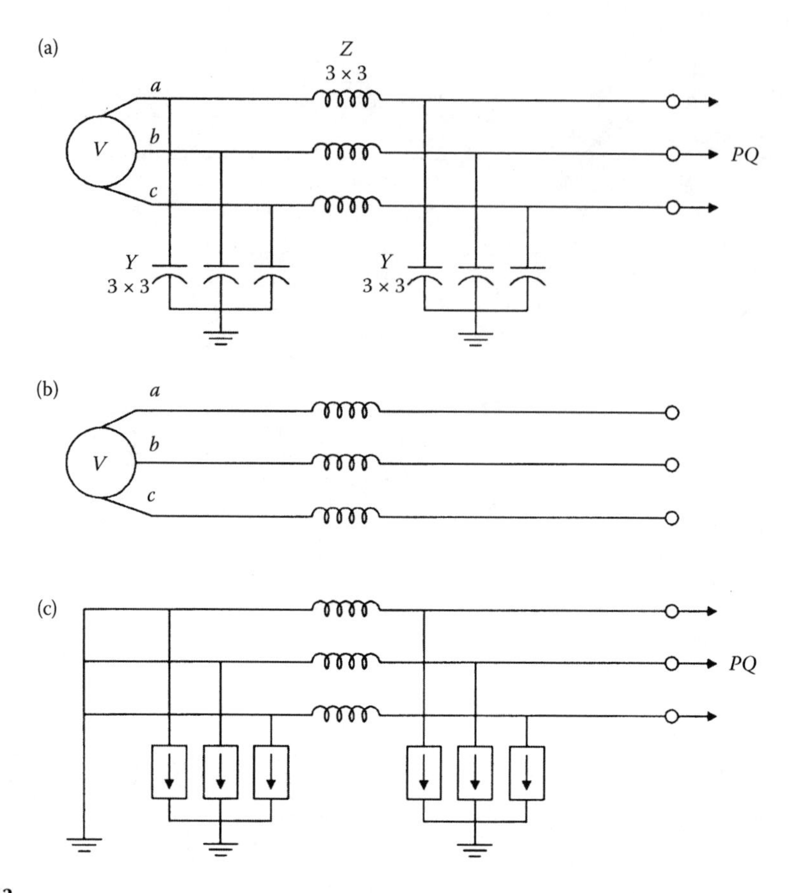

FIGURE 10.13
(a) Original system for load flow, (b) only swing bus activated, and (c) swing bus grounded, only current injections activated.

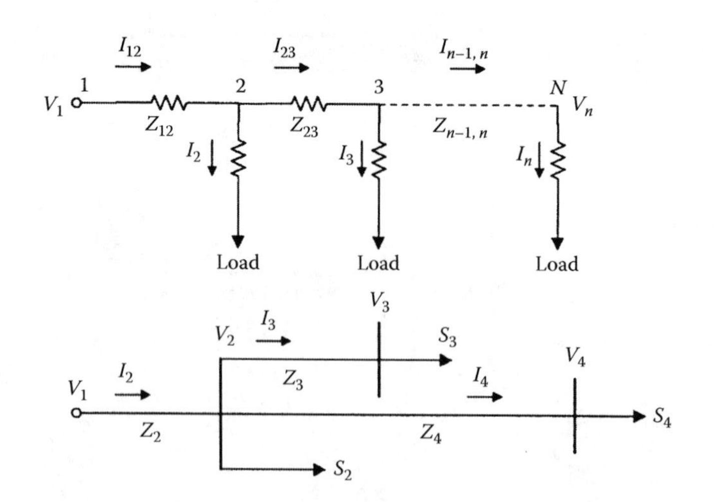

FIGURE 10.14
(a) Distribution system as a ladder network and (b) load flow with a teed-off lateral.

Actually, the node current must be calculated on the basis of complex power at the load:

$$I_{\text{node}} = (S_{\text{node}}/V_{\text{node}})^* \tag{10.52}$$

Starting from the last node, the sending-end voltage is calculated in the *forward sweep*, as in the linear case. This voltage will be different from the sending-end voltage. Using this voltage, found in the first iteration, a *backward sweep* is performed, i.e., the voltages are recalculated starting from the first node to the nth node. This new voltage is used to recalculate the currents and voltages at the nodes in the second forward sweep. The process can be repeated until the required tolerance is achieved.

A lateral circuit, Figure 10.14b, can be handled as follows [9, 10]:

- Calculate the voltage at node 2, starting from node 4, ignoring the lateral to node 3. Let this voltage be V_2.
- Consider that the lateral is isolated and is an independent ladder. Now, the voltage at node 3 can be calculated and therefore current I_3 is known.
- The voltage at node 2 is calculated back, i.e., voltage drop I_3Z_3 is added to voltage V_3. Let this voltage be V_2'. The difference between V_2 and V_2' must be reduced to an acceptable tolerance. The new node 3 voltage is $V_{3(\text{new})} = V_3 - (V_2 - V_2')$. The current I_3 is recalculated and the calculations iterated until the desired tolerance is achieved.

10.4 Optimal Capacitor Locations

The optimal location of capacitors in a distribution system is a complex process [11, 12]. The two main criteria are voltage profiles and system losses. Correcting the voltage profile will require capacitors to be placed toward the end of the feeders, while emphasis on loss reduction will result in capacitors being placed near load centers. An automation strategy based on intelligent customer meters, which monitor the voltage at consumer locations and communicate this information to the utility, can be implemented [13].

We will examine the capacitor placement algorithm based on loss reduction and energy savings using dynamic programming concepts [14]. A reader may first peruse Chapter 11 before going through this section. We can define the objective functions as follows:

- Peak power loss reduction
- Energy loss reduction
- Voltage and harmonic control
- Capacitor cost

The solved variables are fixed and switched capacitors, and their number, size, location, and switched time. Certain assumptions in this optimization process are the loading conditions of the feeders, type of feeder load, and capacitor size based on the available standard ratings and voltages. Consider the placement of a capacitor at node K (Figure 10.15). The peak power loss reduction of the segment is

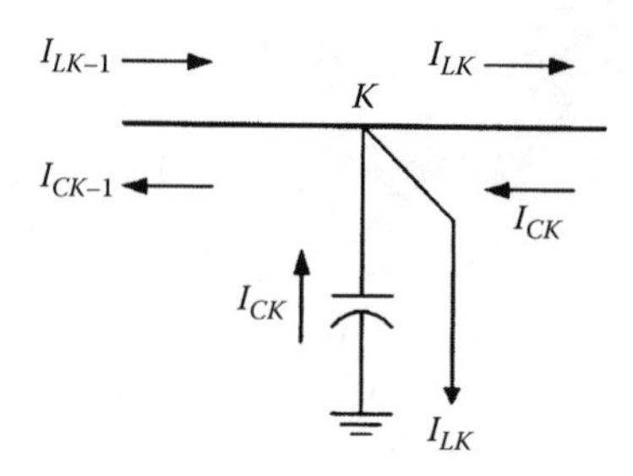

FIGURE 10.15
Load and capacitance current flows at node k with a capacitor bank.

$$PL_k = 3R_k(2I_{LK}I_{CK} - I_{CK}^2) \tag{10.53}$$

where R_k is the resistance of segment K, I_{LK} is the peak reactive current in segment K before placement of the capacitor, and I_{CK} is the total capacitor current flowing through segment K.

The power loss reduction under average load can be written as follows:

$$PLA_K = 3R_K(2I_{CK}I_{LK}LF - I_{CK}^2) \tag{10.54}$$

where all the terms are as defined before and LF is the load factor. The overall objective function for loss reduction can then be written as follows:

$$F = \sum_{k=1}^{N}[F_P PL_K + F_{av} PLA_K] \tag{10.55}$$

where F_p is the monetary conversion factor for power loss reduction under peak load in $/kW/year, F_{av} is the monetary conversion factor for power loss under average load in $/kW/year, and N is the number of line segments.

An optimal value function can be defined as follows:

$$S(x, y_x) = \text{Max}\left[\sum_{k=1}^{N}[F_P PL_K + F_{av}\text{PLA}_K]\right] \tag{10.56}$$

where $S(x, y_x)$ is the maximum value of the objective function calculated from the substation to the line segment if the capacitive current flowing through segment x is y_s. Also

$$I_{CX} = I_{\text{base}} y_x \tag{10.57}$$

We can define some constraints in the optimization process.

The power factor at the substation outlet is to be maintained lagging, i.e., the load current is greater than the capacitive current:

$$I_{L1} > I_{C1} \tag{10.58}$$

This requires that the total capacitive current flowing through $(k-1)$ segment be at least equal to the total capacitive current flowing into node k:

$$I_{CK-1} \geq I_{CK} \tag{10.59}$$

The values of y_x are defined over an adequate range of discrete values.

A dynamic programming formulation seeks the minimum (or maximum) path subject to constraints and boundary conditions. The recursive relation is given by

$$S(x, y_x) = MAX_{y_{x-1} \geq y_x} [F_p PL_X (I_{\text{base}} y_x) + F_{\text{av}} PLA_x (I_{\text{base}} y_x) + S(x-1, y_{s-1})] \tag{10.60}$$

The boundary condition is given by

$$S(0, y_0) = 0 \quad \text{for all } y_0 \tag{10.61}$$

i.e., stage 0 means that the substation is encountered and no capacitors are installed beyond this point. The recursive procedure is illustrated by an example.

Consider the three-segment system in Figure 10.16. Assume that the value of $I_{L1} = 1.0$ per unit and the smallest capacitor size is 0.3 per unit with an upper limit of 0.9 per unit. In Figure 10.17, stage 1 shows that a target node can be constructed with every possible value of I_{C1}. A path is added for each target node and the cost is calculated using Equation 10.51 for $N = 1$:

$$\cos t = 3R_1(2 \times 1.0 \times 0.9 - 0.9^2) F_p \times I_{\text{base}}^2 + 3R_1(2 \times 1.0 \times 0.9 - 0.9^2) F_{\text{av}} I_{\text{base}}^2$$

The graph can be constructed as shown in Figure 10.17 for the other two stages also, and the cost of the path connecting the two nodes can be calculated in an identical manner. Once the cost associated with every path is calculated, the problem reduces to finding the maximum cost path from one of the initial nodes to the target nodes [14]. The voltage and switching constraints can be added to the objective function.

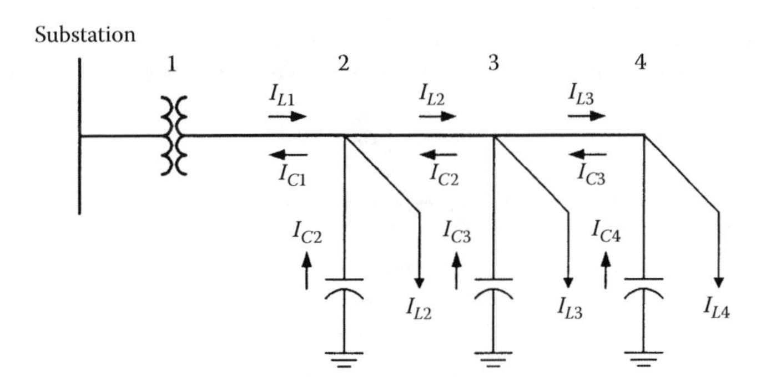

FIGURE 10.16
A three-section distribution system with capacitor compensation.

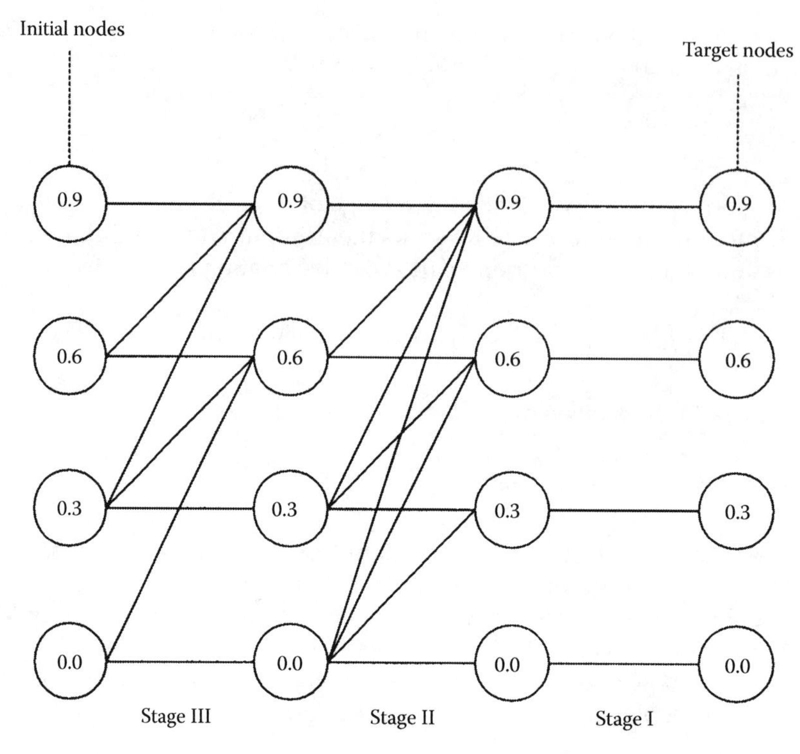

FIGURE 10.17
Three-stage flowchart of dynamic programming.

References

1. L Roy. Generalized polyphase fault analysis program: Calculation of cross country faults. *Proc IEE (Part B)*, 126(10), 995–1000, 1979.
2. MA Loughton. The analysis of unbalanced polyphase networks by the method of phase coordinates, Part 1. System representation in phase frame reference. *Proc Inst Electr Eng*, 115(8), 1163–1172, 1968.
3. MA Loughton. The analysis of unbalanced polyphase networks by the method of phase coordinates, Part 2. System fault analysis. *Proc Inst Electr Eng*, 1165(5), 857–865, 1969.
4. TH Chen. *Generalized Distribution Analysis System*. PhD dissertation, University of Texas at Arlington, 1990.
5. M Chen, WE Dillon. Power system modeling. *Proc IEEE Power App Syst*, 62, 901–915, 1974.
6. G Kron. *Tensor Analysis of Networks*, McDonald, London, 1965.
7. RB Adler, CC Mosher. *Steady State Voltage Power Characteristics for Power System Loads in Stability of Large Power Systems*. ed. Richard T. Byerly and EW Kimbark. IEEE Press, New York, 1974.
8. HL Willis, H Tram, MV Engel, L Finley. Selecting and applying distribution optimization methods. *IEEE Comp Appl Power*, 9(1), 12–17, 1996.
9. WH Kersting, DL Medive. *An Application of Ladder Network to the Solution of Three-Phase Radial Load Flow Problems. IEEE PES Winter Meeting*, New York, 1976.

10. WH Kersting. *Distribution System Modeling and Analysis*, 2nd ed. CRC Press, Boca Raton, FL, 2006.
11. BR Williams, David G. Walden. Distribution automation strategy for the future. *IEEE Comp Appl Power*, 7(3), 16–21, 1994.
12. JJ Grainger, SH Lee. Optimum size and location of shunt capacitors for reduction of losses on distribution feeders. *IEEE Trans Power App Syst*, 100(3), 1105–1118, 1981.
13. SH Lee, JJ Gaines. Optimum placement of fixed and switched capacitors of primary distribution feeders. *IEEE Trans Power App Syst*, 100, 345–352, 1981.
14. SK Chan. Distribution System Automation. PhD dissertation, University of Texas at Arlington, 1982.

11

Optimization Techniques

The application of computer optimization techniques in power systems is reaching new dimensions with improvements in algorithm reliability, speed, and applicability. Let us start with a simple situation. Optimization can be aimed at reducing something undesirable in the power system, e.g., the system losses or cost of operation, or maximizing a certain function, e.g., efficiency or reliability. Such maxima and minima are always subject to certain constraints, i.e., tap settings on transformers, tariff rates, unit availability, fuel costs, etc. The problem of optimization is thus translated into the problem of constructing a reliable mathematical model aimed at maximizing or minimizing a certain function, within the specified constraints.

It is possible to model a wide range of problems in planning, design, control, and measurement. Traditionally, optimization has been used for the economic operation of fossil-fueled power plants, using an economic dispatch approach. In this approach, inequality constraints on voltages and power flows are ignored and real power limits on generation and line losses are accounted for. A more complicated problem is system optimization over a period of time.

The optimization techniques are often applied offline. For many power system problems, an off-line approach is not desirable, because optimal solution is required for immediate real-time implementation and there is a need for efficient and reliable methods. Table 11.1 shows the interaction of various levels of system optimization.

Linear programming [1–3] deals with situations where a maximum or minimum of a certain set of linear functions is desired. The equality and inequality constraints define a region in the multidimensional space. Any point in the region or boundary will satisfy all the constraints; thus, it is a region enclosed by the constraints and not a discrete single-value solution. Given a meaningful mathematical function of one or more variables, the problem is to find a maximum or minimum, when the values of the variables vary within some certain allowable limits. The variables may react with each other or a solution may be possible within some acceptable violations, or a solution may not be possible at all.

Mathematically, we can minimize

$$f(x_1, x_2, \ldots, x_n) \tag{11.1}$$

subject to

$$\begin{aligned} g_1(x_1, x_2, \ldots, x_n) &\leq b \\ g_2(x_1, x_2, \ldots, x_n) &\leq b_2 \\ &\cdots \\ g_m(x_1, x_2, \ldots, x_n) &\leq b_n \end{aligned} \tag{11.2}$$

TABLE 11.1
Various Levels of System Optimization

Time Duration	Control Process	Optimized Function
Seconds	Automatic generation control	Minimize area control error, subject to system dynamic constraints
Minutes	Optimal power flow	Minimize instantaneous cost of operation or other indexes, e.g., pollution
Hours and days	Unit commitment, hydrothermal	Minimize cost of operation
Weeks	Grid interchange coordination	Minimize cost with reliability constraints
Months	Maintenance scheduling	Minimize cost with reliability constraints
Years	Generation planning	Minimize expected investment and operational costs

The linear programming is a special case of an objective function (Equation 11.1) when all the constraints (Equation 11.2) are linear.

11.1 Functions of One Variable

A function $f(x)$ has its global minima at a point x^* if $f(x) \leq f(x^*)$ for all values of x over which it is defined. Figure 11.1 shows that the function may have relative maxima or minima. A *stationary point*, sometimes called a *critical point*, is defined where

$$f'(x) = \mathrm{d}f(x)/\mathrm{d}x = 0 \tag{11.3}$$

The critical point and stationary point are used interchangeably. Sometimes, the critical point is defined as any point which could be a global optimum. This is discussed further. The function $f(x)$ is said to have a weak relative maximum at x_0 if there exists $\in$, $0 < \in < \delta$, such

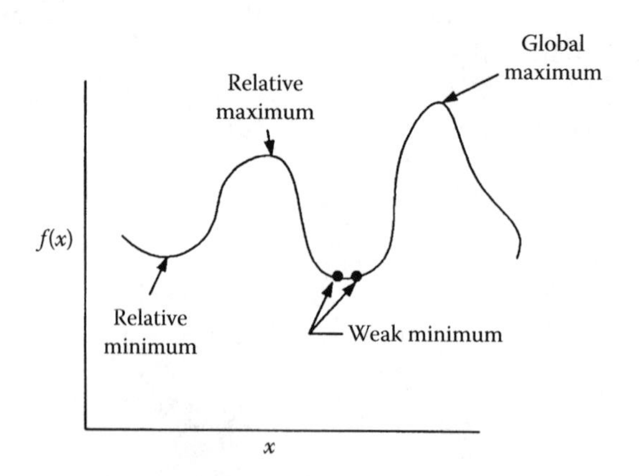

FIGURE 11.1
A single-value function showing relative maxima and minima.

that $f(x) \le f(x^0)$ and there is at least one point x in the interval $[x^0 - \in, x^0 + \in]$ such that $f(x) = f(x^0)$. The relative minimum of a function occurs at a point where its derivative is zero. For a vector

$$\begin{aligned} g_1 &= \left.\frac{\partial f}{\partial x_1}\right|_{x=x_0} = 0 \\ &\ldots \\ g_n &= \left.\frac{\partial f}{\partial x_n}\right|_{x=x_0} = 0 \end{aligned} \tag{11.4}$$

The derivative condition is necessary, but not sufficient, as the derivative can occur at maxima or saddle points. Additional conditions are required to ascertain that a minimum has been obtained.

11.2 Concave and Convex Functions

Important characteristics of the functions related to existence of maxima and minima are the convexity and concavity. A function $f(x)$ is convex for some interval in x, if for any two points x_1 and x_2 in the interval and all values of λ, $0 \le \lambda \le 1$

$$f\left[\lambda x_2 + (1-\lambda)x_1\right] \le \lambda f(x_2) + (1-\lambda) f(x_1) \tag{11.5}$$

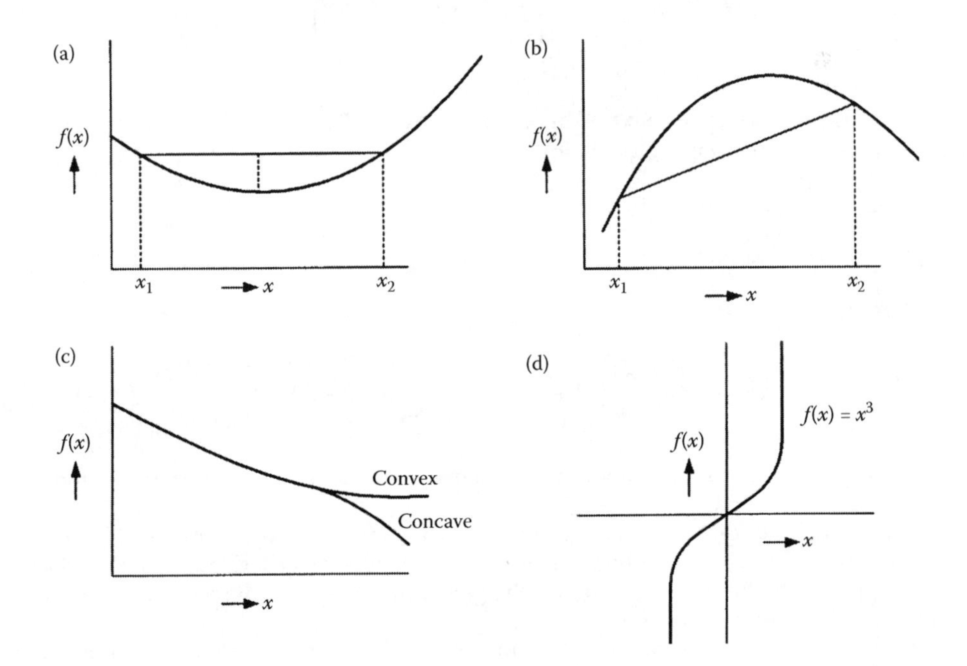

FIGURE 11.2
(a) A convex function, (b) concave function, (c) convex and concave function, and (d) graph of function $f(x) = x^3$.

The definition of the concave function is

$$f\left[\lambda x_2+(1-\lambda)x_1\right]\geq f(x_2)+(1-\lambda)f(x_1) \tag{11.6}$$

Figure 11.2 shows the convexity and concavity. Some functions may not be definitely convex or concave. The function $f(x) = x^3$ in Figure 11.2d is concave in the interval $(-\infty, 0)$ and convex in the interval $(0, \infty)$.

The convexity plays an important role. If it can be shown that the objective function is convex and the constraint set is convex, then the problem has a unique solution. It is not easy to demonstrate this. The optimal power flow problem (Chapter 12) is generally non-convex. Therefore, multiple minima may exist which may differ substantially.

11.3 Taylor's Theorem

If $f(x)$ is continuous and has a first derivative, then for any two points x_1 and x_2, where $x_2 = x_1 +$ incremental h, there is θ, $0 \leq \theta \leq 1$, *so that*

$$f(x_2)=f(x_1)+hf'\left[\theta x_1+(1-\theta)x_2\right] \tag{11.7}$$

Extending

$$f(x_2)=f(x_1)+hf'(x_1)+\frac{h^2}{2!}f'(x_1)+\cdots+\frac{h^n}{n!}f^{(n)}\left[\theta x_1+(1-\theta)x_2\right] \tag{11.8}$$

A function $f(x)$ has a relative maximum or minimum at x^* only if n is even, where n is the order of the first nonvanishing derivative at x^*. It is a maximum if $f^{(n)}(x^*)<0$ and a minimum if $f^{(n)}(x^*)>0$.

Example 11.1

Consider $f(x) = (x - 1)^4$

$$\frac{df(x)}{dx}=4(x-1)^3,\quad \frac{d^2f(x)}{dx^2}=12(x-1)^2,\quad \frac{d^3f(x)}{dx^3}=24(x-1),\quad \frac{d^4f(x)}{dx^4}=24$$

The fourth derivative is even, and is the first nonvanishing derivative at $x=1$; therefore, $x=1$ is the minimum.

For a continuous differentiable function, in a small interval, the maximum and minimum can be determined as shown above. These are relative or local, and most of the time, we will be interested in finding the global maxima or minima for $a \leq x \leq b$. A procedure for this is as follows:

Compute f(a) and f(b), compute $f'(x)$, find roots of $f'(x) = 0$. If there are no roots in [a, b], then z^* is larger of $f(a)$ and $f(b)$. If there are roots in [a, b], then z^* is the largest of $f(a), f(b), f(x_1),\ldots, f(x_k)$.

Example 11.2

Find global maxima of

$$f(x)=x^3+x^2-x+4, \quad 0\le x\le 2$$

Here

$$f(0)=4$$

$$f(2)=14$$

Compute $f'(x)$:

$$f'(x)=3x^2+2x-1$$

The roots are 1/3 and –1. The root –1 is not in interval [0, 2]; therefore, ignore it. Then $z^* = \text{Max}[f(0), f(2), f(1/3)]$; $f(2)$ gives the maxima = 14.

11.3.1 Optima of Concave and Convex Functions

- The convexity of a function assumes great importance for minimization. If $f(x)$ is convex over a closed interval $a \le x \le b$, then any relative minimum of $f(x)$ is also the global minima.
- The global maximum of a convex function $f(x)$ over a closed interval $a \le x \le$ b is taken on either $x = a$ *or* $x = b$ or both.
- If the functions $f_k(\overline{x})$, $k = 1, 2, \ldots, p$ are convex functions over some convex set X in Euclidean space E^n, then function $f(\overline{x})=\sum_{k=1}^{n} f_k(\overline{x})$ is also convex over X.
- The sum of convex functions is a convex function. The sum of concave functions is a concave function.
- If $f(\overline{x})$ is a convex function over a closed convex set X in E^n, then any local minimum of $f(\overline{x})$ is also the global minimum of $f(\overline{x})$ over X.

11.3.2 Functions of Multivariables

Equation 11.8 can be written as follows:

$$f(\overline{x}_2)=f(\overline{x}_1)+\overline{\nabla} f\left[\theta\overline{x}_1+(1-\theta)\overline{x}_2\right]h \tag{11.9}$$

where $\overline{\nabla} f$ is the gradient vector:

$$\overline{\nabla} f=\left(\frac{\partial f}{\partial x_1},\frac{\partial f}{\partial x_2},\ldots,\frac{\partial f}{\partial x_n}\right) \tag{11.10}$$

Matrix $\bar{H}$ is of $m \times n$ dimensions and is called a Hessian. It consists of second partial derivatives of $f(x)$:

$$H_{ij} = \frac{\partial^2 f(x_1, \ldots, x_n)}{\partial x_i \, \partial x_j} \tag{11.11}$$

A sufficient condition for $f(x)$ to have a relative minimum at point x^* is that $\bar{H}$ be positive definite. Also, if the solution to the gradient set of equations is unique, then the solution is a global minimum.

Example 11.3

Consider the solution of a function

$$3x_1^2 - 9x_1 + 4x_2^2 - 3x_1x_2 - 3x_2$$

The gradient vector is

$$g = \begin{vmatrix} 6x_1 - 9 - 3x_2 \\ 8x_2 - 3x_1 - 3 \end{vmatrix}$$

Setting it to zero and solving for x_1 and x_2 gives

$$\begin{vmatrix} x_1 \\ x_2 \end{vmatrix} = \begin{vmatrix} 3/13 \\ -1/13 \end{vmatrix}$$

This solution is unique. The Hessian is given by

$$\bar{H} = \begin{vmatrix} \dfrac{\partial^2 f}{\partial x_1^2} & \dfrac{\partial^2 f}{\partial x_1 \, \partial x_2} \\ \dfrac{\partial^2 f}{\partial x_2 \, \partial x_1} & \dfrac{\partial^2 f}{\partial x_2^2} \end{vmatrix} = \begin{vmatrix} 6 & -3 \\ -3 & 8 \end{vmatrix}$$

Thus, $\bar{H}$ is a positive definite, and therefore the above solution is a global solution.

A property of a function can be defined as *unimodality*. A function is unimodal if there is a path from every point x to the optimal point along which the function continuously increases or decreases. Figure 11.3a shows a strongly unimodal function, while Figure 11.3b shows a nonunimodal function. A strictly unimodal function will have just one local optimum that corresponds to the global optimum.

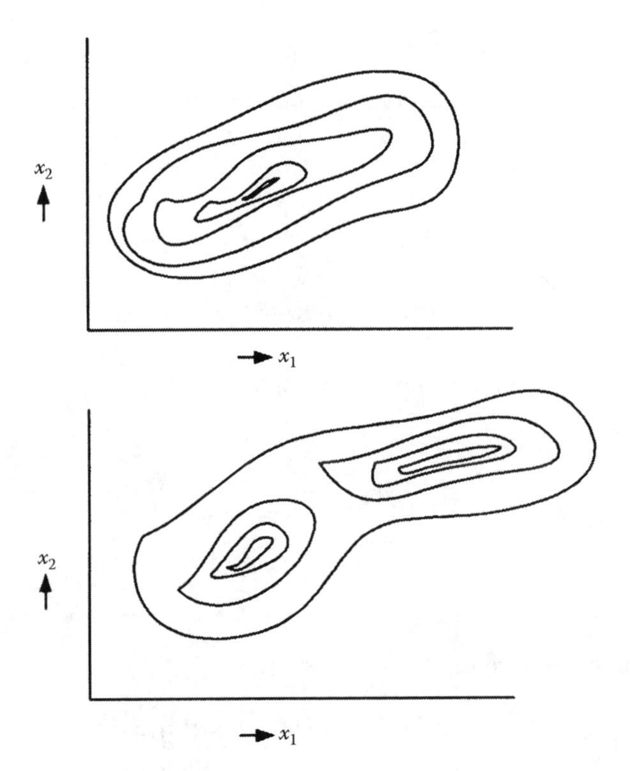

FIGURE 11.3
(a) Function showing strong unimodality and (b) nonunimodal function.

11.4 Lagrangian Method: Constrained Optimization

Suppose a function

$$f(x_1x_2) = k \tag{11.12}$$

has to be minimized subject to constraint that

$$g(x_1x_2) = b \tag{11.13}$$

The function $f(x_1, x_2)$ increases until it just touches the curve of $g(x_1, x_2)$. At this point, the slopes of f and g will be equal. Thus,

$$\begin{aligned} \frac{dx_1}{dx_2} &= -\frac{\partial f/\partial x_2}{\partial f/\partial x_1} \text{ slope of } f(x_1, x_2) \\ \frac{dx_1}{dx_2} &= -\frac{\partial g/\partial x_2}{\partial g/\partial x_1} \text{ slope of } g(x_1, x_2) \end{aligned} \tag{11.14}$$

Therefore,

$$\frac{\partial f/\partial x_2}{\partial f/\partial x_1}=\frac{\partial g/\partial x_2}{\partial g/\partial x_1} \tag{11.15}$$

or

$$\frac{\partial f/\partial x_2}{\partial f/\partial x_1}=\frac{\partial g/\partial x_2}{\partial g/\partial x_1}=\lambda \tag{11.16}$$

This common ratio λ is called the Lagrangian multiplier. Then

$$\frac{\partial f}{\partial x_1}-\lambda\frac{\partial g}{\partial x_1}=0 \tag{11.17}$$

$$\frac{\partial f}{\partial x_2}-\lambda\frac{\partial g}{\partial x_2}=0 \tag{11.18}$$

The Lagrangian function is defined as follows:

$$F(x_1,x_2,\lambda)=f(x_1,x_2)+\lambda\left[b-g(x_1,x_2)\right] \tag{11.19}$$

Differentiating Equation 11.19 with respect to x_1, x_2, and λ and equating to zero will give Equations 11.17 and 11.18. These are the same conditions as if a new unconstrained function h of three variables is minimized:

$$h(x_1x_2\lambda)=f(x_1x_2)-\lambda g(x_1x_2) \tag{11.20}$$

Thus,

$$\begin{aligned}\frac{\partial h}{\partial x_1}&=0\\ \frac{\partial h}{\partial x_2}&=0\\ \frac{\partial h}{\partial \lambda}&=0\end{aligned} \tag{11.21}$$

Example 11.4

Minimize the function

$$f(x_1x_2)=x_1+2x_1x_2+9.5x_2=k$$

for

$$g(x_1x_2)=x_1^2+x_2-12$$

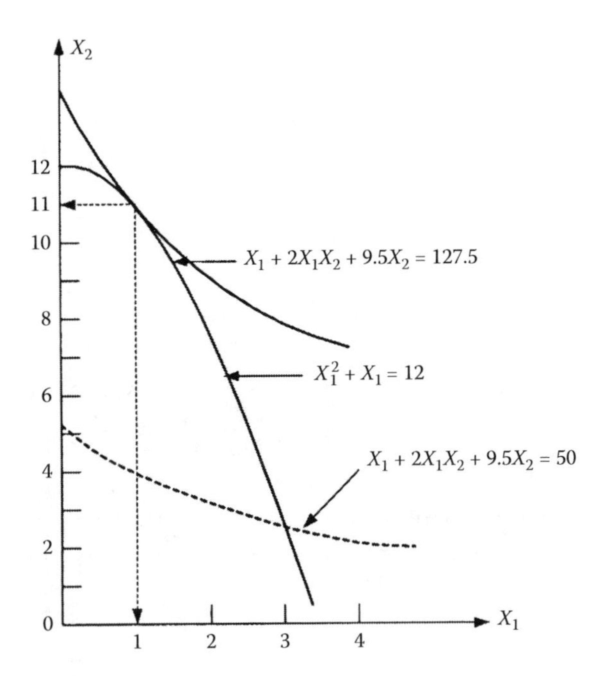

FIGURE 11.4
Functions in Example 11.3.

Form the function with the Lagrangian multiplier as in Equation 11.20

$$h(x_1x_2) = x_1 + 2x_1x_2 + 9.5x_2 - \lambda(x_1^2 + x^2 - 12)$$

$$\frac{\partial h}{\partial x_1} = 1 + 2x^2 - 2\lambda x_1 = 0$$

$$\frac{\partial h}{\partial x_2} = 2x_1 + 9.5 - \lambda = 0$$

$$\frac{\partial h}{\partial \lambda} = -x_1^2 - x^2 + 12 = 0$$

This gives $x_1 = 1$, $x_2 = 11$, and $\lambda = 11.5$. The value of the function is 127.5. However, this is not the global minimum.

Care has to be exercised in applying this method. A convex and a concave function may be tangential to each other, while the absolute minimum may be somewhere else, as in this example. Figure 11.4 shows that the calculated value is not really the minimum.

11.5 Multiple Equality Constraints

The function

$$f(x_1, x_2, \ldots, x_n) \tag{11.22}$$

subject to m equality constraints

$$\begin{aligned} &g_1(x_1, x_2, \ldots, x_n) \\ &g_2(x_1, x_2, \ldots, x_n) \\ &\ldots \\ &g_m(x_1, x_2, \ldots, x_n) \end{aligned} \tag{11.23}$$

can be minimized by forming a function

$$\begin{aligned} h(x_1, x_2, \ldots, x_n, \lambda_1, \lambda_2, \ldots, \lambda_m) &= f(x_1, x_2, \ldots, x_n) + \lambda_1 g_1(x_1, x_2, \ldots, x_n) \\ &+ \cdots + \lambda_m g_m(x_1, x_2, \ldots, x_n) \end{aligned} \tag{11.24}$$

where $n + m$ partial derivatives with respect to x_i, $i = 1$, to n, and λ_j, $j = 1$ to m are obtained. The simultaneous equations are solved and the substitution gives the optimum value of the function.

11.6 Optimal Load Sharing between Generators

We will apply Lagrangian multipliers to the optimal operation of generators, ignoring transmission losses. The generators can be connected to the same bus, without appreciable impedance between these, which will be a valid system for ignoring losses. The cost of fuel impacts the cost of real power generation. This relationship can be expressed as a quadratic equation:

$$C_i = \frac{1}{2} a_n P_{Gn}^2 + b_n P_{Gn} + w_n \tag{11.25}$$

where a_n, b_n, and w_n are constants. w_n is independent of generation.

The slope of the cost curve is the incremental fuel cost (IC):

$$\partial C_1 / \partial P_{Gn} = (IC)_n = a_n P_{Gn} + b_n \tag{11.26}$$

or inversely, the generation can be expressed as a polynomial of the form

$$P_{Gn} = \alpha_n + \beta_n (dC_n / dP_{Gn}) + \gamma_n (dC_n / dP_{Gn})^2 + \cdots \tag{11.27}$$

Considering spinning reserve, the total generation must exceed power demand. The following inequality must be strictly observed:

$$\sum P_G > P_D \tag{11.28}$$

where P_G is the real rated power capacity and P_D is the load demand. The load on each generator is constrained between upper and lower limits:

$$P_{\min} \le P \le P_{\max} \tag{11.29}$$

The operating cost should be minimized, so that various generators optimally share the load:

$$\begin{aligned} &C = \sum_{i=1}^{i=n} C_i P_{Gi} \text{ is minimum when} \\ &\sum_{i=1}^{i=n} P_{Gi} - P_D = 0 \end{aligned} \tag{11.30}$$

Further, the loading of each generator is constrained in Equation 11.29.

This is a nonlinear programming (NLP) problem as the cost index C is nonlinear. If the inequality constraint of Equation 11.28 is ignored, a solution is possible by Lagrangian multipliers:

$$\Gamma = C - \lambda\left[\sum_{i=1}^{i=n} P_{Gi} - P_D\right] \tag{11.31}$$

where λ is the Lagrangian multiplier.

Minimization is achieved by the condition that

$$\partial\Gamma / \partial P_{Gn} = 0 \tag{11.32}$$

Since C is a function of P only, the partial derivative becomes a full derivative:

$$\begin{aligned} &\frac{dC_n}{dP_{Gn}} = \lambda, \text{ i.e.,} \\ &\frac{dC_1}{dP_{G1}} = \frac{dC_2}{dP_{G2}} = \cdots = \frac{dC_n}{dP_{Gn}} = \lambda \end{aligned} \tag{11.33}$$

i.e., all units must operate at the same incremental cost. Figure 11.5 shows the graphic iteration of λ starting from an initial value of λ_0. The three different curves for C represent three different polynomials given by Equation 11.25. An initial value of $IC_0 = \lambda_0$ is assumed and the outputs of the generators are computed. If

$$\sum_{i=1}^{n} P_{Gi} = P_D \tag{11.34}$$

the optimum solution is reached; otherwise, increment λ by $\Delta\lambda$ and recalculate the generator outputs.

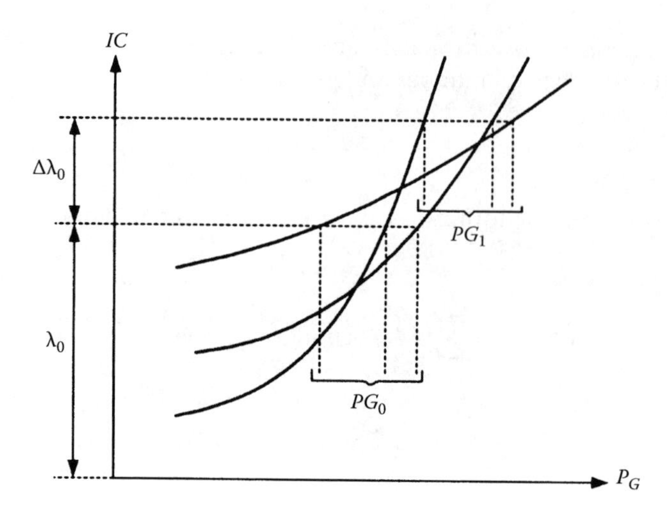

FIGURE 11.5
Graphical representation of Lagrangian multiplier iteration.

Example 11.5

Consider that two generators are required to share a load of 500 MW. The incremental costs of these two generators are given by

$$\frac{dC_a}{dP_{Ga}} = \lambda_a = 0.003P_a + 3.0\$/\text{MW h}$$

$$\frac{dC_b}{dP_{Gb}} = \lambda_b = 0.005P_b + 2.2\$/\text{MW h}$$

If P_a and P_b are the loads on each generator, then

$$P_a + P_b = 500$$

and from Equation 11.33

$$\lambda_a = \lambda_b$$

Solution of these equations gives

$$P_a = 175 \text{ MW}$$
$$P_b = 235 \text{ MW}$$
$$\lambda = \lambda_a = \lambda_b = 3.825(\$/\text{MW h})$$

11.7 Inequality Constraints

A function $f(x_1, x_2)$ subject to inequality constraint $g(x_1, x_2) \geq 0$ can be minimized by adding a nonnegative Z^2 to the inequality constraints; $g(x_1, x_2)$ takes a value other than zero, only if the constraint $g(x_1,x_2) \geq 0$ is violated. If $g(x_1, x_2) < 0$, Z^2 takes a value required to satisfy the equation $g(x_1,x_2) + Z^2 = 0$. If $g(x_1, x_2) \geq 0$, then $Z^2 = 0$:

$$h(x_1, x_2) = f(x_1, x_2) - \mu\left[g(x_1, x_2) + Z^2\right] = 0 \tag{11.35}$$

The function h now has four variables, x_1 x_2, μ, and Z. The partial derivatives of h are obtained with respect to these variables and are equated to zero:

$$\begin{aligned} \frac{\partial h}{\partial x_1} &= \frac{\partial f}{\partial x_1} - \mu \frac{\partial g}{\partial x_1} = 0 \\ \frac{\partial h}{\partial x_2} &= \frac{\partial f}{\partial x_2} - \mu \frac{\partial g}{\partial x_2} = 0 \\ \frac{\partial h}{\partial \mu} &= -g(x_1, x_2) - Z^2 = 0 \\ \frac{\partial h}{\partial Z} &= -2\mu Z = 0 \end{aligned} \tag{11.36}$$

The last condition means that either μ or Z or both μ and Z must be equal to zero.

Example 11.6

Minimize

$$f(x_1, x_2) = 2x_1 + 2x_1x_2 + 3x_2 = k$$

subject to the inequality constraint

$$g(x_1, x_2) = x_1^2 + x_2 \geq 0$$

First form the unconstrained function:

$$\begin{aligned} h &= f(x_1, x_2) - \mu\left[g(x_1, x_2) + Z^2\right] \\ &= 2x_1 + 2x_1x_2 + 3x_2 - \mu\left(x_1^2 + x_2 - 3 + Z^2\right) = k \end{aligned}$$

This gives

$$\begin{aligned} \frac{\partial h}{\partial x_1} &= 2 + 2x_2 - 2\mu x_1 = 0 \\ \frac{\partial h}{\partial x_2} &= 2x_1 + 3 - \mu = 0 \\ \frac{\partial h}{\partial \mu} &= x_1^2 + x_2 - 3 + Z^2 = 0 \\ \frac{\partial h}{\partial Z} &= 2\mu Z \end{aligned}$$

From the last equation, either Z or μ or both are zero. Assume Z = 0, then solving the equations gives $x_1 = 0.76$ and $x_2 = 2.422$. Again, it can be shown that this is the relative minimum and not the global minimum.

11.8 Kuhn–Tucker Theorem

This theorem makes it possible to solve an NLP problem with several variables, where the variables are constrained to satisfy certain equality and inequality constraints. The minimization problem with constraints for control variables can be stated as follows:

$$\min f(\bar{x},\bar{u}) \tag{11.37}$$

subject to the equality constraints

$$g(\bar{x},\bar{u},\bar{p})=0 \tag{11.38}$$

and inequality constraints

$$\bar{u}-\bar{u}_{max}\leq 0 \tag{11.39}$$

$$\bar{u}_{min}-\bar{u}\leq 0 \tag{11.40}$$

Assuming convexity of the functions defined above, the gradient

$$\bar{\nabla}\Gamma=0 \tag{11.41}$$

where Γ is the Lagrangian formed as follows:

$$\Gamma=f(\bar{x},\bar{u})+\lambda^t g(\bar{x},\bar{u},\bar{p})+\alpha_{max}^t(\bar{u}-\bar{u}_{max})+\alpha_{min}^t(\bar{u}_{min}-\bar{u}) \tag{11.42}$$

and

$$\begin{aligned}\alpha_{max}^t(\bar{u}-\bar{u}_{max})&=0\\ \alpha_{min}^t(\bar{u}_{min}-\bar{u})&=0\\ \alpha_{max}&\geq 0\\ \alpha_{min}&\geq 0\end{aligned} \tag{11.43}$$

Equations 11.43 are *exclusion* equations. Multiples α_{max} and α_{min} are dual variables associated with the upper and lower limits on control variables, somewhat similar to λ. The superscript t stands for transpose, see Volume 1. If u_i violates a limit, it can be either the upper or lower limit and not both. Thus, either of the two inequality constraints, Equation 11.39 or 11.40, is active at any one time, i.e., either α_{max} or α_{min} exists at one time and not both.

The gradient equation can be written as follows:

$$\frac{\partial \Gamma}{\partial x} = \frac{\partial f}{\partial x} + \left(\frac{\partial g}{\partial x}\right)^t \lambda = 0 \tag{11.44}$$

$$\frac{\partial \Gamma}{\partial u} = \frac{\partial f}{\partial u} + \left(\frac{\partial g}{\partial u}\right)^t \lambda + \alpha_i = 0 \tag{11.45}$$

In Equation 11.45

$$\begin{aligned} &\alpha_i = \alpha_{i,\max} \text{ if } u_i - u_{i,\max} > 0 \\ &\alpha_i = -\alpha_{i,\min} \text{ if } u_{i,\min} - u_i > 0 \\ &\frac{\partial \Gamma}{\partial \lambda} = g(\bar{x}, \bar{u}, \bar{p}) = 0 \end{aligned} \tag{11.46}$$

Thus, α given by Equation 11.45 for any feasible solution, with λ computed from Equation 11.44, is a negative gradient with respect to $\bar{u}$:

$$\alpha = -\frac{\partial \Gamma}{\partial u} \tag{11.47}$$

At the optimum, α must also satisfy the exclusions equations:

$$\begin{aligned} &\alpha_i = 0 \text{ if } u_{i(\min)} < u_i < u_{i(\max)} \\ &\alpha_i = \alpha_{i(\max)} \geq 0 \text{ if } u_i < u_{i(\max)} \\ &\alpha_i = -\alpha_{i(\min)} \leq 0 \text{ if } u_i < u_{i(\min)} \end{aligned} \tag{11.48}$$

Using Equation 11.47, these equations can be written as follows:

$$\begin{aligned} &\frac{\partial \Gamma}{\partial u_i} = 0 \text{ if } u_{i(\min)} < u_i < u_{i(\max)} \\ &\frac{\partial \Gamma}{\partial u_i} \leq 0 \text{ if } u_i = u_{i(\max)} \\ &\frac{\partial \Gamma}{\partial u_i} \geq 0 \text{ if } u_i = u_{i(\min)} \end{aligned} \tag{11.49}$$

11.9 Search Methods

In the above discussions, the functions must be continuous, differentiable, or both. In practice, very little is known about a function to be optimized. Many functions may defy simple characterization, convex or concave. The search methods can be classified as follows:

1. The unconstrained one-dimensional search methods, which can be simultaneous or sequential.

   The simultaneous search method can be again subdivided into the following:

   Exhaustive search

   Random search

   The random methods can be divided into the following:

   Dichotomous search

   Equal interval search

   Fibonacci search

2. Multidimensional search methods, which can also be simultaneous or sequential. Again, the simultaneous search methods are exhaustive search and random search, while sequential methods are as follows:

   Multivariate grid search

   Univariate search

   Powell's method

   Method of steepest descent

   Fletcher–Powell method

   Direct search

One-dimensional search methods place no constraints on the function, i.e., continuity or differentiability. An exhaustive one-dimensional search method, e.g., subdivides the interval [0,1] into $\Delta x/2$ equally spaced intervals, and the accuracy of the calculations will depend on the selection of Δx. In practice, the functions have more than one variable, maybe thousands of variables, which may react with each other, and, hence, multidimensional methods are applied. We will discuss and analyze all the search methods stated above.

11.9.1 Univariate Search Method

A univariate search changes one variable at a time so that the function is maximized or minimized. From a starting point, with a reasonable estimate of the solution x_0, find the next point x_1 by performing a maximization (or minimization) with respect to the variable:

$$\bar{x}_1 = \bar{x}_0 + \lambda_1 \bar{e}_1 \tag{11.50}$$

where $e_1 = [1, 0, \ldots, 0]$ and λ_1 is a scalar. This can be generalized so that

$$f(\bar{x}_k + \lambda_{k+1}\bar{e}_{k+1}) \quad k = 0, 1, \ldots, n-1 \tag{11.51}$$

is maximized. The process is continued until $|\lambda_k|$ is less than some tolerance value.

Table 11.2 for minimization of function: $3x_1^2 + 4x_2^2 - 5x_1x_2 - 2x_1$ illustrates this procedure and Figure 11.6 shows the search method. The advantage is that only one function is minimized at a time, while the others are held constant. The new values of the other functions

TABLE 11.2

Univariate Search Method–Function of Two Variables

Calculated Points	Minimized Function	Value Found	Best Current Estimate
(6, 6)	$f(x_1, 6)$	$x_1 = 5.2$	5.2, 6
(5.2, 6)	$f(5.2, x_2)$	$x_2 = 2.9$	5.2, 2.9
(5.2, 2.9)	$f(x_1, 2.9)$	$x_1 = 1.8$	1.8, 2.9
(1.8, 2.9)	$f(1.8, x_2)$	$x_2 = 0.8$	1.8, 0.8

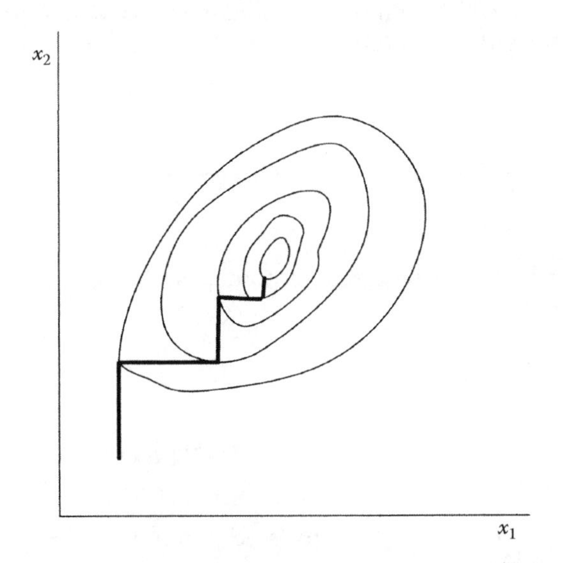

FIGURE 11.6
Convergence in gradient and optimal gradient methods.

are found, one at a time, by substituting the value of the first function and minimizing. The advantage is that it does not require calculations of derivatives. The search method is ineffective when the variables have interaction, or geometrically there are deep valleys and/or ridges.

11.9.2 Powell's Method of Conjugate Directions

If a quadratic function to be minimized is

$$h(\bar{x}) = \bar{x}^t \ \bar{A}\bar{x} + \bar{b}^t \bar{x} + c \tag{11.52}$$

The directions $\bar{P}$ and $\bar{q}$ are then defined as conjugate directions [3] if

$$\bar{p}^t \bar{A}\bar{q} = 0 \tag{11.53}$$

For a unique minimum, it is necessary that matrix $\bar{A}$ is positive definite. Each iterative step begins with a one-dimensional search in n linearly independent directions. If these directions are called

$$\bar{r}_1, \bar{r}_2, \ldots, \bar{r}_n \tag{11.54}$$

and we assume that we start at point x_0, then initially these directions are chosen to be the coordinates

$$r_1 = (x_1, 0, \ldots, 0),\ r_2 = (0, x_2, \ldots, 0), \ldots, r_n = (0, 0, \ldots, x_n) \tag{11.55}$$

The first iteration corresponds to the univariate method, in which one variable is changed at a time. Each iteration develops a new direction. If a positive definite quadratic function is being minimized, then after n iterations all the directions are mutually conjugate.

11.10 Gradient Methods

Starting with an initial value, a sequence of points can be generated so that each subsequent point makes

$$f(x^0) > f(x^1) > f(x^2) \cdots \tag{11.56}$$

$$f(x^0 + \Delta x) = f(x^0) + \left[\nabla f(x^0)\right]^t \Delta x + \cdots \tag{11.57}$$

Close to x^0, vector $\nabla f\left(x^0\right)$ is in a direction to increase $f(x)$. Thus, to minimize it, $-\nabla f\left(X^0\right)$ is used. If a gradient vector is defined as

$$\bar{g}_k = \bar{\nabla} f(X^k) \tag{11.58}$$

Then

$$\bar{x}^{k+1} = \bar{x}^k + (h_k - \bar{g}_k) \tag{11.59}$$

Example 11.7

Consider a function

$$f(x_1, x_2) = 2x_1^2 + 2x_1x_2 + 3x_2^2$$

Gradient vector g is

$$\nabla f = \begin{vmatrix} \dfrac{\partial f}{\partial x_1} \\ \dfrac{\partial f}{\partial x_2} \end{vmatrix} = \begin{vmatrix} 4x_1 + 2x_2 \\ 2x_1 + 6x_2 \end{vmatrix}$$

TABLE 11.3

Example 11.6: Minimization of a Function with Gradient Method

K	x_1, x_2	$f(x_1, x_2)$	g_k	$\lvert g_k \rvert$	$g_k/\lvert g_k \rvert$
0	4.000	32.000	16.000	17.888[a]	0.895
	0.000		8.000		0.447
1	3.105[b]	17.105	11.526	12.053	0.956
	−0.447		3.528		0.293
2	2.149	7.699	7.116	7.117	1.0
	−0.740		−0.142		−0.020
3	1.149	2.505	3.156	3.748	0.842
	−0.720		−2.022		−0.539
4	0.307	0.21	0.866	0.986	0.878
	−0.181		−0.472		−0.545
5	−0.571	0.213	–	–	–
	0.364				

[a] $\lvert g_k \rvert = \sqrt{32^2 + 16^2} = 17.888.$

[b] $4.000 - 0.895 = 3.105 \quad (h_0 = h_1 = h_2 = \cdots = 1).$

The successive calculations to $k = 5$ are shown in Table 11.3. Also see the flowchart in Figure 11.7. The initial assumed values are $x_1 = 4.000$ and $x_2 = 0.000$. Compare with Table 11.2 in which values fluctuate.

11.10.1 Method of Optimal Gradient

The method is also called the method of steepest descent [1, 4]. Determine h_k so that

$$h_k = \text{minimum} \quad f(\bar{x}^k - h\bar{g}_k) \tag{11.60}$$

The procedure is as follows:

- Set initial value of $x(= x^0)$
- Iteration count $k = 0$
- Calculate gradient vector g_k
- Find h_k to minimize Equation 11.60

Then

$$\bar{x}^{k+1} = \bar{x}^k - h_k\bar{g}_k \tag{11.61}$$

The convergence is reached if

$$\left| f\left(\bar{x}^{k+1}\right) - f\left(\bar{x}^k\right) \right| < \in \tag{11.62}$$

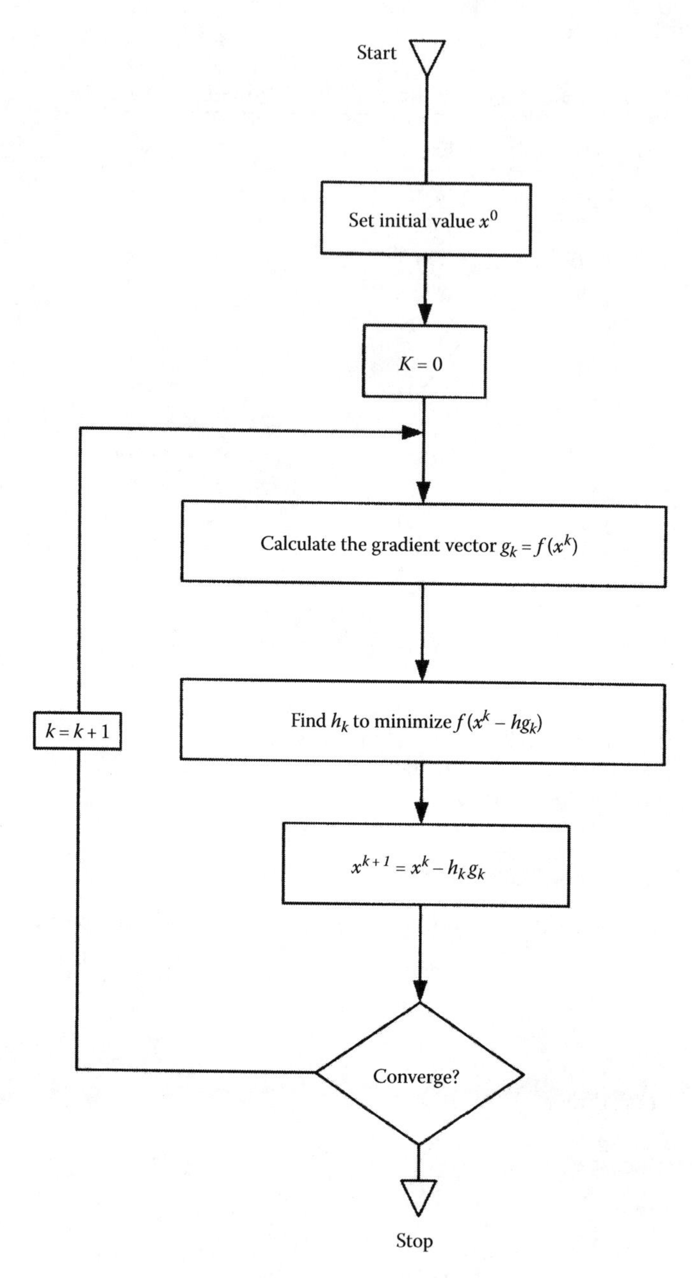

FIGURE 11.7
Flow Chart for the gradient method.

Example 11.8

The previous example is solved by the optimal gradient method to five iterations. The results of the calculations are shown in Table 11.4. The minimum is being approached faster in lesser iterations. A comparison between the gradient method and the optimal gradient method is shown in Figure 11.8.

All iterative methods of minimization, whether quadratically convergent or not, locate h as the limit of sequence $x_0, x_1, x_2, \ldots,$ where x_0 is the initial approximation to the position of the minimum and where for each subsequent iteration, x_i is the position

TABLE 11.4

Example 11.7: Minimization with Optimal Gradient Method

k	x^k	f_k	g_k	$f(x^k - hg_k)$	h_k
0	4.000 0.000	32	16.000 8.000	960.00 h^2 – 320.00 h + 32[a]	0.1758
1	1.1872 –1.4064	5.4176	1.936 –6.064	94.330 h^2 – 40.5196 h + 5.4176	0.2148
2	0.7713 –0.101	1.0646	2.883 0.937	16.623 h^2 – 5.4027 h + 1.0646	0.1863
3	0.2342 –0.2755	0.208	0.386 –1.185	3.593 h^2 – 1.5524/2 h + 0.208	0.2160
4	0.1508 –0.0196	0.0407	0.564 0.184	0.9543 h^2 – 0.3519 h + 0.0407	0.18615
5	0.0458 0.0538	0.01780	–	–	–

[a] $2x_1^2 + 2x_1x_2 + 3x^29 = 960(\text{for } x_1 = 16, x_2 = 8)$

$$|g_k|^2 = (17.888)^2 = 320$$

$$x^{k+1} = x^k - h_k g_k = 4 - (0.1758)(16) = 1.1872$$

and $0 - (0.1758)(8) = -1.4064$

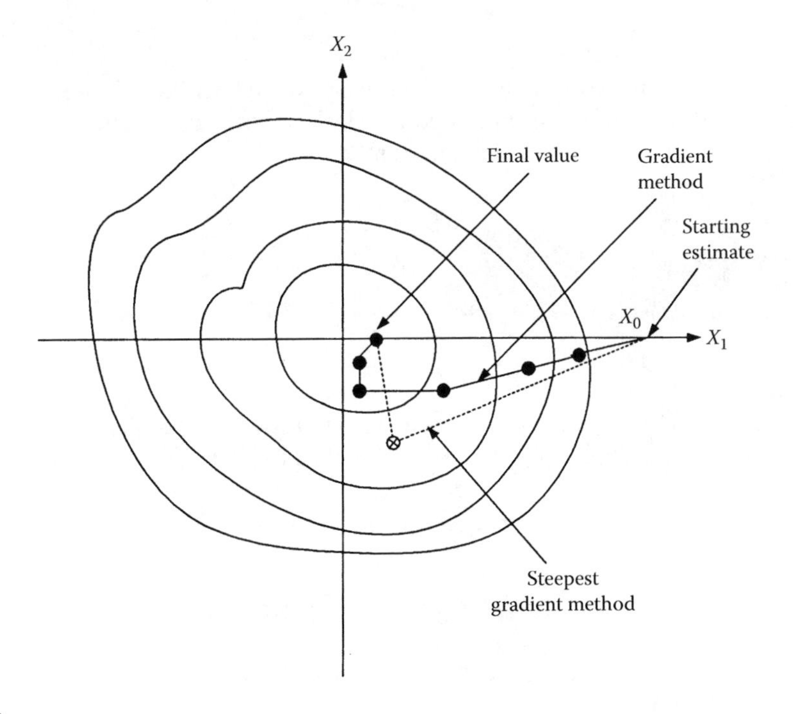

FIGURE 11.8
Convergence in gradient and optimal gradient methods.

of the minimum with respect to variations along the line through x_i in some specified direction p_i. The method of steepest descent uses the directions of the negative gradient of $f(x)$ at x_i, and the method of alternate directions uses cyclically the directions of the n coordinate axes. Methods that calculate each new direction as a part of iteration cycles are more powerful than those in which the directions are assigned in advance [5].

11.11 Linear Programming—Simplex Method

A linear programming problem [1] can be defined as follows:

$$\text{Max } z = \sum_{j=1}^{n} c_j x_j$$

$$\text{st}$$

$$\sum_{j=1}^{i=n} \alpha_{ij} x_j = b_i \quad i = 1,2,\ldots,m \tag{11.63}$$

$$x_j \geq 0 \quad j = 1,2,\ldots,n$$

where "st" stands for "subject to." The coefficients αc_j, and a_{ij} are real scalars and may vary. The objective function

$$f(\bar{x}) = \sum_{j=1}^{n} c_j x_j \tag{11.64}$$

is in linear form in variables x_j, whose values are to be determined. The inequalities can be converted by addition of either nonnegative slack or surplus variables to equalities. In matrix form, Equation 11.63 can be written as follows:

$$\begin{aligned} max\ z &= \bar{c}^t \bar{x} \\ \bar{A}\bar{x} &= \bar{b} \\ \bar{x} &\geq \bar{0} \end{aligned} \tag{11.65}$$

Define the following:

A *feasible solution* to the linear programming is a vector which satisfies all the constraints of the problem. The X feasible solutions are defined as follows:

$$X = \left\{ \bar{x} \middle| A\bar{x} = \bar{b}, \bar{x} \geq \bar{0} \right\} \tag{11.66}$$

A *basic feasible solution* is a feasible solution with no more than m positive x values. These positive x values correspond to linearly independent columns of matrix $\bar{A}$. A nondegenerate basic feasible solution is a basic feasible solution with exactly m positive x_j values. It has fewer than m positive x_j values.

The following can be postulated:

- Every basic feasible solution corresponds to an extreme point of the convex set of feasible solutions of X.
- Every extreme point of X has m linearly independent vectors (columns of matrix $\bar{A}$) associated with it.

- There is some extreme point at which the objective function z takes its maximum or minimum value.

The matrix $\bar{A}$ can be partitioned into a basic matrix $\bar{B}$ and a nonbasic matrix $\bar{N}$. Similarly, the vector $\bar{x}$ can be partitioned and the following equation can be written:

$$\bar{A}\bar{x} = (\bar{B}, \bar{N})\begin{bmatrix} \bar{x}_B \\ \bar{x}_N \end{bmatrix} = \bar{b} \tag{11.67}$$

Therefore,

$$\bar{B}\bar{x}_B + \bar{N}\bar{x}_N = \bar{b} \tag{11.68}$$

For the basic solution, $\bar{x}_N$ should be zero; $\bar{x}_B = B^{-1}\bar{b}$ is the basic solution. Variables $\bar{x}_B$ are called the basic variables. If the vector $\bar{c}$ is partitioned as $\bar{c}_B, \bar{c}_N$, then the optimum function $z = \bar{c}_B\bar{x}_B$. From the basic properties of vectors in E^n, any combination of the columns of the $\bar{A}$ matrix can be written as a linear combination of columns in the matrix $\bar{B}$:

$$\bar{a}_j = \sum_{i=1}^{m} y_{ij}\bar{a} = B\bar{y}_i \tag{11.69}$$

or

$$\bar{y}_j = B^{-1}\bar{a}_j \tag{11.70}$$

If a new basic feasible solution is required, then one or more vectors $\bar{a}_i$ should be removed from $\bar{B}$, and substituted with some other vectors $\bar{a}_j$ from $\bar{N}$. In the simplex method of Dantzig, only one vector at a time is removed from the basis matrix and replaced with the new vector which enters the basis [6]. As each feasible solution corresponds to an *extreme point* of the convex set of solutions, by replacing one vector at a time, *the movement is from one extreme point to an adjacent extreme point of the convex set*. The interior point method basically differs in this concept.

Consider a vector $\bar{a}_r$ from the set $\bar{a}_j$ which is to enter the basis, so that $\bar{y}_{rj} \neq 0$; then from Equation 11.69

$$\bar{a}_r = \frac{1}{y_{rj}}\bar{a}_j - \sum_{i=1}^{m} \frac{y_{ij}}{y_{rj}}\bar{a}_i \tag{11.71}$$

The substitution gives the new basic solution. One more condition that needs to be satisfied is that x_{B_r} is nonnegative. The column r to be removed from the basis can be chosen by the following equation:

$$\frac{x_{Br}}{y_{rj}} = \min\left\{\frac{x_{Bi}}{y_{ij}}, y_{ij} > 0\right\} \tag{11.72}$$

The minimum in Equation 11.72 may not be unique. In this case, one or more variables in the new basic solution will be zero and the result will be a degenerate basic solution. The vector to enter the basis can be selected so that

1. The new objective function = old objective function $+(x_{Br}/y_{rj})(c_j - z_j)$.
2. In the absence of degeneracy, $(x_{Br}/y_{rj}) > 0$. Therefore, $(c_j - z_j)$ should be selected to be greater than zero.

Example 11.9

Maximize

$$z = 2x_1 - 5x_2$$
$$\text{st}$$
$$2x_1 + 4x_2 \geq 16$$
$$3x_1 + 9x_2 \leq 30$$
$$x_1, x_2 \geq 0$$

Convert inequalities to equalities by adding slack or surplus variables:

$$2x_1 + 4x_2 - x_3 = 16$$
$$3x_1 + 9x_2 + x_4 = 30$$
$$x_1, x_2, x_3, x_4 \geq 0$$

Define matrix $\bar{A}$ and its vectors $\bar{a}_p, j = 1,2,3,$ and 4

$$\bar{A} = \begin{vmatrix} 2 & 4 & -1 & 0 \\ 3 & 9 & 0 & 1 \end{vmatrix} \quad \bar{b} = \begin{vmatrix} 16 \\ 30 \end{vmatrix}$$

Therefore,

$$\bar{a}_1 = \begin{vmatrix} 2 \\ 3 \end{vmatrix} \bar{a}_2 = \begin{vmatrix} 4 \\ 9 \end{vmatrix} \bar{a}_3 = \begin{vmatrix} -1 \\ 0 \end{vmatrix} \bar{a}_4 = \begin{vmatrix} 0 \\ 1 \end{vmatrix}$$

Consider that $\bar{a}_1$ and $\bar{a}_4$ form the initial basis; then

$$x_1\bar{a}_1 + x_4\bar{a}_4 = \bar{b}$$

$$x_1 \begin{vmatrix} 2 \\ 3 \end{vmatrix} + x_4 \begin{vmatrix} 0 \\ 1 \end{vmatrix} = \begin{vmatrix} 16 \\ 30 \end{vmatrix}$$

This gives $x_1 = 8$, $x_2 = 6$, and $x_3 = x_4 = 0$. Therefore, the initial feasible solution is

$$\bar{x}_B = \begin{vmatrix} 8 \\ 6 \end{vmatrix}$$

Express vectors $\bar{a}_2$ and $\bar{a}_3$ not in the basis in terms of vectors $\bar{a}_1$ and $\bar{a}_4$:

$$\bar{a}_2 = y_{12}\bar{a}_1 + y_{42}\bar{a}_4$$
$$\bar{a}_3 = y_{13}\bar{a}_1 + y_{43}\bar{a}_4$$

Therefore,

$$\left|\begin{matrix} 4 \\ 9 \end{matrix}\right| = y_{12}\left|\begin{matrix} 2 \\ 3 \end{matrix}\right| + y_{42}\left|\begin{matrix} 0 \\ 1 \end{matrix}\right|$$

Solving: $y_{12} = y_{42} = 3$; therefore, $\bar{y}_2 = |2,3|$.
Similarly,

$$\left|\begin{matrix} -1 \\ 0 \end{matrix}\right| = y_{13}\left|\begin{matrix} 2 \\ 3 \end{matrix}\right| + y_{43}\left|\begin{matrix} 0 \\ 1 \end{matrix}\right|$$

Solving, $y_{13} = -1/2$, $y_{42} = 3/2$; therefore, $\bar{y}_3 = |-1/2, 3/2|$.
Compute z for $\bar{y}_2 = |2,3|$ and $\bar{y}_3 = |-1/2, 3/2|$:

$$z_j = \bar{c}_B' \bar{y}_j$$

Here, $\bar{c}_B^t = |2,0|$. Therefore,

$$z_2 = \left|\begin{matrix} 2 & 0 \end{matrix}\right| \left|\begin{matrix} 2 \\ 3 \end{matrix}\right| = 4$$

$$z_3 = \left|\begin{matrix} 2 & 0 \end{matrix}\right| \left|\begin{matrix} -1/2 \\ 3/2 \end{matrix}\right| = -1$$

where $z_2 - c_2 = 4 + 5 = 9$, and $z_3 - c_3 = -1 - 0 = -1$.
Since $z_3 - c_3 < 0$, $\bar{a}_3$ will enter the basis. The vector to leave the basis can only be $\bar{a}_1$ or $\bar{a}_4$. The criterion is given by Equation 11.72:

$$\min\left\{\frac{x_{Bi}}{y_{13}}, y_{13} > 0\right\} \text{ i.e., } \frac{x_1}{y_{13}} \text{ or } \frac{x_4}{y_{43}} > 0$$

but $y_{13} = -1/2$; therefore, $\bar{a}_4$ leaves the basis. The original value of the objective function is

$$z = \bar{c}_B' \bar{x}_B = |2,0| \left|\begin{matrix} 8 \\ 6 \end{matrix}\right| = 16$$

The new value is given by

$$\hat{z} = z + \frac{x_4}{y_{43}}(c_3 - z_3) = 16 + \frac{6}{3/2}(1) = 20$$

where $\hat{z}$ denotes the new value.
Compute the new value of $\bar{x}_B$:

$$x_1\bar{a}_1 + x_3\bar{a}_3 = \bar{b}$$

$$x_1\begin{vmatrix} 2 \\ 3 \end{vmatrix} + x_3\begin{vmatrix} -1 \\ 0 \end{vmatrix} = \begin{vmatrix} 16 \\ 30 \end{vmatrix}$$

Solving for $x_1 = 10, x_3 = 4$

$$\hat{z} = |2,0|\begin{vmatrix} 10 \\ 4 \end{vmatrix} = 20$$

$\bar{a}_1$ was in the basis, $\bar{a}_4$ left the basis, and $\bar{a}_3$ entered the basis. Calculate $\bar{a}_2$ and $\bar{a}_4$ in terms of $\bar{a}_1$ and $\bar{a}_3$:

$$\bar{a}_2 = y_{12}\bar{a}_1 + y_{32}\bar{a}_3$$

$$\bar{a}_4 = y_{14}\bar{a}_1 + y_{34}\bar{a}_3$$

$$\begin{vmatrix} 4 \\ 9 \end{vmatrix} = y_{12}\begin{vmatrix} 2 \\ 3 \end{vmatrix} + y_{32}\begin{vmatrix} -1 \\ 0 \end{vmatrix}$$

$$\begin{vmatrix} 0 \\ 1 \end{vmatrix} = y_{14}\begin{vmatrix} 2 \\ 3 \end{vmatrix} + y_{34}\begin{vmatrix} -1 \\ 0 \end{vmatrix}$$

$\bar{y}_2 = |3, 2|$ and $\bar{y}_4 = |1/3, 2/3|$.
Compute Z_j

$$z_2 = |2,0|\begin{vmatrix} 3 \\ 2 \end{vmatrix} = 6 \quad z_4 = |2,0|\begin{vmatrix} 1/3 \\ 2/3 \end{vmatrix} = 2/3$$

$$z_2 - c_2 = 6 + 5 > 0$$

$$z_4 - c_4 = 2/3 - 0 > 0$$

Therefore, the optimal solution is as follows: $x_1 = 8, x_2 = 0, x_3 = 6, x_4 = 0$, and $z = 20$.

11.12 Quadratic Programming

NLP solves the optimization problems that involve a nonlinear objective and constraint functions. Problems with nonlinearity confined to objective function only can be solved as an extension of the simplex method. Sensitivity and barrier methods (Chapter 16) are considered fairly generalized to solve NLP problems. Quadratic programming (QP) is a special case where the objective function is quadratic, involving square or cross-products of two variables. There are solution methods with the additional assumption that the function is convex. The type of problem is referred to as quadratic optimization. It can be

applied to maintaining required voltage profile, maximizing power flow, and minimizing generation costs. The problem is of the form:
maximize $f(\overline{x}) = \overline{c}'\overline{x} + \overline{x}'P\overline{x}$

$$\text{maximize } f(\overline{x}) = \overline{c}^t\overline{x} + \overline{x}^t P\overline{x}$$

$$\text{st} \tag{11.73}$$

$$A\overline{x} \le \overline{b} \text{ and } \overline{x} \ge 0$$

There are m constraints and n variables; $\overline{A}$ is an $m \times n$ matrix, P is an $n \times n$ matrix, $\overline{c}$ is an n component vector, and $\overline{b}$ an m component vector, all with known linear elements. The vector $\overline{x}' = (x_1, x_2, \ldots, x_n)$ is the vector of unknowns. The nonlinear part of the problem is the objective function second term $\overline{x}'P\overline{x}$. Because it is of quadratic form, the matrix P is symmetric. We can postulate the following:

1. The function $f(\overline{x})$ is concave if P is a negative semidefinite or negative definite.
2. The function $f(\overline{x})$ is convex if P is a positive semidefinite or positive definite.
3. There are no computationally feasible methods for obtaining global maximum to QP unless $f(\overline{x})$ is concave. If the function is concave, then Kuhn–Tucker conditions are satisfied by a global maxima:

$$F\left(\overline{x}\overline{\lambda}\right) = \overline{c}^t\overline{x} + \overline{x}^t P\overline{x} + \sum_{i=1}^{m} \lambda_i \left[b_i - \sum_{j=1}^{n} a_{ij} x_j \right] \tag{11.74}$$

Therefore, the global maximum to QP is any point $\overline{x}_0$ which satisfies the following:

$$\overline{x}_{0j} > 0$$

$$\partial F / \partial x_j = c_j + 2\sum_{k=1}^{n} p_{jk} x_k - \sum_{i=1}^{m} \lambda_i a_{ij} = 0 \quad \text{at} \quad \overline{x} = \overline{x}_0 \tag{11.75}$$

$$\text{If} \quad \overline{x}_{0j} = 0$$

$$\partial F / \partial x_j = c_j + 2\sum_{k=1}^{n} p_{jk} x_k - \sum_{i=1}^{m} \lambda_i a_{ij} \le 0 \quad \text{at} \quad \overline{x} = \overline{x}_0 \tag{11.76}$$

$$\text{If} \quad \lambda_{0j} > 0$$

$$\partial F / \partial \lambda_i = b_i - \sum_{j=1}^{n} a_{ij} x_j = 0 \quad \text{at} \quad \overline{x} = \overline{x}_0 \tag{11.77}$$

$$\text{If} \quad \lambda_{0j} = 0$$

$$\partial F / \partial \lambda_i = b_i - \sum_{j=1}^{n} a_{ij} x_j \le 0 \quad \text{at} \quad \overline{x} = \overline{x}_0. \tag{11.78}$$

These relations can be rewritten by adding slack variables $y_j, j=1,2,...,n, x_{si}, i=1,2,...,n$ to Equations 11.75 and 11.77:

$$2\sum_{k=1}^{n} p_{jk}x_k - \sum_{i=1}^{m}\lambda_i a_{ij} + y_j = -c_j$$
$$\sum_{j=1}^{n} a_{ij}x_j + x_{si} = b_i \tag{11.79}$$

Note that

$$x_j y_j = 0 \quad j=1,2,...,n$$
$$\lambda_i x_{si} = 0 \quad i=1,2,...,n \tag{11.80}$$

Thus, a point $\overline{x}_0$ will have a global maximum solution only if there exist nonnegative numbers $\lambda_1, \lambda_2, ..., \lambda_m, x_{s1}, x_{s2}, ..., x_{sm}, y_1, y_2, ..., y_n$ so that Equations 11.78–11.80 are satisfied. Except for constraints in () these are all linear and simplex method can be applied.

A general equation for QP with only inequality constraint is

$$\text{minimize} \quad c^t x + \frac{1}{2}x^t Gx$$
$$\text{st} \tag{11.81}$$
$$Ax > b$$

G and A are n-square symmetric matrix and x and c are n-vectors.

11.13 Dynamic Programming

Dynamic programming (DP) can be described as a computational method of solving optimization problems without reference to particularities, i.e., linear programming or NLP. It can be used to solve problems where variables are continuous and discrete or for optimization of a definite integral. It is a multistage decision process, where the techniques of linear programming and NLP optimization are not applicable.

A system at discrete points can be represented by state vectors:

$$X_i = [x_1(i), x_2(i), ..., x_n(i)] \tag{11.82}$$

A *single* decision (di) is made out of a number of choices to start, and at every stage a single decision is taken, then at stage $i+1$, state vector is $X(i+1)$. DP is applicable when

$$X(i+1) = G[X(i), d(i), i] \tag{11.83}$$

At each stage, a return function is described:

$$R(i) = R\left[X(i), d(i), i\right] \quad (11.84)$$

Certain constraints are imposed at each stage:

$$\varphi[X(i), d(i), i] \leq 0 \geq 0 = 0 \quad (11.85)$$

If there are *N* stages and total return is denoted by *I*, then

$$I = \sum_{i=1}^{N} R(i) \quad (11.86)$$

That is, choose sequences *d*(1), *d*(2), …, d(*N*), which result in optimization of *I*. This is called a discrete time multistage process. If any of the functions, *G*, *R*, … defining such a system does not involve chance elements, the system is referred as deterministic; otherwise, it is called a probabilistic system.

11.13.1 Optimality

It is based on the principle of optimality, which states that a subpolicy of an optimal policy must in itself be an optimal subpolicy. The essential requirement is that the objective function must be separable.

The problem with *n* variables is divided into *n* subproblems, each of which contains only one variable. These are solved sequentially, so that the combined optimal solution of *n* problems yields the optimal solution to the original problem of *n* variables.

Let initial stage be C, and first decision is taken arbitrarily, then corresponding to *d*(*i*):

$$X(2) = G[C, d(1)] \quad (11.87)$$

The maximum return over the remaining $N - 1$ stages is

$$f_{N-1}[G\{C, d(1)\}] \quad (11.88)$$

and the total return is the sum of Equations 11.87 and 11.88. Then using the principle of optimality

$$f_N(C) = R[C, d_N(c)] + f_{N-1}[G\{C, d_N(C)\}] \quad (11.89)$$

Consider that an objective function is separable and can be divided into *n* subfunctions, let us say three, for example. Then

$$f(x) = f_1(x_1) + f_2(x_2) + f_3(x_3) \quad (11.90)$$

The optimal value of $f(x)$ is then the maximum of

$$\max\left[f_1(x_1)+f_2(x_2)+f_3(x_3)\right] \tag{11.91}$$

taken over all the nonnegative integers of the constraint variables. If the optimal solution is described by $\hat{x}_1, \hat{x}_2$, and $\hat{x}_3$, and assuming that $\hat{x}_2$ and $\hat{x}_3$ are known, then the problem reduces to that of a single variable:

$$\max\left\{f_1 x_1 + \left[f_2\hat{x}_2 + f_3\hat{x}_3\right]\right\} \tag{11.92}$$

We do not know $\hat{x}_2$ and $\hat{x}_3$, but these must satisfy the constraints. That is the principle of optimality is applied in stages, till the entire system is optimal.

As an example, consider that the total number of units in a generating station, their individual cost characteristics, and load cycle on the station are known; the unit commitment can be arrived at by DP.

Let the cost function be defined as follows:

$F_N(x)$ = minimum cost of generating x MW from N units

$f_N(y)$ = Cost of generating y MW by Nth unit

$f_{N-1}(x - y)$ = Minimum cost of generating $(x - y)$ MW by $N - 1$ units DP gives the following recursive relationship:

$$F_N(x)=\text{Min}\left[f_N(y)+F_{N-1}(x-y)\right] \tag{11.93}$$

Let t units be generated by N units. As a first step, arbitrarily choose *any one* unit out of the t units. Then, $F_1(t)$ is known from Equation 11.78. Now, $F_2(t)$ is the minimum of

$$\begin{gathered}\left[f_2(0)-F_1(t)\right]\\ \left[f_2(1)-F_1(t-1)\right]\\ \cdots\\ \left[f_2(t)-F_1(0)\right]\end{gathered} \tag{11.94}$$

This will give the most economical *two units* to share the total load. The cost curve of these two units can be combined into a single unit, and the third unit can be added. Similarly, the minima of $F_3(t)$, $F_4(t)$, …, $F_n(t)$ is calculated.

Example 11.10

Consider that three thermal units, characteristics, and cost indexes as shown in Table 11.5 are available. Find the unit commitment for sharing a load of 200 MW. The indexes a and b for each unit in cost Equation 11.25 are specified in Table 11.5 and indexes w are zero. Use a second-degree polynomial as

$$C_i=\frac{1}{2}a_i P_{Gi}^2+b_i P_{Gi}+w_i \ \$/\text{h} \tag{11.95}$$

TABLE 11.5

Example 11.9: Dynamic Programming—A System with Three Thermal Units

	Generation Capability		Cost Indexes		
Unit Number	Minimum	Maximum	a	b	w
1	50	200	0.02	0.50	0
2	50	200	0.04	0.60	0
3	50	200	0.03	0.70	0

We select unit 1 as the first unit:

$$F_1(200) = f_1(200) = \frac{1}{2}(0.02)(200)^2 + (0.50)(200) = \$500/\text{h}$$

We will consider a step of 50 MW to illustrate the problem. Practically, it will be too large. From Equation 11.79

$$F_2(200) = \min\{[f_2(0) + F_1(200)], [f_2(50) + F_1(150)], [f_2(100) + F_1(100)]$$
$$\times [f_2(150) + F_1(50)], [f_2(200) + F_1(0)]\}$$

The minimum of these expressions is given by

$$[f_2(50) + F_1(150)] = \$380/\text{h}$$

Similarly, calculate

$$F_2(150) = \min\{[f_2(0) + F_1(150)], [f_2(50) + F_1(100)], [f_2(100)$$
$$+ F_1(50)], [f_2(150) + F_1(0)]\}$$

$$F_2(100) = \min\{[f_2(0) + F_1(100)], [f_2(50) + F_1(50)], [f_2(100) + F_1(0)]\}$$

$$F_2(50) = \min\{[f_2(0) + F_1(50)], [f_2(50) + F_1(0)]\}$$

The minimum values are as follows:

$$F_2(150) = [f_2(50) + F_1(100)] = \$230/\text{h}$$

$$F_2(100) = [f_2(50) + F_1(50)] = \$130/\text{h}$$

$$F_2(50) = [f_2(0) + F_1(50)] = \$50/\text{h}$$

Bring the third unit in

$$F_3(200) = \min\{[f_3(0) + F_2(200)], [f_3(50) + F_2(150)][f_3(100) + F_2(100)]$$
$$\times [f_3(150) + F_2(50)][f_3(200) + F_2(0)]\}$$

The steps as for unit 2 are repeated; these are not shown. The minimum is given by

$$F_3(200) = [f_3(50) + F_2(150)] = \$302.50/\text{h}$$

Therefore, the optimum load sharing on units 1, 2, and 3 is 100, 50, and 50 MW, respectively. If we had started with any of the three units as the first unit, the results would have been identical. The dimensions for large systems are of major consideration, and often DP is used as a subprocess within an optimization process.

A final value problem can be converted into an initial value problem.

Example 11.11

Maximize

$$x_1^2 + x_2^2 + x_3^2$$

$$\text{st}$$

$$x_1 + x_2 + x_3 \geq 10$$

Let

$$u_3 = x_1 + x_2 + x_3$$

$$u_2 = x_1 + x_2 = u_3 - x_3$$

$$u_1 = x_1 = u_2 - x_2$$

Then

$$\begin{aligned} F_3(u_3) &= \text{Max}_{x3}\left[x_3^2 + F_2(u_2)\right] \\ F_2(u_2) &= \text{Max}_{x_2}\left[x_2^2 + F_1(u_2)\right] \\ F_1(u_1) &= [x_1^2] = (u_2 - x_2)^2 \end{aligned} \tag{11.96}$$

Substituting

$$F_2(u_2) = \text{Max}_{x_2}\left[x_2^2 + (u_2 - x_2)^2\right] \tag{11.97}$$

Take the partial derivative and equate to zero for maxima:

$$\begin{aligned} \partial F_2(u_2)/\partial x_2 &= 2x_2 - 2(u_2 - x_2) = 0 \\ x_2 &= 2u_2 \end{aligned} \tag{11.98}$$

Therefore,

$$\begin{aligned} F_3(u_3) &= \text{Max}_{x_3}\left[x_3^2 + F_2(u_2)\right] \\ &= \text{Max}_{x_2}\left[x_3^2 + 5(u_3 - x_3)^2\right] \end{aligned} \tag{11.99}$$

Taking the partial derivative

$$\partial F_3(u_3)/\partial x_3 = 2x_3 - 10(u_3 - x_3) = 0$$
$$x_3 = (5/6)u_3 \tag{11.100}$$

Hence

$$F_3(u_3) = (5/6)u_3^2 \quad u_3 > 10$$

That is, $F_3(u_3)$ is maximum for $u_3 = 10$. Back substituting gives $x_3 = -8.55$, $x_2 = 3.34$, $x_1 = -1.67$. The maximum value of the function is 83.33.

11.14 Integer Programming

Many situations in power systems are discrete, i.e., a capacitor bank is online or offline. A generator is synchronized: on at some time (status = 1) and off (status 0) at other times. If all the variables are of integer type, the problem is called integer programming. If some variables are of continuous type, the problem is called mixed-integer programming. These problems have a nonconvex feasible region and a linear interpolation between feasible points (status 0 and 1) gives an infeasible solution. Nonconvexity makes these problems more difficult to solve than those of smooth continuous formulation. The two mathematical approaches are as follows:

- Branch and bound methods
- Cutting plane methods

In the branch and bound method, the problem is divided into subproblems based on the values of the variables, i.e., consider three variables and their *relaxed* noninteger solution as $x_1 = 0.85$, $x_2 = 0.45$, and $x_3 = 0.92$. Here, x_2 is far removed from the integer values; thus, a solution is found with $x_2 = 0$ and $x_2 = 1$. This leads to two new problems, with alternate possibilities of $x_2 = 1$ or $x_2 = 0$. The process can be continued, where the possible solutions with all integer values will be displayed. Bounding is used to cut off whole sections of the tree, without examining them and this requires an incumbent solution, which can be the best solution found so far in the process, or an initial solution using heuristic criteria.

In the cutting plane method, additional constraints, called cutting planes, are introduced, which create a sequence of continuous problems. The solution of these continuous problems is driven toward the best integer solution.

Optimization applied to power systems is a large-scale problem. Given a certain optimization problem, is there a guarantee that an optimal solution can be found and the method will converge? Will the optimized system be implementable? How it can be tested? A study [6] shows that optimal power flow results are sensitive to ULTC (under-load tap changing) operation and load models. The optimized system did not converge on load flow and had transient stability problems. High-fidelity mathematical models and accurate robust methods of solution are required for real solutions to real problems and to avoid the dilemma of real solutions to nonreal problems or nonsolutions to real world problems.

Problems

11.1 Find minimum of the function: $2x_1^2 + 3x_2^2 + 3x_3^2 - 6x_1 - 8x_2 + 20x_3 + 45$.

11.2 Find global maximum of

$$f(\bar{x}) = 5(x_1 - 2)^2 - 7(x_2 + 3)^2 + 7x_1x_2 \quad \bar{a} \leq \bar{x} \leq \bar{b}$$

$$\bar{a} = [0, 0] \quad \bar{b} = [10, 4]$$

11.3 Minimize $z = 3x_1^2 + 2x_2^2$, subject to $3x_1 - 2x_2 = 8$.

11.4 If two functions are convex, is it true that their product will also be convex?

11.5 Find maximum of $x_1^2x_2^2x_3^2x_4^2$, given that $x_1^2 + x_2^2 + x_3^2 + x_4^2 = c$.

11.6 Minimize: $f(x_1, x_2) = 2.5x_1^2 + 3x_1x_2 + 2.5x_2^2$ by gradient method.

11.7 Repeat Problem 6, with optimal gradient method.

11.8 Maximize: $z = 4x_1 + 3x_2$ st

$$2x_1 + 3x_2 \leq 16$$

$$4x_1 + 2x_2 \leq 10$$

$$x_1, x_2 \geq 0$$

11.9 There are three generators, each having a minimum and maximum generation capability of 10 and 50 MW, respectively. The cost functions associated with these three generators can be represented by Equation 11.95. Assume that the cost factors of generators 1, 2, and 3 are the same as in Table 11.5. Consider a step of 10 MW.

11.10 The incremental costs of two generators in \$/MWh is given by

$$dC_1/dP_{G1} = 0.02P_{G1} + 4$$

$$dC_1/dP_{G1} = 0.015P_{G1} + 3$$

Assuming that the total load varies over 100–400 MW, and the maximum and minimum loads on each generator are 250 and 50 MW, respectively, how should the load be shared between the two units as it varies?

References

1. PE Gill, W Murray, MH Wright. *Practical Optimization*. Academic Press, New York, 1984.
2. DG Lulenberger. *Linear and Non-Linear Programming*. Addison Wesley, Reading, MA, 1984.
3. RJ Vanderbei. *Linear Programming: Foundations and Extensions*, 2nd ed. Kluwer, Boston, MA, 2001.

4. R Fletcher, MJD Powell. A rapidly convergent descent method for minimization. *Comput J*, 5(2), 163–168, 1962.
5. R Fletcher, CM Reeves. Function minimization by conjugate gradients. *Comput J*, 7(2), 149–153, 1964.
6. GB Dantzig. *Linear Programming and Extensions*. Princeton University Press, Princeton, NJ, 1963.
7. E Vaahedi, HMZ El-Din. Considerations in applying optimal power flow to power system operation. *IEEE Trans Power App Syst*, 4(2), 694–703, 1989.

12

Optimal Power Flow

In the load flow problem, the system is analyzed in symmetrical steady state. The specified variables are real and reactive power at *PQ* buses, real powers and voltages at *PV* buses, and voltages and angles at slack buses. The reactive power injections can also be determined, based on the upr and lower limits of the reactive power control. However, this does not immediately lead to optimal operating conditions, as infinitely variable choices exist in specifying a balanced steady-state load flow situation.

The constraints on the control variables are not arbitrary, i.e., the power generation has to be within the constraints of load demand and transmission losses. These demands and limits are commonly referred to as equality and inequality constraints. There are a wide range of control values for which these constraints may be satisfied, and it is required to select a performance, which will minimize or maximize a desired performance index (Chapter 11).

12.1 Optimal Power Flow

The optimal power flow (OPF) problem was defined in early 1960, in connection with the economic dispatch of power [1]. Traditionally, the emphasis in performance optimization has been on the cost of generation; however, this problem can become fairly complex when the hourly commitment of units, hourly production of hydroelectric plants, and cogeneration and scheduling of maintenance without violating the needs for adequate reserve capacity are added. The OPF problem can be described as the cost of minimization of real power generation in an interconnected system where real and reactive power, transformer taps, and phase-shift angles are controllable and a wide range of inequality constraints are imposed. It is a static optimization problem of minute-by-minute operation. It is a nonlinear optimization problem. In load flow, we linearize the network equations in terms of given constraints about an assumed starting point and then increment it with Δ, repeating the process until the required tolerance is achieved. Today, any problem involving the steady state of the system is referred to as an OPF problem. The time span can vary (Table 12.1). In the OPF problem, the basic definitions of state variable, control vector, and input demand vector are retained. OPF requires solving a set of nonlinear equations, describing optimal operation of the power system:

$$\begin{aligned} &\text{Minimize} \quad F(\bar{x},\bar{u}) \\ &\text{with constraints} \quad g(\bar{x},\bar{u}) = 0 \qquad (12.1) \\ &h(\bar{x},\bar{u}) \leq 0 \end{aligned}$$

TABLE 12.1

Constraints, Controls, and Objective Functions in OPF

Inequality and Equality Constraints	Controls	Objectives
Power flow equations	Real and reactive power generation	Minimize generation cost
Limits on control variables		Minimize transmission losses
Circuit loading active and reactive	Voltage profiles and Mvar generation at buses	Minimize control shifts
Net area active and reactive power generation	LTC transformer tap positions	Minimize number of controls rescheduled
		Optimize voltage profile
Active and reactive power flow in a corridor	Transformer phase shifts	Minimize area active and reactive power loss
	Net interchange	
Unit Mvar capability	Synchronous condensers	Minimize shunt reactive power compensation
Active and reactive resverve limits	SVCs, capacitors, and reactor banks	
		Minimize load shedding
Net active power export	Load transfer	Minimize air pollution
Bus voltage magnitudes and angle limits	HVDC line MW flows	
	Load shedding	
Spinning reserve	Line switching	
Contingency constraints	Standby start-up units	
Environmental and security constraints		

Note: Dependent variables are all variables that are not control functions.

where $g(\bar{x},\bar{u})$ represent nonlinear equality constraints and power flow constraints, and $h(\bar{x},\bar{u})$ are nonlinear inequality constraints, limits on the control variables, and the operating limits of the power system. Vector $\bar{x}$ contains dependent variables. The vector $\bar{u}$ consists of control variables; $F(\bar{x},\bar{u})$ is a scalar objective function.

Constraints, controls, and objectives are listed in Table 12.1.

12.1.1 Handling Constraints

The constraints are converted into equality constraints by including slack variables or including constraints at the binding limits. Let $B(\bar{x},\bar{u})$ be the binding constraint set, comprising both equalities and enforced inequalities $h(\bar{x},\bar{u})$ with $u_{\min} \le u \le u_{\max}$.

The Lagrangian function of Equation 12.1 is then

$$\Gamma(\bar{x},\bar{u}) = f(\bar{x},\bar{u}) + \lambda^t B(\bar{x},\bar{u}) \quad (12.2)$$

The binding constraints cannot be definitely known beforehand and identification during optimization becomes necessary. A back-off mechanism is necessary to free inequality constraints if later on these become nonbinding.

Another approach is the penalty modeling, which penalizes the cost function if the functional inequality constraint is violated. Its Lagrangian function is

$$\Gamma(\bar{x},\bar{u}) = f(\bar{x},\bar{u}) + \lambda^t B(\bar{x},\bar{u}) + W(\bar{x},\bar{u}) \tag{12.3}$$

where

$$W(\bar{x},\bar{u}) = \sum_i r_i h_i^2(\bar{x},\bar{u}) \tag{12.4}$$

where i is the violated constraint, r_i is the penalty weight, and $h_i(\bar{x},\bar{u})$ is the ith constraint. Kuhn–Tucker conditions should be met for the final $\bar{x},\bar{u}$, and λ. Squaring the constraints increases the nonlinearity and decreases the sparsity.

The augmented Lagrangian function [2] is a convex function with suitable values of ρ and is written as

$$\Gamma(\bar{x},\bar{u},\lambda,\rho) = f(\bar{x},\bar{u}) - \lambda^t B(\bar{x},\bar{u}) + \frac{1}{2}\rho B(\bar{x},\bar{u})^t B(\bar{x},\bar{u}) \tag{12.5}$$

where

$$\begin{gathered} B(\bar{x},\bar{u}) = 0 \\ (x,u)_{\min} \le (x,u) \le (x,u)_{\max} \end{gathered} \tag{12.6}$$

A large value of ρ may mask the objective function and a small value of ρ may make the function concave. Thus, the augmented Lagrangian method minimizes Equation 12.5, subject to constraints, Equation 12.6.

12.2 Decoupling Real and Reactive Power OPF

It was shown in Chapter 4 that the real and reactive subsets of variables and constraints are weakly coupled. For high X/R systems, the effect of real power flow on voltage magnitude and of reactive power flow on voltage phase-angle change is relatively negligible. Thus, during a real power OPF subproblem, the reactive power control variables are kept constant, and in reactive power OPF, the real power controls are held constant at their previously set values [3]. A summary of control and constraints of decoupled subproblem is shown in Table 12.2. The advantages are as follows:

- Decoupling greatly improves computational efficiency.
- Different optimization techniques can be used.
- A different optimization cycle for each subproblem is possible.

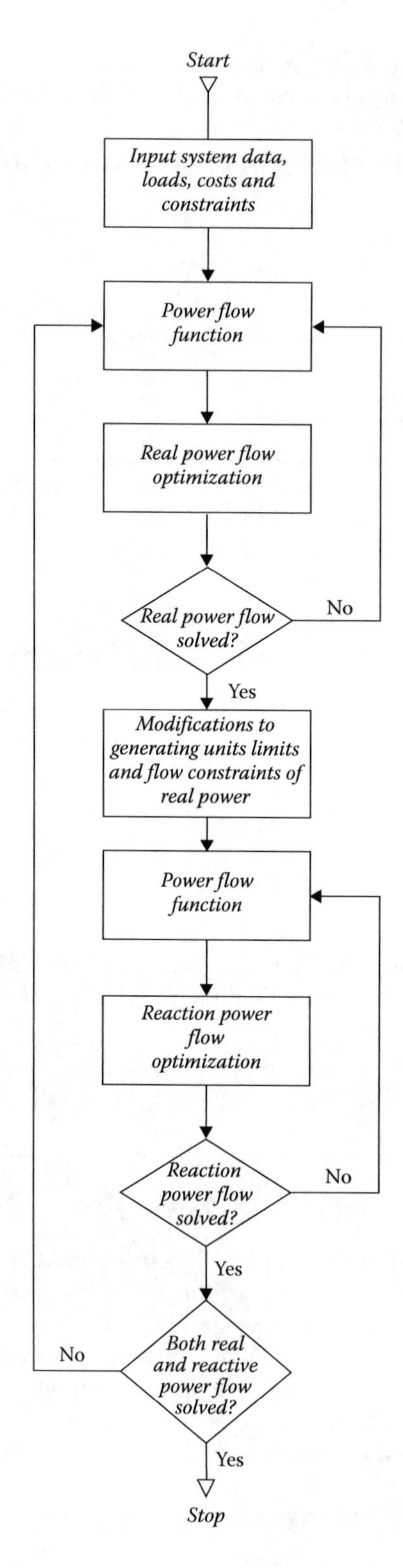

FIGURE 12.1
Decoupled optimal power flow.

TABLE 12.2
Controls and Constraints of Decoupled Active and Reactive Power OPF

OPF	Constraints	Controls
Active power OPF	Network power flow Bus voltage angles Circuit loading MW MW branch flow MW reserve margins Area MW interchange MW flow on a corridor Limits on controls	MW generation Transformer phase shifter positions Area MW Interchange HVDC line MW flows Load shedding Line Switching Load transfer Fast start-up units
Reactive power OPF	Network reactive power flow Bus voltage magnitude Mvar loading Unit Mvar capability Area Mvar generation Corridor Mvar flow Mvar reserve margins Limits on controls	Generator voltages and Mvars LTC tap positions Capacitor, reactor, SVC synchronous condenser statuses

Real power controls are set to optimize operating costs, while reactive power controls are optimized to provide secure postcontingency voltage level or reactive power dispatch. The solution method for the real and reactive power OPF subproblem can be different. Figure 12.1 shows the flowchart for the decoupled solution.

12.3 Solution Methods of OPF

OPF methods can be broadly classified into two optimization techniques:

- Linear programming (LP)-based methods
- Nonlinear programming-based methods

LP is reliable for solving specialized OPF problems characterized by linear separable objectives. The nonlinear programming techniques are as follows:

- Sequential quadratic programming
- Augmented Lagrangian methods
- Generalized reduced gradient methods
- Projected augmented Lagrangian
- Successive LP
- Interior point (IP) methods

The new projective scaling Karmarkar's algorithm, see Section 12.9, for LP has the advantage of speed for large-scale problems by as much as 50:1 when compared to simplex methods. The variants of inheritor-point methods are dual affine, barrier, and primal affine algorithm.

12.4 Generation Scheduling Considering Transmission Losses

In Chapter 11, we considered optimizing generation, ignoring transmission losses. Generation scheduling, considering transmission line losses, can be investigated as follows.

The objective is to minimize the total cost of generation at any time and meeting the constraints of load demand with transmission losses, that is, to minimize the function,

$$C = \sum_{i=1}^{i=k} C_i(P_{Gi}) \tag{12.7}$$

Total generation = total demand plus total transmission losses. Therefore,

$$\sum_{i=1}^{k} P_{Gi} - P_D - P_L = 0 \tag{12.8}$$

where k is the number of generating plants, P_L is the total transmission loss, and P_D is the system load (total system demand). To minimize the cost, solve for the Lagrangian as

$$\Gamma = \sum_{i=1}^{k} C_i(P_{Gi}) - \lambda\left[\sum_{i=1}^{k} P_{Gi} - P_D - P_L\right] \tag{12.9}$$

If power factor at each bus is assumed to remain constant, the system losses are a function of generation and the power demand is unpredictable; thus, the power generation is the only control function. For optimum real power dispatch,

$$\frac{\partial \Gamma}{\partial P_{Gi}} = \frac{dC_i}{dP_{Gi}} - \lambda + \lambda \frac{\partial P_L}{\partial P_{Gi}} = 0, \quad i = 1, 2, \ldots, k \tag{12.10}$$

Rearranging

$$\lambda = \frac{dC_i}{dP_{Gi}} L_i \quad i = 1, 2, \ldots, k \tag{12.11}$$

where

$$L_i = \frac{1}{(1 - \partial P_L / \partial P_{Gi})} \tag{12.12}$$

is called the penalty factor of the ith plant. This implies that the minimum fuel cost is obtained when the incremental fuel cost of each plant multiplied by its penalty factor is the same for all the plants. The incremental transmission loss (ITL) associated with the ith plant is defined as $\partial P_L / \partial P_{Gi}$. Thus,

$$\frac{dC_i}{dP_{Gi}} = \lambda[1-(\text{ITL})_i] \tag{12.13}$$

This equation is referred to as the *exact coordination equation*.

12.4.1 General Loss Formula

The exact power flow equations should be used to account for transmission loss. Commonly, the transmission loss is expressed in terms of active power generation only. This is called the *B*-coefficient method. The transmission losses using *B* coefficients are given by

$$P_L = \sum_{m=1}^{k}\sum_{n=1}^{k} P_{Gm}B_{mn}P_{Gn} \tag{12.14}$$

where P_{Gm}, P_{Gn} is the real power generation at *m*, *n* plants, and B_{mn} are loss coefficients, which are constant under certain assumed conditions.

In matrix form,

$$\overline{P}_L = \overline{P}_G^t \overline{B}\, P_G^t \tag{12.15}$$

$$\overline{P}_G = \begin{vmatrix} P_{G1} \\ P_{G2} \\ - \\ P_{Gk} \end{vmatrix} \quad \overline{B} = \begin{vmatrix} B_{11} & B_{12} & \cdot & B_{1k} \\ B_{21} & B_{22} & \cdot & B_{2k} \\ \cdot & \cdot & \cdot & \cdot \\ B_{k1} & B_{k2} & \cdot & B_{kk} \end{vmatrix} \tag{12.16}$$

The *B* coefficients are given by

$$B_{mn} = \frac{\cos(\theta_m - \theta_n)}{|V_m||V_n|\cos\phi_m \cos\phi_n} \sum_p I_{pm} I_{pm} R_p \tag{12.17}$$

where θ_m and θ_n are the phase angles of the generator currents at *m* and *n* with respect to a common reference; cos ϕ_m and cos ϕ_n are the power factors of the load currents at *m* and *n* plants; I_{pm} and I_{pn} are the current distribution factors, i.e., the ratio of load current to total load current; and R_p is the resistance of the *p*th branch.

In order that B_{mn} do not vary with the load, the assumptions are as follows:

- Ratios I_{pm} and I_{pn} remain constant.
- Voltage magnitudes remain constant.
- The power factor of loads does not change, i.e., the ratio of active to reactive power remains the same.
- Voltage phase angles are constant.

From Equations 12.15 and 12.16, for a three-generator system, the loss equation will be

$$P_L = B_{11}P_{G1}^2 + B_{22}P_{G2}^2 + B_{33}P_{G3}^2 + 2B_{12}P_{G1}P_{G2} + 2B_{23}P_{G2}P_{G3} + 2B_{31}P_{G3}P_{G1} \tag{12.18}$$

Due to simplifying assumptions, a more accurate expression for loss estimation may be required. The loss equation based on bus impedance matrix and current vector is

$$\bar{I}_{\text{bus}} = \bar{I}_p + j\bar{I}_q = \begin{vmatrix} I_{p1} \\ I_{p2} \\ \cdot \\ I_{pn} \end{vmatrix} + j \begin{vmatrix} I_{q1} \\ I_{q2} \\ \cdot \\ I_{qn} \end{vmatrix}$$

$$\begin{aligned} \bar{P}_L + j\bar{Q}_L &= \bar{I}^t_{\text{bus}} \bar{Z}_{\text{bus}} \bar{I}^*_{\text{bus}} \\ &= (\bar{I}_p + j\bar{I}_q)^t (\bar{R} + j\bar{X})(\bar{I}_p - j\bar{I}_q) \end{aligned} \tag{12.19}$$

The $\bar{Z}_{\text{bus}}$ is symmetrical; therefore, $\bar{Z}_{\text{bus}} = \bar{Z}^t_{\text{bus}}$. Expanding and considering the real part only,

$$\bar{P}_L = \bar{I}^t_p \bar{R} \bar{I}_p + \bar{I}^t_p \bar{X} \bar{I}_q + \bar{I}^t_q \bar{R} \bar{I}_q - \bar{I}^t_q \bar{X} \bar{I}_p$$

Again, as $\bar{X}$ is symmetrical, $\bar{I}^t_p \bar{X} \bar{I}_q - \bar{I}^t_q \bar{X} \bar{I}_p = 0$, which gives

$$\bar{P}_L = \bar{I}^t_p \bar{R} \bar{I}_p + \bar{I}^t_q \bar{R} \bar{I}_q \tag{12.20}$$

The currents at a bus in terms of bus voltage and active and reactive power are as follows:

$$I_{pi} = \frac{1}{|V_i|}(P_i \cos\theta_i + Q_i \sin\theta_i) \tag{12.21}$$

$$I_{qi} = \frac{1}{|V_i|}(P_i \sin\theta_i - Q_i \cos\theta_i) \tag{12.22}$$

Substituting these current values into Equation 12.20 and simplifying, the loss equation is

$$P_L = \sum_{j=1}^{n} \sum_{k=1}^{n} \left[\frac{R_{ki} \cos(\theta_i - \theta_k)}{|V_k||V_i|}(P_k P_i + Q_k Q_i) + \frac{R_{ki} \sin(\theta_i - \theta_k)}{|V_k||V_i|}(P_k Q_i - Q_k P_i) \right] \tag{12.23}$$

For small values of θ_k and θ_i, the second term of Equation 12.23 can be ignored:

$$P_L = \sum_{j=1}^{n} \sum_{k=1}^{n} \left[\frac{R_{ki} \cos(\theta_i - \theta_k)}{|V_k||V_i|}(P_k P_i + Q_k Q_i) \right] \tag{12.24}$$

Equation 12.24 can be put in matrix form. Let

$$C_{ki} = \frac{R_{ki} \cos(\theta_i - \theta_k)}{|V_k||V_i|} \tag{12.25}$$

$$D_{ki} = \frac{R_{ki} \sin(\theta_i - \theta_k)}{|V_k||V_i|} \tag{12.26}$$

Then Equation 12.24 is

$$P_{\text{loss}} = \sum_{k=1}^{n} \sum_{i=1}^{n} (P_k C_{ki} P_i + Q_k C_{ki} Q_i + P_k D_{ki} Q_i - Q_k D_{ki} P_i) \tag{12.27}$$

or in matrix form

$$P_{\text{loss}} = |P_1, P_2, \ldots, Pn \quad Q_1, Q_2, \ldots, Q_n|$$

$$\times \begin{vmatrix} C_{11} & C_{12} & . & C_{1n} & D_{11} & D_{12} & . & D_{1n} \\ C_{21} & C_{22} & . & C_{2n} & D_{21} & D_{22} & . & D_{2n} \\ . & . & . & . & . & . & . & . \\ -D_{11} & -D_{12} & . & -D_{1n} & C_{11} & C_{12} & . & C_{1n} \\ -D_{21} & -D_{22} & . & -D_{2n} & C_{21} & C_{22} & . & C_{2n} \\ . & . & . & . & . & . & . & . \\ -D_{n1} & -D_{n2} & . & -D_{nn} & C_{n1} & C_{n2} & . & C_{nn} \end{vmatrix} \cdot \begin{vmatrix} P_1 \\ P_2 \\ . \\ P_n \\ Q_1 \\ Q_2 \\ . \\ Q_n \end{vmatrix} \tag{12.28}$$

or in partitioned form

$$\bar{P}_{\text{loss}} = |\bar{P}_l \bar{Q}_l| \begin{vmatrix} \bar{C} & \bar{D} \\ -\bar{D} & \bar{C} \end{vmatrix} \begin{vmatrix} \bar{P} \\ \bar{Q} \end{vmatrix} \tag{12.29}$$

where $\bar{D}$ is zero for the approximate loss equation.

12.4.2 Solution of Coordination Equation

The solution of the coordination equation is an iterative process. For the nth plant,

$$\frac{dC_n}{dP_n} + \lambda \frac{\partial P_L}{\partial P_n} = \lambda \tag{12.30}$$

As C_n is given by Equation 12.25,

$$(\text{IC})_n = \frac{dC_n}{dP_n} = a_n P_n + b_n \quad \$/\text{MWh} \tag{12.31}$$

From Equation 12.14,

$$\frac{\partial P_L}{\partial P_n} = 2 \sum_{m=1}^{k} P_m B_{mn} \tag{12.32}$$

or

$$\frac{\partial P_L}{\partial P_k} = 2\sum_{i=1}^{n}(C_{ki}P_i + D_{ki}Q_i) \tag{12.33}$$

In the approximate form,

$$\frac{\partial P_L}{\partial P_k} = 2\sum_{i=1}^{n} C_{ki}P_i \tag{12.34}$$

Thus, substituting in (12.31),

$$a_n P_n + b_n + 2\lambda \sum_{m=1}^{k} 2B_{mn}P_m = \lambda$$

$$P_n = \frac{1 - \dfrac{b_n}{\lambda} - \displaystyle\sum_{m=1\ m\neq n}^{k} 2B_{mn}P_m}{\dfrac{a_n}{\lambda} + 2B_{mn}} \tag{12.35}$$

The iterative process is enumerated as follows:

1. Assume generation at all buses except the swing bus and calculate bus power and voltages based on load flow.
2. Compute total power loss P_L. Therefore, ITL can be computed based on Equation 12.33.
3. Estimate initial value of λ, and calculate $P_1, P_2, \ldots, P_n$ based on equal incremental cost.
4. Calculate generation at all buses; Equation 12.34 or polynomial 12.26 can be used.
5. Check for ΔP at all generator buses:

$$\Delta P = \left| \sum_{i=1}^{n} P_{Gi}^{k} - P_D - P_L^{k} \right| \leq \epsilon_1 \tag{12.36}$$

6. Is $\Delta P < \epsilon$?

If no, update λ. If yes, is the following inequality satisfied?

$$P_{Gi}^{k} - P_{Gi}^{k-1} \leq \epsilon_2 \tag{12.37}$$

7. If no, advance iteration count by 1 to $k+1$.

The flowchart is shown in Figure 12.2.

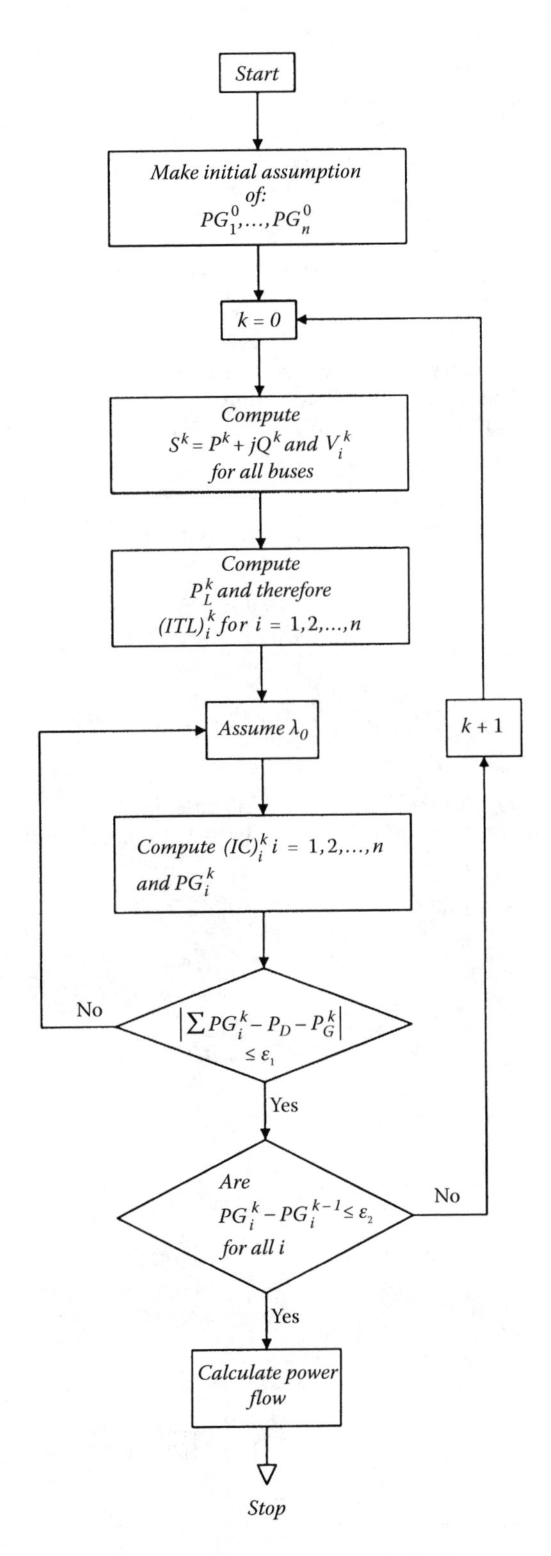

FIGURE 12.2
Flowchart: iterative generator scheduling considering transmission losses.

Example 12.1

Consider two generators of Example 11.4. These supply load through transmission lines as shown in Figure 12.3. The line impedances are converted to a load base of 500 MVA for convenience. The power factor of the load is 0.9 lagging. From Equation 12.34 and ignoring coefficients D_{ki},

$$\frac{\partial P_L}{\partial P_1} = 2(C_{11}P_1 + C_{12}P_2) = 2C_{11}P_1$$

where coefficient C_{12} represents the resistance component of the mutual coupling between the lines. As we are not considering coupled lines, $C_{12}=0$. If the sending-end and receiving-end voltages are all assumed as equal to the rated voltage and $\theta_1=\theta_2=0$, then

$$\frac{\partial P_L}{\partial P_1} = 2C_{11}P_1 = 2R_{11}P_1 = 2\times 0.005P_1 = 0.01P_1$$

Similarly,

$$\frac{\partial P_L}{\partial P_2} = 2C_{22}P_2 = 2\times 0.0065P_2 = 0.013P_2$$

As an initial estimate, assume a value of λ, as calculated for load sharing without transmission line losses, as equal to 3.825. Set both $\lambda_1=\lambda_2=3.825$ and calculate the load sharing:

$$\lambda = \lambda_1 \frac{1}{1-\partial P_L/\partial P_1}$$

$$= \lambda_2 \frac{1}{1-\partial P_L/\partial P_2}$$

Here,

$$\lambda_1 = 0.003(500)P_1 + 3.3 = 1.5P_1 + 3.3$$

$$\lambda_2 = 0.005(500)P_2 + 2.2 = 2.5P_2 + 2.2$$

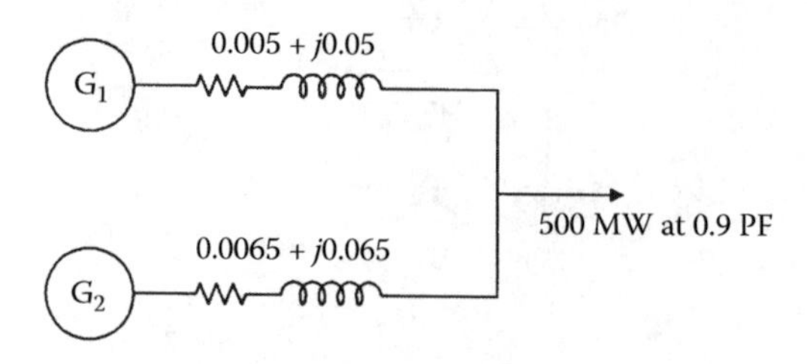

FIGURE 12.3
Configuration for Example 12.1.

Thus,

$$3.825(1-0.01P_1) = 1.5P_1 + 3.3$$

$$3.825(1-0.013P_2) = 2.5P_2 + 2.2$$

This gives

$$P_1 = 0.341, \quad P_2 = 0.6373, \quad P_1 + P_2 = 0.9783s$$

The sum of powers should be >1.0, as some losses occur in transmission. The initial value of λ is low. Consider a demand of approximately 1.02 (2% losses) and adjust the value of λ. A value of 3.87 gives $P_1 = 0.37$ (= 185 MW) and $P_2 = 0.655$ (= 327.5 MW). The total demand is, therefore, 1.025, representing 2.5% losses.

A load flow is now required with outputs of generators 1 and 2 limited to the values calculated above. The load flow results give a load voltage of $0.983 < -0.93°$ at the load bus, and a reactive power input of 156 and 106 Mvar, respectively, from generators 1 and 2 (reactive power loss of 30 Mvar and an active power loss of 3 MW, which is approximately 0.6%). Thus, the total system demand is 1.006. New values of loss coefficients and λ values are found by the same procedure.

Practically, the generator scheduling considers the following constraints in the optimization problem [4–7].

- System real power balance
- Spinning reserve requirements
- Unit generation limits
- State of the unit, i.e., it may be on or off
- Thermal unit minimum and maximum starting up/down times
- Ramp rate limits as the unit ramps up and down
- Fuel constraints
- System environmental (emission) limits, i.e., SO_2 and NO_x limits
- Area emission limits

In addition, the following reactive power limits are imposed:

- Reactive power operating reserve requirements
- Reactive power generation limits and load bus balance
- System voltage and transformer tap limits

12.5 Steepest Gradient Method

The independent variables are represented by u. These are the variables controlled directly, say voltage on a *PV* bus or generation. Dependent, state, or basic variables, which depend on the independent variables, are donated by x, i.e., the voltage and angle on a *PQ* bus. Fixed, constant, or nonbasic variables may be in either of the above two classes and have reached an upper or lower bound and are being held at that bound, i.e., when the voltage on a *PV* bus hits its limit. These are donated by p.

The optimization problem can be stated as minimizations:

$$\min \ c(\bar{x},\bar{u})$$
$$\text{st} \tag{12.38}$$
$$f(\bar{x},\bar{u},\bar{p})=0$$

Define the Lagrangian function:

$$\Gamma(\bar{x},\bar{u},\bar{p})=c(\bar{x},\bar{u})+\lambda^t f(\bar{x},\bar{u},\bar{p}) \tag{12.39}$$

The conditions for minimization of the unconstrained Lagrangian function are as follows:

$$\frac{\partial\Gamma}{\partial x}=\frac{\partial C}{\partial x}+\left[\frac{\partial f}{\partial x}\right]^t\lambda=0 \tag{12.40}$$

$$\frac{\partial\Gamma}{\partial u}=\frac{\partial C}{\partial u}+\left[\frac{\partial f}{\partial u}\right]^t\lambda=0 \tag{12.41}$$

$$\frac{\partial\Gamma}{\partial\lambda}=f(\bar{x},\bar{u},\bar{p})=0 \tag{12.42}$$

Equation 12.42 is the same as the equality constraint, and $\partial f/\partial x$ is the same as the Jacobian in the Newton–Raphson method of load flow.

Equations 12.40–12.42 are nonlinear and can be solved by the steepest gradient method. The control vector $\bar{u}$ is adjusted so as to move in the direction of steepest descent. The computational procedure is as follows:

An initial guess of $\bar{u}$ is made, and a feasible load flow solution is found. The method iteratively improves the estimate of x:

$$\bar{x}^{k+1}=\bar{x}^k+\bar{\Delta}x \tag{12.43}$$

where $\bar{\Delta}x$ is obtained by solving linear Equation 12.22

$$\bar{\Delta}x=-\left[J(\bar{x}^k)\right]^{-1}f(\bar{x}^k) \tag{12.44}$$

Equation 12.40 is solved for λ:

$$\lambda=-\left[\left(\frac{\partial f}{\partial x}\right)^t\right]^{-1}\frac{\partial C}{\partial x} \tag{12.45}$$

λ is inserted into Equation 12.41 and the gradient is calculated:

$$\bar{\nabla}\Gamma = \frac{\partial C}{\partial u} + \left[\frac{\partial f}{\partial u}\right]^t \lambda \tag{12.46}$$

If $\nabla\Gamma \to 0$ is within the required tolerances, then the minimum is reached, otherwise find a new set of variables:

$$\bar{u}_{\text{new}} = \bar{u}_{\text{old}} + \bar{\Delta}u \tag{12.47}$$

where

$$\bar{\Delta}u = -\alpha\bar{\nabla}\Gamma \tag{12.48}$$

Here, Δu is the step in the negative direction of the gradient. The choice of α in Equation 12.48 is important. The step length is optimized. Too small a value slows the rate of convergence and too large a value may give rise to oscillations.

12.5.1 Adding Inequality Constraints on Control Variables

The control variables are assumed to be unconstrained in the above discussions. Practically, these will be constrained, i.e., generation has to be within certain upper and lower bounds:

$$\bar{u}_{\min} \le \bar{u} \le \bar{u}_{\max} \tag{12.49}$$

If Δ_u in Equation 12.48 causes u_i to exceed one of the limits, then it is set corresponding to that limit.

$$\begin{aligned} u_{i,\text{new}} &= u_{i,\max} \quad \text{if } u_{i,\text{old}} + \Delta u_i > u_{i,\max} \\ u_{i,\text{new}} &= u_{i,\min} \quad \text{if } u_{i,\text{old}} + \Delta u_i < u_{i,\min} \end{aligned} \tag{12.50}$$

Otherwise, $u_{i,\text{new}} = u_{i,\text{old}} + \Delta u_i$.

The conditions for minimizing Γ under constraint are as follows (Chapter 11):

$$\begin{aligned} &\frac{\partial \Gamma}{\partial u_i} = 0 \quad \text{if } u_{i\min} < u_i < u_{i,\max} \\ &\frac{\partial \Gamma}{\partial u_i} \le 0 \quad \text{if } u_i = u_{i,\max} \\ &\frac{\partial \Gamma}{\partial u_i} \ge 0 \quad \text{if } u_i < u_{i,\min} \end{aligned} \tag{12.51}$$

Thus, the gradient vector must satisfy the optimality constraints in Equation 12.51.

12.5.2 Inequality Constraints on Dependent Variables

The limits on dependent variables are an upper and lower bound, e.g., on a PQ bus the voltage may be specified within an upper and a lower limit. Such constraints can be handled by the penalty function method (Chapter 11). The objective function is augmented by penalties for the constraint violations. The modified objective function can be written as

$$C' = C(\bar{x}, \bar{u}) + \sum W_j \tag{12.52}$$

where W_j is added for each violated constraint. A suitable penalty function is described by

$$\begin{aligned} W_j &= \frac{S_i}{2}(x_j - x_{j\max})^2 \quad \text{when } x_j > x_{j\max} \\ W_j &= \frac{S_i}{2}(x_j - x_{j\min})^2 \quad \text{when } x_j < x_{j\min} \end{aligned} \tag{12.53}$$

The plot is shown in Figure 12.4. This shows how the rigid bounds are replaced by soft limits. The higher the value of S_i, the more rigidly the function will be enforced [8].

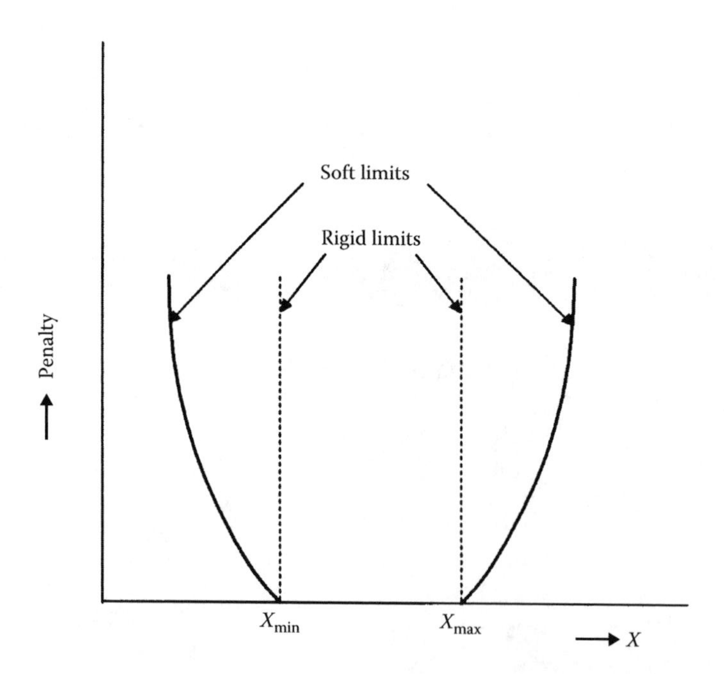

FIGURE 12.4
Penalty function: rigid and soft limits.

12.6 OPF Using Newton Method

We will examine OPF using the Newton method, through an example [9,10]. Consider a four-bus system with the following variables:

- Voltage magnitudes v_1, v_2, v_3, and v_4
- Voltage phase angles θ_1, θ_2, θ_3, and θ_4
- Real power injections p_1 and p_2 from controllable generators 1 and 2
- t_{23} and t_{34}, the tap ratios

The optimally ordered vector $\bar{y}$ for all variables is

$$\bar{y}^t = [p_1, p_2, t_{23}, t_{34}, \theta_1, \upsilon_1, \theta_2, \upsilon_2, \theta_3, \upsilon_3, \theta_4, \upsilon_4] \tag{12.54}$$

The objective function is the sum of cost functions of the active power outputs of the generators:

$$F_1(y) = \frac{1}{2} G_1 p_1^2 + b_1 p_1 + w_1 \tag{12.55}$$

$$F_2(y) = \frac{1}{2} G_2 p_2^2 + b_2 p_2 + w_2 \tag{12.56}$$

$$F(y) = F_1(y) + F_2(y) \tag{12.57}$$

The inequality constraints are the upper and lower bounds on the following:

- Power outputs of generators 1 and 2
- Phase-shift angle
- Voltages of the four buses
- Tap ratios of two transformers

12.6.1 Functional Constraints

The functional constraints consist of equalities that are always active and inequalities that are made active only when necessary to maintain feasibility. Real and active power loads are examples of functional equality constraints that are always active. The upper and lower bounds on the real and reactive power outputs of the generators are examples of functional inequality constraints that may be active or inactive.

The given equalities for buses 3 and 4 are as follows:

$$CP_i = P_i(y) - p_i \tag{12.58}$$

$$CQ_i = Q_i(y) - q_i \quad \text{for } i = 3, 4 \tag{12.59}$$

where P is the real power mismatch, CQ_i is the reactive power mismatch, P_i is the schedule real power load, and q_i is the schedule reactive power load.

The real power injection at buses 1 and 2 is controlled. The sum of real power flow at buses 1 and 2, including injections, is zero. The equation for this constraint is

$$CP_i = P_i(y) - p_i, \text{ for } i = 1, 2 \tag{12.60}$$

The reactive power output of two generators is constrained:

$$q_{i\min} \le Q_i(y) \le q_{i\max} \tag{12.61}$$

When $Q_i(y)$ is feasible, the mismatch in Equation 12.62 for CQ_i is made inactive. When $Q_i(y)$ needs to be enforced, CQ_i is made active:

$$CQ_i = [Q_i(y) - q_i] \tag{12.62}$$

12.6.2 Lagrangian Function

The Lagrangian for any problem similar to the example can be written in the form

$$\Gamma = F - \sum \lambda_{pi} CP_i - \sum \lambda_{qi} CQ_i \tag{12.63}$$

where F is the objective function, λ_{pi} is the Lagrangian multiplier for CP_i, and λ_{qi} is the Lagrangian for CQ_i. Equation 12.63 implies active constraints for real and reactive power injections.

The matrix equation of the linear system for minimizing the Lagrangian for any OPF by Newton's method is

$$\begin{vmatrix} \bar{H}(y,\lambda) & -\bar{J}^t(y) \\ -\bar{J}(y) & \bar{0} \end{vmatrix} \begin{vmatrix} \bar{\Delta y} \\ \bar{\Delta\lambda} \end{vmatrix} = \begin{vmatrix} -\bar{g}(y) \\ -\bar{g}(\lambda) \end{vmatrix} \tag{12.64}$$

where $\bar{H}(y,\lambda)$ is the Hessian matrix, $\bar{J}(y)$ is the Jacobian matrix, $\bar{g}(y)$ is the gradient with respect to y, and $\bar{g}(\lambda)$ is the gradient with respect to λ.

Equation 12.64 can be written as

$$\bar{W}\Delta z = -\bar{g} \tag{12.65}$$

The diagonal and upper triangle of the symmetric $\bar{W}$ matrix is shown in Figure 12.5, when all bus mismatch constraints are active. Some explanations are as follows: (1) the row and column for $\Delta\theta_1$ is inactive as bus 1 is a swing bus, (2) reactive power injections at buses 1 and 2 can take any value, and, thus, (3) row/column $\Delta\lambda_{q1}$ and $\Delta\lambda_{q2}$ are inactive.

	0	1	2	3	4	Δz		$-g$
0	G1	J 1				$\Delta P1$		$-\partial\Gamma/\partial p1$
	G2		J 1			$\Delta P2$		$-\partial\Gamma/\partial p2$
	H		H H J J	H H J J		$\Delta t\,23$		$-\partial\Gamma/\partial t23$
	H			H H j J	H H J J	$\Delta t\,34$		$-\partial\Gamma/\partial t34$
1		H H J J	H H J J	H H J J		$\Delta\theta 1$		$-\partial\Gamma/\partial\theta 1$
		H J J	H H J J	H H J J		$\Delta v1$		$-\partial\Gamma/\partial V1$
		0 0	J J 0 0	J J 0 0		$\Delta\lambda p1$		$-\partial\Gamma/\partial\lambda_{p1}$
		0	J J 0 0	J J 0 0		$\Delta\lambda q1$		$-\partial\Gamma/\partial\lambda_{q1}$
2			H H J J	H H J J	H H J J	$\Delta\theta 2$		$-\partial\Gamma/\partial\theta 2$
			H J J	H J J	H H J J	$\Delta v2$	=	$-\partial\Gamma/\partial V2$
			0 0	0 0	J J 0 0	$\Delta\lambda p2$		$-\partial\Gamma/\partial\lambda_{p2}$
			0	0	J J 0 0	$\Delta\lambda q2$		$-\partial\Gamma/\partial\lambda_{q2}$
3				H H J J	H H J J	$\Delta\theta 3$		$-\partial\Gamma/\partial\theta 3$
				H J J	H H J J	$\Delta v3$		$-\partial\Gamma/\partial V3$
				0 0	J J 0 0	$\Delta\lambda p3$		$-\partial\Gamma/\partial\lambda_{p3}$
				0	J J 0 0	$\Delta\lambda q3$		$-\partial\Gamma/\partial\lambda_{q3}$
4					H H J J	$\Delta\theta 4$		$-\partial\Gamma/\partial\theta 4$
					H J J	$\Delta v4$		$-\partial\Gamma/\partial V4$
					0 0	$\Delta\lambda p4$		$-\partial\Gamma/\partial\lambda_{p4}$
					0	$\Delta\lambda q4$		$-\partial\Gamma/\partial\lambda_{q4}$

FIGURE 12.5
Optimal load flow: matrix W.

Gradient vector $\bar{g}$ is composed of first derivatives of the form $\partial\Gamma/\partial yi$ or $\partial\Gamma/\partial\lambda i$. As an example,

$$\frac{\partial\Gamma}{\partial p_2} = \frac{\partial}{\partial p_2}\left[F_2 - \lambda_{p2}CF_2\right] \tag{12.66}$$

$$\frac{\partial\Gamma}{\partial p_2} = \frac{\partial}{\partial p_2}\left[\frac{1}{2}G_2 p_2^2 + b_2 p_2 + w_2 - \lambda_{p2}(y) - p_2\right]$$

$$\frac{\partial\Gamma}{\partial p_2} = G_2 p_2 + b_2 + \lambda_{p2} \tag{12.67}$$

and

$$\frac{\partial\Gamma}{\partial\theta_2} = \frac{\partial}{\partial\theta_2}\{-\lambda_{pi}CP_i - \lambda_{qi}CQ_i - \lambda_{p2}CP_{2i} - \lambda_{q2}CQ_2\} \tag{12.68}$$

The Jacobian matrix is dispersed throughout the matrix W. Its elements are as follows:

$$\frac{\partial^2\Gamma}{\partial y_i\,\partial\lambda_j} = \frac{\partial^2\Gamma}{\partial\lambda_j\,\partial y_i} \tag{12.69}$$

where y_i could be θ_1, υ_i t_{ij}, or p_i and λ_j can be λ_{pi} or λ_{qi}. These second-order partial derivatives are also first partial derivatives of the form

$$\frac{\partial P_i}{\partial \theta_i} = \frac{\partial Q_i}{\partial \upsilon_i} \tag{12.70}$$

These are elements of the Jacobian matrix J.

12.6.3 Hessian Matrix

Each element of H is a second-order partial derivative of the form

$$\frac{\partial^2 \Gamma}{\partial y_i \partial y_j} = \frac{\partial^2 \Gamma}{\partial y_j \partial y_i} \tag{12.71}$$

The elements of the Hessian, for example, objective function are as follows:

$$\begin{aligned} \frac{\partial^2 F_1}{\partial p_1^2} &= G_1 \\ \frac{\partial^2 F_1}{\partial p_1 \partial \lambda_{p1}} &= 1 \\ \frac{\partial^2 F_1}{\partial p_2^2} &= G_2 \end{aligned} \tag{12.72}$$

Other elements of H are sums of several second partial derivatives, i.e.,

$$\frac{\partial^2 \Gamma}{\partial \theta_2 \partial \upsilon_4} = -\lambda_{p2} \frac{\partial^2 CP_2}{\partial \theta_2 \partial \upsilon_4} - \lambda_{q2} \frac{\partial^2 CQ_2}{\partial \theta_2 \partial \upsilon_4} - \lambda_{p4} \frac{\partial^2 CP_4}{\partial \theta_2 \partial \upsilon_4} - \lambda_{q4} \frac{\partial^2 CQ_4}{\partial \theta_2 \partial \upsilon_4} \tag{12.73}$$

The representative block 3×3 of the matrix W in Figure 12.5 is

$$\left| \begin{matrix} \frac{\partial^2 \Gamma}{\partial \theta_3^2} & \frac{\partial^2 \Gamma}{\partial \theta_3 \partial v_3} & \frac{-\partial^2 \Gamma}{\partial \theta_3 \partial \lambda_{p3}} & \frac{-\partial^2 \Gamma}{\partial \theta_3 \partial \lambda_{q3}} \\ 0 & \frac{\partial^2 \Gamma}{\partial v_3^2} & \frac{-\partial^2 \Gamma}{\partial v_3 \partial \lambda_{p3}} & \frac{-\partial^2 \Gamma}{\partial v_3 \partial \lambda_{q3}} \\ 0 & 0 & 0 & 0 \\ 0 & 0 & 0 & 0 \end{matrix} 0 \right| \tag{12.74}$$

and block 3×4 is

$$\left| \begin{matrix} \frac{\partial^2 \Gamma}{\partial \theta_3 \partial \theta_4} & \frac{\partial^2 \Gamma}{\partial \theta_3 \partial v_4} & \frac{-\partial^2 \Gamma}{\partial \theta_3 \partial \lambda_{p4}} & \frac{-\partial^2 \Gamma}{\partial \theta_3 \partial \lambda_{q4}} \\ \frac{\partial^2 \Gamma}{\partial v_3 \partial \theta_4} & \frac{\partial^2 \Gamma}{\partial v_3 \partial v_4} & \frac{-\partial^2 \Gamma}{\partial v_3 \partial \lambda_{p4}} & \frac{-\partial^2 \Gamma}{\partial v_3 \partial \lambda_{q4}} \\ \frac{-\partial^2 \Gamma}{\partial \lambda_{3_p} \partial \theta_4} & \frac{-\partial^2 \Gamma}{\partial \lambda_{3_p} \partial v_4} & 0 & 0 \\ \frac{-\partial^2 \Gamma}{\partial \lambda_{q3} \partial \theta_4} & \frac{-\partial^2 \Gamma}{\partial \lambda_{q3} \partial v_4} & 0 & 0 \end{matrix} \right| \tag{12.75}$$

12.6.4 Active Set

The active set are the variables that must be enforced for a solution. This set includes unconditional variables and the functions that would violate the constraints if the bounds were not enforced. The following variables are enforced unconditionally:

$\theta_1 = 0$

Power mismatch equations for buses 3 and 4, i.e., CP_i and CQ_i

The mismatch equations for controllable real power, Equation 12.60

The values of the following variables and functions have inequality constraints:

Variables that will violate their bounds, p_1, p_2, t_{23}, t_{34}, and all bus voltages

Mismatch equations for reactive power, where reactive power injections would violate one of their bounds. In this case, CQ_1 and CQ_2

12.6.5 Penalty Techniques

A penalty can be modeled as a fictitious controllable quadratic function, aimed at increasing the optimized cost if the constraint is violated. A penalty is added when a lower or upper bound is violated. If the penalty function is added to Γ and taken into account for evaluation of $\bar{W}$ and $\bar{g}$, it effectively becomes a part of the objective function and creates a high cost of departure of the variable in the solution $\bar{W}\Delta z = -\bar{g}$. For large values of S_i, the function will be forced close to its bound. To modify W and g, the first and second derivatives of the function are computed, in a similar way to the other function in F:

$$\frac{dEy_i}{dy_i} = S_i(y_i^0 - \bar{y}_i)$$
$$\frac{d^2Ey_i}{d^2y_i} = S_i \tag{12.76}$$

where $\bar{y}_1$ is the upper bound and y_1^0 is the current value of y_1. The first derivative is added to $\partial\Gamma/\partial y_1$ of g and the second derivative to $\partial^2\Gamma/\partial y_{12}$. It is assumed that $\bar{W}$ and $\bar{g}$ are not factorized. However, practically, $\bar{W}$ will be factorized and its factors must be modified to reflect addition of penalties. The imposition of a penalty on a variable then requires a change in factors to reflect a change in one diagonal element of $\bar{W}$; $\bar{g}$ must also be modified. When a penalty is added or removed by modifying factors of $\bar{W}$, this effects irritative correction of the variable S_i depending on the computer word size. The larger the value of S_i, the more accurate the enforcement of the penalty.

To test whether a penalty is still needed, it is only necessary to test the sign for its Lagrange multiplier μ:

$$\mu_i = -S_i(y_i - \bar{y}_i) \tag{12.77}$$

If $\bar{y}_i$ is the upper bound and $\mu < 0$ or if $\bar{y}_i$ is the lower bound and μ is >0, the penalty is still needed. Otherwise, it can be removed.

12.6.6 Selecting Active Set

The problem is to find a good active set for solving

$$\bar{W}^k \Delta z^{k+1} = -\bar{g}^k \tag{12.78}$$

An active set which is correct at z^k may be wrong at z^{k+l}, because some inequality constraints will be violated and some enforced inequality constraints will not be necessary. The active set has to be adjusted in the iteration process until an optimum is found.

12.6.7 Algorithm for the Coupled Newton OPF

Initialize:

Vector $\bar{z}$ of variables y and Lagrange multipliers λ are given initial values; λ_{pi} can be set equal to unity and λ_{qi} equal to zero. All other variables can have initial values as in a load flow program.

Select active set.

Evaluate g_k.

Test for optimum, if all the following conditions are satisfied:

- $\bar{g}_k = 0$
- λ_i and μ_i for all inequalities pass the sign test
- The system is feasible

If all these conditions are satisfied, the optimum has been reached. Exit.

Primary iteration. Evaluate and factorize $W(z_k)$.

Solve:

$$\bar{W}(z^k)\Delta z^k + 1 = -\bar{g}(z^k) \tag{12.79}$$

Compute new state. The flowchart is shown in Figure 12.6.

Some of the characteristics of this method are as follows:

- Each iteration is a solution for the minimum of a quadratic approximation of the Lagrangian function.
- Corrections in variables and Lagrangian multipliers that minimize the successive quadratic approximations are obtained in one simultaneous solution of the sparse linear matrix equation.
- All variables (control, state, or bounded) are processed identically.
- Penalty techniques are used to activate and deactivate the inequality constraints. The use of penalties is efficient and accuracy is not sacrificed.
- Solution speed is proportional to network size and is not much affected by the number of free variables or inequality constraints.

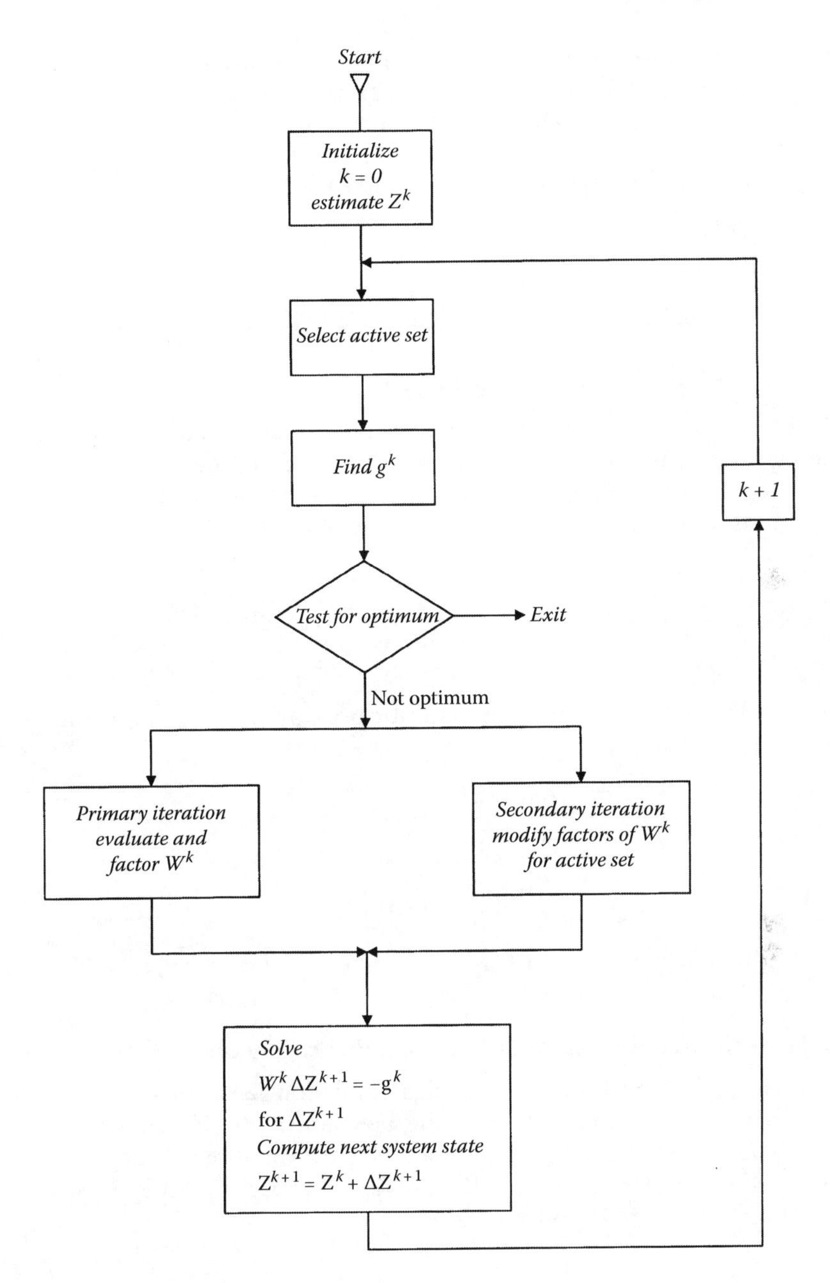

FIGURE 12.6
Flowchart of the Newton method: optimal power flow.

12.6.8 Decoupled Formation

From Equation 12.65, a decoupled formation of the Newton OPF can be written as

$$\bar{W}'\Delta z' = -\bar{g}'$$

$$\bar{W}''\Delta z'' = \bar{g}''$$

where $\bar{W}', \bar{W}''$ pertain to $P\theta$ and Pv subsystems, and similarly $\bar{g}', \bar{g}''$. The decoupling divides the problem into $P\theta$ and Pv *subsystems*. The elements of H omitted in decoupling are relatively small, while the elements of J may not be negligible. The decoupled W matrices have 2×2 formation in Equations 12.74 and 12.75. The factorization of $\bar{W}', \bar{W}''$ requires approximately the same computational effort as the nonfactorized version of Newton OPF [9].

12.7 Sequential Quadratic Programming

The method solves the problem of Equation 12.1 by repeatedly solving a quadratic programming approximation, which is a special case of nonlinear programming [11,12]. The objective function is quadratic and the constraints are linear (Chapter 11). The objective function $f(\bar{x}, \bar{u})$ is replaced by a quadratic approximation:

$$q^k(\bar{D}) = \nabla f(\bar{x}, \bar{u})^k \bar{D} + \frac{1}{2}\bar{D}^t \nabla^2 \Gamma\left[(\bar{x}, \bar{k})^k, \lambda^k\right]\bar{D} \tag{12.80}$$

The step D^k is calculated by solving the quadratic programming subproblem:

$$\begin{aligned} &\min q^k(\bar{D}) \\ &\text{st} \\ &G(\bar{x}, \bar{u})^k + J(\bar{x}, \bar{u})^k \bar{D} = 0 \\ &H(\bar{x}, \bar{u})^k + I(\bar{x}, \bar{u})^k \bar{D} = 0 \end{aligned} \tag{12.81}$$

where J and I are the Jacobian matrices corresponding to constraints G and H. The Hessian of the Lagrangian, $[\nabla^2 \Gamma(\bar{x}, \bar{k})^k, \lambda^k]$ appearing in Equation 12.80, is calculated using quasi-Newton approximation. (Computation of the Hessian in the Newton method is time-consuming. Quasi-Newton methods provide an approximation of the Hessian at point k, using the gradient information from the previous iterations.) After D^k is computed by solving Equation 12.81, (x, u) is updated using

$$(\bar{x}, \bar{u})^{k+1} = (\bar{x}, \bar{u})^k + \alpha^k \bar{D}^k \tag{12.82}$$

where α^k is the step length. Ascertaining the step length in constrained systems must be chosen to minimize the objective function as well as to constrain violations. These two criteria are often conflicting, and a merit function is employed that reflects the relative importance of these two aims. There are several approaches to select merit functions. Sequential quadratic programming has been implemented in many commercial packages.

12.8 Successive Linear Programming

Successive LP can accommodate all types of constraints quite easily and offers flexibility, speed, and accuracy for specific applications. The approach uses linearized programming solved by using different variants of the simplex method. The dual relaxation method and primal method upper bounds are more successfully implemented.

The primal approach tries to obtain an initial value of the objective function based on all the constraints and then modify the objective function by sequentially exchanging constraints on limits with those not on limits. The dual relaxation method finds an initial value of the objective function, which optimally satisfies n number of control variables. It then satisfies the remaining violated constraints by relaxing some binding constraints [13]. The method has simpler, less time-consuming initialization and can detect infeasibility at an early stage of the optimization process, when only a few of the constraints are considered binding in the solution.

Figure 12.7 shows the flowchart [14]. The method allows one critical violated constraint into tableau form at each step. One of the existing binding constraints must leave the basis. The currently violated constraint may enter the basis arbitrarily; however, the constraint to leave the basis must be optimally selected.

The constraints in the basis are tested for eligibility. This test is the *sensitivity* between the incoming constraint and existing binding constraint k:

$$S_k = \frac{\Delta r_{in}}{\Delta r_k} \tag{12.83}$$

where Δr_{in} is the amount by which the violating branch flow or generation will be corrected by entering its constraint into the basis, and Δr_k is the amount by which existing binding constraint k will change when freed. Constraint k is eligible,

- Both this constraint and the incoming constraint are on the upper or lower limit and S_k is positive.
- Both the constraints are on the opposite limits and S_k is negative.

The *ratio test* involving the incremental cost yk of each binding constraint is stated as follows:

The binding constraint to be freed from the basis *among all the eligible constraints* is the one for which

$$\left|\frac{y_k}{S_k}\right| = \text{minimum} \tag{12.84}$$

If no eligible constraints are found, the problem is infeasible.

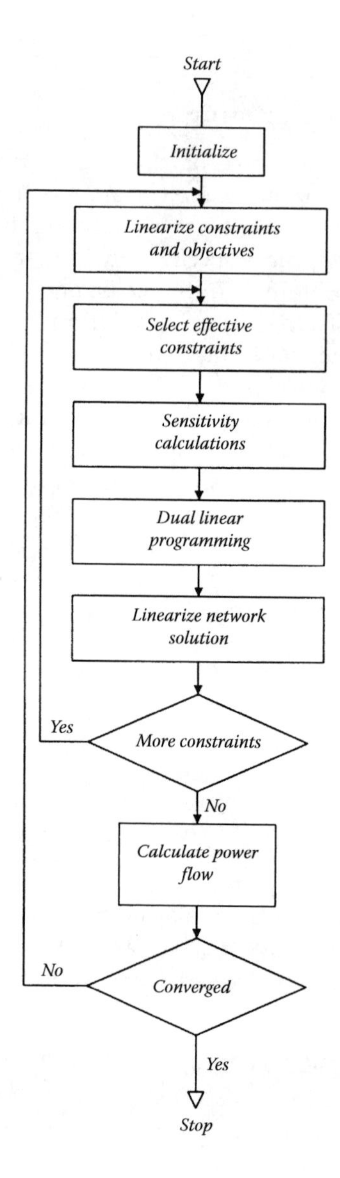

FIGURE 12.7
Flowchart: dual relaxation method.

12.9 Interior Point Methods and Variants

In 1984, Karmarkar [15] announced the polynomially bounded algorithm, which was claimed to be 50 times faster than the simplex algorithm. Consider the LP problem defined as

$$\begin{aligned} &\min \bar{c}^t \bar{x} \\ &\text{st } \bar{A}\bar{x} = \bar{b} \\ &\bar{x} \geq 0 \end{aligned} \tag{12.85}$$

where $\bar{c}$ and $\bar{x}$ are n-dimensional column vectors, $\bar{b}$ is an m-dimensional vector, and $\bar{A}$ is an $m \times n$ matrix of rank m, $n \geq m$. The conventional simplex method requires 2^n iterations to find the solution. The polynomial-time algorithm is defined as an algorithm that solves the LP problem in $O(n)$ steps. The problem of Equation 12.85 is translated into

$$\begin{aligned} &\min \bar{c}^t \bar{x} \\ &\text{st } \bar{A}\bar{x} = 0 \\ &\bar{e}^t \bar{x} = 1 \\ &\bar{x} \geq 0 \end{aligned} \tag{12.86}$$

where $n \geq 2$, $\bar{e} = (1, 1, \ldots, 1)^t$ and the following holds:

- The point $x^0 = (1/n, 1/n, \ldots, 1/n)^t$ is feasible in Equation 12.86.
- The objective value of Equation 12.86 = 0.
- Matrix $\bar{A}$ has full rank of m.

The solution is based on projective transformations followed by optimization over an inscribed sphere, which creates a sequence of points converging in polynomial time. A projective transformation maps a polytope $P \subseteq R^n$ and a strictly IP $a \in P$ into another polytope P' and a point $a' \in P'$. The ratio of the radius of the largest sphere contained in P' with the same center a' is $O(n)$. The method is commonly called an IP method due to the path it follows during solution. The number of iterations required is not dependent on the system size.

We noted in Chapter 11 that, in the simplex method, we go from one extreme point to another extreme point (Example 11.8). Figure 12.8 shows a comparison of the two methods. The simplex method solves an LP problem, starting with one extreme point on the boundary of the feasible region and then goes to a better neighboring extreme point along the boundary, finally stopping at the optimum extreme point. The IP method stays in the interior of the polytope.

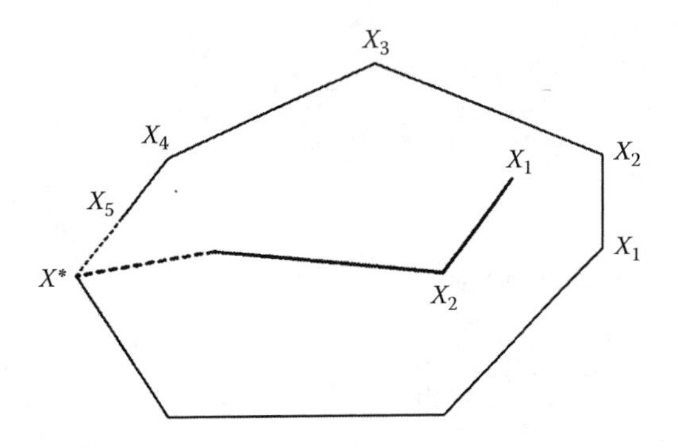

FIGURE 12.8
Iterations in simplex and interior point method.

12.9.1 Karmarkar Interior Point Algorithm

The algorithm creates a sequence of points $x^0, x^1, \ldots, x^k$ in the following steps:

1. Let x^0=center of simplex.
2. Compute next point $x^{k+l}=h(x^k)$. The function $\phi = h(a)$ is defined by the following steps.
3. Let $\bar{D}$=diag$(a_1, a_2, \ldots, a_n)$ be the diagonal matrix.
4. Augment $\bar{A}\bar{D}$ with rows of 1s:

$$\bar{B} = \begin{vmatrix} \bar{A}\bar{D} \\ \bar{e}^t \end{vmatrix}$$

5. Compute orthogonal projection of Dc into the null space of B:

$$\bar{c}_p = \left[1 - \bar{B}^t(\bar{B}\bar{B}^t)^{-1}\bar{B}\right]\bar{D}c$$

6. The unit vector in the direction of $\bar{c}_p$ is

$$\bar{c}_u = \frac{\bar{c}_p}{|c_p|}$$

7. Take a step of length ωr in the direction of c_u:

$$Z = a - \omega r c_u \quad \text{where } r = \frac{1}{\sqrt{n(n-1)}}$$

8. Apply reverse protective transformation to z:

$$\bar{\phi} = \frac{\bar{D}\bar{z}}{\bar{e}^t\bar{D}\bar{z}}$$

Return ϕ to x^{k+l}.
The potential function is defined as

$$f(\bar{x}) = \sum_{i=1}^{n} \text{In}\left(\frac{c^t x}{x_i}\right) \tag{12.87}$$

12.9.1.1 Check for Infeasibility

There should be certain improvement in the potential function at each step. If the improvement is not achieved, it can be concluded that the objective function must be positive. This forms a test of the feasibility.

12.9.1.2 Check for Optimality

The check involves going from the current IP to an extreme point without increasing the value of the objective function and testing the extreme point for optimality. The check is carried out periodically.

It can be shown that in $O[n(q+\log n)]$ steps, a feasible point x is found such that

$$c^{-t}\overline{x}=0 \text{ or } \frac{c^{-t}\overline{x}}{c^{-t}a_0}\leq 2^{-q} \text{ where } a_0=(1/n)\overline{e} \tag{12.88}$$

12.9.2 Barrier Methods

One year after Karmarkar announced his method, Gill et al. presented an algorithm based on projected Newton logarithmic barrier methods [16]. Gill's work [16] showed that Karmarkar work can be viewed as a special case of a barrier-function method for solving nonlinear programming problems. One drawback of the Karmarkar algorithm is that it does not generate dual solutions, which are of economic significance. Todd's work [17], an extension of Karmarkar's algorithm, generates primal and dual solutions with objective values converging to a common optimal primal and dual value. Barrier-function methods treat inequality constraints by creating a barrier function, which is a combination of original objective function and weighted sum of functions with positive singularity at the boundary. As the weight assigned to the singularities approaches zero, the minimum of barrier-function approaches the minimum of original function.

The following are variants of Karmarkar's IP method:

- Projective scaling methods
- Primal and dual affine methods [18]
- Barrier methods
- Extended quadratic programming using IP

12.9.3 Primal–Dual IP Method

The algorithms based on Karmarkar's projective method have polynomial-time complexity requiring $O(nL)$ iterations. These algorithms do not appear to perform well in practice. The algorithms based on dual affine scaling methods exhibit good behavior to real-world problems. Most primal–dual barrier path following methods have been shown to require $O(\sqrt{n}L)$ iterations at the most. We will discuss a primal–dual IP method for LP [19,20].

The algorithm is based on the following:

- Newton method for solving nonlinear equations
- Lagrange method for optimization with inequality constraints
- Fiacco and McCormick barrier method for optimization with inequality constraints

Consider an LP problem (Equation 12.1), and let it be a primal problem. The dual problem is

$$\begin{aligned}
&\max \bar{b}^t \bar{y} \\
&\text{st } \bar{A}^t \bar{y} + \bar{z} = \bar{c} \\
&\bar{z} \geq 0
\end{aligned} \tag{12.89}$$

First, Lagranges are formed with barriers:

$$\begin{aligned}
&\Gamma_p(\bar{x}, \bar{y}) = \bar{c}^t \bar{x} - \mu \sum_{j=1}^{n} \text{In}(x_j) - y^t(\bar{A}\bar{x} - \bar{b}) \\
&\Gamma_p(\bar{x}, \bar{y}, \bar{z}) = \bar{b}^t \bar{y} - \mu \sum_{j=1}^{n} \text{In}(z_j) - \bar{x}^T(\bar{A}^t \bar{y} + \bar{z} - \bar{c})
\end{aligned} \tag{12.90}$$

The first-order necessary conditions for Equation 12.90 are as follows:

$$\begin{aligned}
&\bar{A}\bar{x} = \bar{b} \text{ (primal feasibility)} \\
&\bar{A}^t \bar{y} + \bar{z} = \bar{c} \text{ (dual feasibility)} \\
&\bar{x}_j \bar{z}_j = \mu \text{ for } j = 1, 2, \ldots, n
\end{aligned} \tag{12.91}$$

If $\mu = 0$, then the last expression in Equation 12.91 corresponds to ordinary complimentary slackness. In barrier methods, μ starts at some positive value and approaches zero, as $x(\mu) \to x^*$ (the constrained minima). Using Newton's method to solve Equation 12.91, we have

$$\begin{aligned}
&\bar{A}\bar{\Delta}x = \bar{b} - \bar{A}x^0 = d\bar{P} \\
&\bar{A}^t \bar{\Delta}y + \bar{\Delta}z = \bar{c} - \bar{z}^0 - \bar{A}^t y^0 = -d\bar{D} \\
&\bar{Z}\bar{\Delta}x + \bar{X}\bar{\Delta}z = \mu\bar{e} - \bar{X}\bar{Z}\bar{e}
\end{aligned} \tag{12.92}$$

where

$$\begin{aligned}
&\bar{X} = \text{diag } (x_1^0, \ldots, x_n^0) \\
&\bar{Z} = \text{diag } (z_1^0, \ldots, z_n^0) \\
&\bar{e}^t = (1, 1, \ldots, 1)
\end{aligned} \tag{12.93}$$

From Equation 12.92, the following equations are obtained:

$$\begin{aligned}
&\bar{\Delta}y = (\bar{A}\bar{Z}^{-1}\bar{X}\bar{A}^t)^{-1}(\bar{b} - \mu\bar{A}\bar{Z}^{-1}\bar{e} - \bar{A}\bar{Z}^{-1}\bar{X}d\bar{D}) \\
&\bar{\Delta}z = -d\bar{D} - \bar{A}^t \bar{\Delta}y \\
&\bar{\Delta}x = \bar{Z}^{-1}\left[\mu\bar{e} - \bar{X}\bar{Z}\bar{e} - \bar{X}\bar{\Delta}z\right]
\end{aligned} \tag{12.94}$$

x, y, and z are then updated:

$$x^1 = x^0 + \alpha_p \Delta x$$
$$y^1 = y^0 + \alpha_d \Delta y \tag{12.95}$$
$$z^1 = z^0 + \alpha_d \Delta z$$

where α_p and α_d are step sizes for primal and dual variables, chosen to preserve $x > 0$ and $z > 0$.

This completes one iteration. Instead of taking several steps to converge for a fixed value of μ, μ is reduced from step to step. Monteriro and Adler [20] proposed a scheme for updating μ, as follows:

$$\mu^{k+1} = \mu^k \left(1 - \frac{t}{\sqrt{n}}\right) \quad 0 < t < \sqrt{n} \tag{12.96}$$

and proved that convergence is obtained in a maximum of $O(\sqrt{n})$ iterations. The convergence criterion is

$$\frac{\bar{c}^t \bar{x} - \bar{b}^t \bar{y}}{1 + \left|\bar{b}^t \bar{y}\right|} < \in \tag{12.97}$$

12.10 Security- and Environmental-Constrained OPF

Security-constrained OPF, written as SCOPF for abbreviation, considers outage of certain equipment or transmission lines [21,22]. The security constraints were introduced in early 1970 and online implementation became a new challenge. In conventional OPF, the insecurity of the system during contingency operations is not addressed. Traditionally, security has relied on preventive control, i.e., the current-operating point is feasible in the event of occurrence of a given subset of the set at all possible contingencies. This means that the base case variables are adjusted to satisfy postcontingency constraints, which can be added to Equation 12.1:

$$\min f(\bar{z})$$
$$\text{st } g(\bar{z}) = 0$$
$$h(\bar{z}) = 0, \tag{12.98}$$
$$\phi_i(\bar{u}_i - \bar{u}_0) \le \Theta_i$$

Here, $f(\bar{z}) = f(\bar{x}, \bar{u})$ and the last constraint, called the coupling constraint, reflects the rate of change in the control variables of the base case. Θ_i is the vector of upper bounds reflecting ramp rate limits. Without the coupling constraints, the problem is separable into $N + 1$ subproblems.

TABLE 12.3

Power System Security Levels

Security Level	Description	Description	Control Actions
1	Secure	All loads supplied and no operating limits violated; the system is adequate secure and economical	
2	Correctly secure	All loads supplied, no operating limits violated, except under contingency conditions which can be corrected without loss of load by contingency-constrained OPF	
3	Alert	All loads supplied, no operating limits violated, except that some violations caused by contingencies cannot be corrected without loss of load	Preventive rescheduling can raise it to levels 1 and 2
4	Correctable emergency	All loads supplied but operating limits are violated	Can be corrected by remedial actions and moves to levels 2 or 3
5	Noncorrectable emergency	All loads supplied, but operating limits violated; can be corrected by remedial actions with some loss of load	Remedial actions result in some loss of load
6	Restorative	No operating limits violated, but there is loss of load	–

Source: Monticelli, A.J. et al., *IEEE Trans. Power Syst.*, 2(1), 175–182, 1987, shows various levels of SCOPF.
Note: Research works [23–32] provide further reading.

The environmental constraints are implemented. The Clean Air Act requires utilities to reduce SO_2 and NO_x emissions. These are expressed as separable quadratic functions of the real power output of individual generating units.

The environmental constraints $e(Z) \leq 0$ can be added to Equation 12.98. This makes the problem of SCOPF rather difficult to solve by conventional methods, and decomposition strategies are used in the solution of OPF. Talukdar–Giras decomposition [11] is an extension of the sequential programming method. Benders decomposition [27] is a variable partitioning method, where certain variables are held fixed while the others are being solved. The values of the complicating variables are adjusted by the master problem, which contains the basic set and a subproblem which has the extended OPF problem. Iterations between the master problem and subproblem continue until the original problem is solved. Table 12.3 adapted from Reference [22] shows various levels of SCOPF. Studies [23–32] provide further reading.

References

1. J Carpentier. Contribution á létude du dispatching. *Economique de la Societe Francaise des Electriciens*, 3, 431–447, 1962.
2. A Santos Jr. A dual augmented Lagrangian approach for optimal power flow. *IEEE Trans Power Syst*, 3(3), 1020–1025, 1988.

3. RR Shoults, DI Sun. Optimal power flow based upon P–Q decomposition. *IEEE Trans Power App Syst*, 101, 397–405, 1982.
4. M Piekutowski, IR Rose. A linear programming method for unit commitment incorporating generation configurations, reserve and flow constraints. *IEEE Trans Power App Syst*, PAS-104(12), 3510–3516, 1982.
5. DI Sun. Experiences with implementing optimal power flow for reactive scheduling in Taiwan power system. *IEEE Trans Power Syst*, 3(3), 1193–1200, 1988.
6. GA Maria, JA Findlay. A Newton optimal flow program for Ontario hydro EMS. *IEEE Trans Power Syst*, 2(3), 576–584, 1987.
7. C Wang, SM Shahidehpour. A decomposition approach to nonlinear multi-area generation scheduling with tie-line constraints using expert systems. *IEEE Trans Power Syst*, 7(4), 1409–1418, 1992.
8. HW Dommel, WF Tinney. Optimal power flow solutions. *IEEE Trans Power App Syst*, PAS-87, 1866–1876, 1968.
9. DI Sun, BT Ashley, BJ Brewer, BA Hughes, WF Tinney. Optimal power flow by Newton approach. *IEEE Trans Power App Syst*, 103, 2864–2880, 1984.
10. WF Tinney, DI Sun. Optimal power flow: Research and code development. EPRI Research Report EL-4894, February 1987.
11. SN Talukdar, TC Giras. A fast and robust variable metric method for optimal power flows. *IEEE Trans Power App Syst*, 101(2), 415–420, 1982.
12. RC Burchett, HH Happ, KA Wirgau. Large scale optimal power flow. *IEEE Trans Power App Syst*, 101(10), 3722–3732, 1982.
13. B Sttot, JL Marinho. Linear programming for power system network security applications. *IEEE Trans Power App Syst*, 98, 837–848, 1979.
14. DS Kirschen, HP Van Meeten. MW/voltage control in a linear programming based optimal load flow. *IEEE Trans Power Syst*, 3, 782–790, 1988.
15. N Karmarkar. A new polynomial-time algorithm for linear programming. *Combinatorica*, 4(4), 373–395, 1984.
16. PE Gill, W Murray, MA Saunders, JA Tomlin, MH Wright. On projected Newton barrier methods for linear programming and an equivalence to Karmarkar's projective method. *Math Program*, 36, 183–209, 1986.
17. MJ Todd, BP Burrell. An extension of Karmarkar's algorithm for linear programming using dual variables. *Algorithmica*, 1, 409–424, 1986.
18. RE Marsten. Implementation of dual affine interior point algorithm for linear programming. *ORSA Comput*, 1, 287–297, 1989.
19. IJ Lustig, RE Marsen, NDF Shanno. IP methods for linear programming computational state of the art technical report. Computational Optimization Center, School of Industrial and System Engineering, Georgia Institute of Technology, Atlanta, GA, 1992.
20. RDC Monteiro, I Adler. Interior path following primal–dual algorithms. Part 1: Linear programming, Part II: Convex quadratic programming. *Math Program*, 44, 27–66, 1989.
21. B Stott, O Alsac, AJ Monticelli. Security analysis and optimization. *Proc IEEE*, 75(2), 1623–1644, 1987.
22. AJ Monticelli, MVF Pereira, S Granville. Security constrained optimal power flow with post-contingency corrective rescheduling. *IEEE Trans Power Syst*, 2(1), 175–182, 1987.
23. RC Burchett, HH Happ, DR Vierath. Quadratically convergent optimal power flow. *IEEE Trans Power App Syst*, 103, 3267–3275, 1984.
24. AM Sasson. Optimal power flow solution using the Hessian matrix. *IEEE Trans Power App Syst*, 92, 31–41, 1973.
25. IJ Lustig. Feasibility issues in an interior point method for linear programming. *Math Program*, 49, 145–162, 1991.
26. IEEE Power Engineering Society. IEEE Tutorial Course–Optimal Power Flow, Solution Techniques, Requirements and Challenges–IEEE. Document Number 96TP 111–0, 1996.

27. JF Benders. Partitioning for solving mixed variable programming problems. *Numerische Mathematik*, 4, 283–252, 1962.
28. SN Talukdar, VC Ramesh. A multi-agent technique for contingency constrained optimal power flows. *IEEE Trans Power Syst*, 9(2), 885–861, 1994.
29. I Adler. An implementation of Karmarkar's algorithm for solving linear programming problems. *Math Program*, 44, 297–335, 1989.
30. IEEE Committee Report. IEEE reliability test system. *IEEE Trans Power App Syst*, PAS-98, 2047–2054, 1979.
31. IEEE Working Group Report. Description and bibliography of major economy-security functions Part I and Part II. *IEEE Trans Power App Syst*, 100, 211–235, 1981.
32. JA Momoh. Application of quadratic interior point algorithm to optimal power flow. EPRI Final Report, RP 2473–36 II, 1992.

13

Heuristic Optimization Techniques

We have discussed the classical method of optimization in Chapters 11 and 12. These become fairly involved and have limitations. Heuristic methods have evolved in recent years for solving optimization problems that were difficult and even impossible to solve with classical methods. These may offer advantages, for example, these systems are very robust and solution time is much shorter than with the classical approach. These find applications in a number of electrical engineering problems, not limited to optimal power flow. For example,

- Economic dispatch and unit commitment
- Maintenance and generation scheduling
- Constrained load flow problems
- Power system controls

Modern Heuristic Optimization Techniques—Theory and Applications to Power Systems is a good reference with respect to applications to electrical power systems [1].

Living organisms exhibit complex behavior—the biological evolution has solved problems typified by chance, chaos, and nonlinear activities. These are also the problems that are difficult or impossible to handle with classical methods of optimization, and these problems do appear in the field of power systems. The term *evolutionary computation* is used to cover all evolutionary algorithms (EAs), which offer practical advantages to the difficult optimization problems.

An EA consists of initialization; this may be a pure random sampling of possible solutions. This is followed by a selection process in light of a performance index. A numerical value is assigned to two equally possible solutions—that is, the criteria need not be specified with precision—which is required in classical optimization. The randomly varying individuals are subjected to a fitness function. A difference equation can be written as

$$\mathrm{x}[t+1] = s(v(|\,\mathrm{x}(t)\,|)) \tag{13.1}$$

where $|\mathrm{x}(t)|$ is the population at time t, v is the random variation operator, and s is the selection operator [2].

13.1 Genetic Algorithm

13.1.1 Background

Computer simulation of evolution started as early as in 1954 with work of Nils Barr Celli, followed by work of Rechenburg and Hans-Paul in the 1960s and early 1970s. Genetic algorithms (GAs) became popular through work of John Holland in the 1970s and particularly his book: *Adaptation in Natural and Artificial Systems* (1975). His work originated with studies of cellular automata, conducted at the University of Michigan. He formalized a framework for predicting the quality of next generation, known as "Holland's Schema Theorem." (This theorem has since been proved erroneous. Premature convergence occurs because the population is not infinite, which is the basic hypothesis of the theorem.) In the late 1980s, General Electric started selling the world's first GA product for mainframe computers. In 1989, Axcelis, Inc. released Evolver, the world's first commercial GA product. It is currently in its sixth version. See References [3–7].

13.1.2 Methodology

The GA is based on conjecture of natural selection and genetics. It is a technique that searches parallel peaks in a multipath approach. This reduces the possibility of local minimum trapping. The parameters are coded—that helps to evolve current state into the next with minimum computations. It uses a fitness function for evaluating the fitness of each string to guide the search. There is no need to calculate the derivates or other auxiliary functions.

GAs belong to a class of EAs that generate solutions using techniques inspired by natural evolution, such as inheritance, mutation, selection, and crossover.

A population of candidates' solutions called individuals, creatures, or phenotypes is evolved for better solutions. Each candidate has a set of properties (its chromosomes or genotypes) that can be mutated and altered. The solutions are presented in binary, as strings of zeros and ones. However, other encoding is possible.

13.1.3 Initialization

The population size depends on the nature of the problem and may contain several hundred to thousands of solutions. Often the initial population is generated randomly, allowing a wide range of possible solutions (the search space).

During each successive generation, a proportion of the existing population is selected to breed a new generation. Individual solutions are selected through a fitness-based process, where fitter solutions are more likely to be selected.

The fitness function—which is problem dependent—is defined over the genetic representation and measures the quality of the represented solution. In some problems, it may be hard to define the fitness criteria, and in these cases a simulation may be used to determine the fitness function value of a phenotype.

13.1.4 Genetic Operators

The next step is to generate a second-generation population of solutions from those selected through a combination of genetic operators. Chromosomes in a generation are forced to evolve into better ones in the next generation by four basic GA operators:

- Reproduction
- Crossover
- Mutation
- And the problem-specified fitness function also called the objective function

(a)

Current generation		Next generation
01011101	→	01011101
11000110	→	11000110

(b)

Current generation		Next generation
01011101	→	01000101
11000110	→	01011101

(c)

Current generation		Next generation
01011101	→	01000101
11000110	→	11000111

FIGURE 13.1
Basic operations of GA (a) reproduction, (b) crossover, and (c) mutation.

In reproduction, a number of selected exact copies of chromosomes in the current population become the offsprings. In crossover, randomly selected crossovers of two individual chromosomes are swapped to produce the offspring. In mutation, randomly selected genes in chromosomes are altered by a probability equal to the mutation rate. For binary coding, it means that digit 1 becomes 0 and vice versa. Figure 13.1 shows (a) reproduction, (b) crossover, and (c) mutation.

The fitness function (objective function) plays the role of an environment to distinguish between good and bad chromosomes. The conceptual procedure from one generation to another is shown in Figure 13.2. Each chromosome is evaluated by objective function and some good chromosomes are selected [7,8].

13.1.5 Termination

The generational process is continued till a termination condition is reached, which may be the following:

- When a solution is found which satisfies minimum criteria
- Fixed number of generations is reached
- The successive iterations do not produce a better result
- Combination of the above

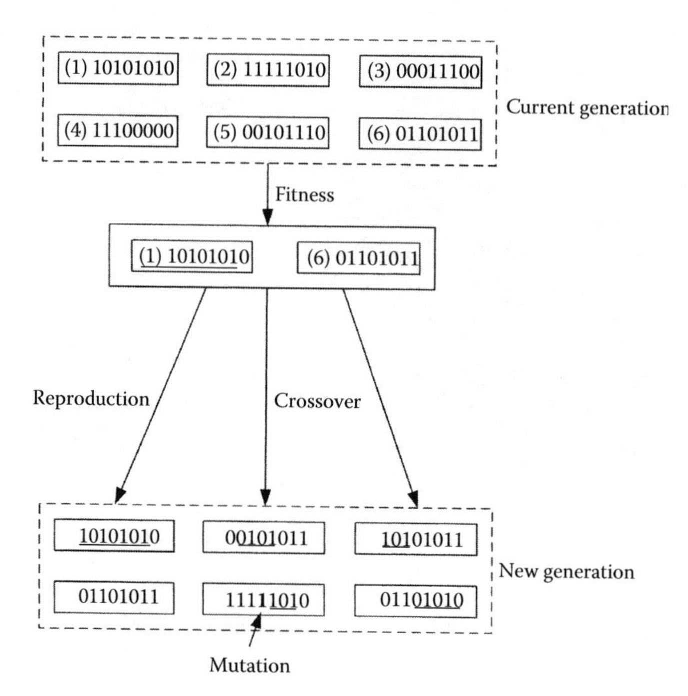

FIGURE 13.2
Procedure for GA operations from one generation to another.

13.1.6 Limitations

Some limitations are as follows:

- Repeated fitness function evaluation for complex problems is the most limiting segment. Finding solutions to complex high-dimensional multimodal problems often requires very expensive fitness function evaluations.
- GA does not scale well with complexity. Where the numbers of elements which are exposed to mutation are large, there is an exponential increase in search space size. The complex problems must be broken down into simplest representation possible.
- GA may have a tendency to converge toward local optima or even arbitrary points rather than the global optima of the problem. The likelihood of this depends on the shape of the four fitness landscapes—the problem may be alleviated by using different fitness functions, increasing rate of mutation, or using selective techniques that maintain a diverse population of solutions.
- Dynamic data sets are difficult as the genomes begin to converge early on toward solutions that may not be valid for later data. Several methods have been proposed to remedy this, for example, increasing genetic diversity.

13.1.7 Adaptive GA

An adaptive GA (AGA) is a significant variant of GAs. The possibilities of crossover (pc) and mutation (pm) greatly determine the solution accuracy and convergence speed. Instead of using fixed values of pc and pm, AGAs utilize the population information in each generation and adjust pc and pm in order to maintain population diversity and also sustain the convergence capacity. The adjustments of pc and pm depend on the fitness values of the solutions. In clustering-based adaptive GA, through the use of clustering analysis to judge the optimization states of the population, the adjustments of pc and pm depend on these optimization states. Also see References [9–13].

13.2 Particle Swarm Optimization

Particle swarm optimization (PSO) is originally attributed to Kennedy, Eberhart, and Shi and was first intended for simulating social behavior as a representation of movement of organisms in bird flock or fish pond. PSO was aimed to treat nonlinear problems with continuous variables. Treatment of mixed integer nonlinear optimization is difficult, which PSO can handle and can be organized with small algorithm. It is somewhat similar to GA—initially the system is popularized with a population of random solutions. Each random solution called a particle is flown through the problem hyperspace with a random velocity. It may take lesser time to converge as compared to GA. It is a population-based stochastic search algorithm. It mimics the natural process of group communication of swarms of animals, e.g., insects or birds. If one member finds a desirable path, the rest of the swarm follows it. In PSO, the behavior is imitated by particles with certain positions and velocities in a search space, wherein the population is called a swarm, and each member of the swarm is called a particle. Each particle flies through the search space and remembers the best position it has seen. Members of the swarm communicate these positions to each other and adjust their positions and velocities based on these best positions. The velocity adjustment is based on historical behavior of the particles as well as their companions. In this way, the particles fly to the optimum position.

Let a function *f*, where the gradient is not known, be minimized. The function takes a candidate solution as argument in the form of a vector of real numbers and produces a real number as output, which indicated the objective function value of the given candidate solution. The goal is to find a solution *a* for which

$$f(a) \leq f(b) \tag{13.2}$$

for all values of *b* in the search space. This means that *a* is the global minimum.

Let *S* be the number of particles in the swarm, each having a position

$$x_i \in R^n \tag{13.3}$$

in the search space and a velocity

$$v_i \in R^n \tag{13.4}$$

Let P_i be the best known position of the particle 1 and let *g* be the best known position of the swarm. A basic PSO program is then as follows.

For each particle $i = 1,\ldots, S$ do:

- Initiate the particle position with a uniformly distributed random vector:

$$x_i \approx U(b_{lo}, b_{up}) \tag{13.5}$$

where b_{lo} and b_{up} are the lower and upper boundaries of the search space.

- Initialize the particle best known position to its initial position:

$$p_i \leftarrow x_i \tag{13.6}$$

If

$$f(p_i) < f(g) \tag{13.7}$$

- Update the swarm best known position:

$$g \leftarrow p_i \tag{13.8}$$

- Initialize the particle velocity:

$$v_i \approx U\left(-\left|b_{up} - b_{lo}\right|, \left|b_{up} - b_{lo}\right|\right) \tag{13.9}$$

Until a termination criterion is met, for example, the number of iterations performed, or a solution with objective function value is found, *repeat:*

- For each particle $i = 1,\ldots, S$ do:
- For each dimension $d = 1,\ldots, n$ do:
- Pick random numbers: $r_p, r_g, \ldots, U(0, 1)$
- Update the particle velocity:

$$v_{i,d} \leftarrow \omega v_{id} + \phi_p r_p (p_{i.d} - x_{i,d}) + \phi_g r_g (g_d - x_{id}) \tag{13.10}$$

- Update the particle position:
  $x_i \leftarrow x_i + v_i$ If,
  $(f(x) < f(p_i))$, do:
- Update the particle best known position:

$$p_i \leftarrow x_i \tag{13.11}$$

If $(f(p_i) < f(g))$

- Update the swarm's best known position

$$g \leftarrow p_i \tag{13.12}$$

Now g holds the best found position. A flowchart is shown in Figure 13.3. The choices of PSO parameters have a large impact on optimization performance. This has been the subject of much research. The PSO parameters can also be tuned by using another overlaying optimizer, a concept known as meta-optimization.

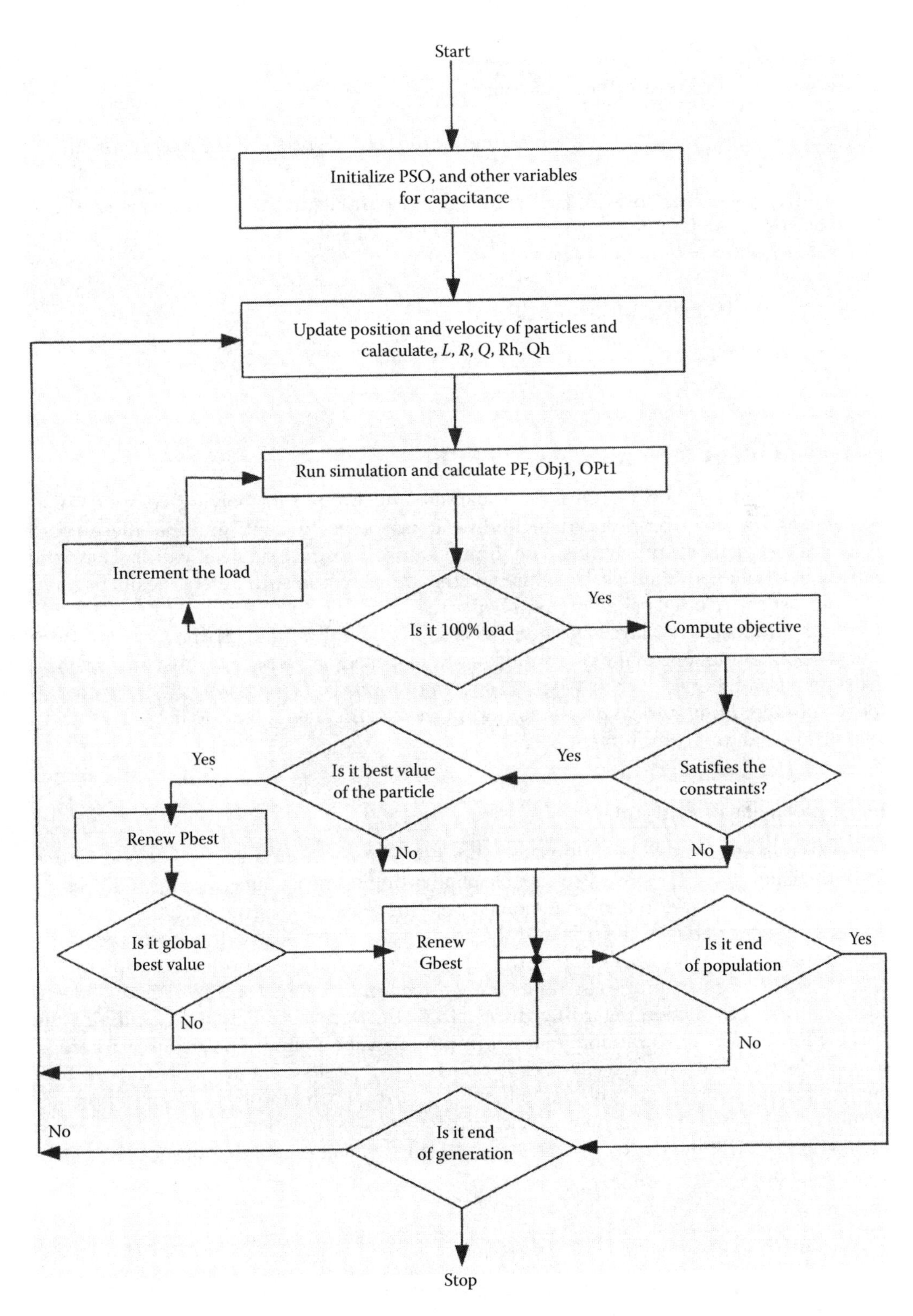

FIGURE 13.3
PSO, flowchart for optimization.

13.2.1 Convergence in PSO

Convergence in PSO means the following:

- All particles have converged to a point in the search space, which may not be the optimum.
- Convergence to a local optimum where all particles best p or, alternately, the swarm's best known position g, approaches a local optimum of the problem, regardless of how the swarm behaves.

See References [14–19] for further reading.

13.3 Ant Colony Optimization Algorithms

Ant colony optimization (ACO) is a probabilistic technique for solving computational problems, which can be reduced to finding good paths through graphs—these come from metaheuristic optimizations. The term is derived from two Greek words. Heuristic derives from the verb heuriskein, which means to find while suffix meta means beyond. Metaheuristics are often called modern heuristics.

Initially proposed by Marco Dorigo in 1992 in his PhD thesis [20,21], the first algorithm was to search for an optimal path through a graph, based on behavior of real ants seeking the shortest path between their nest and source of food. The idea has since diversified to solve a number of numerical problems, and in the electrical engineering field; generation scheduling and unit commitment [22,23].

13.3.1 Behavior of Real Ants

The real ants are capable of finding the shortest path from the nest to the food source without visual clues. They are also capable of adapting to the changes. In Figure 13.4, suppose there was no horizontal obstruction (shown shaded in this figure) initially, and there is a straight path between the nest and the food source. If an obstruction occurs, initially some ants may take the longer dotted route till all will follow the shortest route along the obstruction. These capabilities are essentially due to pheromone trails, which ants use to communicate information regarding their path or the decision where to go. Initially, the ants explore the area in a random manner [24]. As soon as an ant finds a source, it carries some of the found food to the nest. While doing so it deposits a chemical pheromone trail on the ground, which guides others to the food source. Each ant probabilistically prefers to follow a direction rich in pheromone rather than a poorer one. In turn, the shortest path will have a higher deposit on average, and this will cause a higher number of ants to choose the path.

13.3.2 Ant System

The ant system (AS) was the first ant colony search algorithm, proposed in the early 1990s. After the initialization process, m ants are put on their initial vertex on the construction graph, and all pheromones are initiated to be a small value (range 0–1), for all the edges.

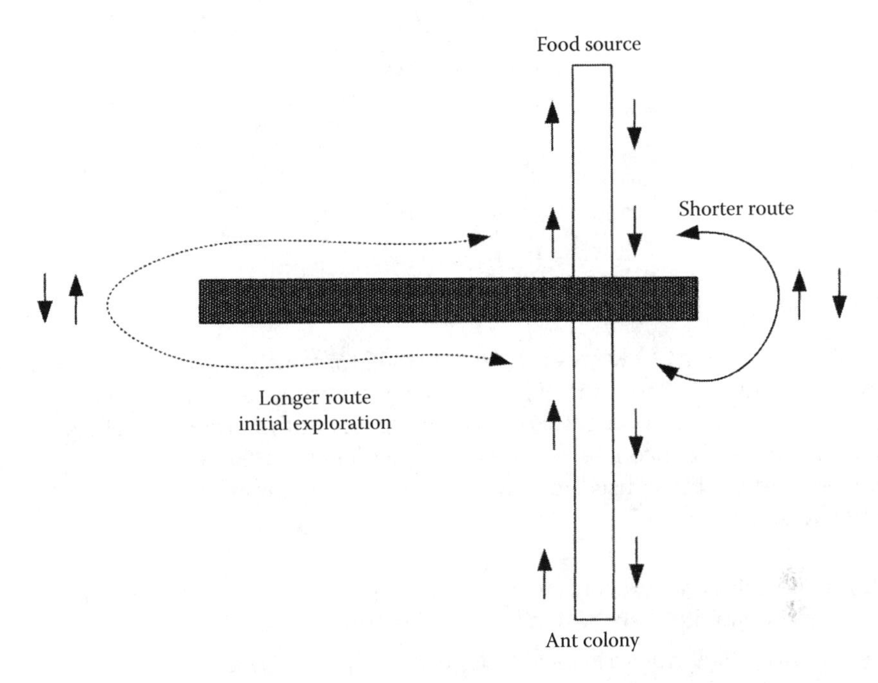

FIGURE 13.4
Behavior of real ants in search of food: from ant colony to food source and back depositing pheromone trail.

Suppose kth ant is at vertex c_i in the construction step t. The kth ant performs a randomized walk from vertex c_i to the next vertex to construct a feasible solution incrementally in such a way that the next vertex is chosen stochastically. The probability the kth ant chooses to move from i to j can be determined by the random-proportional state transition rule [25].

$$P_k(i,j) = \frac{[\tau(i,j)]^{\alpha} \cdot [\eta(i,j)]^{\beta}}{\sum_{1}^{t} [\tau(i,u)]^{\alpha} \cdot [\eta(i,u)]^{\beta}} \quad j, \quad u \in N_{k,i} \tag{13.13}$$
$$= 0 \quad \text{otherwise}$$

where $\tau(i, j)$ represents the pheromone trail associated with $l_{i,j}$, which is the connection between i and j, and $\eta(i, j)$ is the desirability of adding connection $l_{i,j}$ to solution, and can be determined according to optimization problem under consideration. It is usually set to be the inverse of the cost $J_{i,j}$ associated with edge $l_{i,j}$.

$N_{k,i}$ is the feasible neighbor components of the kth ant at vertex c_i with respect to problem constraints $\Omega, N_{k,i} \subseteq N_i$. α, β determine the relative importance of pheromone versus heuristic value: $\alpha, \beta > 0$.

Once all ants have completed their solutions, the pheromone trails are updated:

$$\tau(i,j) \leftarrow (1-\alpha) \cdot \tau(i,j) + \sum_{k=1}^{m} \Delta\tau_k(i,j) \tag{13.14}$$

where α is a pheromone decay parameter with $0 < \alpha \leq 1$, and

$$\Delta\tau_k(i,j) = \frac{1}{J_{i,j}} \quad (i,j) \in \phi$$
$$= 0 \quad \text{otherwise} \tag{13.15}$$

where ϕ are the set of moves done by the kth ant.

13.3.3 Ant Colony System

The ant colony system (ACS) was developed after AS and outperforms AS. The basic working of ACS algorithm is shown in Figure 13.5. Given an optimization problem, derive a finite set c of solution components. Second, define a set of pheromone values τ—this set of values is commonly called pheromone model, which is fundamentally a parameterized probabilistic model. This is used to generate solutions to the problem by iteration in the following two steps:

- Candidate solutions are constructed using a pheromone model, that is, a parameterized probability distribution over the solution space.
- The candidate solutions are used to modify the pheromone values in a way that is deemed to bias future sampling toward high-quality solutions.

The reinforcement of solution components depending on the solution quality is an important ingredient of ACO algorithms. It assumes that good solutions consist of good solution components. To learn which components contribute to good solutions can help assembling them into better solutions.

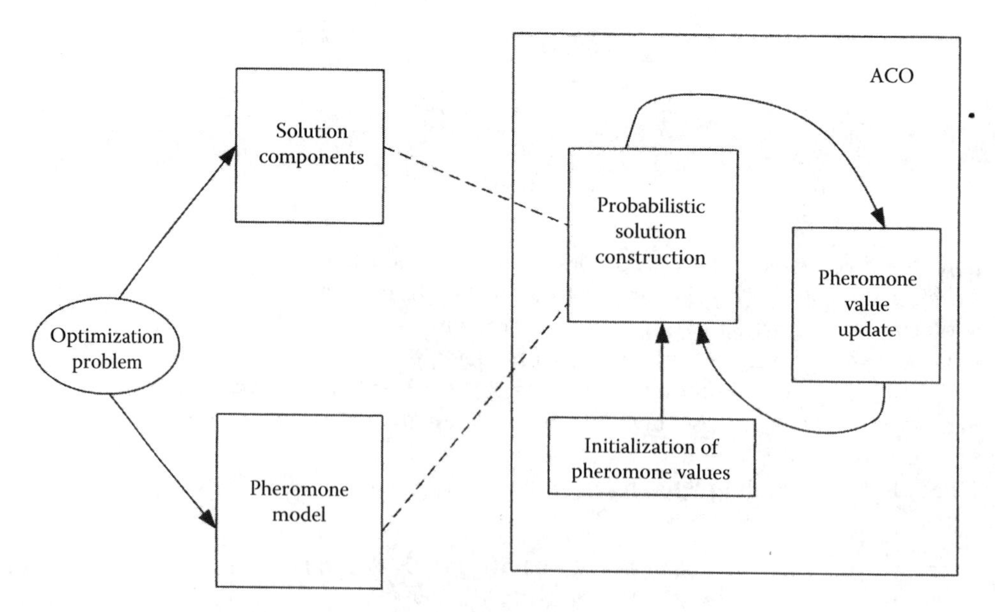

FIGURE 13.5
Building blocks of ACO algorithm.

13.3.4 Other Extensions

The other extensions of ACO variations are as follows:

Max–Min Ant System (MMAS)

This adds maximum and minimum pheromone amounts [τ_{max}, τ_{min}]. All edges are reinitialized to τ_{max} when nearing stagnation.

Rank-Based Ant System (ASrank)

All solutions are ranked according to their length. The amount of pheromone deposited is then weighted for each solution, so that solutions with shorter paths deposit more pheromone than the solutions with longer paths.

Continuous Orthogonal Ant Colony (COAC)

The pheromone deposit of COAC is to enable ants to search for solutions collaboratively and effectively. By using an orthogonal design method, ants in feasible domain can explore their chosen regions effectively, with enhanced global search capability and accuracy.

Recursive ACO

The whole search domain is divided into several subdomains and solves the objectives on these domains [6]. The results from all subdomains are compared and the best few of them are promoted for the next level. The subdomains corresponding to the selected results are further subdivided and the process is repeated till an output of desired precision is obtained [7].

13.3.5 Convergence

Like most metaheuristics, it is difficult to estimate the speed of convergence. For some versions, it is possible to prove that it is convergent.

A flowchart of ACS and MMAS is depicted in Figure 13.6.

13.4 Tabu Search

The word tabu comes from Tongan, a Polynesian language, and is used by aborigines of Tonga to indicate something which is holy or sacred and cannot be touched. It was created by Fred W. Glover in 1986 [26] and formalized in 1989 [27,28]. It is a metaheuristic search method employing local search methods used in mathematical optimization. Its origin is not related to biological or physical optimization processes. It has been applied to network problems in electrical power systems, including optimum capacitor installations, long-term transmission network extension, and distribution planning problems, and also other scientific problems.

Local searches take a potential solution to a problem and check its immediate neighbors, that is, solutions that are similar except for some minor details, in the hope of finding better solutions. Local search methods may get stuck up in suboptimal regions or on plateaus where many solutions may fit.

The local search is enhanced by *relaxation* of basic rule, that is, at each step a worsening move is acceptable, if no improving move is available. In addition, prohibitions called tabu are introduced to discourage the search from coming back to previously visited solutions.

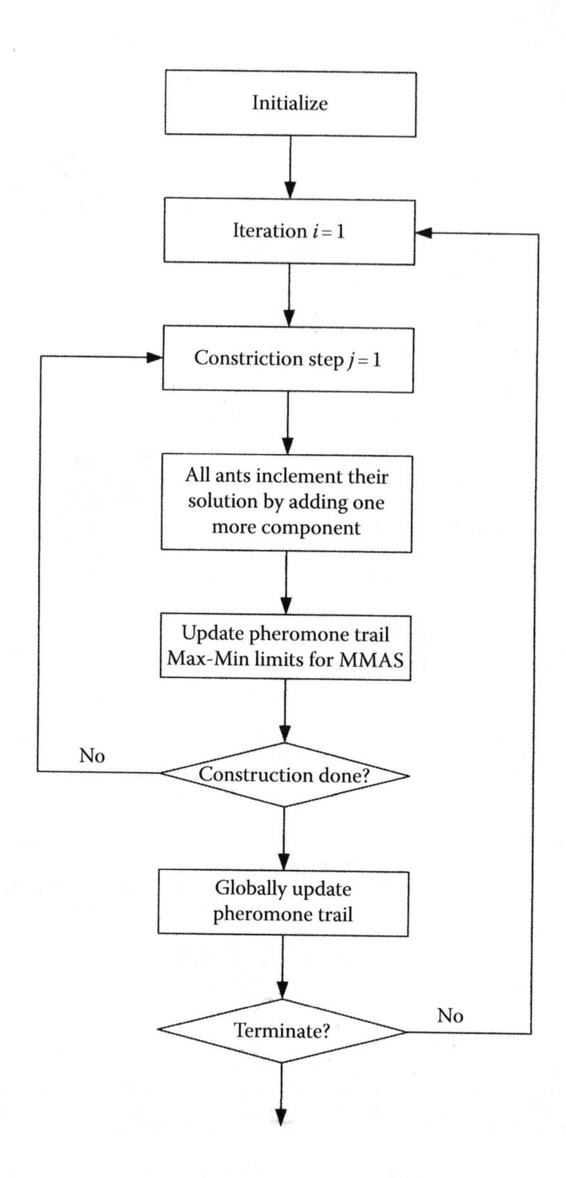

FIGURE 13.6
Flowchart of ACO.

The implementation of tabu search uses memory structures that describe the visited solutions or user-defined set of rules. If a solution has been previously visited within a certain short-term period, or if it has violated a rule, it is marked tabu, so that the algorithm no longer considers it.

Tabu search uses a local or neighborhood search procedure to move from one solution to an improved solution say from x to x' in the neighborhood of x, until some stopping criterion has been satisfied. It explores search space that might have been left unexplored by other local search procedures. The solutions admitted to the new neighborhood $N^*(x)$ are determined through use of memory structures. Using these structures, the search progresses iteratively from current solution x to an improved solution x' in the neighborhood $N^*(x)$. These memory structures form a list called *tabu list*, which is essentially a set of rules and banned solutions used to filter which solutions will be admitted to neighborhood

$N^*(x)$ to be further explored by the search. More commonly, a tabu list may consist of solutions that have changed by process of moving one solution to another.

13.4.1 Types of Memory

- Short-term memory contains a list of solutions recently considered. If a potential solution appears on the tabu list, it cannot be revisited until it reaches an expiration point.
- Intermediate-term memory has the intensification rules to bias the search to promising areas of search space.
- Long-term memory has diversification rules that drive the search into new regions.

In practice, the three memories can overlap.

The fitness function is a mathematical function, which returns a score when the aspiration criterion is satisfied—for example, an aspiration criterion could be considered as a new search space is found. If a candidate has a higher fitness value than the current best, it is set as the new best. The local best candidate is always added to the tabu list, and if this list is full, some elements will be allowed to expire. Generally items that expire from the list are of the same order as the elements added.

The core of Tabu search is embedded in the short-term memory process. Figure 13.7 shows that short-term memory search constitutes a form of aggressive exploration to make the best move possible, to satisfy laid constraints. Without these restrictions, the method could take a step away from the local optimum. Figure 13.8 shows the logic of choosing the best admissible candidate. For each move, the list is evaluated. The evaluation of a move can be based on the change produced in the objective function value or the evaluation can be based on generating approximate solutions or may utilize local measures of attractiveness. It is usually preferable to check first whether a given move has a higher evaluation than its preceding move before checking tabu status.

Intermediate- and long-term memories operate primarily as a basis of strategies for intensifying and diversifying the search. In fact, the fundamental elements of diversification and intensification are present in the short-term memory. In some cases, only short-term memory has produced solutions superior to those produced by other means, and intermediate- and long-term memories are bypassed. However, the long-term memory can be an improvement for hard problems. The modular form of a process makes it easy to create and test short-term memory component first and then to incorporate remaining components if additional refinement is required.

Also see References [29–31]

13.5 Evolution Strategies

Evolution strategies (ESs) have evolved to share many features with GA. Both maintain a population of possible solutions and use a selection mechanism for selecting the best individuals from the population. Evolutionary programming (EP) is similar to GA. It relies on behavior linkage between parents and their offsprings, rather than specific genetic operators.

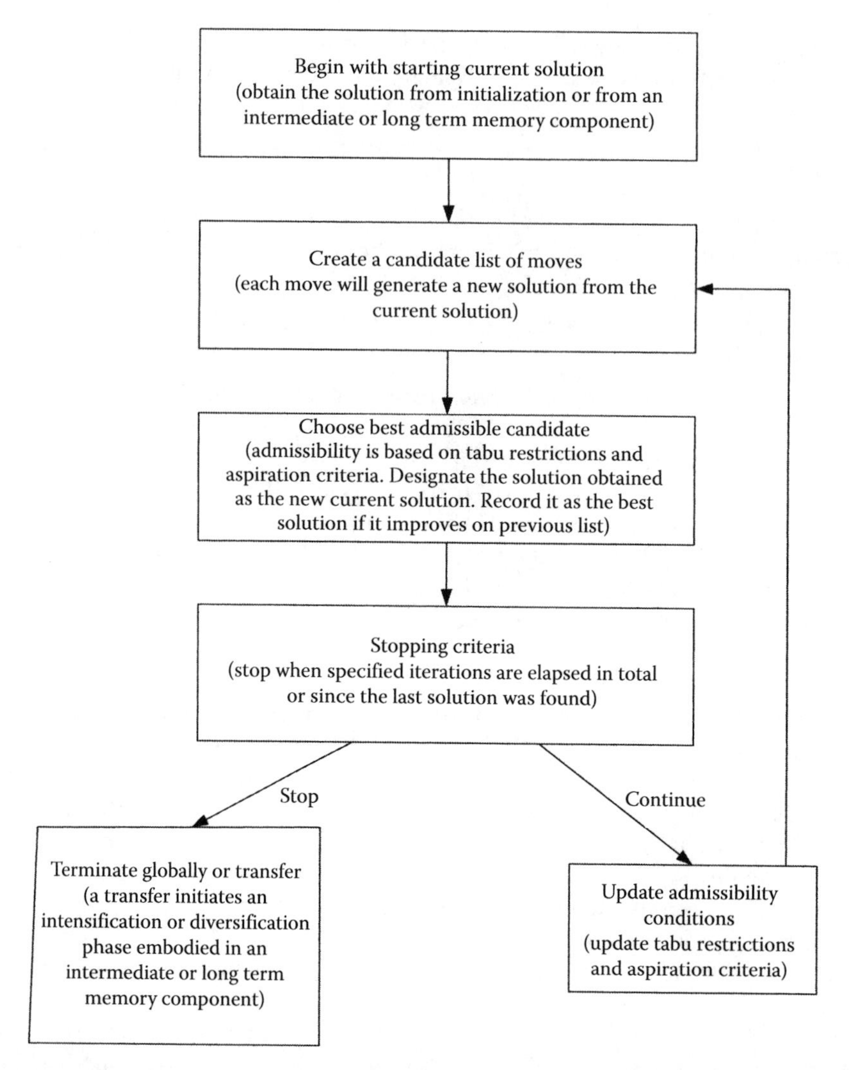

FIGURE 13.7
Block circuit and operations for short-term memory, tabu search.

The difference is that the GA operates on binary strings while the ES operates on floating point vectors and uses mutation as the dominant operator.

It was first used by Lawrence J. Fogel in the USA in 1960 in order to simulate evolution as a learning process to generate artificial intelligence. Currently, it is a wide evolutionary computing dialect with no fixed structure in contrast with some other dialects. It is becoming harder to distinguish from EP. EP algorithms perform well approximating solutions to all types of problems, because ideally they do not make any assumptions about underlying fitness landscape and have applications in the field of engineering, biology, economics, marketing, etc.

Classical optimization requires approximate settings of exogenous variables, and also EP, but EP can optimize these parameters as a part of the search for optimal solutions [32]. Suppose a search requires real-valued vectors that minimize a function $f(x)$, where x is a

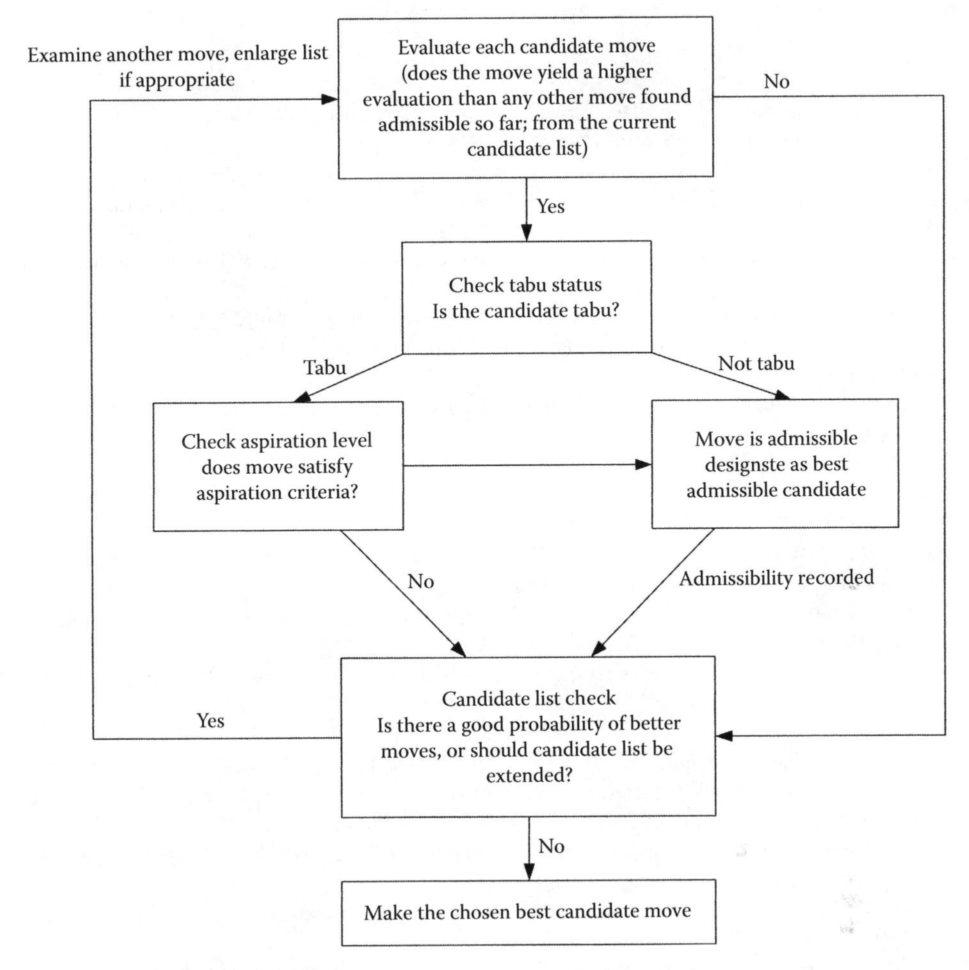

FIGURE 13.8
Best candidate selection through tabu search.

vector in n dimensions. A typical EP can use Gaussian random variation on current parent solutions to generate offspring:

$$x_i' = x_i + \sigma_i N(0,1) \tag{13.16}$$

where σ_i is the standard deviation of a random variable denoted by $N(0, 1)$. Setting the step size is important, see [33].

$$x_i' = \sigma_i \exp(\tau_i N(0,1) + \tau' N_i(0,1)) \tag{13.17}$$

where $\tau \propto (2n)^{0.5}$, $\tau' \propto (2n^{0.5})^{0.5}$

In this manner, the standard deviations are subject to variations and selection to guide the search to optima. Also see References [34,35]

This chapter forms an introduction, and the references provide further reading.

References

1. KY Lee, AE Mohmed. *Modern Heuristic Optimization Techniques: Theory and Applications to Power Systems*. IEEE Press, Piscataway, NJ, 2008.
2. DB Fogel, A Ghozil. Using fitness distributions to design more efficient evolutionary computations. *Proceedings of IEEE Conference on Evolutionary Computation*, Nagoya, Japan, pp. 11–19, May 1996.
3. NA Barricelli. Symbiogenetic evolution process realized in artificial methods, *Methodos*, 9, 143–182, 1957.
4. I Rechenberg. *Evolutions straegie*. Holzmann-Froboog, Stuttgart. ISBN 3-7728-0373-3, 1973.
5. Hans-Paul Schwefel. *Numerical Optimization of Computer Models*. Wiley, New York. ISBN 0-471-09988-0, 1981.
6. DE Goldberg. *Genetic algorithms in Search, Optimization and Machine Learning*. Addison-Wesley, Reading, MA. 1989.
7. JF Frenzel. Genetic algorithms, *IEEE Potentials*, 12, 21–24, 1998.
8. J Kennedy, R Eberhart. Particle swarm optimization. *Proceedings of International Conference on Neural Networks*, Perth, Australia, vol. 4, pp. 1942–1948, 1995.
9. M Srinivas, L Patnaik. Adaptive probabilities of crossover and mutation in genetic algorithms. *IEEE Trans Syst, Man Cybern*, 24(4), 656–667, 1994.
10. J Chung Zhang, WL Lo. Clustering based adaptive crossover and mutation probabilities for genetic algorithms, *IEEE Trans Evol Comput*, 11(3), 326–335, 2007.
11. Y-M Chen. Passive filter design using genetic algorithms, *IEEE Trans Ind Electron*, 50(1), 202–207, 2003.
12. ML Schmitt. Theory of genetic algorithms, *Theor Comput Sci*, 259, 1–61, 2001.
13. ML Schmitt. Theory of genetic algorithms II: Models for genetic operators over the string-tensor representation of population and convergence to global optima for arbitrary fitness function under scaling, *Theor Comput Sci*, 310, 181–231, 2004.
14. R Poli. Analysis of the publications on applications of PSO, *Journal of Artificial Evolution and Applications*, 1–8, 2008.
15. Y Shi, RC Eberhart. Parameter selection in particle swarm optimization, *Proc Evol Program*, 7(EP98), 591–600, 1998.
16. M Clerc, J Kennedy. The particle swarm-explosion, stability and convergence in a multi-dimensional complex space, *IEEE Trans Evol Comput*, 1(6), 58–73, 2002.
17. IC Trela. The particle swarm optimization algorithm: Convergence analysis and parameter selection, *Inf Process Lett*, 85(6), 317–325, 2003.
18. M Clerc, J Kennedy. The particle swarm-explosion, stability and convergence in a multidimensional complex space, *IEEE Trans Evol Comput*, 1(6), 58–73, 2002.
19. MR Bonyadi, MZ Michalewicz. A locally convergent rotationally invariant PSO algorithm, *Swarm Intell*, 8(3), 159–198, 2014.
20. A Colorni, M Dorigo, V Maniezzo. *Distributed Optimization by Ant Colonies, actes de la premiére conference europeéne sur la vie artificielle*. France Elsevier Publishing, Paris, pp. 133–142, 1999.
21. M Dorigo. Optimization learning and natural algorithms, PhD thesis, Politecnico di Milano, 1992.
22. MR Irving, YH Song. Optimization techniques for electrical power systems, *IEEE Power Eng J*, 14(5), 245–254, 2000.
23. YH Song, CS Chou, Y Min. Large scale economic dispatch by artificial ant colony search algorithm, *Electr Mach Power Syst*, 27(5), 679–690, 1999.
24. C Blum, M Dorigo. The hyper-cube framework for ant colony optimization, *IEEE Trans Syst, Man Cybern*, 34(2), 1161–1172, 2004.
25. M Dorigo, LM Gambardella. Ant colony system: A cooperative learning approach to the traveling salesman problem, *IEEE Trans Evol Comput*, 1(1), 53–66, 1997.

26. F Glover. Future paths for integer programming and links to artificial intelligence, *Comput Oper Res*, 13(5), 533–549, 1986.
27. F Glover. Tabu search—Part 1, *ORSA J Comput*, 1(2), 190–206, 1989.
28. F Glover. Tabu search—Part II, *ORSA J Comput*, 2(1), 4–32, 1990.
29. M Laguna, R Marti. *Scatter Search: Methodology and Implementation*. Kluwer Academic Publishers, Boston, 2003.
30. AH Mentawy, YL Abdel-Magid, SZ Selim. Unit commitment by Tabu search, *IEEE Proc Gene Transm Distrib*, 145(1), 56–64, 1998.
31. YC Wang, HT Yang, CL Huang. Solving the capacitor placement problem in radial distribution system using Tabu search, *IEEE Trans Power Syst*, 11(4), 1868–1873, 1996.
32. R Rosenberg. Simulation of genetic populations with biochemical properties, PhD dissertation, University of Michigan, 1967.
33. T Black, HP Schwefel. An overview of evolutionary programming for parameter optimization, *Evol Comput*, 1(1), 1–24, 1993.
34. HP Schwefel. *Evolution and Optimum Seeking*. John Wiley, New York, 1995.
35. HJ Bremermann. Evolution through evolution and recombination. In: *Self-Organizing Sysems* (Ed. MC Jacobi), Spartan Books, Washington, DC, pp. 93–106, 1962.

Appendix A: Calculation of Line and Cable Constants

This appendix presents an overview of calculations of line and cable constants with an emphasis on three-phase models and transformation matrices. Practically, the transmission or cable system parameters will be calculated using computer-based subroutine programs. For simple systems, the data are available in tabulated form for various conductor types, sizes, and construction [1–4]. Nevertheless, the basis of these calculations and required transformations are of interest to a power system engineer. The models described are generally applicable to steady-state studies. Frequency-dependent models are required for transient analysis studies, which are not discussed in this volume [5].

A.1 AC Resistance

As we have seen, the conductor ac resistance is dependent upon frequency and proximity effects, temperature, spiraling, and bundle conductor effects, which increase the length of wound conductor in spiral shape with a certain pitch. The ratio R_{ac}/R_{dc} considering proximity and skin effects is given in Volume 1, Chapter 10. The resistance increases linearly with temperature and is given by the following equation:

$$R_2 = R_1\left(\frac{T+t_2}{T+t_1}\right) \tag{A.1}$$

where R_2 is the resistance at temperature t_2, R_1 is the resistance at temperature t_1, and T is the temperature coefficient, which depends on the conductor material. It is 234.5 for annealed copper, 241.5 for hard drawn copper, and 228.1 for aluminum. The resistance is read from manufacturers' data, databases in computer programs, or generalized tables. The internal resistance and the ac-to-dc resistance ratios using Bessel's functions are provided in Reference [6]. The frequency-dependent behavior can be simulated by dividing the conductor into n hollow cylinders with an internal reactance and conductivity, and the model replicated by an equivalent circuit consisting of series of connections of hollow cylinders [7]. Table 10.14 shows ac-to-dc resistance of conductors at 60 Hz.

A.2 Inductance

The *internal* inductance of a solid, smooth, round metallic cylinder of infinite length is due to its internal magnetic field when carrying an alternating current and is given by

$$L_{int} = \frac{\mu_0}{8\pi}\ \text{H/m (Henry per meter)} \tag{A.2}$$

where μ_0 is the permeability $= 4\pi \times 10^{-7}$ (H/m). Its *external* inductance is due to the flux outside the conductor which is given by

$$L_{ext} = \frac{\mu_0}{2\pi} \ln\left(\frac{D}{r}\right) \text{H/m} \tag{A.3}$$

where D is any point at a distance D from the surface of the conductor and r is the conductor radius. In most inductance tables, D is equal to 1 ft and adjustment factors are tabulated for higher conductor spacings. The total reactance is

$$L = \frac{\mu_0}{2\pi}\left[\frac{1}{4} + \ln\frac{D}{r}\right] = \frac{\mu_0}{2\pi}\left[\ln\frac{D}{e^{-1/4}r}\right] = \frac{\mu_0}{2\pi}\left[\ln\frac{D}{\text{GMR}}\right] \text{H/m} \tag{A.4}$$

where GMR is called the geometric mean radius and is $0.7788r$. It can be defined as the radius of a tubular conductor with an infinitesimally thin wall that has the same external flux out to a radius of 1 ft as the external and internal flux of a solid conductor to the same distance.

A.2.1 Inductance of a Three-Phase Line

We can write the inductance matrix of a three-phase line in terms of flux linkages λ_a, λ_b, and λ_c:

$$\begin{vmatrix} \lambda_a \\ \lambda_b \\ \lambda_c \end{vmatrix} = \begin{vmatrix} L_{aa} & L_{ab} & L_{ac} \\ L_{ba} & L_{bb} & L_{bc} \\ L_{ca} & L_{cb} & L_{cc} \end{vmatrix} \begin{bmatrix} I_a \\ I_b \\ I_c \end{bmatrix} \tag{A.5}$$

The flux linkages λ_a, λ_b, and λ_c are given by

$$\begin{aligned} \lambda_a &= \frac{\mu_0}{2\pi}\left[I_a \ln\left(\frac{1}{\text{GMR}_a}\right) + I_b \ln\left(\frac{1}{D_{ab}}\right) + I_c \ln\left(\frac{1}{D_{ac}}\right)\right] \\ \lambda_b &= \frac{\mu_0}{2\pi}\left[I_a \ln\left(\frac{1}{D_{ba}}\right) + I_b \ln\left(\frac{1}{\text{GMR}_b}\right) + I_c \ln\left(\frac{1}{D_{bc}}\right)\right] \\ \lambda_c &= \frac{\mu_0}{2\pi}\left[I_a \ln\left(\frac{1}{D_{ca}}\right) + I_b \ln\left(\frac{1}{D_{cb}}\right) + I_c \ln\left(\frac{1}{\text{GMR}_c}\right)\right] \end{aligned} \tag{A.6}$$

where D_{ab}, D_{ac},... are the distances between conductor of a phase with respect to conductors of b and c phases; L_{aa}, L_{bb}, and L_{cc} are the self-inductances of the conductors; and L_{ab}, L_{bb}, and L_{cc} are the mutual inductances. If we assume a symmetrical line, i.e., the GMR of all three conductors is equal and also the spacing between the conductors is equal. The equivalent inductance per phase is

$$L = \frac{\mu_0}{2\pi} \ln\left(\frac{D}{\text{GMR}}\right) \text{H/m} \tag{A.7}$$

The phase-to-neutral inductance of a three-phase symmetrical line is the same as the inductance per conductor of a two-phase line.

A.2.2 Transposed Line

A transposed line is shown in Figure A.1. Each phase conductor occupies the position of two other phase conductors for one-third of the length. The purpose is to equalize the phase inductances and reduce unbalance. The inductance derived for a symmetrical line is still valid and the distance D in Equation A.7 is substituted by GMD (geometric mean distance). It is given by

$$\text{GMD} = (D_{ab}D_{bc}D_{ca})^{1/3} \tag{A.8}$$

A detailed treatment of transposed lines with rotation matrices is given in Reference [8].

A.2.3 Composite Conductors

A transmission line with composite conductors is shown in Figure A.2. Consider that group X is composed of n conductors in parallel and each conductor carries $1/n$ of the line current. The group Y is composed of m parallel conductors, each of which carries $-1/m$ of the return current. Then L_x, the inductance of conductor group X, is

$$L_x = 2\times10^{-7}\ln\frac{\sqrt[nm]{(D_{aa'}D_{ab'}D_{ac'}\ldots D_{am})\ldots(D_{na'}D_{nb'}D_{nc'}\ldots D_{nm})}}{\sqrt[n^2]{(D_{aa}D_{ab}D_{ac}\ldots D_{an})\ldots(D_{na}D_{nb}D_{nc}\ldots D_{nn})}} \tag{A.9}$$

Henry per meter.

We write Equation A.9 as

$$L_x = 2\times10^{-7}\ln\left(\frac{D_m}{D_{sx}}\right)\text{H/m} \tag{A.10}$$

Similarly,

$$L_y = 2\times10^{-7}\ln\left(\frac{D_m}{D_{sy}}\right)\text{H/m} \tag{A.11}$$

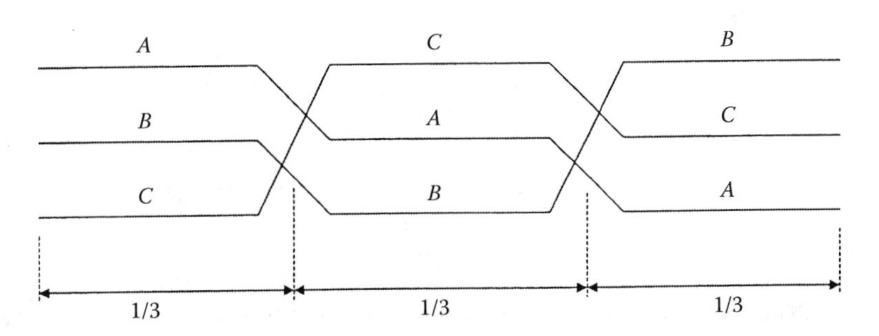

FIGURE A.1
Transposed transmission line.

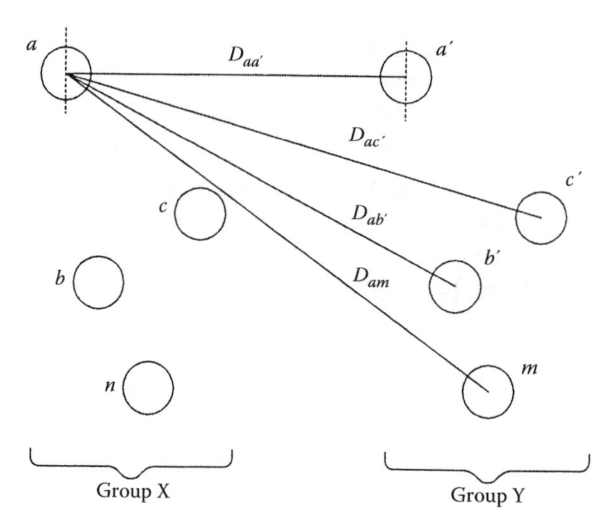

FIGURE A.2
Inductance of composite conductors.

The total inductance is

$$L = (L_x + L_y)\text{H/m} \tag{A.12}$$

A.3 Impedance Matrix

In Chapter 1, we decoupled a symmetrical three-phase line 3×3 matrix having equal self-impedances and mutual impedances (see Equation 1.37). We showed that the off-diagonal elements of the sequence impedance matrix are zero. In high-voltage transmission lines, which are transposed, this is generally true and the mutual couplings between phases are almost equal. However, the same cannot be said of distribution lines and these may have unequal off-diagonal terms. In many cases, the off-diagonal terms are smaller than the diagonal terms and the errors introduced in ignoring these will be small. Sometimes, an equivalence can be drawn by the following equations:

$$\begin{aligned} Z_s &= \frac{Z_{aa} + Z_{bb} + Z_{cc}}{3} \\ Z_m &= \frac{Z_{ab} + Z_{bc} + Z_{ca}}{3} \end{aligned} \tag{A.13}$$

i.e., an average of the self-impedances and mutual impedances can be taken. The sequence impedance matrix then has only diagonal terms (see Example A.1).

A.4 Three-Phase Line with Ground Conductors

A three-phase transmission line has couplings between phase-to-phase conductors and also between phase-to-ground conductors. Consider a three-phase line with two ground conductors, as shown in Figure A.3. The voltage V_a can be written as follows:

$$\begin{aligned} V_a &= R_a I_a + j\omega L_a I_a + j\omega L_{ab} I_b + j\omega L_{ac} I_c + j\omega L_{aw} I_w + j\omega L_{av} I_v \\ &\quad - j\omega L_{an} + V_a' + R_n I_n + j\omega L_n I_n - j\omega L_{an} I_a - j\omega L_{bn} I_b \\ &\quad - j\omega L_{cn} I_c - j\omega L_{wn} I_w - j\omega L_{vn} I_v \end{aligned} \tag{A.14}$$

where,

$R_a, R_b, \ldots, R_n$ are the resistances of phases $a, b, \ldots, n$

$L_a, L_b, \ldots, L_n$ are the self-inductances

$L_{ab}, L_{ac}, \ldots, L_{an}$ are the mutual inductances.

This can be written as follows:

$$\begin{aligned} V_a &= (R_a + R_n) I_a + R_n I_b + R_n I_c + j\omega(L_a + L_n - 2L_{an}) I_a \\ &\quad + j\omega(L_{ab} + L_n - L_{an} - L_{bn}) I_b + j\omega(L_{ac} + L_n - L_{an} - L_{cn}) I_c + R_n I_w \\ &\quad + j\omega(L_{aw} + L_n - L_{an} - L_{wn}) I_w + R_n I_\upsilon + j\omega(L_{a\upsilon} + L_n - L_{an} - L_{\upsilon n}) I_\upsilon + V_a' \\ &= Z_{aa-g} I_a + Z_{ab-g} I_b + Z_{ac-g} I_c + Z_{aw-g} I_w + Z_{a\upsilon-g} I_\upsilon + V_a' \end{aligned} \tag{A.15}$$

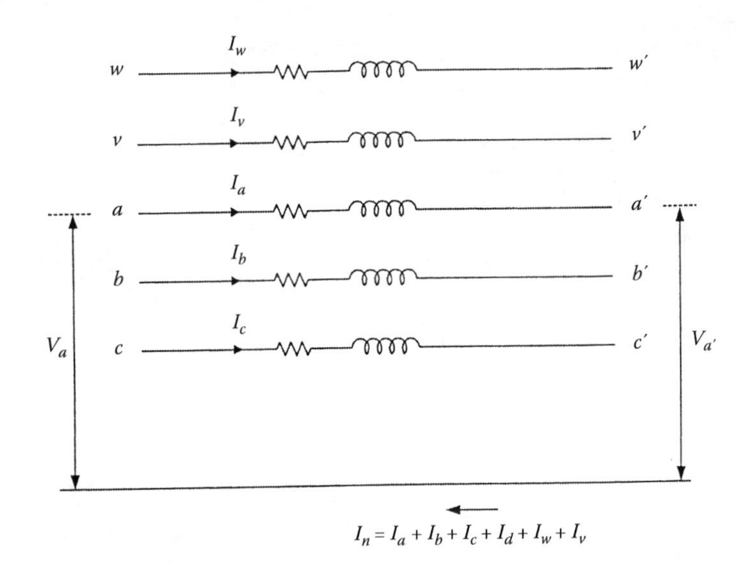

FIGURE A.3
Transmission line section with two ground conductors.

where Z_{aa-g} and Z_{hb-g} are the self-impedances of a conductor with ground return, and Z_{ab-g} and Z_{ac-g} are the mutual impedances between two conductors with common earth return. Similar equations apply to the voltages of other phases and ground wires. The following matrix then holds for the voltage differentials between terminals marked *w*, *v*, *a*, *b*, and *c*, and *w*′, *v*′, *a*′, *b*′, and *c*′:

$$\left|\begin{matrix} \Delta V_a \\ \Delta V_b \\ \Delta V_c \\ \Delta V_w \\ \Delta V_w \end{matrix}\right| = \left|\begin{matrix} Z_{aa-g} & Z_{ab-g} & Z_{ac-g} & Z_{aw-g} & Z_{av-g} \\ Z_{ba-g} & Z_{bb-g} & Z_{bc-g} & Z_{bw-g} & Z_{bv-g} \\ Z_{ca-g} & Z_{cb-g} & Z_{cc-g} & Z_{cw-g} & Z_{cv-g} \\ Z_{wa-g} & Z_{wb-g} & Z_{wc-g} & Z_{ww-g} & Z_{wv-g} \\ Z_{va-g} & Z_{vb-g} & Z_{vc-g} & Z_{vw-g} & Z_{vv-g} \end{matrix}\right| \left|\begin{matrix} I_a \\ I_b \\ I_c \\ I_w \\ I_v \end{matrix}\right| \tag{A.16}$$

In the partitioned form, this matrix can be written as follows:

$$\left|\begin{matrix} \Delta \overline{V}_{abc} \\ \Delta \overline{V}_{wv} \end{matrix}\right| = \left|\begin{matrix} \overline{Z}_A & \overline{Z}_B \\ \overline{Z}_C & \overline{Z}_D \end{matrix}\right| \left|\begin{matrix} \overline{I}_{abc} \\ \overline{I}_{wv} \end{matrix}\right| \tag{A.17}$$

Considering that the ground wire voltages are zero:

$$\begin{aligned} \Delta \overline{V}_{abc} &= \overline{Z}_A \overline{I}_{abc} + \overline{Z}_B \overline{I}_{wv} \\ 0 &= \overline{Z}_c \overline{I}_{abc} + \overline{Z}_D \overline{I}_{wv} \end{aligned} \tag{A.18}$$

Thus,

$$\overline{I}_{wv} = -\overline{Z}_D^{-1} \overline{Z}_C \overline{I}_{abc} \tag{A.19}$$

$$\Delta \overline{V}_{abc} = (\overline{Z}_A - \overline{Z}_B \overline{Z}_D^{-1} \overline{Z}_C) \overline{I}_{abc} \tag{A.20}$$

This can be written as follows:

$$\Delta \overline{V}_{abc} = \overline{Z}_{abc} \overline{I}_{abc} \tag{A.21}$$

$$\overline{Z}_{abc} = \overline{Z}_A - \overline{Z}_B \overline{Z}_D^{-1} \overline{Z}_C = \left|\begin{matrix} Z_{aa'-g} & Z_{ab'-g} & Z_{ac'-g} \\ Z_{ba'-g} & Z_{bb'-g} & Z_{bc'-g} \\ Z_{ca'-g} & Z_{cb'-g} & Z_{cc'-g} \end{matrix}\right| \tag{A.22}$$

The five-conductor circuit is reduced to an equivalent three-conductor circuit. The technique is applicable to circuits with any number of ground wires provided that the voltages are zero in the lower portion of the voltage vector.

A.5 Bundle Conductors

Consider bundle conductors, consisting of two conductors per phase (Figure A.4). The original circuit of conductors a, b, c and a', b', c' can be transformed into an equivalent conductor system of a'', b'', and c''.

Each conductor in the bundle carries a different current and has a different self-impedance and mutual impedance because of its specific location. Let the currents in the conductors be I_a, I_b, and I_c, and I'_a, I'_b, and I'_c, respectively. The following primitive matrix equation can be written as follows:

$$\begin{vmatrix} V_a \\ V_b \\ V_c \\ V'_a \\ V'_b \\ V'_c \end{vmatrix} = \begin{Vmatrix} Z_{aa} & Z_{ab} & Z_{ac} & Z_{aa'} & Z_{ab'} & Z_{ac'} \\ Z_{ba} & Z_{bb} & Z_{bc} & Z_{ba'} & Z_{bb'} & Z_{bc'} \\ Z_{ca} & Z_{cb} & Z_{cc} & Z_{ca'} & Z_{cb'} & Z_{cc'} \\ Z_{a'a} & Z_{a'b} & Z_{a'c} & Z_{a'a'} & Z_{a'b'} & Z_{a'c'} \\ Z_{b'a} & Z_{b'b} & Z_{b'c} & Z_{b'a'} & Z_{b'b'} & Z_{b'c'} \\ Z_{c'a} & Z_{c'b} & Z_{c'c} & Z_{c'a'} & Z_{c'b'} & Z_{c'c'} \end{Vmatrix} \begin{vmatrix} I_a \\ I_b \\ I_c \\ I_{a'} \\ I_{b'} \\ I_{c'} \end{vmatrix} \tag{A.23}$$

This can be partitioned so that

$$\begin{vmatrix} \bar{V}_{abc} \\ \bar{V}_{a'b'c'} \end{vmatrix} = \begin{Vmatrix} \bar{Z}_1 & \bar{Z}_2 \\ \bar{Z}_3 & \bar{Z}_4 \end{Vmatrix} \begin{vmatrix} \bar{I}_{abc} \\ \bar{I}_{a'b'c'} \end{vmatrix} \tag{A.24}$$

for symmetrical arrangement of bundle conductors $\bar{Z}_1 = \bar{Z}_4$.

Modify so that the lower portion of the vector goes to zero. Assume that

$$\begin{aligned} V_a &= V'_a = V''_a \\ V_b &= V'_b = V''_b \\ V_c &= V'_c = V''_c \end{aligned} \tag{A.25}$$

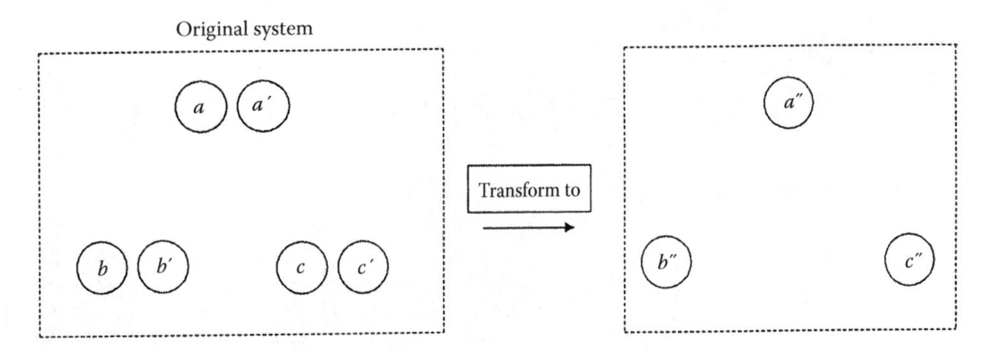

FIGURE A.4
Transformation of bundle conductors to equivalent single conductors.

The upper part of the matrix can then be subtracted from the lower part:

$$\left|\begin{array}{c} V_a \\ V_b \\ V_c \\ \hdashline 0 \\ 0 \\ 0 \end{array}\right| = \left|\begin{array}{ccc:ccc} Z_{aa} & Z_{ab} & Z_{ac} & Z_{aa'} & Z_{ab'} & Z_{ac'} \\ Z_{ba} & Z_{bb} & Z_{bc} & Z_{ba'} & Z_{bb'} & Z_{bc'} \\ Z_{ca} & Z_{cb} & Z_{cc} & Z_{ca'} & Z_{cb'} & Z_{cc'} \\ \hdashline Z_{a'a}-Z_{aa} & Z_{a'b}-Z_{ab} & Z_{a'c}-Z_{ac} & Z_{a'a'}-Z_{aa'} & Z_{a'b'}-Z_{ab'} & Z_{a'c'}-Z_{ac'} \\ Z_{b'a}-Z_{ba} & Z_{b'b}-Z_{bb} & Z_{b'c}-Z_{bc} & Z_{b'a'}-Z_{ba'} & Z_{b'b'}-Z_{bb'} & Z_{b'c'}-Z_{bc'} \\ Z_{c'a}-Z_{ca} & Z_{c'b}-Z_{cb} & Z_{c'c}-Z_{cc} & Z_{c'a'}-Z_{ca'} & Z_{c'b'}-Z_{cb'} & Z_{c'c}-Z_{cc'} \end{array}\right| \left|\begin{array}{c} I_a \\ I_b \\ I_c \\ \hdashline I_{a'} \\ I_{b'} \\ I_{c'} \end{array}\right| \tag{A.26}$$

We can write it in the partitioned form as follows:

$$\left|\begin{array}{c} \bar{V}_{abc} \\ 0 \end{array}\right| = \left|\begin{array}{cc} \bar{Z}_1 & \bar{Z}_2 \\ \bar{Z}_2^t - \bar{Z}_1 & \bar{Z}_4 - \bar{Z}_2 \end{array}\right| \left|\begin{array}{c} \bar{I}_{abc} \\ \bar{I}_{a'b'c'} \end{array}\right| \tag{A.27}$$

$$\begin{aligned} I_a'' &= I_a + I_a' \\ I_b'' &= I_b + I_b' \\ I_c'' &= I_c + I_c' \end{aligned} \tag{A.28}$$

The matrix is modified as shown in the following:

$$\left|\begin{array}{cccc:cc} Z_{aa} & Z_{ab} & Z_{ac} & Z_{aa'}-Z_{aa} & Z_{ab'}-Z_{ab} & Z_{ac'}-Z_{ac} \\ Z_{ba} & Z_{bb} & Z_{bc} & Z_{ba'}+Z_{ba} & Z_{bb'}+Z_{bb} & Z_{bc'}-Z_{bc} \\ Z_{ca} & Z_{cb} & Z_{cc} & Z_{ca'}-Z_{ca} & Z_{cb'}-Z_{cb} & Z_{cc'}-Z_{cc} \\ \hdashline Z_{a'a}-Z_{aa} & Z_{a'b}-Z_{ab} & Z_{a'c}-Z_{ac} & Z_{a'a'}-Z_{aa'}-Z_{a'a}+Z_{aa} & Z_{a'b'}-Z_{ab'}-Z_{a'b}+Z_{ab} & Z_{a'c'}-Z_{ac'}-Z_{a'c}+Z_{ac} \\ Z_{b'a}-Z_{ba} & Z_{b'b}-Z_{bb} & Z_{b'c}-Z_{bc} & Z_{b'a'}-Z_{ba'}-Z_{b'a}+Z_{ba} & Z_{b'b'}-Z_{bb'}-Z_{b'b}+Z_{bb} & Z_{b'c'}-Z_{bc'}-Z_{b'c}+Z_{bc} \\ Z_{c'a}-Z_{ca} & Z_{c'b}-Z_{cb} & Z_{c'c}-Z_{cc} & Z_{c'a'}-Z_{ca'}-Z_{c'a}+Z_{ca} & Z_{c'b'}-Z_{cb'}-Z_{c'b}+Z_{cb} & Z_{c'c'}-Z_{cc'}-Z_{c'c}+Z_{cc} \end{array}\right| \left|\begin{array}{c} I_a + I_a^! \\ I_b + I_b^! \\ I_c + I_c^! \\ \hdashline I_a' \\ I_b' \\ I_c' \end{array}\right| \tag{A.29}$$

or in the partitioned form

$$\left|\begin{array}{c} \bar{V}_{abc} \\ 0 \end{array}\right| = \left|\begin{array}{cc} \bar{Z}_1 & \bar{Z}_2 - \bar{Z}_1 \\ \bar{Z}_2^t - \bar{Z}_1 & (\bar{Z}_4 - \bar{Z}_2) - (\bar{Z}_2^t - \bar{Z}_1) \end{array}\right| \left|\begin{array}{c} \bar{I}_{abc}'' \\ \bar{I}_{a',b',c'}' \end{array}\right| \tag{A.30}$$

This can now be reduced to the following 3×3 matrix as before:

$$\left|\begin{array}{c} V_a'' \\ V_b'' \\ V_c'' \end{array}\right| = \left|\begin{array}{ccc} Z_{aa}'' & Z_{ab}'' & Z_{ac}'' \\ Z_{ba}'' & Z_{bb}'' & Z_{bc}'' \\ Z_{ca}'' & Z_{cb}'' & Z_{cc}'' \end{array}\right| \left|\begin{array}{c} I_a'' \\ I_b'' \\ I_c'' \end{array}\right| \tag{A.31}$$

A.6 Carson's Formula

The theoretical value of Z_{abc-g} can be calculated by Carson's formula (c. 1926). This is of importance even today in calculations of line constants. For an n-conductor configuration, the earth is assumed as an infinite uniform solid with a constant resistivity. Figure A.5 shows image conductors in the ground at a distance equal to the height of the conductors above ground and exactly in the same formation, with the same spacing between the conductors. A flat conductor formation is shown in Figure A.5:

$$Z_{ii} = R_i + 4\omega P_{ii}G + j\left[X_i + 2\omega G \ln\frac{S_{ii}}{r_i} + 4\omega Q_{ii}G\right]\Omega/\text{mi} \tag{A.32}$$

$$Z_{ij} = 4\omega P_{ii}G + j\left[2\omega G \ln\frac{S_{ij}}{D_{ij}} + 4\omega Q_{ij}G\right]\Omega/\text{mi} \tag{A.33}$$

where,

Z_{ii}=self-impedance of conductor i with earth return (Ω/mi)
Z_{ij}=mutual impedance between conductors i and j (Ω/mi)
R_i=resistance of conductor in Ω/mi
S_{ii}=conductor-to-image distance of the ith conductor to its own image
S_{ij}=conductor-to-image distance of the ith conductor to the image of the jth conductor

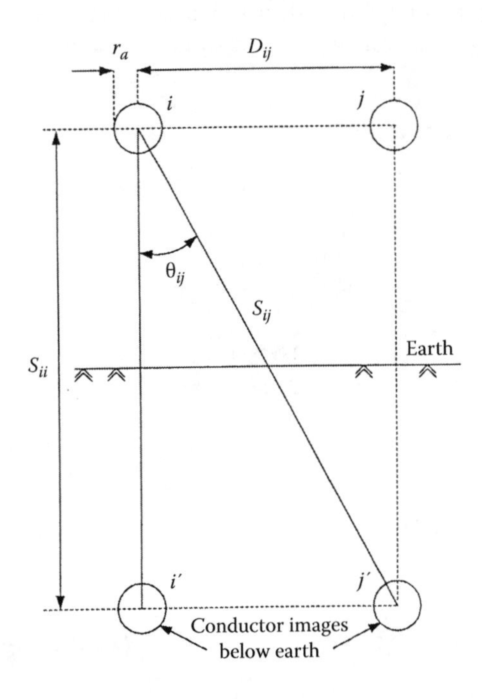

FIGURE A.5
Conductors and their images in the earth (Carson's formula).

D_{ij}=distance between conductors i and j
r_i=radius of the conductor (in ft)
ω=angular frequency
$G=0.1609347\times10^{-7}$ Ω-cm
GMR_i=geometric mean radius of the conductor i
ρ=soil resistivity
θ_{ij}=angle as shown in Figure A.5
Expressions for P and Q are

$$P=\frac{\pi}{8}-\frac{1}{3\sqrt{2}}k\cos\theta+\frac{k^2}{16}\cos 2\theta\left(0.6728+\ln\frac{2}{k}\right)+\frac{k^2}{16}\theta\sin\theta+\frac{k^3\cos 3\theta}{45\sqrt{2}}-\frac{\pi k^4\cos 4\theta}{1536} \tag{A.34}$$

$$Q=-0.0386+\frac{1}{2}\ln\frac{2}{k}+\frac{1}{3\sqrt{2}}\cos\theta-\frac{k^2\cos 2\theta}{64}+\frac{K^3\cos 3\theta}{45\sqrt{2}}-\frac{k^4\sin 4\theta}{384}-\frac{k^4\cos 4\theta}{384}\left(\ln\frac{2}{k}+1.0895\right) \tag{A.35}$$

where

$$k=8.565\times10^4 S_{ij}\sqrt{f/p} \tag{A.36}$$

in which S_{ij} is in ft and ρ is the soil resistivity in Ω-m, and f is the system frequency. This shows dependence on frequency as well as on soil resistivity.

A.6.1 Approximations to Carson's Equations

These approximations involve P and Q and the expressions are given by the following:

$$P_{ij}=\frac{\pi}{8} \tag{A.37}$$

$$Q_{ij}=-0.03860+\frac{1}{2}\ln\frac{2}{k_{ij}} \tag{A.38}$$

Using these assumptions, f=60 Hz and soil resistivity=100 Ω-m, the equations reduce to

$$Z_{ii}=R_i+0.0953+j0.12134\left(\ln\frac{1}{\text{GMR}_i}+7.93402\right)\Omega/\text{mi} \tag{A.39}$$

$$Z_{ij}=0.0953+j0.12134\left(\ln\frac{1}{D_{ij}}+7.93402\right)\Omega/\text{mi} \tag{A.40}$$

Equations A.39 and A.40 are of practical significance for calculations of line impedances. (Equations A.32 through A.40 are not available in SI units.)

Example A.1

Consider an unsymmetrical overhead line configuration, as shown in Figure A.6. The phase conductors consist of 556.5 KCMIL (556,500 circular mils) of ACSR conductor consisting of 26 strands of aluminum, two layers and seven strands of steel. From the properties of ACSR conductor tables, the conductor has a resistance of 0.1807 Ω at 60 Hz and its GMR is 0.0313 ft at 60 Hz; the conductor diameter is 0.927 in. The neutral consists of 336.4 KCMIL, ACSR conductor, resistance 0.259 Ω/mi at 60 Hz and 50°C and GMR 0.0278 ft, and conductor diameter 0.806 in. It is required to form a primitive Z matrix, convert it into a 3×3 Z_{abc} matrix, and then to sequence impedance matrix Z_{012}.

Using Equations A.39 and A.40:

$$Z_{aa} = Z_{bb} = Z_{cc} = 0.2760 + j1.3831$$

$$Z_{nn} = 0.3543 + j1.3974$$

$$Z_{ab} = Z_{ba} = 0.0953 + j0.8515$$

$$Z_{bc} = Z_{cb} = 0.053 + j0.7654$$

$$Z_{ca} = Z_{ac} = 0.0953 + j0.7182$$

$$Z_{an} = Z_{na} = 0.0953 + j0.7539$$

$$Z_{bn} = Z_{nb} = 0.0953 + j0.7674$$

$$Z_{cn} = Z_{nc} = 0.0953 + j0.7237$$

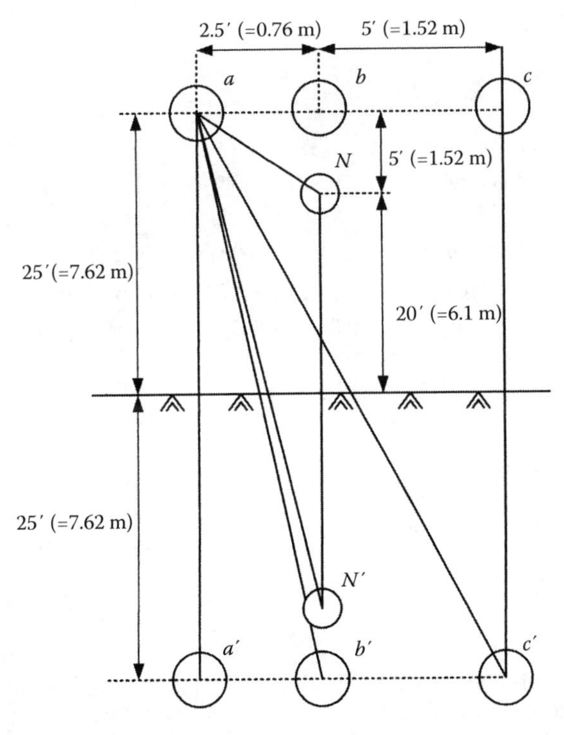

FIGURE A.6
Distribution line configuration for calculations of line parameters (Examples A.1 and A.2).

Therefore, the primitive impedance matrix is

$$\bar{Z}_{\text{prim}} = \begin{vmatrix} 0.2760+j1.3831 & 0.0953+j0.8515 & 0.0953+j0.7182 & 0.0953+j0.7539 \\ 0.0953+j0.8515 & 0.2760+j1.3831 & 0.0953+j0.7624 & 0.0953+j0.7674 \\ 0.0953+j0.7182 & 0.0953+j0.7624 & 0.2760+j1.3831 & 0.0953+j0.7237 \\ 0.0953+j0.7539 & 0.0953+j0.7674 & 0.0953+j0.7237 & 0.3543+j1.3974 \end{vmatrix}$$

Eliminate the last row and column using Equation A.22:

$$\bar{Z}_{abc} = \begin{vmatrix} 0.2747+j0.9825 & 0.0949+j0.4439 & 0.0921+j0.3334 \\ 0.0949+j0.4439 & 0.2765+j0.9683 & 0.0929+j0.3709 \\ 0.0921+j0.3334 & 0.0929+j0.3709 & 0.2710+j1.0135 \end{vmatrix}$$

Convert to Z_{012} by using the transformation equation, Volume 1:

$$\bar{Z}_{912} = \begin{vmatrix} 0.4606+j1.7536 & 0.0194+j0.0007 & -0.0183+j0.0055 \\ -0.0183+j0.0055 & 0.1808+j0.6054 & -0.0769-j0.0146 \\ 0.0194+j0.0007 & 0.0767-j0.0147 & 0.1808+j0.6054 \end{vmatrix}$$

This shows the mutual coupling between sequence impedances. We could average out the self-impedances and mutual impedances according to Equation A.13:

$$Z_s = \frac{Z_{aa}+Z_{bb}+Z_{cc}}{3} = 0.2741+j0.9973$$

$$Z_m = \frac{Z_{ab}+Z_{bc}+Z_{ca}}{3} = -0.0127-j0.0044$$

The matrix Z_{abc} then becomes

$$\bar{Z}_{abc} = \begin{vmatrix} 0.2741+j0.9973 & -0.0127-j0.0044 & -0.0127-j0.0044 \\ -0.0127-j0.0044 & 0.2741+j0.9973 & -0.0127-j0.0044 \\ -0.0127-j0.0044 & -0.0127-j0.0044 & 0.2741+j0.9973 \end{vmatrix} \Omega/\text{mi}$$

and this gives

$$\bar{Z}_{012} = \begin{vmatrix} 0.2486+j0.9885 & 0 & 0 \\ 0 & 0.2867+j1.0017 & 0 \\ 0 & 0 & 0.2867+j1.0017 \end{vmatrix} \Omega/\text{mi}$$

Example A.2

Figure A.7 shows a high-voltage line with two 636,000 mils ACSR bundle conductors per phase. Conductor GMR=0.0329 ft, resistance=0.1688 Ω/mi, diameter=0.977 in., and spacing are as shown in Figure A.7. Calculate the primitive impedance matrix and reduce it to a 3×3 matrix, then convert it into a sequence component matrix.

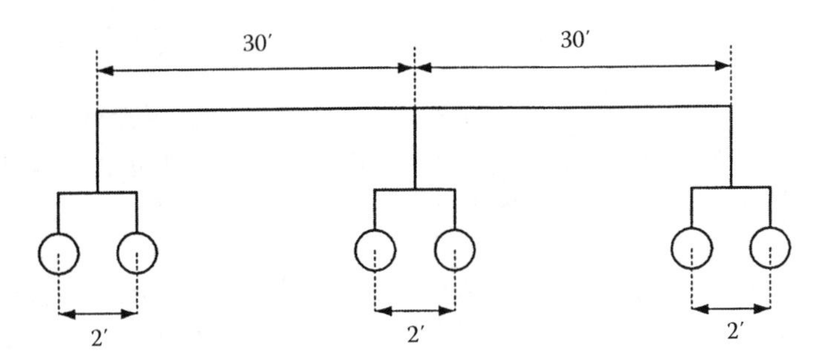

FIGURE A.7
Configuration of bundle conductors (Example A.2).

From Equations A.39 and A.40 and the specified spacings in Figure A.7, matrix Z_1 is

$$\bar{Z}_1 = \begin{vmatrix} 0.164+j1.3770 & 0.0953+j0.5500 & 0.0953+j0.4659 \\ 0.0953+j0.5500 & 0.164+j1.3770 & 0.0953+j0.5500 \\ 0.0953+j0.4659 & 0.0953+j0.5500 & 0.164+j1.3770 \end{vmatrix}$$

This is also equal to Z_4, as the bundle conductors are identical and symmetrically spaced. Matrix Z_2 of Equation A.27 is

$$\bar{Z}_2 = \begin{vmatrix} 0.0953+j0.8786 & 0.0953+j0.5348 & 0.0953+j0.4581 \\ 0.0953+j0.5674 & 0.0953+j0.8786 & 0.0953+j0.5348 \\ 0.0953+j0.4743 & 0.0953+j0.8786 & 0.0953+j0.8786 \end{vmatrix}$$

The primitive matrix is 6×6 given by Equation A.23 formed by partitioned matrices according to Equation A.24. Thus, from $\bar{Z}_1$ and $\bar{Z}_2$, the primitive matrix can be written. From these two matrices, we will calculate matrix Equation A.30:

$$\bar{Z}_1 - \bar{Z}_2 = \begin{vmatrix} 0.069+j0.498 & j0.0150 & j0.0079 \\ -j0.0171 & 0.069+j0.498 & j0.0150 \\ -j0.00847 & -j0.0170 & 0.069+j0.498 \end{vmatrix}$$

and

$$\bar{Z}_k = (\bar{Z}_1 - \bar{Z}_2) - (\bar{Z}_2^t - \bar{Z}_1) = \begin{vmatrix} 0.138+j0.997 & -j0.0022 & -j0.0005 \\ -j0.0022 & 0.138+j0.997 & -j0.0022 \\ -j0.0005 & -j0.0022 & 0.138+j0.997 \end{vmatrix}$$

The inverse is

$$\bar{Z}_k^{-1} = \begin{vmatrix} 0.136-j0.984 & 0.000589-j0.002092 & 0.0001357-j0.0004797 \\ 0.0005891-j0.002092 & 0.136-j0.981 & 0.0005891-j0.002092 \\ 0.0001357-j0.0004797 & 0.0005891-j0.002092 & 0.136-j0.981 \end{vmatrix}$$

then the matrix $(\bar{Z}_2 - \bar{Z}_1)\bar{Z}_k^{-1}(\bar{Z}_2^t - \bar{Z}_1)$ is

$$\begin{vmatrix} 0.034 + j0.2500 & -0.000018 - j0.000419 & 0.0000363 - j0.0003871 \\ -0.000018 - j0.000419 & 0.034 + j0.2500 & -0.000018 - j0.000419 \\ 0.0000363 - j0.000387 & -0.000018 - j0.000419 & 0.034 + j0.2500 \end{vmatrix}$$

Note that the off-diagonal elements are relatively small as compared to the diagonal elements. The required 3×3 transformed matrix is then Z_1 minus the above matrix:

$$\bar{Z}_{\text{transformed}} = \begin{vmatrix} 0.13 + j1.127 & 0.095 + j0.55 & 0.095 + j0.466 \\ 0.095 + j0.55 & 0.13 + j1.127 & 0.095 + j0.55 \\ 0.095 + j0.466 & 0.095 + j0.55 & 0.13 + j1.127 \end{vmatrix} \Omega/\text{mi}$$

Using Equation A.35, the sequence impedance matrix is

$$\bar{Z}_{0.12} = \begin{vmatrix} 0.32 + j2.171 & 0.024 - j0.014 & -0.024 - j0.014 \\ -0.024 - j0.014 & 0.035 + j0.605 & -0.048 + j0.028 \\ 0.024 - j0.014 & 0.048 + j0.028 & 0.035 + j0.605 \end{vmatrix} \Omega/\text{mi}$$

A.7 Capacitance of Lines

The shunt capacitance per unit length of a two-wire, single-phase transmission line is

$$C = \frac{\pi\varepsilon_0}{\ln(D/r)} \text{ F/m (Farads per meter)} \tag{A.41}$$

where ε_0 is the permittivity of free space=8.854×10^{-12} F/m, and other symbols are as defined before. For a three-phase line with equilaterally spaced conductors, the line-to-neutral capacitance is

$$C = \frac{2\pi\varepsilon_0}{\ln(D/r)} \text{ F/m} \tag{A.42}$$

For unequal spacing, D is replaced with GMD from Equation A.7. The capacitance is affected by the ground and the effect is simulated by a mirror image of the conductors exactly at the same depth as the height above the ground. These mirror-image conductors carry charges which are of opposite polarity to conductors above the ground (Figure A.8). From this figure, the capacitance to ground is

$$C_n = \frac{2\pi\varepsilon_0}{\ln(\text{GMD}/r) - \ln\left(\sqrt[3]{S_{ab'}S_{bc'}S_{ca'}} \,/\, \sqrt[3]{S_{aa'}S_{bb'}S_{cc'}}\right)} \tag{A.43}$$

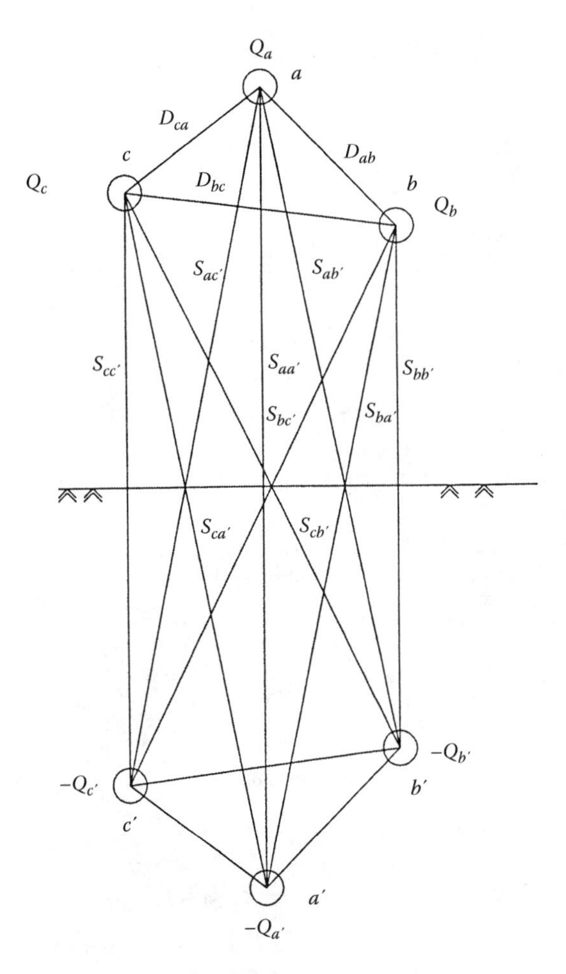

FIGURE A.8
Calculations of capacitances, conductors, mirror images, spacing, and charges.

Using the notations in Equation A.10, this can be written as follows:

$$C_n = \frac{2\pi\varepsilon_0}{\ln(D_m / D_s)} = \frac{10^{-9}}{18\ln(D_m / D_s)}\,\text{F/m} \tag{A.44}$$

A.7.1 Capacitance Matrix

The capacitance matrix of a three-phase line is

$$\bar{C}_{abc} = \begin{vmatrix} C_{aa} & -C_{ab} & -C_{ac} \\ -C_{ba} & C_{bb} & -C_{bc} \\ -C_{ca} & -C_{cb} & C_{cc} \end{vmatrix} \tag{A.45}$$

This is diagrammatically shown in Figure A.9a. The capacitance between the phase conductors a and b is C_{ab} and the capacitance between conductor a and ground is $C_{aa}-C_{ab}-C_{ac}$.

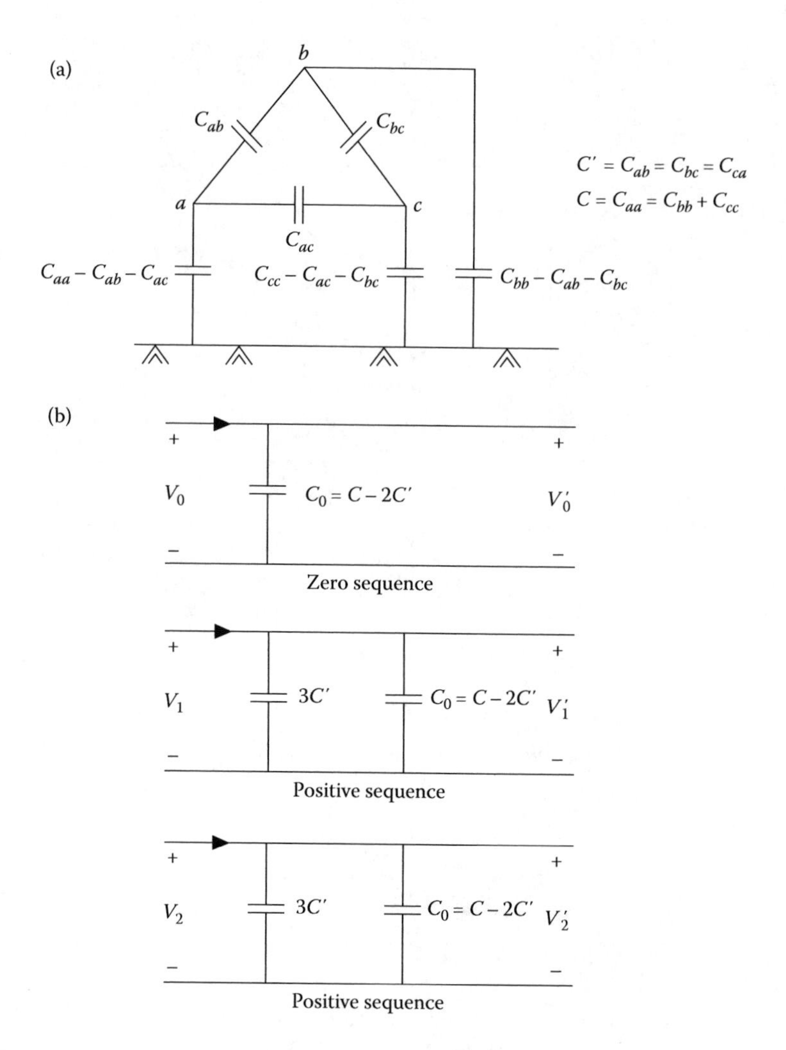

FIGURE A.9
(a) Capacitance of three-phase line and (b) equivalent positive, negative, and zero sequence capacitances.

If the line is perfectly symmetrical, all the diagonal elements are the same and all the off-diagonal elements of the capacitance matrix are identical:

$$\bar{C}_{abc} = \begin{vmatrix} C & -C' & -C' \\ -C' & C & -C' \\ -C' & -C' & C \end{vmatrix} \tag{A.46}$$

Symmetrical component transformation is used to diagonalize the matrix:

$$\bar{C}_{012} = \bar{T}_s^{-1}\bar{C}_{abc}\bar{T}_s = \begin{vmatrix} C-2C' & 0 & 0 \\ 0 & C+C' & 0 \\ 0 & 0 & C+C' \end{vmatrix} \tag{A.47}$$

The zero, positive, and negative sequence networks of capacitance of a symmetrical transmission line are shown in Figure A.9b. The eigenvalues are $C-2C'$, $C+C'$, and $C+C'$. The capacitance $C+C'$ can be written as $3C'+(C-2C')$; i.e., it is equivalent to the line capacitance of a three-conductor system plus the line-to-ground capacitance of a three-conductor system.

In a capacitor, $V=Q/C$. The capacitance matrix can be written as follows:

$$\bar{V}_{abc} = \bar{P}_{abc}\bar{Q}_{abc} = \bar{C}_{abc}^{-1}\bar{Q}_{abc} \tag{A.48}$$

where $\bar{P}$ is called the potential coefficient matrix, i.e.,

$$\begin{vmatrix} V_a \\ V_b \\ V_c \end{vmatrix} = \begin{vmatrix} P_{aa} & P_{ab} & P_{ac} \\ P_{ba} & P_{bb} & P_{bc} \\ P_{ca} & P_{cb} & P_{cc} \end{vmatrix} \begin{vmatrix} Q_a \\ Q_b \\ Q_c \end{vmatrix} \tag{A.49}$$

where

$$P_{ii} = \frac{1}{2\pi\varepsilon_0}\ln\frac{S_{ii}}{r_i} = 11.17689\ln\frac{S_{ii}}{r_i} \tag{A.50}$$

and

$$P_{ii} = \frac{1}{2\pi\varepsilon_0}\ln\frac{S_{ii}}{D_{ij}} = 11.17689\ln\frac{S_{ij}}{D_{ij}} \tag{A.51}$$

where,

S_{ij}=conductor-to-image distance below ground (ft)
D_{ij}=conductor-to-conductor distance (ft)
r_i=radius of the conductor (ft)
ε_0=permittivity of the medium surrounding the conductor=1.424×10^{-8}
For sine-wave voltage and charge, the equation can be expressed as follows:

$$\begin{vmatrix} I_a \\ I_b \\ I_c \end{vmatrix} = j\omega \begin{vmatrix} C_{aa} & -C_{ab} & -C_{ac} \\ -C_{ba} & -C_{bb} & -C_{bc} \\ -C_{ca} & -C_{cb} & C_{cc} \end{vmatrix} \begin{vmatrix} V_a \\ V_b \\ V_c \end{vmatrix} \tag{A.52}$$

The capacitance of three-phase lines with ground wires and with bundle conductors can be addressed as in the calculations of inductances. The primitive P matrix can be partitioned and reduces to a 3×3 matrix.

Example A.3

Calculate the matrices P and C for Example A.1. The neutral is 30 ft (9.144 m) above ground and the configuration of Figure A.6 is applicable.

The mirror images of the conductors are drawn in Figure A.6. This facilitates calculation of the spacings required in Equations A.50 and A.51 for the P matrix. Based on the geometric distances and conductor diameter, the primitive P matrix is

$$\bar{P} = \begin{vmatrix} P_{aa} & P_{ab} & P_{ac} & P_{an} \\ P_{ba} & P_{bb} & P_{bc} & P_{bn} \\ P_{ca} & P_{cb} & P_{cc} & P_{cn} \\ P_{na} & P_{nb} & P_{nc} & P_{nn} \end{vmatrix}$$

$$= \begin{vmatrix} 80.0922 & 33.5387 & 21.4230 & 23.3288 \\ 33.5387 & 80.0922 & 25.7913 & 24.5581 \\ 21.4230 & 25.7913 & 80.0922 & 20.7547 \\ 23.3288 & 24.5581 & 20.7547 & 79.1615 \end{vmatrix}$$

This is reduced to a 3×3 matrix:

$$P = \begin{vmatrix} 73.2172 & 26.3015 & 15.3066 \\ 26.3015 & 72.4736 & 19.3526 \\ 15.3066 & 19.3526 & 74.6507 \end{vmatrix}$$

Therefore, the required $\bar{C}$ matrix is inverse of $\bar{P}$, and $\bar{Y}_{abc}$ is

$$\bar{Y}_{abc} = j\omega\bar{P}^{-1} = \begin{vmatrix} j6.0141 & -j1.9911 & -j0.7170 \\ -j1.9911 & j6.2479 & -j1.2114 \\ -j0.7170 & -j1.2114 & j5.5111 \end{vmatrix} \mu \text{ siemens/mi}$$

A.8 Cable Constants

The construction of cables varies widely; it is mainly a function of insulation type, method of laying, and voltage of application. For high-voltage applications above 230 kV, oil-filled paper-insulated cables are used, though recent trends see the development of solid dielectric cables up to 345 kV. A three-phase solid dielectric cable has three conductors enclosed within a sheath and because the conductors are much closer to each other than those in an overhead line and the permittivity of an insulating medium is much higher than that of air, the shunt capacitive reactance is much lower as compared to an overhead line. Thus, use of a T or Π model is required even for shorter cable lengths.

The inductance per unit length of a single-conductor cable is given by

$$L = \frac{\mu_0}{2\pi} \ln \frac{r_1}{r_2} \text{ H/m} \tag{A.53}$$

where r_1 is the radius of the conductor and r_2 is the radius of the sheath, i.e., the cable outside diameter divided by 2.

When single-conductor cables are installed in magnetic conduits, the reactance may increase by a factor of 1.5. Reactance is also dependent on conductor shape, i.e., circular or sector, and on the magnetic binders in three-conductor cables.

A.8.1 Zero Sequence Impedance of the OH Lines and Cables

The zero sequence impedance of the lines and cables is dependent upon the current flow through a conductor and return through the ground, or sheaths, and encounters the impedance of these paths. The zero sequence current flowing in one phase also encounters the currents arising out of that conductor self-inductance, from mutual inductance to other two phase conductors, from the mutual inductance to the ground and sheath return paths, and the self-inductance of the return paths. Tables and analytical expressions are provided in Reference [9]. As an example, the zero sequence impedance of a three-conductor cable with a solidly bonded and grounded sheath is given by

$$z_0 = r_c + r_e + j0.8382\frac{f}{60}\log_{10}\frac{D_e}{\text{GMR}_{3c}} \quad \text{(A.54)}$$

where,

r_c = ac resistance of one conductor Ω/mi

r_e = ac resistance of earth return (depending upon equivalent depth of earth return, soil resistivity, taken as 0.286 Ω/mi)

D_e = distance to equivalent earth path (see Reference [9])

GMR_{3c} = geometric mean radius of conducting path made up of three actual conductors taken as a group (in in.):

$$\text{GMR}_{3c} = \sqrt[3]{\text{GMR}_{1c}S^2} \quad \text{(A.55)}$$

where GMR_{1c} is the geometric mean radius of individual conductor and $S = (d + 2t)$, where d is the diameter of the conductor and t is the thickness of the insulation.

A.8.2 Concentric Neutral Underground Cable

We will consider a concentric neutral construction as shown in Figure A.10a. The neutral is concentric to the conductor and consists of a number of copper strands that are wound helically over the insulation. Such cables are used for underground distribution, directly buried or installed in ducts. Referring to Figure A.10a, d is the diameter of the conductor, d_0 is the outside diameter of the cable over the concentric neutral strands, and d_s is the diameter of an individual neutral strand. Three cables in flat formation are shown in Figure A.10b. The GMR of a phase conductor and a neutral strand are given by the following expression:

$$\text{GMR}_{cn} = \sqrt[n]{\text{GMR}_s n R^{n-1}} \quad \text{(A.56)}$$

where GMR_{cn} is the equivalent GMR of the concentric neutral, GMR_S is the GMR of a single neutral strand, n is the number of concentric neutral strands, and R is the radius of a circle passing through the concentric neutral strands (see Figure A.10a) = $(d_0 - d_s)/2$ (in ft).

The resistance of the concentric neutral is equal to the resistance of a single strand divided by the number of strands.

The GMD between concentric neutral and adjacent phase conductors is

$$D_{ij} = \sqrt[n]{D_{mn}^n - R^n} \quad \text{(A.57)}$$

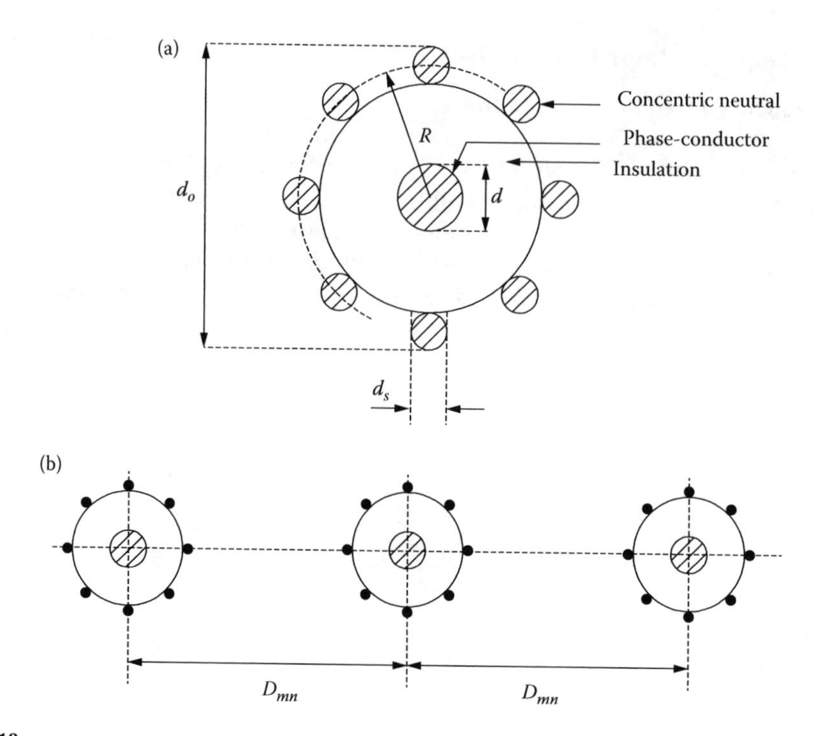

FIGURE A.10
(a) Construction of a concentric neutral cable and (b) configuration for cable series reactance (Example A.4).

where D_{ij} is the *equivalent* center-to-center distance of the cable spacing. Note that it is less than D_{mn}, the center-to-center spacing of the adjacent conductors, Figure A.10b. Carson's formula can be applied and the calculations are similar to those in Example A.1.

Example A.4

A concentric neutral cable system for 13.8 kV has a center-to-center spacing of 8 in. The cables are 500 KCMIL, with 16 strands of #12 copper wires. The following data are supplied by the manufacturer:

GMR phase conductor = 0.00195 ft
GMR of neutral strand = 0.0030 ft
Resistance of phase conductor = 0.20 Ω/mi
Resistance of neutral strand = 10.76 Ω/mi. Therefore, the resistance of the concentric neutral = 10.76/16 = 0.6725 Ω/mi.
Diameter of neutral strand = 0.092 in.
Overall diameter of cable = 1.490 in.
Therefore, $R = (1.490 - 0.092)/24 = 0.0708$ ft.

The effective conductor phase-to-phase spacing is approximately 8 in., from Equation A.55.

The primitive matrix is a 6×6 matrix, similar to Equation A.16. In the partitioned form, Equation A.17, the matrices are

$$\bar{Z}_a = \begin{vmatrix} 0.2953+j1.7199 & 0.0953+j1.0119 & 0.0953+j0.9278 \\ 0.0953+j1.0119 & 0.2953+j1.7199 & 0.0953+j1.0119 \\ 0.0953+j0.9278 & 0.0953+j1.0119 & 0.2953+j1.7199 \end{vmatrix}$$

The spacing between the concentric neutral and the phase conductors is approximately equal to the phase-to-phase spacing of the conductors. Therefore,

$$\bar{Z}_B = \begin{vmatrix} 0.0953+j1.284 & 0.0953+j1.0119 & 0.0953+j0.9278 \\ 0.0953+j1.0119 & 0.0953+j1.284 & 0.0953+j1.0119 \\ 0.0953+j0.9278 & 0.0953+j1.0119 & 0.0953+j1.284 \end{vmatrix}$$

Matrix $\bar{Z}_c = \bar{Z}_B$ and matrix $\bar{Z}_D$ is given by

$$\bar{Z}_D = \begin{vmatrix} 0.7678+j1.2870 & 0.0953+j1.0119 & 0.0953+j0.9278 \\ 0.0953+j1.0119 & 0.7678+j1.2870 & 0.0953+j1.0119 \\ 0.0953+j0.9278 & 0.0953+j1.0119 & 0.7678+j1.2870 \end{vmatrix} z$$

This primitive matrix can be reduced to a 3×3 matrix, as in other examples.

A.8.2 Capacitance of Cables

In a single-conductor cable, the capacitance per unit length is given by

$$C = \frac{2\pi\varepsilon\varepsilon_0}{\ln(r_1/r_2)} \text{F/m} \tag{A.58}$$

By change of units, this can be written as follows:

$$C = \frac{7.35\varepsilon}{\log(r_1/r_2)} \text{pF/ft} \tag{A.59}$$

Note that ε is the permittivity of the dielectric medium relative to air. The capacitances in a three-conductor cable are shown in Figure A.11. This assumes a symmetrical construction and the capacitances between conductors and from conductors to the sheath are equal. The circuit in Figure A.11a is successively transformed and Figure A.11d shows that the net capacitance per phase $= C_1 + 3C_2$.

By change of units, Equation A.59 can be expressed as follows:

$$C = \frac{7.35\varepsilon}{\log(r_1/r_2)} \text{pF/ft} \tag{A.60}$$

This gives the capacitance of a single-conductor shielded cable. Table A.1 gives the values of ε for various cable insulation types.

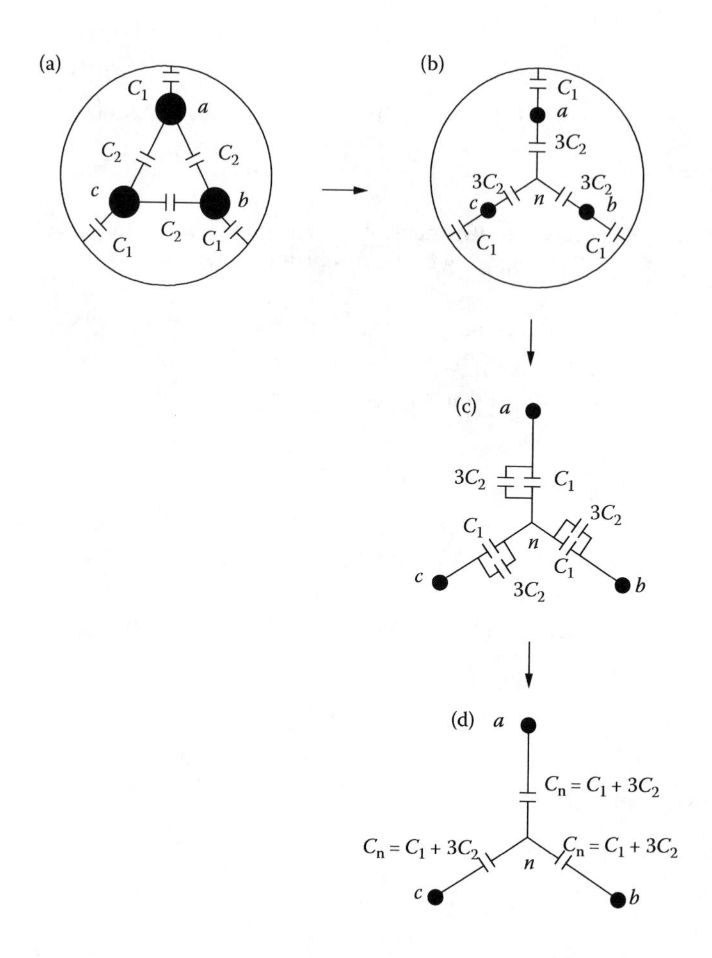

FIGURE A.11
(a) Capacitances in a three-conductor cable, (b) and (c) equivalent circuits, and (d) final capacitance circuit.

TABLE A.1
Typical Values for Dielectric Constants of Cable Insulation

Type of Insulation	Permittivity (ε)
Polyvinyl chloride (PVC)	3.5–8.0
Ethylene-propylene (EP) insulation	2.8–3.5
Polyethylene insulation	2.3
Cross-linked polyethylene	2.3–6.0
Impregnated paper	3.3–3.7

References

1. DG Fink (Ed.). *Standard Handbook for Electrical Engineers* (10th ed.). McGraw-Hill, New York, 1969.
2. JH Watt. *American Electrician's Handbook* (9th ed.). McGraw-Hill, New York, 1970.

3. Central Station Engineers. *Electrical Transmission and Distribution Reference Book* (4th ed.). Westinghouse Corp., East Pittsburgh, PA, 1964.
4. The Aluminum Association. *Aluminum Conductor Handbook* (2nd ed.) The Aluminum Association, Washington, DC, 1982.
5. JC Das. *Transients in Electrical Systems*, McGraw-Hill, New-York, 2010.
6. WD Stevenson. *Elements of Power System Analysis* (2nd ed.). McGraw-Hill, New York, 1962.
7. CIGRE, Working Group 33.02(Internal Voltages). Guide Lines for Representation of Network Elements when Calculating Transients, Paris.
8. PM Anderson. *Analysis of Faulted Systems*, Iowa State University Press, Ames, IA, 1973.
9. Westinghouse. *Transmission and Distribution Reference Book*, Westinghouse, East Pittsburg, PA, 1964.

Appendix B: Solution to the Problems

B.2 Solution to the Problems: Chapter 2

B.2.1 Problem 2.1

As $D = 1$, $K = 1$, and $M = 10$ and

$$T = \frac{M}{D} = 10 \text{ s}$$

The Laplace transform of the step change in the load is

$$\Delta P_L(s) = \frac{-0.02}{s}$$

Then

$$\Delta \omega_r(s) = \left(\frac{-0.02}{s}\right)\left(\frac{K}{1+sT}\right)$$

Here, $K = 1$.

Therefore, taking inverse transform

$$\Delta \omega_r(t) = -0.02e^{-t/T} + 0.02$$

$$= -0.02e^{-0.1t} + 0.02$$

This can be plotted as shown in Figure B.1.

B.2.2 Problem 2.2

The load is 1500 and 50 MW load is suddenly applied; then based on 1500 MVA base, the load change is 0.033 pu. Again $D = 1$; therefore, the steady-state frequency deviation is

$$\Delta \omega_{ss} = -\frac{\Delta P_L}{D} = 0.033$$

Therefore, the steady-state frequency deviation is

$$0.033 \times 60 = 1.98 \text{ Hz}$$

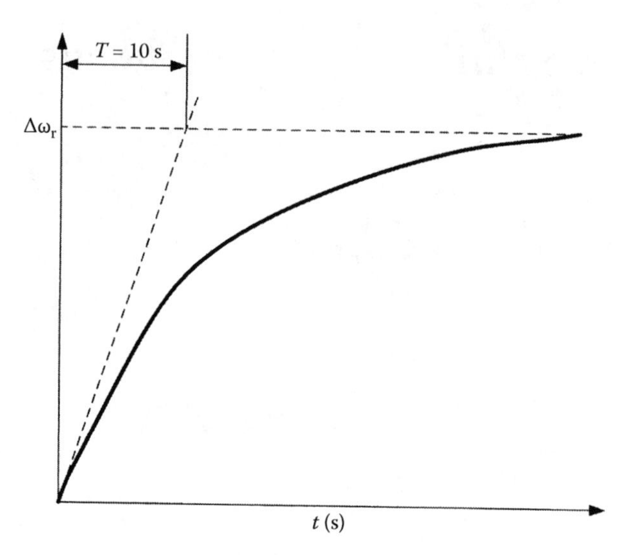

FIGURE B.1
Speed deviation, the time constant is 10 s, Problem 2.1.

B.2.3 Problem 2.3

A 5% regulation in area 1, with 5000 MW of generation, means

$$\frac{1}{R_1} = \frac{1}{0.05} \times \frac{5000}{60} = 1667\,\text{MW/Hz}$$

Similarly, for area 2

$$\frac{1}{R_2} = \frac{1}{0.05} \times \frac{15,000}{60} = 5000\,\text{MW/Hz}$$

The total regulation is

$$\frac{1}{R_1} + \frac{1}{R_2} = \frac{1}{R} = 6667\,\text{MW/Hz}$$

Load damping due to 4000 MW load in area 1

$$D_1 = 1 \times \frac{4000}{100} \times \frac{100}{60} = 66.67\,\text{MW/Hz}$$

Similarly, for area 2

$$D_2 = 1 \times \frac{13,000}{100} \times \frac{100}{60} = 216.67\,\text{MW/Hz}$$

Then

$$D = D_1 + D_2 = 283.34\,\text{MW/Hz}$$

Change in frequency due to loss of 500 MW in area 1

$$\Delta f = \frac{-\Delta P_L}{1/R+D} = \frac{-(-500)}{6667+283.34} = 0.0719\,\text{Hz}$$

Load changes in the two areas due to increase in frequency

$$\Delta P_{D1} = D_1 \Delta f = 4.79\,\text{MW}$$

and

$$\Delta P_{D2} = D_2 \Delta f = 15.58\,\text{MW}$$

Generation change in two areas due to speed regulation

$$\Delta P_{G1} = -\frac{1}{R1}\Delta f = -119.85\,\text{MW}$$

$$\Delta P_{G2} = -\frac{1}{R2}\Delta f = -359.5\,\text{MW}$$

Table B.1 is constructed for the load and generation changes in areas 1 and 2.

With supplementary control

$$ACE_1 = B_1 \Delta f + \Delta P_{12} = 0$$

$$ACE_2 = B_2 \Delta f - \Delta P_{12} = 0$$

Therefore,

$$\Delta f = 0$$

$$\Delta P_{12} = 0$$

There is no change in the tie-line flow, the generation and load in area 1 is reduced by 500 MW.

TABLE B.1

Problem 2.3—Load and Generation Change

Area 1		Area 2	
Load	**Generation**	**Load**	**Generation**
4000 – 500 + 4.79 = 3504.79	5000 – 119.85 = 4880.15	13,000 + 15.58 = 13,015.58	15,000 – 359.5 = 14,640.5

B.3 Solution to the Problems: Chapter 3

B.3.1 Problem 3.1

From Figure 3.4

$$V_S = V_r + I_r\frac{Z}{2} + (I_c + I_r)\frac{Z}{2}$$

$$= V_r + I_r Z + V_c Y\frac{Z}{2}$$

$$= V_r + I_r Z + \left(V_r + I_r\frac{Z}{2}\right)Y\frac{Z}{2}$$

i.e.,

$$A = 1 + \frac{1}{2}YZ$$

$$B = Z\left[1 + \frac{1}{4}YZ\right]$$

$$I_s = I_r + YV_c$$

$$= I_r + \left(V_r + I_r\frac{Z}{2}\right)Y$$

Therefore,

$$C = Y, D = 1 + \left(\frac{1}{2}\right)YZ$$

B.3.2 Problem 3.2

Short line

Ignore Y.

Receiving-end current = 590.6 A < -31.78 = sending-end current. $A = D = 1$, $C = 0$, $B = Z$.

Therefore,

$$V_s = 132.79 < 0° + (0.35 + j0.476)(500)(0.5906 < -31.78° = 294.72 + j65.01 = 301 < 12.44°$$

$$V_s I_s^* = 128.883 + j124.26$$

Sending-end active power = 386.65 MW (three-phase).

Also sending-end reactive power (lagging) = 372.78 Mvar.

Π *Model*

$$A = 1 + \left(\frac{YZ}{2}\right)$$
$$Z = 175 + j238$$
$$Y = j0.0026$$
$$A = D = 0.691 + j0.227$$

$$C = Y\left(1 + \frac{1}{4}YZ\right) = -0.0002957 + j0.002198$$

$$V_s = (0.691 + j0.227)132.79(< 0°) + (175 + j238)(0.5906 < -31.78°) = 253.69 + j95.15$$
$$= 270.94 < 20.56°$$

$$I_s = CV_r + DI_r = (-0.0002957 + j0.002198)132.79 < 0 + (0.691 + j0.227)(0.5906 < -31.78°)$$
$$= 0.378 + j0.191$$

The sending-end current has changed from lagging to leading. Sending-end power

$$V_s I_s^* = (253.69 + j95.15)(0.378 - j0.191) = 114.07 - j12.49$$

Total active power supplied = 342.21 MW
Power received = 200 MW
Loss = 142.21 MW
Total reactive power = 37.47 Mvar leading into the source
Load reactive power = 123.95 Mvar.

The distributed capacitance of the line *generates* more reactive power than that demanded by the load.

Infinite Line

$$ZY = -0.619 + j0.455 = 0.768 < 143.68°$$

$$\gamma = \sqrt{ZY} = 0.876 < 71.84° = \alpha + j\beta$$

$$\alpha = 0.27302 \text{ Np}$$

$$\beta = 0.8323 \text{ rad}$$

$$\cosh\gamma = \cosh\alpha\cos\beta + j\sinh\alpha\sin\beta$$
$$= (1.0367)(0.673) + j(0.2733)(0.74) = 0.698 + j0.204$$

$$\sinh\gamma = \sinh\alpha\cos\beta + j\cosh\alpha\sin\beta$$
$$= (0.2733)(0.673) + j(1.0367)(0.74) = 0.186 + j0.767$$

The sending-end voltage

$$V_s = \cosh \gamma V_r + Z_0 \sinh \gamma I_r$$

$$Z_0 = \sqrt{\frac{Z}{Y}} = 320.284 - j105.08$$

$$V_s = (0.698 + j0.204)132.79 < 0° + (320.284 - j105.08)(0.5906 < -31.78°)$$
$$= 233.38 + j96.97$$

(Compare this with π line model calculations.)

$$I_s = \left(\frac{1}{Z_0}\right) \sinh \gamma V_r + \cosh \gamma I_r$$
$$= (0.00282 + j0.000925)(0.186 + j0.767)(132.79 < 0°)$$
$$+ (0.698 + j0.204)(0.5906 < -31.78° = 0.389 + j0.195)$$

Then

$V_s I_s^* = 109.693 - j7.87$ per phase in MW and Mvar, the reactive power leading 23.61 Mvar into the source.

The SIL loading is given by Equation 3.65. The surge impedance is 337; therefore, at 230 kV, SIL = 157 MW.

B.3.3 Problem 3.3

$$A = 0.9 < 2^{o} = 0.9 < \alpha$$

$$B = 120 < 75^{o} = 120 < \beta$$

From Equation 3.107, the center of receiving-end circle is

$$\left|\frac{A}{B}\right| V_r^2 = \left(\frac{0.9}{120}\right)(400)^2 = 1200\,\text{MVA}$$

$$< \beta - \alpha = 73^{\circ}$$

Referring to Figure B.2, the line OC can be drawn.

Receiving-end load = 200 MW at 0.85 power factor.

$$\cos^{-1} 0.85 = 31.8°$$

OP can be located in Figure B.2 to scale.

Then CP by measurements to scale = 1380 MVA

$$CP = 1380 = \frac{|V_s||V_r|}{B}$$

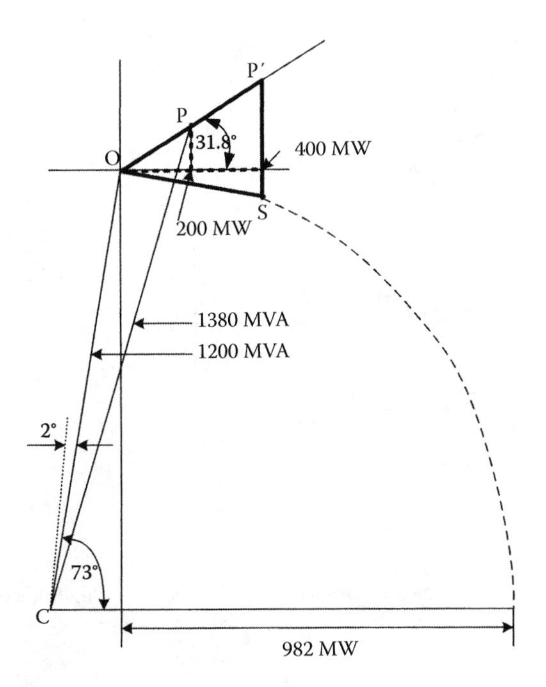

FIGURE B.2
Circle diagram for Problem 3.3.

This gives $V_s = 414$ kV.

For maintaining the receiving-end voltage = 400 kV = sending-end voltage, the radius of the circle diagram is

$$\frac{(400)^2}{120} = 1333$$

Therefore, the maximum MW that can be transferred is 982 MW by measurement.

The load of 400 MW at 0.85 corresponds to 470.6 MVA. Locate this as OP′ at the required power factor angle of 31.8°.

Then, the line *P′S* cuts the circle drawn earlier at S. Read *P′S* = 280 Mvar, which must be supplied at the receiving end.

B.3.4 Problem 3.4

The current reflection coefficients are negative of voltage reflection constants. Thus, at the sending end, $\rho_s = 1$. At the receiving end, $\rho_r = -0.33$. The lattice diagram can be drawn as shown in Figure B.3a.

Then, the sending-end current can be expressed as follows:

$$I_s = \frac{1}{Z_0}\left(1 - 2\rho_r + \rho_r^2 - 2\rho_r^2 + \cdots\right)$$

$$= \frac{1}{R_L}$$

The sending-end current profile is shown in Figure B.3b.

Similarly, the profiles at any point of the line can be drawn.

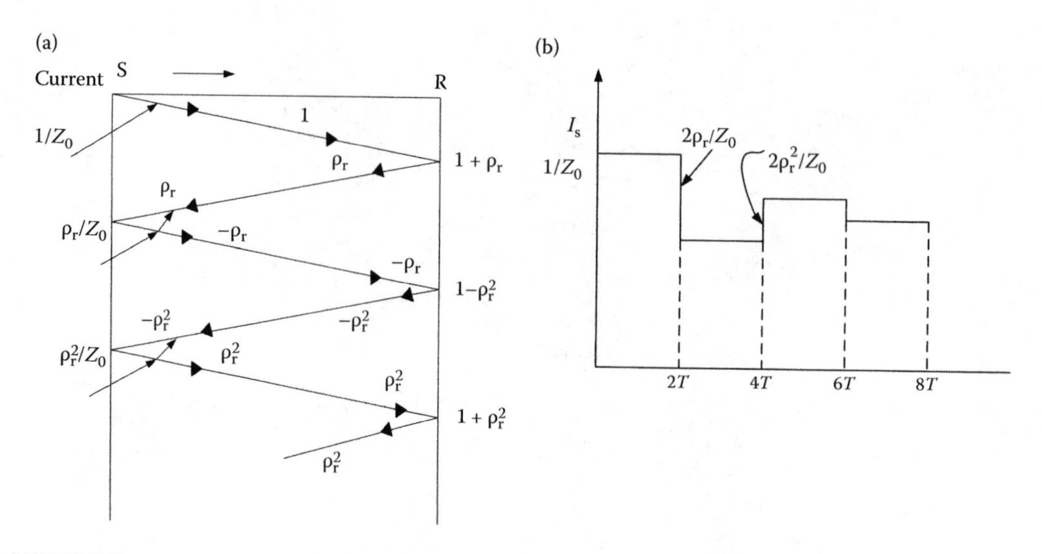

FIGURE B.3
(a) Lattice diagram of current reflections and (b) sending-end current profile, Problem 3.4.

B.3.5 Problem 3.5

$$A = 0.85 < 4°, B = 180 < 65°$$

The sending- and receiving-end voltages are equal $|V_s| = |V_r| = 230\,\text{kV}$ and the load is at unity power P factor, thus reactive power supplied = 0.

From Equation 3.98

$$0 = \frac{230 \times 230}{180} \sin(65° - \delta) - \left(\frac{0.85}{180}\right) \times (230)^2 \sin(65° - 4°)$$

This gives $\delta = 17.6°$.

Then, from Equation 3.97, the active power received is

$$P_r = \frac{(230)^2}{180} \cos(65° - 17.6°) - \left(\frac{0.85}{180}\right) \times (230)^2 \cos(65° - 4°) = 77.80\ \text{MW}$$

If the load is increased to 200 MW, voltages maintained at their rated voltages, then from Equation 3.97, to find δ

$$200 = \frac{(230)^2}{180} \cos(65° - \delta) - \left(\frac{0.85}{180}\right) \times (230)^2 \cos(65° - 4°)$$

This will give a value of δ much higher than β (=65°). The maximum power transfer occurs when $\delta = \beta$.

Therefore, let us try to solve for a reduced power of 150 MW. Then, substituting the values in the same Equation 3.97, $\delta = 42.3°$.

The reactive power supplied is given by Equation 10.98:

$$Q_r = \frac{230 \times 230}{180}\sin(65^\circ - 42.3^\circ) - \left(\frac{0.85}{180}\right) \times (230)^2 \sin(65^\circ - 4^\circ) = -105.08 \text{ Mvar}$$

In addition, the load is specified at a lagging power factor of 0.9. 150 MW of load at 0.9 lagging requires 72.65 Mvar. All this reactive power must also be supplied at the receiving end. That is, a capacitor bank of 105.08 + 72.65 = 177.73 Mvar is required.

B.3.6 Problem 3.6

The general voltage profile of the symmetrical line is shown in Figure B.4.

See Equation 8.29 and Example 8.2 for calculation of midpoint voltage under load and on no-load conditions. This problem actually belongs to Chapter 8. Also see the solution to Problem 8.3.

B.3.7 Problem 3.7

From Equation 3.69, velocity of propagation is 0.16667 × 10^9 cm/s. This is approximately 1/18th of the speed of light. The velocity of propagation in cables is much lower compared to OH lines due to higher capacitance.

B.3.8 Problem 3.8

The maximum power Equations 3.101 and 3.102 can be applied to any line model. More simply, from the phasor diagram (Figure B.5)

$$V_s^2 = (V_r \cos\varphi + RI_r)^2 + (V_r \sin\varphi + XI_r)^2$$

$$= V_r^2 + 2I_r V_r (R\cos\varphi + X\sin\varphi) + I_r^2(R^2 + X^2)$$

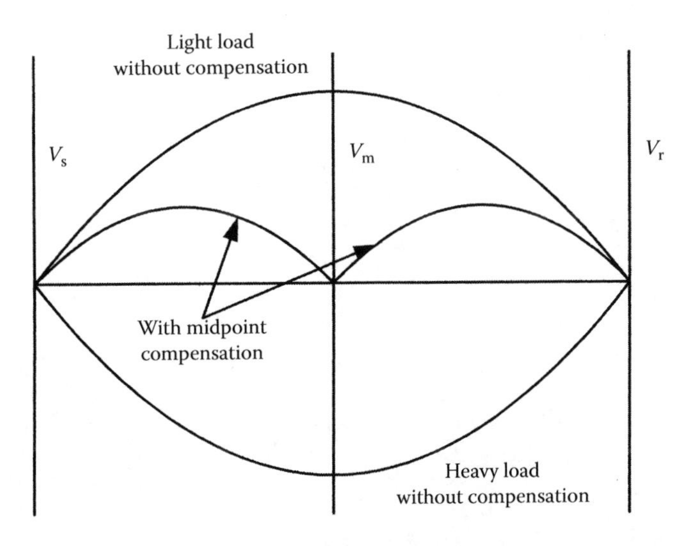

FIGURE B.4
The general voltage profile of a symmetrical line, Problem 3.6.

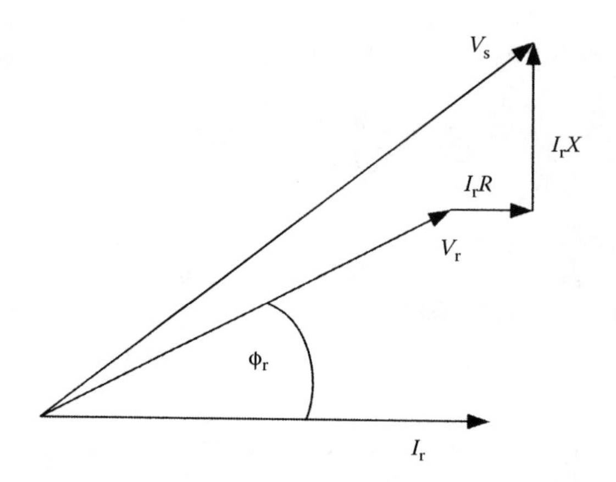

FIGURE B.5
Phasor diagram of a short line.

$$P = VI_r \cos\varphi$$

$$Q = VI_r \sin\varphi$$

The above equations can be rewritten as follows:

$$-V_s^2 + V_r^2 + 2PR + 2QX + \frac{1}{V_r^2}(P^2 + Q^2)(R^2 + X^2) = 0$$

For maximum power $dP/dQ = 0$

$$\frac{dP}{dQ} = -\frac{2X + 2Q\left(\dfrac{Z^2}{V_r^2}\right)}{2R + 2P\left(\dfrac{Z^2}{V_r^2}\right)} = 0$$

This gives

$$2X + 2Q\left(\frac{Z^2}{V_r^2}\right) = 0$$

$$Q = \frac{V_r^2 X}{Z^2}$$

Then substituting

$$P_m = \frac{V_r^2}{Z^2}\left(\frac{ZV_s}{V_r} - R\right)$$

Coming to the numerical part, $V_r = 4160$ V. With 5% regulation $V_s = 4368$ V, $Z = 1.005$. Then, maximum power transfer is 16.367 MW per phase, i.e., 49.10 MW (total).

B.3.9 Problem 3.9

See T-representation of a transmission line (Figure 3.4c) and the phasor diagram (Figure 3.4d). Figure 8.5 shows the circuit and phasor diagram for midpoint compensation for a symmetrical line. Also see Figure 8.5e, which shows the phasor diagram of a symmetrical line.

B.3.10 Problem 3.10

$$Z = 0.1 + j0.86\ \Omega$$

$$Y = j0.01508\ \Omega$$

$$\gamma = \sqrt{ZY} = 0.0046 + j0.114 = \alpha + j\beta$$

$$Z_0 = \sqrt{\frac{Z}{Y}} = 7.564 - j0.438$$

Then

$$A = D = \cosh \gamma l$$

$$\cosh\gamma = \cos h\alpha \cos\beta + j \sin h\alpha \sin\beta$$

$$\beta = 0.114 \text{ rad} = 6.53°$$

$$\cosh\gamma = (1)(0.9935) + j(0.0046)(0.1137)$$

$$= 0.9935 + j0.00052 = A = D$$

$$\sinh\gamma = \sin h\alpha \cos\beta + j \cos h\alpha \sin\beta$$

$$= (0.0046)(0.9935) + j(1)(0.1137)$$

$$= 0.00457 + j0.1137$$

Then

$$C = \frac{\sinh\gamma l}{Z_0} = (0.00457 + j0.1137)/(7.564 - j0.438)$$

$$= -0.0002654 + j0.015$$

$$B = Z_0 \sinh\gamma = (7.564 - j0.438)(0.00457 + j0.1137)$$

$$= 0.084 + j0.858$$

The velocity of propagation is

$$\frac{1}{\sqrt{LC}} \text{ m/s}$$

$$= \frac{1}{\sqrt{\left(\frac{0.86}{2\pi f}\right) \times 0.04 \times 10^{-6}}} = 104{,}680 \text{ m/s}$$

(The parameters in this example do not correspond to a real-world transmission line. The capacitance per mile is too large.)

Therefore,

$$\frac{104{,}680}{60} = 1744\,\text{m}$$

B.4 Solutions to the Problems: Chapter 4

B.4.1 Problem 4.1

The no-load dc voltage is

$$V_{\text{do}} = \frac{3\sqrt{2}}{\pi} \times 2 \times \frac{230}{220} \times 230 = 649.36\,\text{kV}$$

$$V_{\text{d}} = V_{\text{do}} \frac{\cos\alpha + \cos\delta}{2}$$

$$= 649.36 \times \frac{\cos 20° + \cos 35°}{2} = 585.6\ \text{kV}$$

Calculate

$$\Delta V_{\text{d}} = V_{\text{do}} \frac{\cos\alpha + \cos\delta}{2}$$

$$= 649.36 \times \frac{\cos 20° + \cos 35°}{2} = 31.94\ \text{kV}$$

Then

$$R_{\text{c}} = \frac{\Delta V_{\text{d}}}{2I_{\text{d}}} = \frac{31.94}{4} = 7.99\ \Omega$$

$$X_{\text{c}} = \frac{\pi R_{\text{c}}}{3} = 8.37\ \Omega$$

The fundamental component of an alternating current is

$$I_{L1} = \frac{\sqrt{6}}{\pi} \times 2 \times \frac{230}{220} \times 2 = 3.26\,\text{kA}$$

The power factor is

$$\cos\varphi = \frac{\text{V}_{\text{d}}}{\text{V}_{\text{do}}} = \frac{585.6}{649.36} = 0.902$$

$$P_{ac} = P_{dc} = V_d I_d = 586.6 \times 2 = 1172.4\,\text{MW}$$

(Neglect some losses in the rectifier transformer.)

Therefore, the required reactive power compensation is

$$Q = P_{ac} \tan\varphi = 561.2\,\text{Mvar}$$

B.4.2 Problem 4.2

We know that a bipolar circuit can be analyzed with equivalent monopolar link. Based on that concept, the equivalent circuit can be drawn as shown in Figure B.6. (This figure is similar to Figure 4.6.) As there are four bridges of six-pulse in a bipolar link, these can be represented in monopolar link, with $R_c = 6\ \Omega$/bridge.

$$I_d = \frac{1000}{1000} = 1\,\text{kA}$$

(The bipolar link voltage is given ±500 kV. Therefore, equivalent monopolar voltage is 1000 kV.)

At the inverter end, the ideal no-load voltage is

$$V_{doi} = \frac{V_{di} + BR_{ci}I_d}{\cos\gamma_0} = \frac{1000 + 4 \times 6 \times 1}{\cos 18^\circ} = 1066.34\,\text{kV}$$

(B is the number of bridges = 4 here.)

Therefore, the power factor at inverter end is

$$\cos\phi_i = \frac{1000}{1066.34} = 0.9378$$

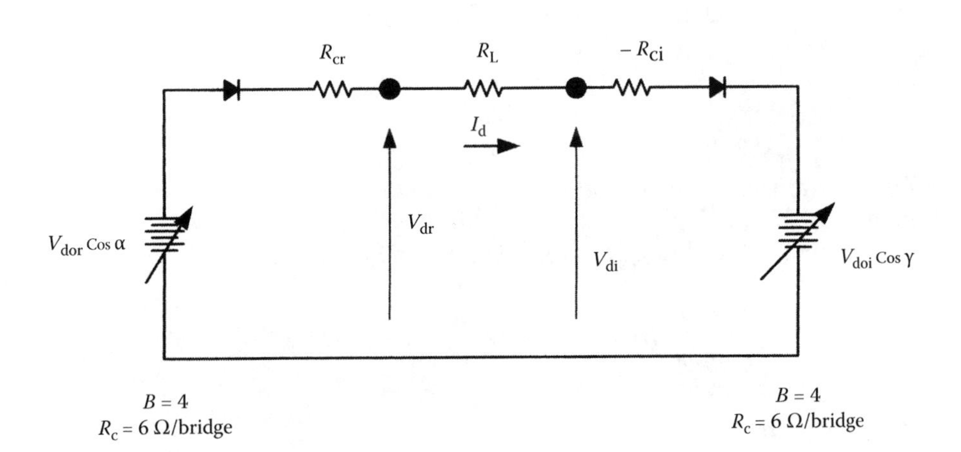

FIGURE B.6
Equivalent circuit of two-terminal dc line, Problem 4.2.

The reactive power required is

$$Q_i = 1000\tan 22.57° = 370.24\,\text{Mvar}$$

The inverter commutation angle can be found as follows:

$$V_{\text{di}} = V_{\text{doi}} \frac{\cos\gamma + \cos\beta}{2}$$

$$1000 = 1066.34 \frac{\cos 18° + \cos\beta}{2}$$

This gives

$$\cos\beta = 0.9152$$

or

$$\beta = 26.39°$$

The inverter commutation angle

$$\mu = \beta - \gamma_0 = 8.39°$$

Similar to inverter end

$$V_{\text{dor}} = \frac{V_{\text{dr}} + BR_{\text{ci}}I_{\text{d}}}{\cos\gamma_0} = \frac{1020 + 4\times 6\times 1}{\cos 18^{\text{o}}} = 1087.2\,\text{kV}$$

(Note that the resistance of each line is given 10 Ω; therefore, the total resistance is 20 Ω, and with 1 kA current, the voltage drop of 20 kV is added to 1000 kV.)

The rms line-to-line voltage on the primary of the transformer is found by assuming transformation ratio of 0.48.

Then rms line-to-line voltage at the rectifier transformers primary side bus is

$$\frac{V_{\text{dor}}}{1.3505BT_{\text{r}}}$$

where T_r is the transformer turns ratio.

This gives 419.2 kV.

The total current of four bridges, reflected on the primary side, is

$$\frac{\sqrt{6}}{\pi}BT_{\text{r}}I_{\text{d}} = \frac{\sqrt{6}}{\pi}\times 4\times 0.48\times 1 = 1.497\,\text{kA}$$

DC power at rectifier is

$$V_{\text{dr}}I_{\text{d}} = 1020\times 1 = 1020\,\text{MW}$$

The rectifier-end power factor is

$$\frac{1020}{1087.2} = 0.939$$

Therefore, the reactive power compensation required at rectifier end is 373.2 Mvar.

B.5 Solution to the Problems: Chapter 5

B.5.1 Problem 5.1

The Y matrix without link between buses 2 and 3 (0.667−j2.00) is

$$\bar{Y}_{\text{bus}} = \begin{vmatrix} 2.176 - j7.673 & -1.1764 + j4.706 & -1.0 + j3.0 & 0 \\ -1.1764 + j4.706 & 2.2764 - j7.731 & 0 & -1 + j3.0 \\ -1.0 + j3.0 & 0 & 1 - j6.30 & j3.333 \\ 0 & -1 + j3.0 & j3.330 & 1.0 - j6.308 \end{vmatrix}$$

Using Equations 5.2

$$y_{33} \to y_{33} + \Delta y_{32} = 1 - j6.30 + 0.667 - j2.0 = 1 - j8.30$$

$$y_{23} \to y_{23} - \Delta y_{23} = 0 - (0.667 - j2.0) = y_{32}$$

$$y_{22} \to y_{22} + \Delta y_{23} = 2.2764 - j7.731 + 0.667 - j2.0 = 2.8434 - j9.681$$

Substituting

$$\bar{Y}_{\text{bus}} = \begin{vmatrix} 2.176 - j7.673 & -1.1764 + j4.706 & -1.0 + j3.0 & 0 \\ -1.1764 + j4.706 & 2.8434 - j9.681 & -0.667 + j2.0 & -1 + j3.0 \\ -1.0 + j3.0 & -0.667 + j2.0 & 1 - j8.30 & j3.333 \\ 0 & -1 + j3.0 & j3.333 & 1.0 - j6.308 \end{vmatrix}$$

B.5.2 Problem 5.2

The admittance values on 100 MVA base are shown in Figure B.7.
Then by examination

$$\bar{Y}_{\text{bus}} = \begin{vmatrix} 2.019 - j45.57 & -2.019 + j45.57 & 0 & 0 \\ -2.019 + j45.7 & 3.112 - j48.50 & -1.094 + j2.93 & 0 \\ 0 & -1.094 + j2.93 & 1.191 - j9.28 & -0.097 + j6.35 \\ 0 & 0 & -0.097 + j6.35 & 0.097 - j6.35 \end{vmatrix}$$

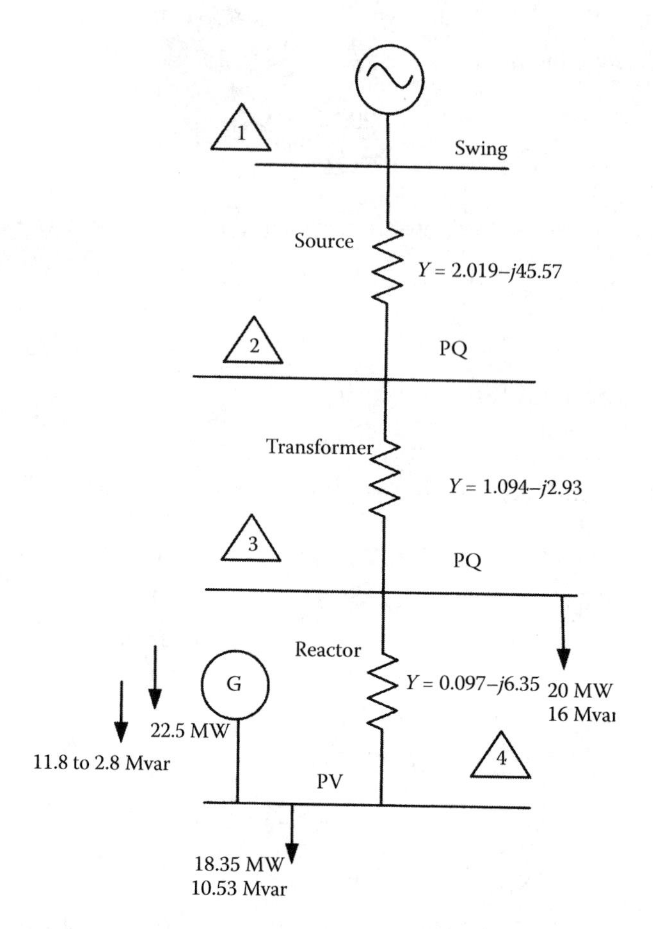

FIGURE B.7
Equivalent circuit of Figure 5P-1.

B.5.3 Problem 5.3

Bus 3 serves a load of 20 MW and 16 Mvar, bus 4 is a PV bus; assume that the voltage required to be maintained = rated voltage = 1 per unit. The motor load is not a constant impedance load, but is considered so in this example. The load types and their characteristics are discussed in Chapter 6. First consider that the generator operates at 22.5 MW and the maximum reactive power output is 11.8 Mvar. Considering that the bus load is directly served from the generator, 3.25 MW and 0.27 Mvar are supplied into the system.

1. *Gauss Method*

Assume $V_1 = V_2 = V_3 = V_4 = 1.0$ per unit as the starting point. Then, using the Gauss method
First iteration

$$V_2^1 = \frac{1}{Y_{22}}\left[\frac{P_2 - jQ_2}{V_2^{0*}} - Y_{21}V_1^0 - Y_{23}V_3^0 - Y_{24}V_4^0\right]$$

$$= \frac{1}{3.112 - j48.50}\left[-(-2.019 - j45.57)1 < 0^\circ - (-1.094 + j2.93)1 < 0^\circ\right]$$

$$= 1.0 < 0^\circ$$

$$V_3^1 = \frac{1}{Y_{33}}\left[\frac{P_3 - jQ_3}{V_3^{0*}} - Y_{31}V_1^0 - Y_{32}V_2^0 - Y_{34}V_4^0\right]$$

$$= \frac{1}{1.191 - j9.28}\left[\frac{-0.2 + j0.16}{1 < 0^\circ} - (-1.094 + j2.93)1 < 0^\circ - (-0.097 + j6.35)1 < 0^\circ\right]$$

$$= 0.98 - j0.019$$

Calculate the bus 4 voltage assuming that the generator operates at its maximum given reactive power output of 11.8 Mvar:

$$V_4^1 = \frac{1}{Y_{44}}\left[\frac{P_4 - jQ_4}{V_4^{0*}} - Y_{41}V_1^0 - Y_{42}V_2^0 - Y_{43}V_3^0\right]$$

$$= \frac{1}{0.097 - j6.35}\left[\frac{0.0325 - j0.0027}{1 < 0^\circ} - (-0.097 + j6.35)1 < 0^\circ\right]$$

$$= 1.001 + j0.005$$

Second iteration

$$V_2^2 = \frac{1}{Y_{22}}\left[\frac{P_2 - jQ_2}{V_2^{0*}} - Y_{21}V_1^1 - Y_{23}V_3^1 - Y_{24}V_4^1\right]$$

$$= \frac{1}{3.112 - j48.50}[-(-2.019 + j45.57)(0.98 - j0.019) - (-1.094 + j2.93)(1.001 + j0.005)]$$

$$= 0.917 - j0.074$$

$$V_3^2 = \frac{1}{Y_{33}}\left[\frac{P_3 - jQ_3}{V_3^{1*}} - Y_{31}V_1^1 - Y_{32}V_2^1 - Y_{34}V_4^1\right]$$

$$= \frac{1}{1.191 - j9.28}\left[\frac{-0.2 + j0.16}{0.98 + j0.019} - (-1.094 + j2.93)1 < 0^\circ - (-0.097 + j6.35)(1.001 + j0.005)\right]$$

$$= 0.981 - j0.016$$

$$V_4^2 = \frac{1}{Y_{44}}\left[\frac{P_4 - jQ_4}{V_4^{1*}} - Y_{41}V_1^1 - Y_{42}V_2^1 - Y_{43}V_3^1\right]$$

$$= \frac{1}{0.097 - j6.35}\left[\frac{0.0325 - j0.0027}{1.001 - j0.005} - (-0.097 + j6.35)(0.98 - j0.019)\right]$$

$$= 0.98 - j0.014$$

The result in the second iteration shows that the generator reactive power output is not enough to maintain the voltage on PV bus 4 to the desired level of 1.0 per unit.

2. *Gauss–Seidel Iteration*

As discussed in the text, the new values of voltages are substituted as soon as these are found, without going through the whole cycle of iteration. For example, in the first iteration, instead of estimating bus 4 voltage with expression

$$V_4^1 = \frac{1}{Y_{44}}\left[\frac{P_4 - jQ_4}{V_4^{0*}} - Y_{41}V_1^0 - Y_{42}V_2^0 - Y_{43}V_3^0\right]$$

$$= \frac{1}{0.097 - j6.35}\left[\frac{0.0325 - j0.0027}{1<0^\circ} - (-0.097 + j6.35)1<0^\circ\right]$$

$$= 1.001 + j0.005$$

we estimate with the following expression

$$V_4^1 = \frac{1}{Y_{44}}\left[\frac{P_4 - jQ_4}{V_4^{0*}} - Y_{41}V_1^1 - Y_{42}V_2^1 - Y_{43}V_3^1\right]$$

$$= \frac{1}{0.097 - j6.35}\left[\frac{0.0325 - j0.0027}{1<0^\circ} - (-0.097 + j6.35)(0.98 - j0.0019)\right]$$

$$= 0.981 - j0.014$$

3. *Gauss-Seidel with acceleration factor*

As in Example 5.6, with 1.6 acceleration factor, the bus 3 voltage is written as follows:

$$V_3^1 = V_3^0 + 1.6(V_3^{1'} - V_3^0)$$

$$= 1<0^\circ + 1.6(0.98 - j0.0019 - 1<0^\circ)$$

$$= 0.986 - j0.003$$

The results of first two iterations with an acceleration factor of 1.6 are shown in Table B.2.

TABLE B.2

Results of Load Flow Calculations, Gauss-Seidel, Acceleration Factor = 1.6

Bus Number	First Iteration	Second Iteration
2	$1<0^\circ$	$0.998 < -1.4^\circ$
3	$0.969 < -1.7^\circ$	$0.936 < -3.6^\circ$
4	$0.956 < -2.2^\circ$	$0.930<3.2^\circ$

B.5.4 Problem 5.4

Omitting the swing bus, the Y matrix is

$$\bar{Y}_{bus} = \begin{vmatrix} 3.112 - j48.50 & -1.094 + j2.93 & 0 \\ -1.094 + j2.93 & 1.191 - j9.28 & -0.097 + j6.35 \\ 0 & -0.097 + j6.35 & 0.097 - j6.35 \end{vmatrix}$$

As per Example 5.7, the pivotal manipulation can be done in any order. Using pivot (2, 2)

$$Y_{21} = \frac{Y_{21}}{Y_{22}} = \frac{-1.094 + j2.93}{1.191 - j9.28} = -0.326 - j0.076$$

$$Y_{22} = \frac{1}{Y_{22}} = \frac{1}{1.191 - j9.28} = 0.014 + j0.06$$

$$Y_{23} = \frac{Y_{23}}{Y_{22}} = \frac{-0.097 + j6.35}{1.191 - j9.28} = -0.674 + j0.076$$

$$Y_{12} = 0.326 + j0.076$$

$$Y_{11} = Y_{11} - \frac{Y_{12}Y_{21}}{Y_{22}} = (3.112 - j48.50) - \frac{(-1.094 + j2.93)(-1.094 + j2.93)}{1.191 - j9.28} = 2.535 - j47.63$$

Therefore, the matrix is

$$\begin{vmatrix} 2.535 - j47.63 & 0.326 + j0.076 & 0 \\ -0.326 - j0.076 & 0.014 + j0.06 & -0.674 + j0.076 \\ 0 & 0.0674 - j0.076 & 0.097 - j6.35 \end{vmatrix}$$

After similar manipulations of pivots (3, 3) and (1, 1), the Z matrix is

$$\begin{vmatrix} 0.00097 + j0.022 & 0.00097 + j0.022 & 0.00097 + j0.022 \\ 0.00097 + j0.022 & 0.113 + j0.321 & 0.113 + j0.321 \\ 0.00097 + j0.022 & 0.113 + j0.321 & 0.115 + j0.479 \end{vmatrix}$$

B.5.5 Problem 5.5

In Figure 5P.1

$$I_2 = \left[\frac{P_2 + jQ_2}{V_2}\right]^* - Y_{21}V = 0 - (-2.019 + j45.57) = 2.019 - j45.57$$

(Initial estimate of all voltages = 1 < 0, and bus 2 does not serve any load.)

$$I_3 = \left[\frac{P_3 + jQ_3}{V_3}\right]^* - Y_{31}V = (-0.2 - j0.16)^* = -0.2 + j0.16$$

Bus 4 has load as well as generation:

$$\text{Load} = 0.1 - j0.075$$

Motor load of 10,000 hp with given efficiency and power factor = 0.0794–j0.029.

Consider an active power generation of 22.5 MW = rating of the generator. Then, active power = 0.047 pu.

Consider that the generator operates at its maximum reactive power output = 11.8 Mvar = 0.118 pu. Therefore, the excess reactive power supplied into the system = 0.014 pu. Then, $I_4 = 0.047 - j0.014$.

Then, the voltages are as follows:

$$\begin{vmatrix} V_2 \\ V_3 \\ V_4 \end{vmatrix} = \bar{Z} \begin{vmatrix} 2.019 - j45.57 \\ -0.2 + j0.16 \\ 0.047 - j0.014 \end{vmatrix} = \begin{vmatrix} 0.997 - j0.0032 \\ 0.936 - j0.033 \\ 0.938 - j0.025 \end{vmatrix}$$

The calculations are repeated with new values of voltages. The reactive power output of the generator is not enough to maintain voltage at bus 4 to the desired level—a result which we found in the solution to Problem 5.3.

B.5.6 Problem 5.6

First consider no taps adjustment, and then the Y matrix is

$$\bar{Y}_{\text{bus}} = \begin{vmatrix} -2j & 2j & 0 & 0 \\ 2j & 0.8 - j3.6 & -0.8 + j1.6 & 0 \\ 0 & -0.8 + j1.6 & 1.6 - j3.2 & -0.8 + j1.6 \\ 0 & 0 & -0.8 + j1.6 & 0.8 - j1.6 \end{vmatrix}$$

The bus 4 has a generator of 23.53 MW, a rated power factor of 0.85, and a rated Mvar of 14.56.

Then, as in Problem 5.3, with Gauss–Seidel iteration, after two iterations, the bus voltages are as follows:

Bus 2 voltage = 0.958 < –3.79°

Bus 3 voltage = 0.977 < –1.08°

Bus 4 voltage = 1.05<4.73°

The Mvar output of the generator is

$$Q_4 = -I_m\left[V_{42}^{*}\{Y_{41}V_1 + Y_{42}V_2 + Y_{43}V_3 + Y_{44}V_4\}\right]$$

$$= -I_m\left[1.05(0.997 + j0.082)\{(-0.8 + j1.6)(0.977)(1.0 - j0.019) + (0.8 - j1.6)(0.958)(0.999 - j0.064)\}\right]$$

$$= 0.049$$

Thus, approximately 4.9 Mvar of the generator is utilized and rest of it remains trapped in the generator.

When a tap of 1:1.1 is provided, modify the Y matrix as shown in the text.

Now modify the matrix for change in tap ratio of the transformer.

$$\bar{Y}_{\text{bus}} = \begin{vmatrix} -2.347 & 2.18j & 0 & 0 \\ 2.18j & 0.8-j3.6 & -0.8+j1.6 & 0 \\ 0 & -0.8+j1.6 & 1.6-j3.2 & -0.8+j1.6 \\ 0 & 0 & -0.8+j1.6 & 0.8-j1.6 \end{vmatrix}$$

Note that in the calculations $V_2/V_2' = (1/1.1) = 0.9$.

The first iteration gives

$$\text{Bus 2 voltage} = 0.916 < -1.95^\circ$$
$$\text{Bus 3 voltage} = 0.959 < -1.96^\circ$$
$$\text{Bus 4 voltage} = 1.05 < 3.2^\circ$$

The generator now supplies 9.40 Mvar into the system.

Practically, when generators are connected in a step-up configuration in utility generating stations, it becomes an important factor to utilize the generator reactive power capability. The reactive power can remain "trapped" in the generator, and cannot be supplied into the high-voltage utility system through a GSU (generator step-up transformer). It requires optimization of the generator operating voltage within steady-state operating limits (maximum of 1.05 per unit according to the ANSI/IEEE standards), impedance of GSU, and the tap settings on GSU.

B.5.7 Problem 5.7

A solution of Example 5.8 is illustrated. Example 5.9 can be similarly solved:

$$\bar{Y} = \begin{vmatrix} 3 & -1 & 0 \\ -1 & 4 & -3 \\ 0 & -3 & 6 \end{vmatrix}$$

Following the procedure detailed in Appendix A, Volume 1

$$L = \begin{vmatrix} 3 & 0 & 0 \\ -1 & 3.667 & 0 \\ 0 & -3 & 3.57 \end{vmatrix}$$

$$U = \begin{vmatrix} 1 & -0.333 & 0 \\ 0 & 1 & -0.814 \\ 0 & 0 & 1 \end{vmatrix}$$

$$\text{Then} \begin{vmatrix} 3 & 0 & 0 \\ -1 & 3.667 & 0 \\ 0 & -3 & 3.57 \end{vmatrix} \begin{vmatrix} V_1' \\ V_2' \\ V_3' \end{vmatrix} = \begin{vmatrix} 3 \\ 4 \\ 2 \end{vmatrix}$$

This gives

$$\begin{vmatrix} V_1' \\ V_2' \\ V_3' \end{vmatrix} = \begin{vmatrix} 1 \\ 1.364 \\ 1.708 \end{vmatrix}$$

Then

$$\begin{vmatrix} 1 & -0.333 & 0 \\ 0 & 1 & -0.814 \\ 0 & 0 & 1 \end{vmatrix} \begin{vmatrix} V_1 \\ V_2 \\ V_3 \end{vmatrix} = \begin{vmatrix} 1 \\ 1.364 \\ 1.708 \end{vmatrix}$$

or

$$\begin{vmatrix} V_1 \\ V_2 \\ V_3 \end{vmatrix} = \begin{vmatrix} 1.916 \\ 2.752 \\ 1.706 \end{vmatrix}$$

The results correspond well with the results arrived in Example 5.8.

B.6 Solution to the Problems: Chapter 6

B.6.1 Problem 6.1

$$\begin{vmatrix} \frac{\partial f_1}{\partial x_1} & \frac{\partial f_1}{\partial x_2} & \frac{\partial f_1}{\partial x_3} \\ \frac{\partial f_2}{\partial x_1} & \frac{\partial f_2}{\partial x_2} & \frac{\partial f_2}{\partial x_3} \\ \frac{\partial f_3}{\partial x_1} & \frac{\partial f_3}{\partial x_2} & \frac{\partial f_3}{\partial x_3} \end{vmatrix} = \begin{vmatrix} 9-x_2 & -x_1 & 0 \\ 1 & 6 & -2x_3 \\ 0 & x_3^2 & 2x_2x_3-10 \end{vmatrix}$$

Then

$$x^1 = \begin{vmatrix} 0 \\ 0 \\ 0 \end{vmatrix} - \begin{vmatrix} 9 & 0 & 0 \\ 1 & 6 & 0 \\ 0 & 0 & -10 \end{vmatrix}^{-1} \begin{vmatrix} -6 \\ -10 \\ 4 \end{vmatrix} = -\begin{vmatrix} 0.111 & 0 & 0 \\ -0.019 & 0.167 & 0 \\ 0 & 0 & -1 \end{vmatrix} \begin{vmatrix} 6 \\ -10 \\ 4 \end{vmatrix} = \begin{vmatrix} -0.667 \\ 1.778 \\ 0.4 \end{vmatrix}$$

Second iteration

$$x^2 = \begin{vmatrix} -0.667 \\ 1.778 \\ 0.4 \end{vmatrix} - \begin{vmatrix} 7.222 & 0.667 & 0 \\ 1 & 6 & -0.8 \\ 0 & 0.16 & -8.58 \end{vmatrix}^{-1} \begin{vmatrix} -10.813 \\ -0.139 \\ 0.283 \end{vmatrix} = \begin{vmatrix} 0.851 \\ 1.755 \\ 0.429 \end{vmatrix}$$

B.6.2 Problem 6.2

The Jacobian is

$$\begin{vmatrix} \Delta P_2 \\ \Delta Q_2 \\ \Delta P_3 \end{vmatrix} = \begin{vmatrix} \partial P_2 / \partial \theta_2 & \partial P_2 / \partial V_2 & \partial P_2 / \partial \theta_3 \\ \partial Q_2 / \partial \theta_2 & \partial Q_2 / \partial V_2 & \partial Q_2 / \partial \theta_3 \\ \partial P_3 / \partial \theta_2 & \partial P_3 / \partial V_2 & \partial P_3 / \partial \theta_3 \end{vmatrix} \begin{vmatrix} \Delta \theta_2 \\ \Delta V_2 \\ \Delta \theta_3 \end{vmatrix}$$

This is as in the solution to Example 6.3.

B.6.3 Problem 6.3

The Y matrix without considering phase shift is

$$Y = \begin{vmatrix} 1.042 - j6.08 & j0.27 & -1.042 + j5.90 \\ j0.27 & 1.042 - j6.08 & -1.042 + j5.90 \\ -1.042 + j5.90 & -1.042 + j5.90 & 2.084 - j11.80 \end{vmatrix}$$

Let us solve first without considering the effect of voltage regulating transformer.

We will demonstrate the solution for the first iteration; the solution can be extended to the second iteration:

$$P_2 = 1.03 \times V_2 \left[0.27 \sin \theta_2\right] + V_2 \times V_2 \left[1.042 \cos(\theta_2 - \theta_2) - 6.08 \sin(\theta_2 - \theta_2)\right]$$

$$+ V_2 \times 1.03 \left[(-1.042) \cos(\theta_2 - \theta_3) + 5.90 \sin(\theta_2 - \theta_3)\right]$$

$$= -0.031$$

$$Q_2 = 1.03 \times V_2 \left[0.27 \cos \theta_2\right] + V_2 \times V_2 \left[1.042 \sin(\theta_2 - \theta_2) + 6.08 \cos(\theta_2 - \theta_2)\right]$$

$$+ V_2 \times 1.03 \left[(-1.042) \sin(\theta_2 - \theta_3) - 5.90 \cos(\theta_2 - \theta_3)\right]$$

$$= +0.278$$

$$P_3 = 1.03 \times 1.03 \left[(-1.042) \cos \theta_3 + 5.91 \sin \theta_3\right] + V_2 \times V_3 \left[(-1.042) \cos(\theta_3 - \theta_2) + 5.90 \sin(\theta_3 - \theta_2)\right]$$

$$+ V_3 \times 1.03 \left[(2.084) \cos(\theta_3 - \theta_3) - 11.80 \sin(\theta_3 - \theta_3)\right]$$

$$= -0.094$$

$$\frac{\partial P_2}{\partial \theta_2} = 1.03V_2\,[0.27\cos\theta_2] + 1.03V_2\,[(1.042)\sin(\theta_2 - \theta_3) + 5.90\cos(\theta_2 - \theta_3)]$$

$$= 6.188$$

$$\frac{\partial P_2}{\partial \theta_3} = 1.03V_2\,[(-1.042)\sin(\theta_2 - \theta_3) - 5.90\cos(\theta_2 - \theta_3)]$$

$$= -6.08$$

$$\frac{\partial P_2}{\partial V_2} = 1.03\,[0.27\sin\theta_2] + 2V_2(1.042) + 1.03\,[(-1.042)\cos(\theta_2 - \theta_3) + 5.90\sin(\theta_2 - \theta_3)]$$

$$= 0.64$$

$$\frac{\partial Q_2}{\partial \theta_2} = 1.03V_2\,[0.27\sin\theta_2] + 1.03V_2\,[(-1.042)\cos(\theta_2 - \theta_3) + 5.90\sin(\theta_2 - \theta_3)]$$

$$= -1.444$$

$$\frac{\partial Q_2}{\partial V_2} = 1.03V_2\,[0.27\cos\theta_2] + 2V_2[6.08] + 1.03V_2\,[(-1.402)\sin(\theta_2 - \theta_3) - 5.90\cos(\theta_2 - \theta_3)]$$

$$= 6.528$$

$$\frac{\partial Q_2}{\partial \theta_3} = 1.03V_2\,[(1.042)\cos(\theta_2 - \theta_3) - 5.90\sin(\theta_2 - \theta_3)]$$

$$= 1.07$$

$$\frac{\partial P_3}{\partial \theta_2} = 1.03V_2\,[(-1.402)\sin(\theta_3 - \theta_2) - 5.90\cos(\theta_3 - \theta_2)]$$

$$= -6.09$$

$$\frac{\partial P_3}{\partial V_2} = -1.03[(-1.402)\sin(\theta_3 - \theta_2) + 5.90\sin(\theta_3 - \theta_2)]$$

$$= -1.07$$

$$\frac{\partial P_3}{\partial \theta_3} = 1.03V_2\,[(2.084)\sin(\theta_3 - \theta_2) + 11.80\cos(\theta_3 - \theta_2)]$$

$$= 11.99$$

Therefore, the Jacobian is

$$\bar{J} = \begin{vmatrix} 6.188 & 0.64 & -6.08 \\ -1.444 & 6.528 & 1.444 \\ -6.09 & -1.07 & 11.99 \end{vmatrix}$$

Thus,

$$\begin{vmatrix} \Delta\theta_2^1 \\ \Delta V_2^1 \\ \Delta\theta_3^1 \end{vmatrix} = \begin{vmatrix} 6.188 & 0.64 & -6.09 \\ -1.444 & 5.722 & 1.444 \\ -6.09 & -1.444 & 11.99 \end{vmatrix}^{-1} \begin{vmatrix} -0.5-(-0.031) \\ -0.5-(+0.278) \\ -0.5-(-0.094) \end{vmatrix} = \begin{vmatrix} -0.214 \\ -0.141 \\ -0.155 \end{vmatrix}$$

This gives $V_2 = 0.859 < 12.07°$

$$V_3 = 1.03 < -8.8°$$

Calculate Q_3 with new voltages:

$$Q_3 = (0.84 - j0.179)(1.03)[-1.042 \sin < -8.8° - 5.91 \cos < -8.8°] + (0.840 - j0.179)(1.018 - j0.155)$$
$$[(-1.402)\sin(-8.8 + 12.07) - 5.91\cos(-8.8 + 12.07)] + (1.018 - j0.155)(1.018 - j0.155)$$
$$[11.64\cos(8.8 - 8.8)] = 0.959 - j0.30$$

That is, approximately 100 Mvar of reactive power injection.

Using a computer solution which goes through a couple of iterations, the converged load flow is shown in Figure B.8a.

Reactive power injection of 110 Mvar is required at bus 3, which is within the specified limits at this bus.

Considering the phase shift, the matrix will be no longer symmetrical as

$$y_{ij} \neq y_{ji}$$

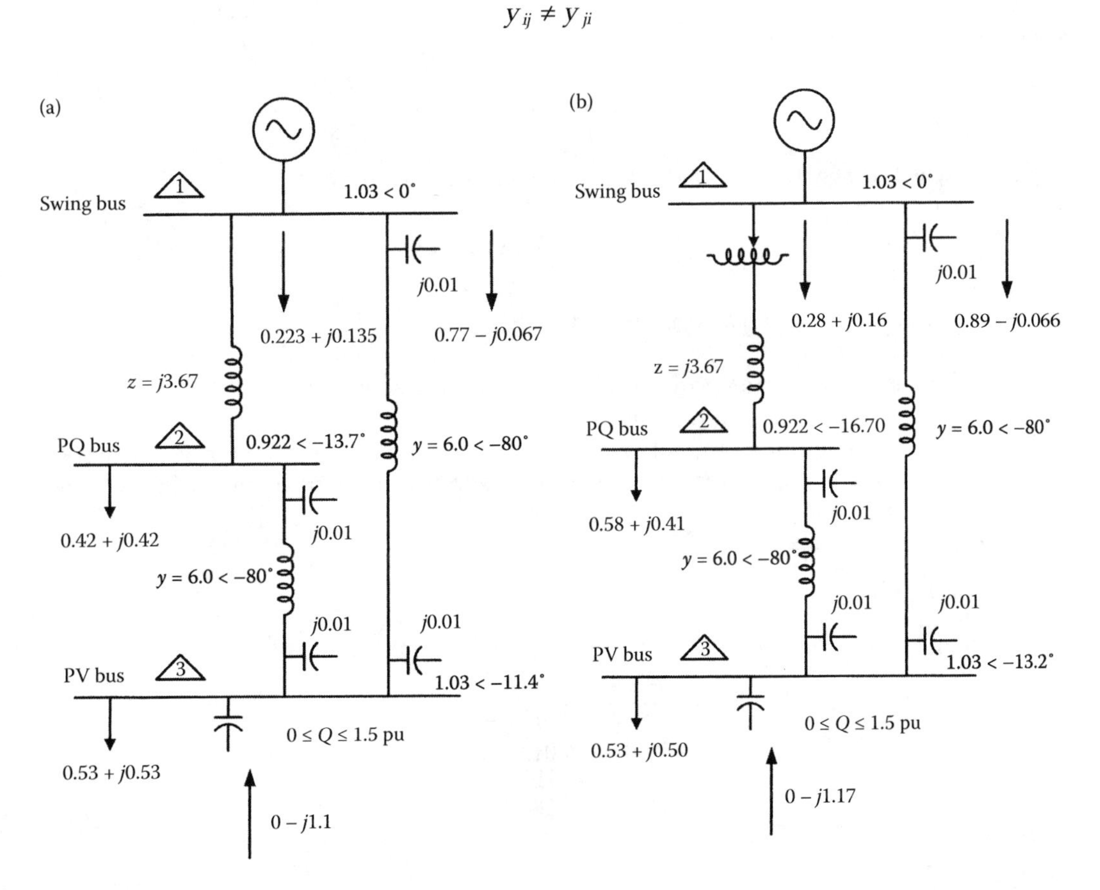

FIGURE B.8
(a) Converged load flow solution and (b) converged load flow solution with phase-shifting transformer, Problem 6.3.

A new row and column can be added to Jacobian, and its dimensions will change from n to $n + 1$.

Using Equations 6.74

$$y_{12} = j0.27e^{j3^\circ} = j0.27(\cos 4^\circ + j\sin 4^\circ) = -0.0188 + j0.269$$

$$y_{21} = j0/27e^{-j3^\circ} = j0.27(\cos 4^\circ - j\sin 4^\circ) = 0.0188 + j0.269$$

Note that a negative real number is involved. Here, we have modeled the line between buses 1 and 2 ignoring resistance, which will always be present.

The modified Y matrix becomes

$$Y = \begin{vmatrix} 1.042 - j6.08 & -0.188 + j0.269 & -1.042 + j5.91 \\ 0.188 + j0.269 & 1.042 - j6.08 & -1.042 + 5.91 \\ -1.042 + j5.91 & -1.042 + j5.91 & 2.084 - j11.64 \end{vmatrix}$$

The complex power flow between nodes i and j is

$$S_{\text{sr}}^* = V_{\text{s}}^* I_{\text{s}}$$

Then, from Equation 12.74

$$S_{\text{sr}}^* = V_{\text{s}}^* \left(|N|^2 yV_{\text{s}} - N^* yV_{\text{r}}\right)$$

Equating real and negative parts, the active power flow becomes

$$P_{\text{sr}} = yV_{\text{s}}\left(NV_{\text{s}} \cos\theta_{\text{L}} - V_{\text{r}} \cos(\theta_{\text{L}} + \delta_{\text{r}} - \delta_{\text{s}} - \varphi)\right)$$

Neglecting resistance impedance angle is 90°.

Therefore, with phase shift, the power transfer is proportional to $\sin(13.67^\circ - 0^\circ) = 0.236$. With phase regulating transformer add 3°, i.e., $\sin(13.67^\circ - 0^\circ - (-3^\circ)) = 0.287$. Thus, the phase shift will increase power flow from 42 to 51.07 MW.

Figure B.8b is a computer simulation with phase shift, also see Problem 11.6. The tap changing mainly impacts the reactive power flow, and the phase shift mainly impacts the active power flow.

B.6.4 Problem 6.4

From Equation 6.65, we can write

$$\begin{vmatrix} \Delta P_2 \\ \Delta P_3 \\ \Delta Q_2 \end{vmatrix} = \begin{vmatrix} \dfrac{\partial P_2}{\partial \theta_2} & \dfrac{\partial P_2}{\partial \theta_3} & 0 \\ \dfrac{\partial P_3}{\partial \theta_2} & \dfrac{\partial P_3}{\partial \theta_3} & 0 \\ 0 & 0 & \dfrac{\partial Q_2}{\partial V_2} \end{vmatrix} \begin{vmatrix} \Delta \theta_2 \\ \Delta \theta_3 \\ \Delta V_2 \end{vmatrix}$$

That is

$$\begin{vmatrix} -0.469 \\ -0.406 \\ -0.778 \end{vmatrix} = \begin{vmatrix} 6.188 & -6.09 & 0 \\ -6.09 & 11.99 & 0 \\ 0 & 0 & 6.528 \end{vmatrix} \begin{vmatrix} \Delta\theta_2 \\ \Delta\theta_3 \\ \Delta V_2 \end{vmatrix}$$

This gives

$$\begin{vmatrix} \Delta\theta_2 \\ \Delta\theta_3 \\ \Delta V_2 \end{vmatrix} = J^{-1} \times \begin{vmatrix} -0.469 \\ -0.406 \\ -0.222 \end{vmatrix} = \begin{vmatrix} -0.218 \\ -0.095 \\ -0.119 \end{vmatrix}$$

Compare with the results arrived in Problem 6.3.

B.6.5 Problem 6.5

The required networks are shown in Figure B.9.

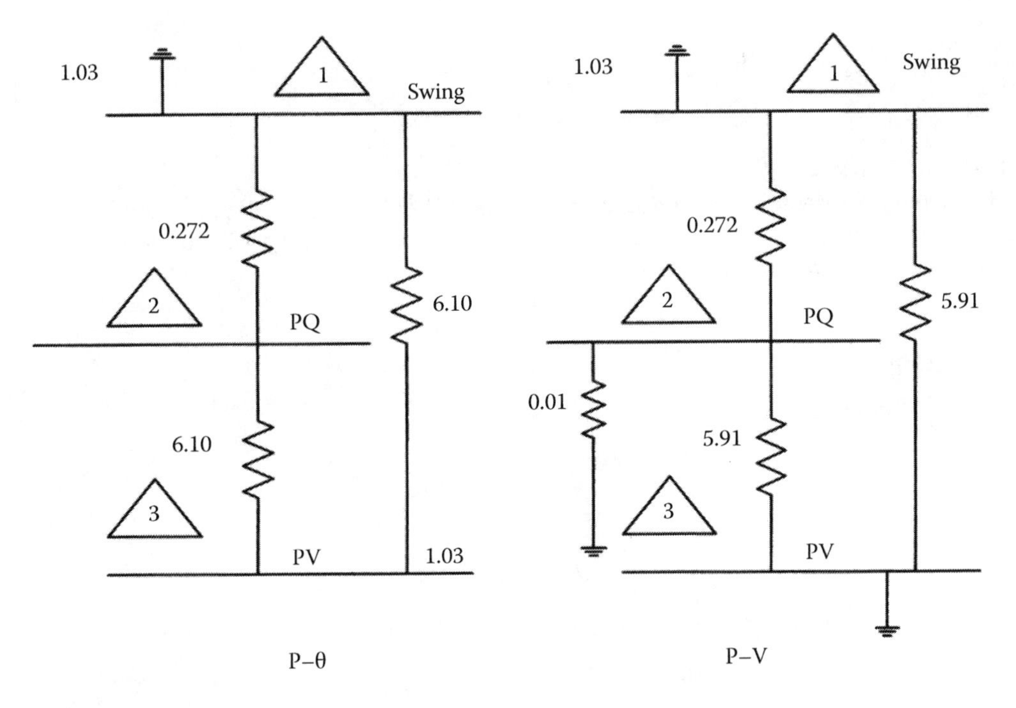

FIGURE B.9
P–θ and P–V networks of Problem 6.5.

B.6.6 Problem 6.6

P–θ network

$$\begin{vmatrix} B_{22} & B_{23} \\ B_{23} & B_{33} \end{vmatrix} \begin{vmatrix} \Delta\theta_2 \\ \Delta\theta_3 \end{vmatrix} = \begin{vmatrix} \Delta P_2 / V_2 \\ \Delta P_3 / V_3 \end{vmatrix}$$

or

$$\begin{vmatrix} 6.372 & -6.10 \\ -6.10 & 12.20 \end{vmatrix} \begin{vmatrix} \Delta\theta_2 \\ \Delta\theta_3 \end{vmatrix} = \begin{vmatrix} -0.469 \\ -0.406 / 1.03 \end{vmatrix}$$

This gives

$$\begin{vmatrix} \Delta\theta_2 \\ \Delta\theta_3 \end{vmatrix} = \begin{vmatrix} -0.2 \\ -0.133 \end{vmatrix}$$

P–V network

$$|B_{22}||\Delta V_2| = |\Delta Q_2 / V_2|$$

$$|6.19||\Delta V_2| = 0.278$$

$$\Delta V_2 = 0.045.$$

That is, $V_2 = 1.045 < -11.4°$.

There is a considerable difference in the results of Problems 3, 4, and 6. It should be noted that these results are for the first iteration only, to demonstrate the procedures, and are not indicative of the final results.

B.6.7 Problem 6.7

From Equation 6.56

$$\begin{vmatrix} \Delta P_2 \\ \Delta Q_2 \\ \Delta P_3 \end{vmatrix} = \begin{vmatrix} \frac{\partial P_2}{\partial e_2} & \frac{\partial P_2}{\partial h_2} & 0 \\ \frac{\partial Q_2}{\partial e_2} & \frac{\partial Q_2}{\partial h_2} & 0 \\ 0 & 0 & \frac{\partial P_3}{\partial e_3} \end{vmatrix} \begin{vmatrix} \Delta e_2 \\ \Delta h_2 \\ \Delta e_3 \end{vmatrix}$$

This requires use of rectangular form of Jacobian. See Equations 6.22 through 6.35 in the text.

Assume that $e_2 = e_3 = 1$ and $h_2 = 0$.

Then

$$\begin{vmatrix} \Delta P_2 \\ \Delta Q_2 \\ \Delta P_3 \end{vmatrix} = \begin{vmatrix} 1.042 & 6.08 & 0 \\ 5.90 & -1.042 & 0 \\ 0 & 0 & 2.084 \end{vmatrix} \begin{vmatrix} \Delta e_2 \\ \Delta h_2 \\ \Delta e_3 \end{vmatrix}$$

In the rectangular form, the $\Delta P, \Delta Q$ can be calculated from the following equations:

$$\Delta P_k = P_{k,\text{ specified}} - \sum_{j=1}^{j=n} \left[e_k (G_{kj} e_j - B_{kj} h_j) + h_k (G_{kj} h_j + B_{kj} e_j) \right]$$

$$\Delta Q_k = Q_{k,\text{ specified}} - \sum_{j=1}^{j=n} \left[h_k (G_{kj} e_j - B_{kj} h_j) - e_k (G_{kj} h_j + B_{kj} e_j) \right]$$

Here, we use the values as calculated before using polar form:

$$\begin{vmatrix} -0.469 \\ -0.778 \\ -0.406 \end{vmatrix} = \begin{vmatrix} 1.042 & 6.08 & 0 \\ 5.91 & -1.042 & 0 \\ 0 & 0 & 2.084 \end{vmatrix} \begin{vmatrix} \Delta e_2 \\ \Delta h_2 \\ \Delta e_3 \end{vmatrix}$$

This gives

$$\begin{vmatrix} \Delta e_2 \\ \Delta h_2 \\ \Delta e_3 \end{vmatrix} = \begin{vmatrix} -0.141 \\ -0.053 \\ -0.195 \end{vmatrix}$$

The new voltage $V_2 = 0.895 - j0.053$.

B.6.8 Problem 6.8

See Figure 6.9 and Table 6.2. Also see Examples 6.7 and 6.8 which calculate the load flow with same loads, but in one case, with constant impedance type (Example 12.8) and in the other case, constant KVA type (Example 12.7). The results of calculation show that the load modeling has profound effect on the load flow and the voltage profiles in the system.

B.6.9 Problem 6.9

The dimensions are 4×4. Ten elements are populated out of sixteen, i.e., the percentage populated = 62.5%:

$$\begin{vmatrix} X & X & 0 & 0 \\ X & X & X & 0 \\ 0 & X & X & X \\ 0 & 0 & X & X \end{vmatrix}$$

B.7 Solution to the Problems: Chapter 7

B.7.1 Problem 7.1

See Section 7.3. Depending upon the details of the study, various methodologies can be used. A load flow study with modeling of starting loads as impact loads will give only initial voltage dip on starting with no idea of the starting time, acceleration, or voltage recovery. A dynamic study can be conducted and the starting characteristics can be plotted as shown in Figure 7.16. A transient stability type of program can be used to study stability and impact of voltage dips on other motor loads. EMTP simulation is another option, which plots transients in much greater detail.

B.7.2 Problem 7.2

The rotor resistance is not directly specified. The total power transferred through the air gap is

$$P_g = I_2^2 \frac{r_2}{s}$$

Rotor copper loss $= I_2^2 r_2$.

The mechanical power output $= (1-s)P_g$.

Assume an operating power factor of 93%. Then, the full load current = 185 A. Motor kVA = 1203, KW = 1119.

That is

$$3(1-s)I_2^2 \frac{r_2}{s} = 1119{,}000\,\text{W}$$

Substituting the values, $r_2 = 0.165\ \Omega$.

For mechanical power output, losses like stator copper loss, windage, and friction must be considered. An efficiency of 94% is specified. That is

$$\eta = \frac{\text{Input} - \text{Losses}}{\text{Input}}$$

This gives losses of 67 kW. Deducting from 1119 kW gives a 1052 KW output.
The rotor speed is (0.985)(1800) = 1773 rpm.
Therefore, the output torque is

$$\frac{1052 \times 10^3 \times 60}{1773} = 35{,}600\,\text{Nm or} = 26{,}237\ \text{lb-ft}$$

The equivalent circuit diagram with numerical values is shown in Figure B.10a and b. Note that R_{mag} is high and can be neglected. In this figure, the stator resistance is considered equal to the rotor resistance—this is only an approximation.

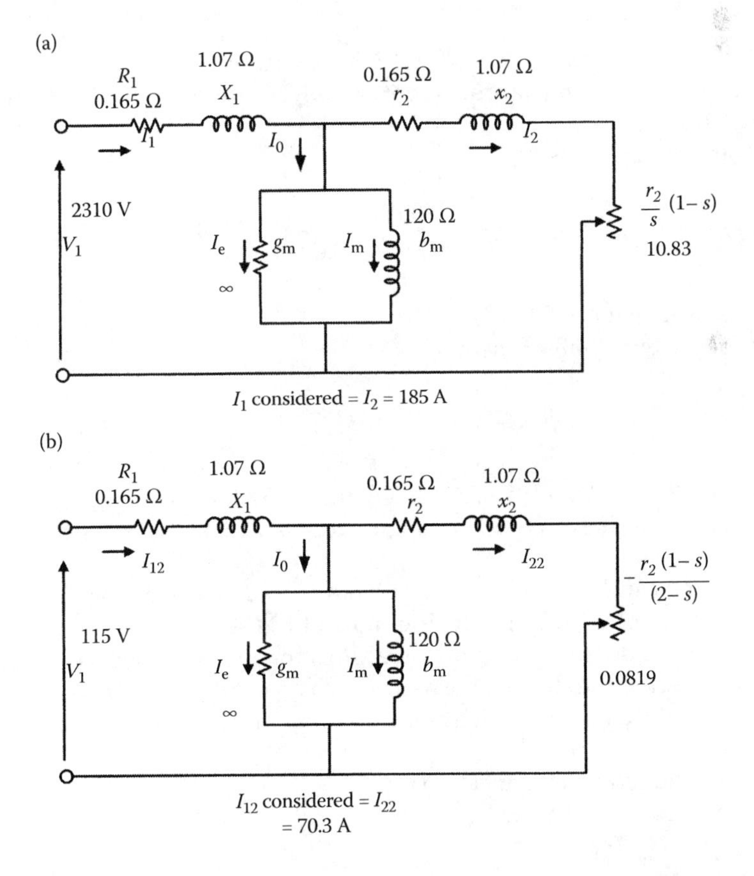

FIGURE B.10
(a) Induction motor equivalent circuit, normal operation and (b) equivalent negative sequence circuit, Problem 7.2.

Text explains that a 5% negative sequence component can give rise to 30%: more conservatively, 38% negative sequence current. Therefore, in Figure B.10b, the negative sequence current can be taken as follows:

$$(0.38)\times(185)=70.3\text{ A}$$

Then, approximately, the negative sequence power output = 1.210 = kW.

The negative sequence torque is 50 N-m.

B.7.3 Problem 7.3

Based on given data, the motor full load current = 406.2 A at a power factor of 0.945.

Therefore, the starting current = 2437 A at 0.15 power factor.

Motor locked rotor impedance is

$$z=\frac{2400<0^\circ}{\sqrt{3}\times 2437<-81.3^\circ}=0.0854+j0.562\ \Omega$$

The external impedance is 2.5 pu on 100 MVA base = 0.144 Ω.

The X/R of external impedance is not given, assume = 10. This gives external impedance = $0.0143 + j0.143$.

Then, the starting current is

$$I_s=\frac{2400<0^\circ}{\sqrt{3}\times(0.0997+j0.605)}=|2259|$$

Ampères at a power factor of 0.163.

The voltage at the terminals of the motor is

$$1385.7<0^\circ-(0.0143+j0.143)(2259)(0.163-j0.987)=1062-j20.77\approx|1062|\text{ V}$$

The line-to-line voltage is 1839 V, i.e., a voltage dip of 23.4% on starting impact.

The locked rotor torque of 88.17% in Figure B.10b will be reduced to approximately 51.76%.

The starting time can be calculated as in Example 7.2. The starting current of 2259 A can be assumed to remain constant during most of the starting period. More accurately, it can be calculated as demonstrated further. Thus, the motor torque–speed curve can be constructed considering the starting current, which on accelerating reduces in magnitude. Also the power factor changes, which improves the terminal voltage, and increases the starting torque.

Consider that motor slip has reduced to 0.3.

The locked rotor current is given by

$$\frac{V}{\sqrt{(R_1+r_2/s)^2+(X_1+x_2)^2}}$$

At start, $s = 1$, consider $R_1 = r_2 = 0.0427\ \Omega$.

Then, at 0.30 slip, $r_2/s = 0.142\ \Omega$.

Then, the starting current is

$$I_{s,0.3\ \text{slip}} = \frac{2400}{(0.0427 + 0.142) + j0.562 + (0.0143 + j0.143)} = |1891|$$

at a power factor of 0.272.

That is, compared to starting instant, the current is reduced form 2259 to 1891 A and its power factor has changed from 0.163 to 0.272.

As the current is reduced, the voltage at the motor terminals is

$$1385.7 < 0° - (0.0143 + j0.143)(1891)(0.272 - j0.96) = 1190 - j47.59 \approx |1119|$$

That is, line-to-line voltage of 1938 V.

Thus, the voltage is now 80.7% of the rated. Then, the motor torque is 65% of the rated.

A similar procedure can be carried out for point-by-point calculation.

Under normal running load, the motor terminal voltage is

$$1385.7 < 0° - (0.0143 + j0.143)(460.2)(0.945 - j0.327) = 1358 - j60 \approx |1358|$$

That is, a line-to-line voltage of 2352 V = 0.98% of the rated. Thus, motor full-load torque need be reduced to 96%.

Based on the above calculations, the motor torque and the starting current are plotted as shown in Figure B.11a and b. To calculate the starting time, Table B.3 can now be constructed.

This gives a starting time of 27.6 s.

B.7.4 Problem 7.4

The voltage dip on starting is calculated as 23.7%. To limit it to desired 10% maximum, a reactor start with 40% tap on the reactor can be used. This reduces the starting current to approximately 904 A, and the starting torque of 88.17% at full voltage will be reduced to approximately 12.07%. This could cause a lockout of motor during starting.

With Krondrofer starting, the starting current becomes proportional to starting torque. This will raise the starting torque of approximately 24%, for the same starting current reduction.

Capacitor start can be considered.

Low-frequency starting can be considered at a considerably higher cost.

B.8 Solution to the Problems: Chapter 8

B.8.1 Problem 8.1

Equation 8.6

From Equations 8.2 and 8.3, we can write

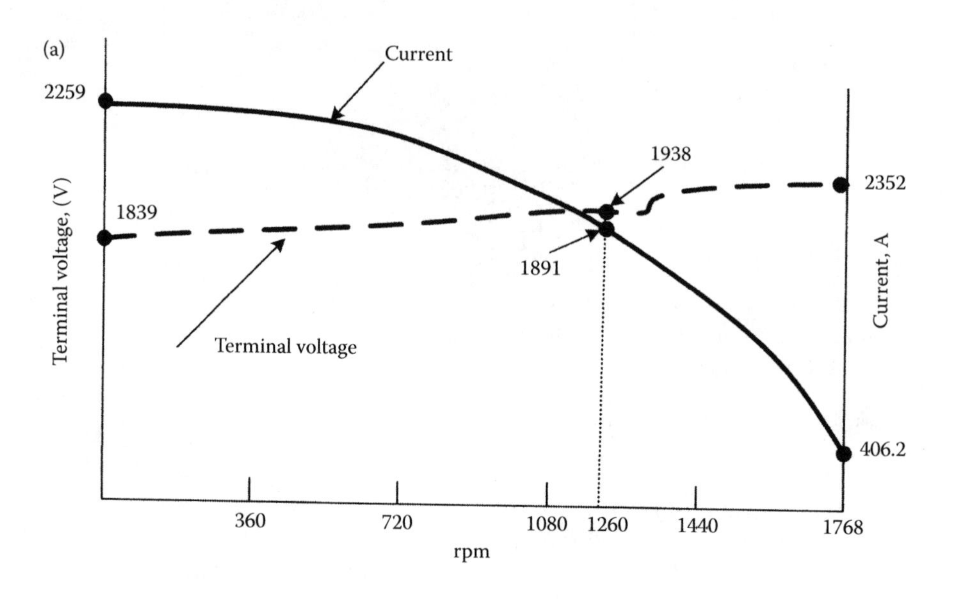

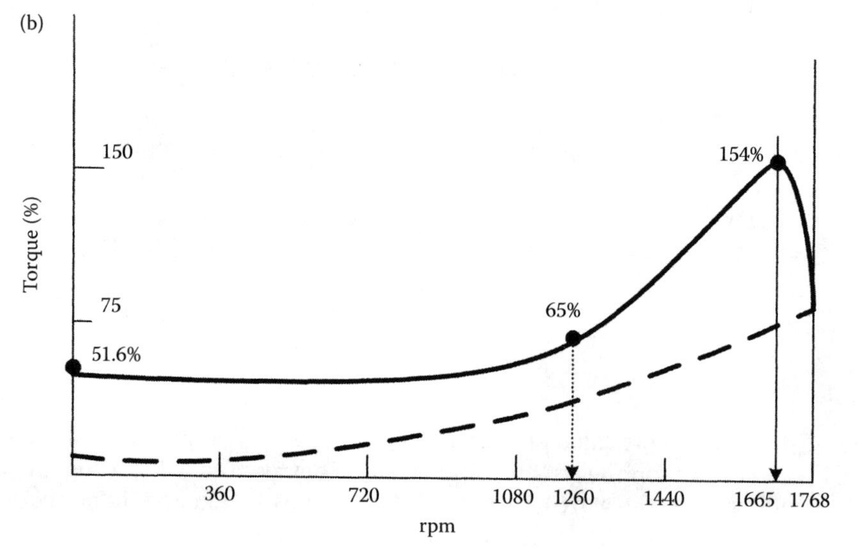

FIGURE B.11

(a) Starting current profile and (b) starting torque–speed characteristics of induction motor, Problem 7.3.

TABLE B.3

Calculations of Starting Time Average Starting Torque during Speed Intervals

Speed Change	Average Torque (%)	Average Torque (lb-ft²)
0–25%	41	2436
25–50%	39	2317
50–75%	43	2555
75% to full load (slip = 1.75%)	45	2673

$$\sin\theta = \frac{PX}{V_s V_r}$$

$$\cos\theta = \frac{QX + V_r^2}{V_s V_r}$$

Squaring and adding gives

$$1 = \left(\frac{PX}{V_s V_r}\right)^2 + \left(\frac{QX + V_r^2}{V_s V_r}\right)^2$$

Thus, Equation 8.6 is derived.

Equation 8.37.

The power equation of a lossless transmission line is

$$P = \frac{V_s V_r}{X_{sF}} \sin\delta$$

(here $\delta = \theta$)

where X_{sF} is the series reactance.

For a midcompensated line, $V_s = V_r = V$ and Figure 8.5d shows that the series reactance is changed to

$$jX_{sF}(1 - s)$$

From Figure 8.5a

$$I_{sh} = I_s' - I_s = j(b_{sh}/2)(1 - k_m)$$

B.8.2 Problem 8.2

The voltage dips by 10%, i.e., 23 kV. Therefore, a compensation of 836.6 Mvar is required.

B.8.3 Problem 8.3

The 200 MVA transformer reactance in ohms = 26.45 at 230 kV. This gives a total line reactance of 76.75 Ω. The capacitance is not specified, and can be assumed to be 5.4×10^{-6} MΩ/mi. Estimated line length is approximately 100 mi.

The specified parameters give $Z_0 = 385\ \Omega$, and $\beta = 2.078$ rad/mi = 0.119°/mi. Thus, $\beta l = 11.9°$.

Then, the electrical length of the line is 11.9° and the maximum power transfer as a function of the natural load is equal to $5.06P_0$.

The midpoint voltage is 1 per unit. The sending-end and receiving-end voltages for a load angle of 30°, from Equation 8.29, are as follows:

$$\left[1 - (\sin 5.7°)^2 (1 - 0.5)^2\right]^{1/2} = 0.862 \text{ pu}$$

The reactive power required from Equation 8.27 is

$$Q_s = -Q_r = P_0 \sin\frac{\theta}{2}\left[\left(\frac{P}{P_0}\right)^2 - 1\right]$$

$$= (0.0098)(0.25 - 1) = 0.00735 \text{ k var/ kW}.$$

With 0.3 series compensation and from Equation 8.34

$$\frac{P_{comp}}{P} = 6.91$$

The surge impedance is $\sqrt{1-0.3}Z_0 = 0.837Z_0$.

The natural loading is now equal to $1.195P_0$.

The new electrical length of the line is

$$\sin^{-1}\left(\frac{1.195P_0}{6.91P_0}\right) = 9.9°$$

Similarly, for 0.3 shunt compensation

$$\frac{P_{comp}}{P} = 6.71$$

The surge impedance is $0.877P_0$.

The natural loading is now equal to $1.14P_0$.

B.8.4 Problem 8.4

A total of 100 MVA at 0.8 power factor at 13.8 kV gives a current of 4183 A $< -36.87°$. V_r is unknown. First consider that $V_r = V_s$.

The line reactance = 0.1 pu = 0.1794 Ω.

Then, the voltage drop through the line is

$$I_r Z = (4184)(0.8 - j0.6)(j0.1794) = 4504 + j600$$

This gives the receiving-end voltage of

$$7967.7 - 4504 - j600 = 3464 < -9.8°$$

The load is considered as a constant impedance load; therefore, the load impedance is

$$Z_l = \frac{13{,}800}{\sqrt{3} \times 4184 < -36.8°} = 1.90 < 36.8°$$

Therefore, in the second iteration, the load current is

$$\frac{3464 < -9.8°}{1.90 < 36.8°} = 1269 - j1346$$

The receiving-end voltage is

$$7967.7-(1269-j1346)(j0.1794)=7726.2-j227$$

The voltage at receiving end can be solved like a load flow problem and after a couple of iterations it is

$$0.941\text{pu}<-4.3°$$

The load supplied is 71 MW and 53.1 Mvar.

The current is

$$\frac{7489<-4.3°}{1.90<36.8°}=2967-j2600\text{ A}$$

The sending-end power $=V_sI_s^*=(7967.7)(2967-j2600)=23.71\times10^6+j20.73\times10^6$ per phase.

Approximately 10 Mvar is lost in the transmission line.

It is a short line; therefore, the line susceptance is neglected. At no-load, the receiving-end voltage is equal to the sending-end voltage. At full load, it dips by 5.9%. The capacitor size to limit the voltage dip to 2% is approximately given by Equation 8.14, which can be written as follows:

$$|V_{\text{raised}}|=|V_{\text{available}}|+\left[\frac{X_{\text{th}}}{|V_{\text{available}}|}\right]Q_{\text{c}}$$

where X_{th} is the Thévenin reactance of the line. Here, $X_{th}=0.1794\ \Omega$.

This gives $Q_c=3.89$ Mvar. If a capacitor bank of this rating is switched at the maximum load, the voltage will improve to 0.98 pu. But this will not meet the requirement of maintaining the receiving-end voltage within 0.98 pu from zero to full load.

When the line is carrying a load of 27 MW, the voltage dip will be 2%. If we switch a capacitor bank of 2.64 Mvar at this point, the voltage will rise to 1.02 pu. Then, as the load develops, the voltage will behave as shown in Figure B.12. Thus, by switching a much lower capacitor bank at partial load of 27 MW, the line regulation can be maintained within plus minus 2%.

B.8.5 Problem 8.5

$$Q_{ij}=|N|\ |y||V_i|\left[|N||V_i|\sin\beta-|V_j|\sin(\beta+\theta_j-\theta_i)\right]$$

Angle β is the angle of the impedance between buses i and j; here, it is equal to 90°.

We could simply use Equation 8.14 as an approximation.

First consider that the transformer is set at the rated taps 500–18 kV. The full load reactive power of the generator is 124 Mvar. The 500 kV voltage is considered as 1<0° per unit. Thus, the problem reduces that what should be the generation voltage to pump 124 Mvar into the utility system.

We know that reactive power flow is more of a function of ΔV. As per ANSI/IEEE standards, a generator can be operated at 1.5 per unit voltage. Consider it as the generation voltage.

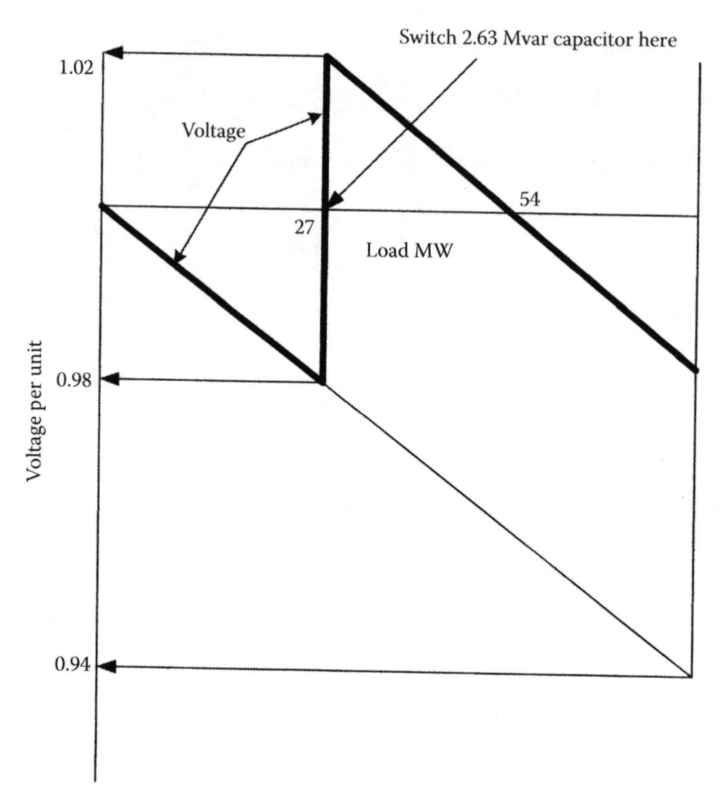

FIGURE B.12
Switching point of capacitor bank for line regulation, Problem 8.4.

If we consider θ_j and $\theta_i = 0$, then

$$Q = 20 \times 1.05[1.05 - 1] = 1.05 \text{ per unit}$$

that is, 105 Mvar can be supplied.

Try a generation voltage of 1.06:

$$Q = 20 \times 1.06[1.06 - 1] = 1.272 \text{ per unit}$$

That is, the generation voltage should be raised to 1.059 per unit.

The reactive current is $-j3977$ at 18 kV multiplied by 1.059, as the generator is operating at this voltage = 4221.6 A. The transformer reactance in ohms = 0.162. The reactive power loss is 8.65 Mvar.

Consider now that the generation voltage is held at 1.0 per unit. The transformer taps are set primary = 525 kV. That means 500 kV primary voltage will give 17.1 kV on the secondary side.

Assume that the generator voltage and utility voltage are both at 1.0 pu.

Then

$$Q_{ij} = |N|\,|y||V_i|\left[|N||V_i|\sin\beta - |V_j|\sin(\beta + \theta_j - \theta_i)\right]$$

$$= (1.05)(20)(1)[(1.05) - (1)] = 1.05$$

The effect of raising primary taps is the same as raising the generation voltage.

It is obvious that if the taps are moved in the other direction, i.e., a tap of say 0.95 is used, meaning that 500 kV primary will raise the 13.8 kV voltage to 14.49 kV, then no reactive power will flow from the generator. On the other hand, the reactive power will flow from the utility into the generator and the generator will act as a reactor to absorb this power. See Figure 8.6 for the generator reactive capability curve.

B.8.6 Problem 8.6

The problem amounts to finding:

- Receiving-end voltage at no-load
- The reactor at the receiving end to restore voltage to 138 kV at no-load
- Receiving-end voltage when supplying the specified load
- Receiving-end capacitor to restore voltage to 138 kV on load

The line length is 100 mi. The parameters are on per mile basis.

All these four steps can be calculated using the load flow methods of Chapters 5 and 6.

With the line at no-load, the receiving-end voltage is 1.085 pu, based upon the source voltage; i.e., it is 149.8 kV. To bring it down to 138 kV, a –15 Mvar inductive power must be supplied.

When the line is loaded to the specified load, the receiving-end voltage is 101.6 kV at an angle of –15.9°. To raise it to 138 kV, 85 Mvar of capacitive power must be supplied. Therefore, the synchronous condensers must be able to supply 85 Mvar capacitive and 15 Mvar lagging, controllable power. It will be large synchronous rotating machine and practically a STATCOM or TCR will be a better choice. Figure B.13 illustrates the reactive power compensation.

B.8.7 Problem 8.7

The equation of the load line is already derived in Equation 8.14. That is,

$$\frac{\partial Q}{\partial V} = -\frac{V_r}{X_{sr}} = -\frac{V_s}{X_{sr}}$$

This can be written as follows:

$$\frac{\Delta V}{V} \approx \frac{\Delta Q}{S_{sc}}$$

where S_{sc} is the system short-circuit MVA.

Per unit voltage change is equal to the ratio of reactive power swing to short-circuit level of the supply system. This can be graphically shown as in Figure B.14. This relation can be called system load line or supply system voltage characteristics.

B.8.8 Problem 8.8

Uncompensated line

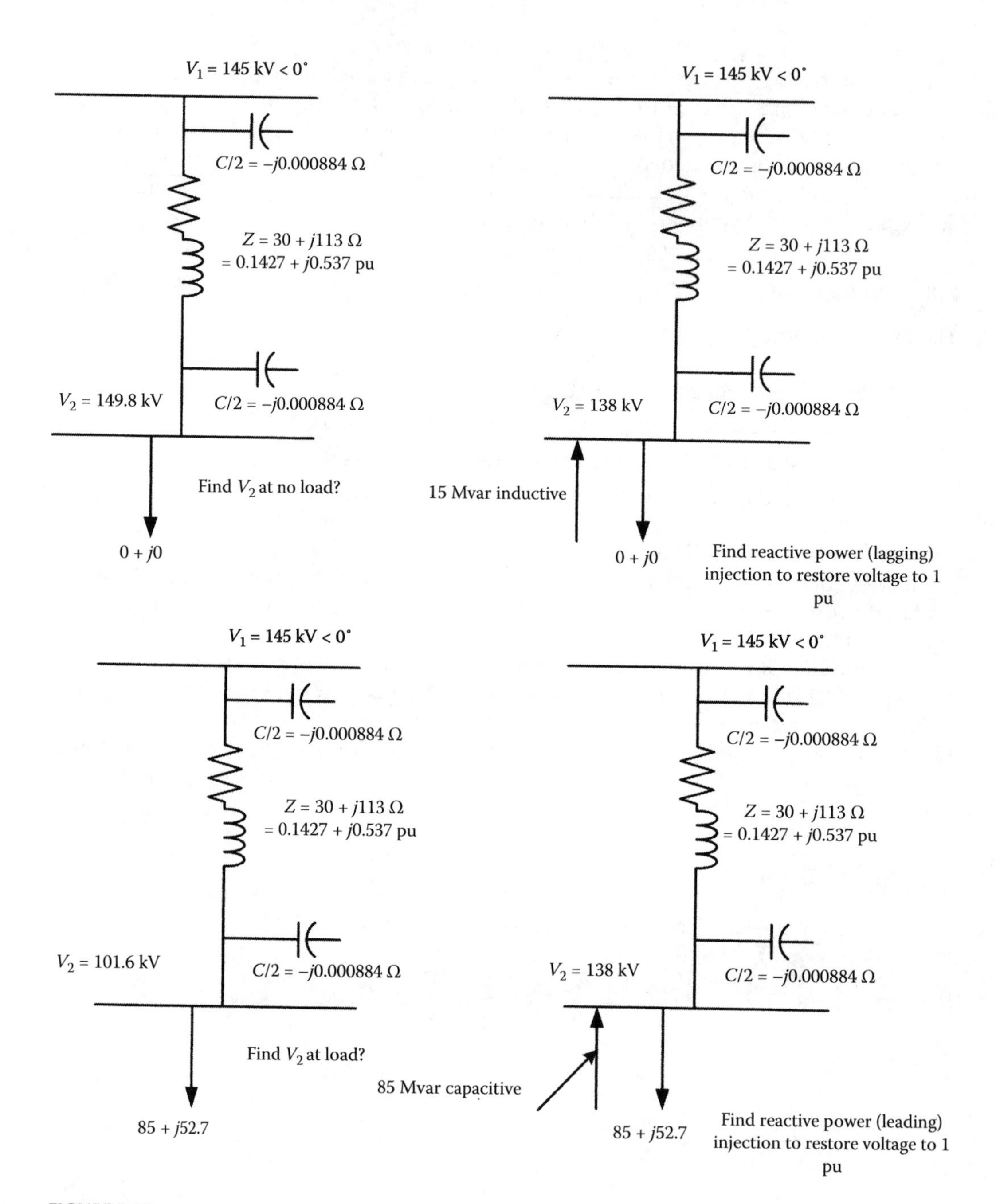

FIGURE B.13
Reactive power compensations of a transmission line, Problem 8.6.

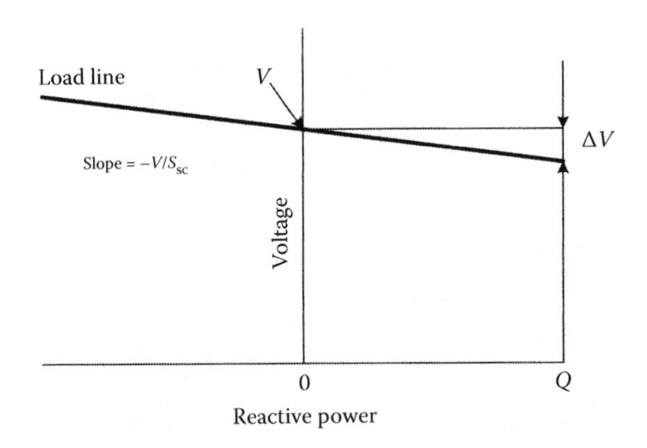

FIGURE B.14
Load line of a synchronous compensator, Problem 8.7.

$$\beta = \sqrt{zy} = 2.078 \times 10^{-3}$$

rad/mi. $\beta l = 23.8°$ for the line length of 200 mi.
Series reactance = 0.302 pu, 100 MVA base:

$$P_m = \frac{V_s V_r}{Z_0 \sin(\beta l)} \sin\delta$$

$$Z_0 = \sqrt{\frac{z}{y}} = 385\ \Omega$$

Thus, the maximum power is 2.478 P_0.

Series compensation
Calculate based on the following equation:

$$P = \frac{V_s V_r \sin\delta}{\left[Z_0 \sin\theta - \frac{X_{Cr}}{2}(1+\cos\theta)\right]}$$

The line series reactance is 160 Ω = 0.302 pu at 230 kV. Therefore, compensation required is 0.181 pu.
Substituting these values, $P = 4.347 \sin\delta$.

Shunt compensation
To completely wipe out the capacitance of the line, the reactors required are as follows:

$$X = Z_0 \frac{\sin(\theta/2)}{1-\cos(\theta/2)}$$

The power transfer is

$$P = \frac{V^2}{2Z_0 \sin(\theta/2)} = 2.43 P_0$$

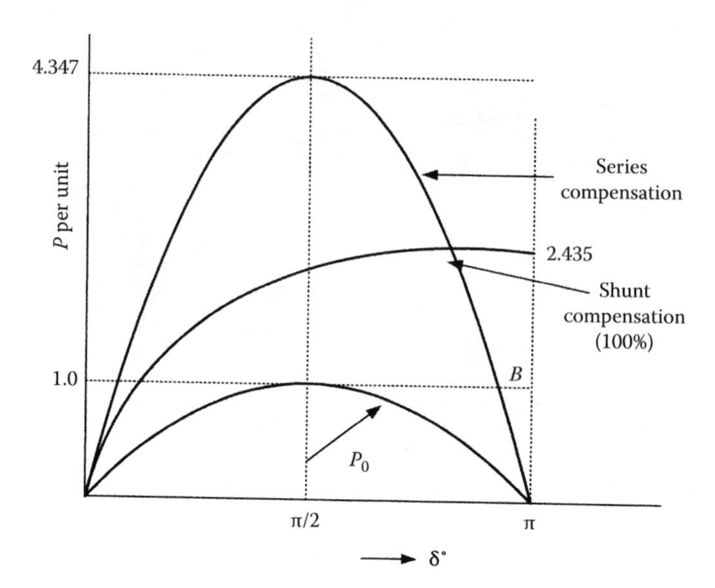

FIGURE B.15
Power angle characteristics with various compensation methods, Problem 8.8.

The maximum power transfer is not much impacted by addition of shunt reactors. The power characteristics are plotted in Figure B.15, also see Figure 8.15.

B.9 Solution to the Problems: Chapter 9

B.9.1 Problem 9.1

- A TCR gives rise to much higher harmonic pollution compared to a STATCOM, though harmonic filters are provided. See Figure 9.5 for the waveform of a STATCOM due to phase multiplication.
- The response of TCR is slower compared to a STATCOM.
- The *V–I* characteristics of STATCOM provides much better voltage support on system undervoltages—see Figure 9.9 for comparison.
- A STATCOM can reduce flicker much better than an SVC—see comparison in Volume 3.
- A TCR has much higher losses, even at zero output compared to that a STATCOM has very low losses.

B.9.2 Problem 9.2

Based on the reactive power flow concepts, a tap of higher than the rated voltage (assume that the transformer is provided with taps on the primary high-voltage windings) increases the turns ratio of the transformer and, thus, decreases the secondary voltage at which the generation is connected. Considering that the generator is operated in the PV mode, to keep the bus voltage constant, set point equal to the rated voltage of the system; the reactive power output of the generator will increase till it hits its maximum.

A tap of voltage lower than the rated voltage of the primary windings raises the secondary side voltage to which the generator is connected, and the generator will back out on its reactive power. Considering that the load demand does not change, more reactive power will be supplied from the utility source.

B.11 Solution of Problems: Chapter 11

B.11.1 Problem 11.1

The gradient vector is

$$g = \begin{vmatrix} 4x_1 - 6 \\ 6x_2 - 8 \\ 6x_3 + 20 \end{vmatrix}$$

Setting it to zero gives

$$\begin{vmatrix} x_1 \\ x_2 \\ x_3 \end{vmatrix} = \begin{vmatrix} \frac{3}{2} \\ \frac{4}{3} \\ \frac{-10}{3} \end{vmatrix}$$

The Hessian is

$$H = \begin{vmatrix} \frac{\partial^2 f}{\partial x_1^2} & \frac{\partial^2 f}{\partial x_1 x_2} & \frac{\partial^2 f}{\partial x_1 x_3} \\ \frac{\partial^2 f}{\partial x_2 x_1} & \frac{\partial^2 f}{\partial x_2^2} & \frac{\partial^2 f}{\partial x_2 x_3} \\ \frac{\partial^2 f}{\partial x_3 x_1} & \frac{\partial^2 f}{\partial x_3 x_2} & \frac{\partial^2 f}{\partial x_3^2} \end{vmatrix} = \begin{vmatrix} 4 & 0 & 0 \\ 0 & 6 & 0 \\ 0 & 0 & 6 \end{vmatrix}$$

This is definite positive. Therefore,

$$\begin{vmatrix} x_1 \\ x_2 \\ x_3 \end{vmatrix} = \begin{vmatrix} \frac{3}{2} \\ \frac{4}{3} \\ \frac{-10}{3} \end{vmatrix}$$

is the minimum point.

B.11.2 Problem 11.2

Find the boundary values:

$$f(0,0) = -43$$

$$f(0,4) = -363$$

$$f(10,0) = -299$$

$$F(10,4) = 257$$

Check for internal stationary points:

$$\frac{\partial f}{\partial x_1} = 10(x_1 - 2) + 7x_2 = 0$$

$$\frac{\partial f}{\partial x_2} = -14(x_2 +) + 7x_1 = 0$$

The solution of these equations is unique:

$$(x_1, x_2) = (0.397, -1.949)$$

x_2 violates the restriction that it is ≥0.
Therefore, clearly, the maxima is given by $f(10,4) = 257$.

B.11.3 Problem 11.3

Form the Lagrangian function:

$$F(x_1, x_2, \lambda) = 3x_1^2 + 2x_2^2 + \lambda(8 - 3x_1 + 2x_2)$$

Then

$$\frac{\partial(x,\lambda)}{\partial x_1} = 6x_1 - 3\lambda = 0$$

$$\frac{\partial(x,\lambda)}{\partial x_2} = 4x_2 + 2\lambda = 0$$

$$\frac{\partial(x,\lambda)}{\partial x_\lambda} = 8 - 3x_1 + 2x_2 = 0$$

The solution of these equations gives

$$x_1 = 8/5, \quad x_2 = -8/5, \quad \lambda = 16/5$$

B.11.4 Problem 11.4

Yes, it is generally true.

B.11.5 Problem 11.5

Form the Lagrangian function:

$$x_1^2 x_2^2 x_3^2 x_4^2 + \lambda(c - x_1^2 - x_2^2 - x_3^2 - x_4^2)$$

$$\frac{\partial(x,\lambda)}{\partial x_1} = 2x_1(x_2^2 x_3^2 x_4^2) + \lambda(-2x_1) = 0$$

$$\frac{\partial(x,\lambda)}{\partial x_2} = 2x_2(x_1^2 x_3^2 x_4^2) + \lambda(-2x_2) = 0$$

$$\frac{\partial(x,\lambda)}{\partial x_3} = 2x_3(x_1^2 x_3^2 x_4^2) + \lambda(-2x_3) = 0$$

$$\frac{\partial(x,\lambda)}{\partial x_4} = 2x_4(x_1^2 x_2^2 x_3^2) + \lambda(-2x_4) = 0$$

$$\frac{\partial(x,\lambda)}{\partial x_\lambda} = (c - x_1^2 - x_2^2 - x_3^2 - x_4^2) = 0$$

Thus, $x_1 = x_2 = x_3 = x_4 = (c/2)$.

B.11.6 Problem 11.6

The problem is then similar to Example 11.7.

The initial values are not given in the problem; let us assume $x_1 = 6$, $x_2 = 0$. Then

$$\nabla f = \left| \begin{array}{c} \dfrac{\partial f}{\partial x_1} \\ \dfrac{\partial f}{\partial x_2} \end{array} \right| = \left| \begin{array}{c} 5x_1 + 3x_2 \\ 3x_1 + 5x_2 \end{array} \right|$$

The results of first five iterations are shown in Table B.4.

B.11.7 Problem 11.7

The results of first five iterations are shown in Table B.5. This shows much rapid convergence. Note the effect of assumptions of initial values. We could have assumed a value of (1, 1) and the convergence would have been faster.

TABLE B.4

Maximization by Gradient Method

K	x_1, x_2	$F(x_1, x_2)$	g_k	$\lvert g_k \rvert$	$g_k / \lvert g_k \rvert$
0	6.000 0.000	90.000	30.000 18.000	34.986	0.857 0.514
1	5.143 –0.514	58.840	24.169 12.855	27.375	0.883 0.470
2	4.260 –0.984	35.207	18.346 7.858	19.958	0.913 0.394
3	3.347 –1.378	18.834	12.568 3.132	12.953	0.970 0.242
4	2.377 –1.620	9.085	6.991 –0.988	7.061	0.990 –0.140
5	1.387 –1.478	4.1806	2.4604 –3.2588	4.0833	0.6025 –0.7980

TABLE B.5

Maximization with Optimal Gradient Method

K	x_1, x_2	$F(x_1, x_2)$	g_k	$f(x^k - hg_k)$	h_k
0	6.000 0.000	90.000	30.000 18.000	$4680h^2 - 1224h + 90$	0.131
1	2.077 –2.354	9.970	3.323 –5.538	$49.08h^2 - 41.72h + 9.70$	0.425
2	0.665 0.000	1.104	3.2230 1.994	$57.42h^2 - 15.02h + 1.104$	0.131
3	0.230 –0.261	0.122	0.368 –0.613	$0.602h^2 - 0.512h + 0.122$	0.425
4	0.074 0.000	0.014	0.368 0.221	$0.405h^2 - 0.184h + 0.014$	0.131

B.11.8 Problem 11.8

The problem is solved similar to Example 11.9. Add slack variables

$$2x_1 + 3x_2 + x_3 = 16$$

$$4x_1 + 2x_2 + x_4 = 10$$

$$x_1, x_2, x_3, x_4 \geq 0$$

Then

$$\bar{A} = \begin{vmatrix} 2 & 3 & 1 & 0 \\ 4 & 2 & 0 & 1 \end{vmatrix}$$

$$\bar{a}_1 = \begin{vmatrix} 2 \\ 4 \end{vmatrix} \quad \bar{a}_2 = \begin{vmatrix} 3 \\ 2 \end{vmatrix} \quad \bar{a}_3 = \begin{vmatrix} 1 \\ 0 \end{vmatrix} \quad \bar{a}_4 = \begin{vmatrix} 0 \\ 1 \end{vmatrix}$$

If a_2 and a_3 are considered to form the basis, then

$$x_2 \begin{vmatrix} 3 \\ 2 \end{vmatrix} + x_3 \begin{vmatrix} 1 \\ 0 \end{vmatrix} = \begin{vmatrix} 16 \\ 10 \end{vmatrix}$$

Solving

$$x_2 = 5 \quad x_3 = 1 \quad x_1 = x_4 = 0$$

The initial feasible solution is

$$x_B = \begin{vmatrix} 5 \\ 1 \end{vmatrix}$$

Following the procedure in Example 11.9, it can be shown that this is also the final solution and the value of the function is $z = 15$.

$$\bar{a}_1 = \begin{vmatrix} 2 \\ 4 \end{vmatrix} \quad \bar{a}_4 = \begin{vmatrix} 0 \\ 1 \end{vmatrix}$$

B.11.9 Problem 11.9

The step is 10 MW, and the total generation is 50 MW.

For unit 1

$$F_1(50) = \$50, F_1(40) = \$36, F_1(30) = 24, F_1(20) = 14, F_1(10) = 6$$

For the second unit

$$f_1(50) = \$80, f_2(40) = 56, f_2(30) = 36, f_2(20) = 20, f_2(10) = 8$$

For $F_2(50)$, calculate the minimum of

$$f_2(0) + F_1(50) = \$50$$

$$f_2(10) + F_1(40) = 44$$

$$f_2(20) + F_1(30) = 44$$

$$f_2(30) + F_1(20) = 50$$

$$f_2(40) + F_1(10) = 62$$

$$f_2(50) + F_1(0) = 80$$

The minimum is \$44/h.

Similarly, for $F_2(40)$

$$f_2(0)+F_1(40)=\$36$$

$$f_2(10)+F_1(30)=32$$

$$f_2(20)+F_1(20)=34$$

$$f_2(30)+F_1(10)=42$$

$$f_2(40)+F_1(0)=56$$

This gives minimum = \$32/h.

Calculate $F_2(30)$

$$f_2(0)+F_1(30)=\$24$$

$$f_2(10)+F_1(20)=22$$

$$f_2(20)+F_1(10)=26$$

$$f_2(30)+F_1(0)=36$$

This gives minimum = \$22/h.

Calculate $F_2(20)$

$$f_2(0)+F_1(20)=\$14$$

$$f_2(10)+F_1(10)=14$$

$$f_2(20)+F_1(0)=20$$

This gives minimum = \$14/h.

Finally, $F_2(10)$

$$f_2(0)+F_1(10)=6$$

$$f_2(10)+F_1(0)=8$$

The minimum is \$6.0/h.

Bring in the third unit

$$f_3(50)=\$72.5,\ f_3(40)=52,\ f_3(30)=34.5,\ f_3(20)=20,\ f_3(10)=8.5$$

Then, the minimum of

$$f_3(0)+F_2(50)\ =0+44=44$$

$$f_3(10)+F_2(40)=8.5+32=40.5$$

$$f_3(20)+F_2(30)\ =20+22=44$$

$$f_3(30)+F_2(20)=34.5+14=48.5$$

$$f_3(40)+F_2(10)=52+6=58$$

$$f_3(50)+F_2(0)=72.5$$

The minimum is 40.5. Thus, unit 1 should operate at 30 MW and units 2 and 3 at 10 MW each.

B.11.10 Problem 11.10

See Example 11.3.

Let the load to be shared = 100 MW.

Then

$$\lambda_a = 0.02P_a + 4 = \lambda_b = 0.015P_b + 3$$

and

$$\lambda_a + \lambda_b = 100$$

This gives $P_b = 86$ MW and $P_a = 14$ MW. However, as the minimum generation of any of the two generators is 50 MW, the entire load should be served from generator 2.

Now consider the maximum load of 400 MW. Following the same equations, generator G_2 should supply 250 MW, and generator 1, 150 MW.

The intermediate load between 100 and 400 MW to be shared can be similarly calculated.

Index

Note: Page numbers followed by "*f*" and "*t*" refer to figures and tables, respectively.

M

N

T